D0327214

MODELLING STOCK MARKET VOLATILITY

Bridging the Gap to Continuous Time

MODELLING STOCK MARKET VOLATILITY

Bridging the Gap to Continuous Time

Edited by

Peter E. Rossi

Graduate School of Business
The University of Chicago
Chicago, Illinois

ACADEMIC PRESS

San Diego London Boston New York Sydney Tokyo Toronto

This book is printed on acid-free paper. ♾

Copyright © 1996 by ACADEMIC PRESS

All Rights Reserved.
No part of this publication may be reproduced or transmitted in any form or by any means, electronic or mechanical, including photocopy, recording, or any information storage and retrieval system, without permission in writing from the publisher.

Academic Press, Inc.
525 B Street, Suite 1900, San Diego, California 92101-4495, USA
http://www.apnet.com

Academic Press Limited
24-28 Oval Road, London NW1 7DX, UK
http://www.hbuk.co.uk/ap/

Library of Congress Cataloging-in-Publication Data

Modelling stock market volatility : bridging the gap to continuous time / edited by Peter Rossi.
p. cm.
Includes index.
ISBN 0-12-598275-5 (alk. paper)
1. Stocks--Prices--Mathematical models. I. Rossi, Peter E. (Peter Eric), date.
HG4636.M63 1996
332.63'222--dc20 96-26267
CIP

PRINTED IN THE UNITED STATES OF AMERICA
96 97 98 99 00 01 BB 9 8 7 6 5 4 3 2 1

CONTENTS

II

CONTINUOUS TIME LIMITS AND OPTIMAL FILTERING FOR ARCH MODELS

III

SPECIFICATION AND ESTIMATION OF CONTINUOUS TIME PROCESSES

CONTRIBUTORS

Department of Agricultural Economics
and Rural Sociology
The Pennsylvania State University
103 Armsby Building
University Park, PA 16802-5600

Numbers in parentheses indicate the pages on which the author's contributions begin.

Yacine Aït-Sahalia (427) Graduate School of Business, University of Chicago, Chicago, Illinois 60637.

Tim Bollerslev (xi) Department of Finance, Northwestern University, Evanston, Illinois 60208.

Phillip A. Braun (65) J. L. Kellogg Graduate School of Management, Northwestern University, Evanston, Illinois 60208.

Robert F. Engle (333) Department of Economics, University of California, San Diego, La Jolla, California 92093.

Dean P. Foster (157, 193, 291) Wharton School, University of Pennsylvania, Philadelphia, Pennsylvania 19104.

A. Ronald Gallant (357) Department of Economics, University of North Carolina, Chapel Hill, North Carolina 27599.

Lars Peter Hansen (385) Department of Economics, University of Chicago, Chicago, Illinois 60637.

Gary G. J. Lee (333) Treasury Quantitative Research, ABN AMRO Bank, N. A., Chicago, Illinois 60602.

Daniel B. Nelson* (3, 17, 37, 65, 79, 129, 157, 193, 241, 291) Graduate School of Business, University of Chicago, Chicago, Illinois 60637.

José Alexandre Scheinkman (385) Department of Economics, University of Chicago, Chicago, Illinois 60637.

Alain M. Sunier† (65) Goldman, Sachs & Co., New York, New York 10004.

George Tauchen (357) Department of Economics, Duke University, Durham, North Carolina 27708.

*Deceased.

†Present Address: Long Term Capital, Greenwich, Connecticut 06830

INTRODUCTION: MODELLING STOCK MARKET VOLATILITY—BRIDGING THE GAP TO CONTINUOUS TIME

Until recent years, there has been a curious dichotomy between the sorts of time-series models favored for empirical work with economic and financial time series and the models typically used in theoretical work on asset pricing. Since its introduction in 1982 by Engle, empirical work with financial time series has been dominated by variants of the autoregressive conditionally heteroskedastic (ARCH) model. On the other hand, much of the theoretical development of contingent claims pricing models has been based on continuous-time models of the sort which can be represented by stochastic differential equations. The reason for these divergent paths is one of mathematical and statistical convenience. Discrete-time ARCH models have been favored over continuous-time models for statistical inference because of the relative ease of estimation. On the theoretical side, application of various "no arbitrage" conditions is most easily accomplished via the Itô differential calculus requires a continuous-time formulation of the problem. The purpose of this volume is to bring together work from a new literature which seeks to forge a closer link between discrete-time and continuous-time models along with new work on practical estimation methods for continuous-time models.

Modelling Stock Market Volatility
Copyright © 1996 by Academic Press, Inc. All rights of reproduction in any form reserved.

A great deal of the seminal work in this area is due to Dan Nelson of the University of Chicago. During an academic career cut short by cancer at 10 years, Dan completed a remarkably comprehensive and coherent research agenda which has provided a complete answer to many of the questions surrounding the relationship between discrete-time ARCH models and continuous-time stochastic volatility models. Thus, this volume serves, in part, to honor Dan and his remarkable achievements (see Bollerslev and Rossi, 1995, for a comprehensive bibliography of Dan's work).

Discrete-time modelling of stock return and other financial time series has been dominated by models formulated to capture various qualitative features of this data. The ARCH model was introduced to capture the predictability of volatility or variance and the so-called volatility clustering as well as the thick-tailed marginal distribution of the time-series process. The single most widely used ARCH formulation is the GARCH(1, 1) model introduced by Bollerslev in 1986:

$$y_t = \mu_t + \sigma_t z_t \qquad \{z_t\} \text{ i.i.d., } \mathrm{E}[Z_t] = 0, \quad \mathrm{Var}(z_t) = 1;$$

$$\sigma_t^2 = \omega + \alpha(\sigma_t z_t)^2 + \beta\sigma_{t-1}^2.$$

This model allows for a smoothly evolving σ_t process as a function of past squared innovations. As a scale mixture model, the GARCH (1, 1) gives rise to a thick-tailed marginal distribution of the data. Application of GARCH models to financial data has produced two common empirical findings:

IGARCH

(1) The variance equation parameter estimates are often close to a unit root (for GARCH(1, 1), $\alpha + \beta = 1$), especially when the models are fit to high frequency data.
(2) In spite of the scale mixing achieved by the ARCH model, nonnormal innovations are required to fit the most stock return series.

The first finding of unit roots caused a good deal of confusion in the early ARCH literature. From a forecasting perspective, the corresponding IGARCH model with $\alpha + \beta \equiv 1.0$ behaves like a random walk; i.e., $\mathrm{E}_t(\sigma_{t+s}^2) = \sigma_{t+1}^2 + (s-1)\omega \to \infty$ almost surely. However, as Nelson shows in "Stationarity and Persistence in the GARCH(1, 1) Model," the persistence of a shock to the conditional variance should be very carefully interpreted. The behavior of a martingale can differ markedly from the behavior of a random walk. Strict stationarity and ergodicity of the GARCH(1, 1) model require geometric convergence of $\{\beta + \alpha z_t^2\}$, or $\mathrm{E}[\ln(\beta + \alpha z_t^2)] < 0$, which is a much less stringent condition than arithmetic convergence, or $\mathrm{E}[\beta + \alpha z_t^2] < 1$, required for the model to be covariance stationary. Thus, IGARCH models can be strictly stationary models in which no second moments exist. This important insight also underlies subsequent work on the consistency and asymptotic normality of maximum-likelihood-based estimators for GARCH-type models.

Another problematic feature of the GARCH formulation is that parameter inequality restrictions are required to keep the variance function positive. This, coupled with the tendency of GARCH model outliers to be negative, led to the development of the EGARCH model introduced in Nelson's two chapters, "Modelling Stock Market Volatility Changes" and "Conditional Heteroskedasticity in Asset Returns: A New Approach." In the EGARCH model, $\ln(\sigma_t^2)$ is parameterized as an ARMA model in the absolute size and the sign of the lagged innovations. For example, the AR(1)-EGARCH model is given by

$$\ln(\sigma_t^2) = \omega + \beta \ln(\sigma_{t-1}^2) + \theta z_{t-1} + \gamma\left[|z_{t-1}| - \mathrm{E}|z_{t-1}|\right].$$

The log formulation enforces variance positivity and the use of both the absolute size and sign of lagged innovations gives the model some robust variance estimation characteristics. As in the GARCH(1, 1) model, γ, $\beta > 0$, large price changes are still followed by large price changes, but with $\theta < 0$ this effect is accentuated for negative price changes, a stylized feature of equity returns often referred to as the "leverage effect." Whether due to changes in leverage or not, a stock market crash is typically followed by a period of much higher volatility than a corresponding upward run in the prices, and the EGARCH model was an instant hit since it parsimoniously captured this phenomenon. In the short time since its introduction more than 100 empirical studies have employed the model, and EGARCH estimation is now also available as a standard procedure in a number of commercial statistical software packages.

While the univariate EGARCH model has enjoyed considerable empirical success, many interesting questions in financial economics necessarily call for a multivariate modelling approach. However, the formulation of multivariate ARCH models poses a number of practical problems, including parameter parsimony and positive definiteness of the conditional covariance matrix estimators. The bivariate version of the EGARCH model in Braun, Nelson, and Sunier ("Good News, Bad News, Volatility, and Betas"), designed explicitly to capture any "leverage effects" in the conditional β's of equity returns, represents a particularly elegant solution to both of these problems.

The empirical finding that the persistence in volatility generally increases as the frequency at which the data is sampled increases suggests that it would be interesting to conduct a limiting experiment with ARCH models (obviously, the fact that stock prices are quoted in eighths means that this limiting experiment will ultimately break down at some intraday sampling frequency). In "ARCH Models as Diffusion Approximations,"

Nelson establishes weak convergence results for sequences of stochastic difference equations (e.g., ARCH models) to stochastic differential equations as the length of the sampling interval between the observations diminishes. For instance, consider the sequence of GARCH(1, 1) models observed at finer and finer time intervals h with conditional variance parameters $\omega_h = \omega h$, $\alpha_h = \alpha(h/2)^{1/2}$, and $\beta_h = 1 - \alpha(h/2)^{1/2} - \theta h$, and conditional mean $\mu_h = hc\sigma_t^2$. Under suitable regularity conditions, the diffusion limit of this process equals

$$dy_t = c\sigma_t^2\, dt + \sigma_t\, dW_{1,t},$$

$$d\sigma_t^2 = \left(\omega - \theta\sigma_t^2\right) dt + \alpha\sigma_t^2\, dW_{2,t},$$

where $W_{1,t}$ and $W_{2,t}$ denote independent Brownian motions. Similarly, the sequence of AR(1)-EGARCH models, with $\beta_h = 1 - \beta h$ and the other parameters as defined in this chapter, converges weakly to the

$$d(\ln(y_t)) = \theta\sigma_t^2\, dt + \sigma_t\, dW_{1,t},$$

$$d\left(\ln\left(\sigma_t^2\right)\right) = -\beta\left(\ln\left(\sigma_t^2\right) - \alpha\right) dt + dW_{2,t},$$

diffusion processes commonly employed in the theoretical options pricing literature.

The continuous-record asymptotics introduced by Nelson can be applied to a number of other issues that arise in the use of ARCH models. For instance, one could regard the ARCH model as merely a device which can be used to perform filtering or smoothing estimation of unobserved volatilities. Nelson's insights are that, even when misspecified, appropriately defined sequences of ARCH models may still serve as consistent estimators for the volatility of the true underlying diffusion, in the sense that the difference between the true instantaneous volatility and the ARCH filter estimates converges to zero in probability as the length of the sampling frequency diminishes. Although the formal proofs for this important result as developed in "Filtering and Forecasting with Misspecified ARCH Models I" are somewhat complex, as with most important insights, the intuition is fairly straightforward. In particular, suppose that the sample path for the instantaneous volatility process, $\{\sigma_t^2\}$, is continuous almost surely. Then for every $\xi > 0$ and every $t > 0$ there exists a $\delta > 0$ such that $\sup_{t-\delta \leq \tau \leq t} |\sigma_\tau^2 - \sigma_t^2| < \xi$. Now, divide this interval into N equal time segments. Given the drift, $\mu_{t-\delta}$, and the instantaneous volatility, $\sigma_{t-\delta}^2$, the N increments $(y_{t-(i-1)\delta/N} - y_{t-i\delta/N})$ $i = 1, 2, \ldots, N$ are then approximately i.i.d. normally distributed with mean $N^{-1}\delta\mu_{t-\delta}$ and variance $N^{-1}\delta\sigma_{t-\delta}^2$. A natural estimate for σ_t^2 is therefore

$$\hat{\sigma}_t^2(\delta, N) = \delta^{-1} \sum_{i=1,N} \left(y_{t-(i-1)\delta/N} - y_{t-i\delta/N}\right)^2,$$

which under suitable moment conditions converges in probability to σ_t^2 as $\delta \to 0$ and $N \to \infty$ by a Law of Large Numbers. Note that the drift term is of second-order importance so that a failure to account for the drift does not affect the consistency of the estimates. Now consider the GARCH(1, 1) estimates for σ_t^2:

$$\hat{\sigma}_t^2 = \omega + \alpha\hat{\varepsilon}_{t-1}^2 + \beta\hat{\sigma}_{t-1}^2 = \omega(1-\beta)^{-1} + \sum_{i=0,\infty} \alpha\beta^i\hat{\varepsilon}_{t-1-i}^2.$$

When applied at increasingly higher sampling frequencies, the corresponding GARCH(1, 1) filter with $\omega_h = \omega h$, $\alpha_h = \alpha(h/2)^{1/2}$, and $\beta_h = 1 - \alpha(h/2)^{1/2} - \theta h$ effectively achieves consistency in the same manner, by estimating σ_t^2 as the average of an increasing number of squared residuals "close" to time t. The AR(1)-EGARCH diffusion approximation discussed above and many other ARCH filters share this smoothing property. Note that the autoregressive coefficients in this consistent GARCH(1, 1) filter, $\alpha_h + \beta_h = 1 - \theta h$, and the AR(1)-EGARCH filter, $\beta_h = 1 - \beta h$, both converge to unity as $h \to 0$. This feature of the consistent filters is important and provides a possible explanation for the widespread empirical findings of apparent IGARCH-type behavior with high-frequency financial data.

Of course, not all consistent ARCH filters will perform equally well in a given situation, so that issues of efficiency become important. The basic framework of continuous-record asymptotics developed in "Asymptotic Filtering Theory for Univariate ARCH Models" and "Asymptotic Filtering Theory for Multivariate ARCH Models" allows for a formal analysis along these lines. It is impossible to do full justice to these path-breaking approaches in a short summary such as this. However, a useful illustration is provided by considering one of the proposed approaches to estimation and filtering for discrete-time stochastic volatility models. In particular, consider the stochastic volatility model

$$y_{t+h} = y_t + h^{1/2}\sigma_t z_{t+h},$$
$$\ln(\sigma_{t+h}^2) = \ln(\sigma_t^2) + h^{1/2}\lambda v_{t+h},$$

where z_t and v_t are assumed to be i.i.d. normally distributed with correlation ρ. The linear Kalman filter advocated by a number of researchers provides a simple way of analyzing this model by transforming the measurement equation to $\ln(y_{t+h} - y_t)^2 = \mu + \ln(\sigma_t^2) + [\ln(z_{t+h}) - E(\ln(z_{t+h}))]$. In fact, Nelson was the first to employ this approach in his 1988 MIT thesis. Of course, since the measurement error is nonnormal, the linear Kalman filter is suboptimal. In particular, the asymptotically optimal linear Kalman filter achieves an asymptotic variance for $h^{-1/4}[\ln(\hat{\sigma}_t^2) - \ln(\sigma_t^2)]$ as $h \to 0$ equal to $\lambda\Psi(1/2)^{1/2}$, where $\Psi(x) \equiv d[\ln(\Gamma(x)]/dx$ denotes the Psi function. Meanwhile, the asymptotically optimal ARCH filter,

$$\ln(\hat{\sigma}_{t+h}^2) = \ln(\hat{\sigma}_t^2) + \rho\lambda(y_{t+h} - y_t)\hat{\sigma}_t^{-1}$$
$$+ \lambda(1-\rho^2)^{1/2}\left[\Gamma(1/2)^{-1/2}\Gamma(3/2)^{1/2}|y_{t+h} - y_t|\hat{\sigma}_t^{-1} - 2^{-1/2}\right],$$

yields an asymptotic variance for $h^{-1/4}[\ln(\hat{\sigma}_t^2) - \ln(\sigma_t^2)]$ as $h \to 0$ of $2^{1/2}\lambda(1 - \rho^2)^{1/2}$. Comparing these two asymptotic variances, it follows that the efficiency loss from using the suboptimal linear Kalman filter in this context may be quite substantial. A number of full information approaches for estimating stochastic volatility models have recently been proposed in the literature. As illustrated by this example, Nelson's work on continuous-record asymptotics is likely to provide invaluable insight on the formulation of filters and moment generating functions in the future analyses of such models.

The Nelson and Foster results on the asymptotic efficiency of various ARCH filters also can be used to characterize the impact of various kinds of misspecification. For instance, the use of absolute innovations rather than squared innovations in the Taylor–Schwert and EGARCH approaches provides filters with superior properties in the presence of thick-tailed innovation densities so often found to be necessary in empirical applications. Similarly, the importance of capturing "leverage effects" or asymmetries in the conditional variance function may also be analyzed by Nelson and Foster's filtering efficiency results.

The practical relevance of these asymptotic filtering formulae is tested in "Variance Filtering with ARCH Models: A Monte Carlo Investigation." ARCH model filters with both known and estimated parameters are found to perform very well in settings calibrated to actual data on stock returns, interest rates, and exchange rates.

Although many different sequences of misspecified ARCH models may provide consistent filtered volatility estimates, the forecasting performance across these filters is likely to be very different. This issue is addressed in Nelson and Foster, "Filtering and Forecasting with Misspecified ARCH Models II: Making the Right Forecast with the Wrong Model." While the higher order moments can be ignored in the continuous-time limit, correctly modelling the first two conditional moments turns out to be crucial for generating accurate volatility forecasts. In practice, some experimentation with alternative flexible functional forms designed to capture the relevant stylized facts for the data set at hand will be important for successful forecasting.

The filtering papers rely on a Markovian assumption for the continuous-time process. In most applications in finance theory, this is not a restrictive assumption, since the existence of a stochastic differential equation representation is typically assumed. However, in applications in which the dimension and nature of the state vector are not known, it may be more appropriate to assume a semi-martingale framework as in "Continuous Record Asymptotics for Rolling Sample Variance Estimators." Utilizing this assumption then allows for a formal analysis of the rolling-regression-type estimators of conditional variances often used by investment professionals.

In many applications, interest focuses on the actual estimation of the parameters of continuous-time diffusion models and corresponding specification testing. The discrete-time ARCH literature has provided a rich summary of the important qualitative features of the discrete-time data which the continuous-time model must be capable of exhibiting. More directly, ARCH models can be used to summarize the joint distribution of the discrete data with continuous-time estimators constructed by minimizing the discrepancy between the joint distribution of the discrete data implied by the continuous-time model and the fitted ARCH density. This is the approach used by Engle and Lee in "Estimating Diffusion Models of Stochastic Volatility."

In "Specification Analysis of Continuous Time Models in Finance," Gallant and Tauchen use their SNP estimator of the conditional density for the discrete data as a score generator and then match moments based on simulation from the assumed continuous-time model. Their SNP density estimator uses an ARCH-style leading term which captures the basic volatility persistence thus allowing for a more efficient nonparametric parameterization. The Gallant and Tauchen procedure also yields a moment-discrepancy diagnostic which can help the investigator determine which features of the data are not adequately captured in the continuous-time specification.

Hansen and Scheinkman take a different approach to the estimation of continuous-time processes with a fully observable state vector but with drift and scale functions that depend nonlinearly on the state. By constructing an appropriate infinitesmal generator, in "Back to the Future: Generating Moment Implications for Continuous-Time Markov Processes," Hansen and Scheinkman are able to construct moment conditions for discretely sampled data. These moment conditions can be used as the basis of a GMM estimation strategy. This procedure therefore allows for the selection of continuous-time parameter estimates to match the observed marginal properties (such as thick-tailness) and serial dependencies (such as volatility persistence) in the discrete data.

In "Nonparametric Pricing of Interest Rate Derivative Securities," Aït-Sahalia takes a semi-parametric approach in which the drift function is parametric and the variance function is estimated to match the marginal density of the discrete data. A nonparametric (often kernel) estimate of the marginal distribution of the discretely sampled data can then be mapped into a nonparametric estimate of the variance function. This procedure selects a diffusion model with the appropriate sort of scale mixing which will reproduce the marginal properties of the data. Aït-Sahalia illustrates this technique through the estimation of the stochastic differential equation followed by the short-term interest rate and computes nonparametric prices for bonds and bond options.

The progress in understanding, formulating, and estimating continuous-time models summarized in this volume will lead to a rapid growth in the empirical application of these models. Until recently, the choice of continuous-time models used in theoretical finance was dictated more by their mathematical tractibility than by the extent to which these models captured important features of the data. It is exciting to contemplate the changes in the class of continuous-time models which will arise as these models are subjected to routine specification testing.

Tim Bollerslev
Peter E. Rossi

PART I

UNDERSTANDING AND SPECIFYING THE DISCRETE TIME MODEL

I

MODELLING STOCK MARKET VOLATILITY CHANGES*

DANIEL B. NELSON

Graduate School of Business
University of Chicago
Chicago, Illinois 60637

1. INTRODUCTION

The crash of October 1987 reminded everyone that stock market volatility is subject to sudden and at best partially predictable shifts. Understanding how and why these shifts occur is important for many areas of economic theory, especially asset pricing, and is important for participants in asset markets, particularly options markets.

Research on the determinants of volatility changes has uncovered several factors that seem to be strongly associated with market volatility changes. These are a few of the more important:

(1) Positive serial correlation in volatility. As Mandelbrot (1963) put it "large changes tend to be followed by large changes—of either sign—and small changes tend to be followed by small changes..." (see also Fama, 1965; French *et al.*, 1987).

(2) Trading and nontrading days. Fama (1965) and French and Roll (1986) have documented that both trading and nontrading days contribute to market volatility. For example, stock market volatility tends to be higher on Mondays than on other days of the week, presumably because the

*Reprinted from Daniel B. Nelson, "Modelling Stock Market Volatility Changes," *ASA 1989 Proceedings of the Business and Economics Statistics Section*, pp. 93–98.

movement of stock prices on a typical Monday reflects information arriving over a 72-hour period, while on most other trading days, price movements reflect information arriving over a 24-hour period.

(3) Leverage effects. Black (1976) noted that when a stock's price drops (rises), the volatility of its return typically rises (falls). Leverage provides at least a partial explanation: A leveraged firm becomes more (less) leveraged when the value of its equity drops (rises). Black argued, however, that the measured effect of stock price changes on volatility was too large to be explained solely by leverage changes (see also Christie, 1982; French *et al.*, 1987).

(4) Recessions and financial crisis. Stock market volatility tends to be high during financial and economical crises such as recessions and bank panics (Schwert, 1988), and market volatility hit an historic high during the depressions of the 1930s (Officer, 1973). Since financial crises and recessions also tend to be accompanied by sharp drops in the stock market, this factor and factor (3) are difficult to distinguish.

(5) Nominal interest rates. Using postwar data, Fama and Schwert (1977), Christie (1982), and Glosten *et al.* (1989) have concluded that high levels of nominal interest rates are associated with high market volatility. Fama and Schwert also found high levels of inflation associated with high market volatility. It seems likely that these conclusions would be modified if prewar data were included, since the stock market volatility hit a historical high during the depression and yet nominal interest rates were very low (although real rates were very high) and prices were falling during the most several parts of the depression.[1]

2. ARCH MODELS

Probably the most important innovation in modelling volatility changes was the introduction by Engle (1982) of ARCH (autoregressive conditionally heteroskedastic) models. Engle's insight was to set the conditional variance of a series of prediction errors $\{\xi_t\}$ equal to some function of lagged errors, time, parameters, and predetermined variables:

$$\sigma_t^2 = \sigma^2(\xi_{t-1}, \xi_{t-2}, \ldots, t, x_t, b), \tag{2.1}$$

$$\xi_t = \sigma_t Z_t, \tag{2.2}$$

where $Z_t \sim$ i.i.d. with $\mathrm{E}(Z_t) = 0$, $\mathrm{E}(Z_t^2) = 1$.

[1]Some other factors may influence volatility: Glosten, *et al.* argued that volatility tends to be lower during December and higher during January. Hardouvelis (1988) argued that initial margin requirements are important determinants of market volatility. See, however, the critique of Schwert (1988) and Hsieh and Miller (1989) of Hardouvelis' methodology.

Next Engle chose a function form for $\sigma^2(\cdot)$:

$$\sigma_t^2 = \omega + \sum_{j=1}^{p} \alpha_j \xi_{t-j}^2, \tag{2.3}$$

where ω and $\{\alpha_j\}$, $j = 1, p$ are nonnegative constants. (This is necessary to keep σ_t^2 nonnegative.) The functional form (2.3) is often called ARCH(p), or just ARCH, though we will reserve this term to refer to any model of the form (2.1)–(2.2). The appeal of the model (2.3) lies in the way it captures the positive serial correlation in ξ_t^2: A high value of ξ_t^2 increases σ_{t+1}^2, which in turn increases the expectation of ξ_{t+1}^2, and so on. In other words, a large (small) value of ξ_t^2 tends to be followed by a large (small) value of ξ_{t+1}^2.

Bollerslev (1986) generalized the ARCH(p) model by introduction GARCH(p, q):

$$\sigma_t^2 = \omega + \sum_{i=1}^{q} \beta_i \sigma_{t-i}^2 + \sum_{j=1}^{p} \alpha_j \xi_{t-j}^2, \tag{2.4}$$

which allows a more parsimonious representation for σ_t^2 as a function of lagged values of ξ_t^2. Again ω, $\{\alpha_j\}$, $j = 1, p$, and $\{\beta_i\}$, $i = 1, q$ are nonnegative constants.

Seasonal and nontrading effects have been incorporated into GARCH models by making the intercept term ω in (2.4) time dependent (see, e.g., Baillie and Bollerslev, 1989).

Leverage effects are harder to accommodate, since in models (2.3) and (2.4) σ_t^2 is driven by past *squared* residuals, so that the signs of past residuals ξ_{t-k} do not influence σ_t^2. This motivated the Exponential ARCH model of Nelson (1988), which uses (2.5)–(2.6) in place of (2.3) or (2.4):

$$\ln(\sigma_t^2) = \alpha_t + \sum_{k=1}^{\infty} \beta_k g(Z_{t-k}), \qquad \beta_1 = 1 \tag{2.5}$$

$$g(Z_t) \equiv \theta Z_t + \gamma[|Z_t| - \mathrm{E}|Z_t|], \tag{2.6}$$

where $\{\alpha_t\}$ is either deterministic or else a function of predetermined variables. This model sets $\ln(\sigma_t^2)$ equal to a distributed lag of a nonlinear transform $g(\cdot)$ of the normalized residuals $\{Z_t\}$. The function $g(\cdot)$ captures both factors (1) and (3) of Section 1: If $\gamma > 0$, the second term in $g(\cdot)$ implies that σ_t^2 tends to rise (fall) when $|Z_t|$ is larger (smaller) than expected. If $\theta < 0$, σ_t^2 tends to rise (fall) when Z_t is negative (positive). Calendar effects and predetermined variables such as interest rates can be accommodated in the $\{\alpha_t\}$ sequence.

A recent study by Pagan and Schwert (1989) compares GARCH and Exponential ARCH models in stock volatility forecasting. Using monthly returns data from 1834–1925, Pagan and Schwert compare the forecasting performance of various parametric and nonparametric models of stock volatility, including GARCH, Exponential ARCH, and the regime-switching model of Hamilton (1989). They find that the nonparametric models are the best predictors in sample, but perform poorly out of sample. Among the parametric models, Exponential ARCH performed the best both in and out of sample.

3. EXAMPLE: VOLATILITY ON THE STANDARD 90 INDEX[2]

Many researchers (e.g., Office, 1973; Pagan and Schwert, 1989) have noted that stock volatility was extraordinarily high during the Great Depression. Volatility was not only high during this period, it was itself quite variable, changing rapidly. This makes the period an especially interesting one to study.

3.1. The Data

From January 1928 through July 1956, Standard Statistics Co. recorded the daily value of an index of 90 heavily traded stocks, the "Standard 90." In this section, we fit an Exponential ARCH model to the volatility on this index and use the model to examine the first three of the factors determining stock volatility discussed in the Introduction (serial correlation, trading and nontrading days, and leverage effects). Since we have no good daily interest rate or production index data, the fourth and fifth factors (recessions, financial crises, and nominal interest rates) are harder to account for separately, and we ignore them.

3.2. The Model

We assume that the conditional variance process follows an Exponential ARCH model with an ARMA representation for $\ln(\sigma_t^2)$:

$$\ln(\sigma_t^2) = \alpha_t + (1 + \Psi_1 L + \cdots + \Psi_q L^q)(1 - \Delta_1 L - \cdots - \Delta_p L^p)\, g(z_{t-1}) \tag{3.1}$$

To account for nontrading days, we let the deterministic factor α_t be given by

$$\alpha_t = \alpha + \ln(1 + N_t \delta), \tag{3.2}$$

[2]The empirical results of this section are reported in a previous working paper, Nelson (1989). Ken French, G. W. Schwert, and Robert Stambaugh kindly provided the data.

where N_t is the number of nontrading days between the current trading day t and the previous trading day $t-1$. δ is the contribution of a nontrading day to σ_t^2 as a function of the contribution of a trading day. For example, until June 1952, the stock markets were open part of Saturday, so that on a typical Monday during the sample period, $N_t = 1$. If $\delta = .2$, then σ_t^2 on a typical Monday would be 20% greater than on a typical Tuesday.[3]

To account for the possibility of a thick-tailed conditional distribution, we assume that $Z_t \sim$ i.i.d. with a GED distribution (Subbotin, 1923). The GED is a family of distributions including the normal, as well as distributions thicker- or thinner-tailed than the normal. The density for a GED with zero mean and unit variance is

$$f(z) = \frac{v \exp\left\{-\frac{1}{2}|z/\lambda|^{v}\right\}}{\lambda 2^{(1+1/v)}\Gamma(1/v)}, \qquad -\infty < z < \infty, \tag{3.3}$$

where $\Gamma(\cdot)$ is the gamma function, and

$$\lambda \equiv \left[2^{(-2/v)}\Gamma(1/v)/\Gamma(3/v)\right]^{1/2}. \tag{3.4}$$

The parameter v is a measure of tail thickness: The fourth moment of z is $\Gamma(5/v)\cdot\Gamma(1/v)/(\Gamma(3/v)^2)$, which is decreasing in v. When $v = 2$, $Z \sim N(0,1)$. When $v > (<) 2$, Z is more thin- (thick-)tailed then the normal.

To account for nonsynchronous trading and the possible effect of σ_t^2 on required returns, we model returns as

$$R_t = a + bR_{t-1} + c\sigma_t^2 + \xi_t, \tag{3.5}$$

where a, b, and c are constants, and R_t is the logarithmic capital gain on the Standard 90 stock index. Intuitively, we might expect c, the "risk premium" term, to be positive. As Glosten *et al.* (1989) show, however, it is possible to construct theoretical models in which c takes either a positive or a negative sign, so there is no strong reason to except c to be positive.

3.3. Estimates

We fit the model by maximum likelihood, using ARMA models for $\ln(\sigma_t^2)$ of order up to ARMA(4, 4). The likelihood values are reported in Nelson (1989). Interestingly, both the Schwarz (1978) Criterion and the AIC selected an ARMA(2, 1) model, the same model selected in Nelson (1989) for the CRSP value-weighted market index from 1962–1987.

[3]Since the empirical work reported here was performed, Mulherin and Gerety (1988) have found that, when the market was open on Saturday, the volatility on was lower on Saturday than other trading days, perhaps because of shorter Saturday trading hours. Including a Saturday dummy in σ_t would probably improve the model.

TABLE 1 Parameter Estimates

Parameter	Coefficient estimate	Standard error
α	−9.7616	0.3060
δ	0.3358	0.0433
γ	0.2173	0.0166
Δ_1	1.8750	0.0197
Δ_2	−0.8751	0.0197
Ψ	−0.9722	0.0063
θ	−0.1138	0.0114
a	$7.1303 \cdot 10^{-4}$	$9.4469 \cdot 10^{-5}$
b	0.0703	0.0104
c	−2.1384	0.8510
v	1.2430	0.0212

Table 1 reports the parameter estimates for the ARMA(2, 1) mode.[4]

The largest AR root in the $\ln(\sigma_t^2)$ process is estimated at about 0.9988, with a standard error of about 0.005, so that the T statistic for a unit root test is approximately -2.313. Although a standard test rejects a unit root at the 1% level, this should be interpreted with caution, since the asymptotic properties of the test statistics in ARCH models (even in the absence of a unit root) are unknown. However the point estimate indicates substantial persistence—i.e., the half-life of a shock associated with the largest AR root is estimated to be about two years.

We find evidence of thick tails in the conditional distribution—i.e., v is significantly below 2, the value for v when $Z \sim N(0,1)$. This is in keeping with the findings of Engle and Bollerslev (1986) in their study of the foreign exchange markets.

3.4. Leverage and Magnitude Effects

Both γ and θ have the expected sign ($\theta < 0$ and $\gamma > 0$) and are significantly different from zero at any standard level, implying that both the magnitude and the sign of market returns are important determinants of conditional variance. The $g(z)$ function looks like Figure 1.

The $g(Z_t)$, the innovation in $\ln(\sigma_{t+1}^2)$, is negative (positive) when the surprise component of returns is very small (large). Nevertheless, $g(Z_t)$ is very sensitive to the sign of Z_t, since it takes a $1\frac{1}{2}$ standard deviation positive Z_t to increase σ_{t+1}^2 but only a $\frac{1}{2}$ standard deviation negative Z_t.

[4]The estimated correlation matrix of the parameter estimates is reported in Nelson (1989).

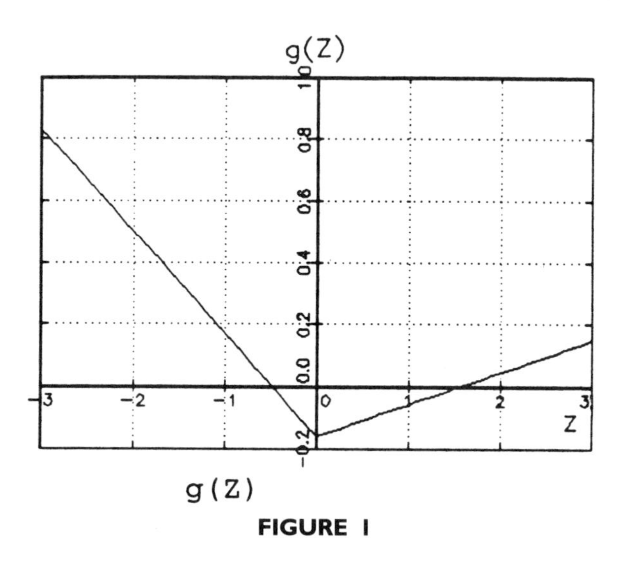

FIGURE 1

To see the importance of leverage-related effects, it is useful to examine the interrelation of σ_t^2 and changes in the level of the market through the sample period. Figures 2 and 3 plot the log of the Standard 90 stock index and the daily conditional standard deviation σ_t.

Casual inspection of Figures 2 and 3 confirms the insight provided by Figure 1: Market drops (rises) are strongly associated with increases (decreases) in σ_t^2. This is even more apparent when we examine some of the episodes of greatest market volatility during the period. Figure 4 plots

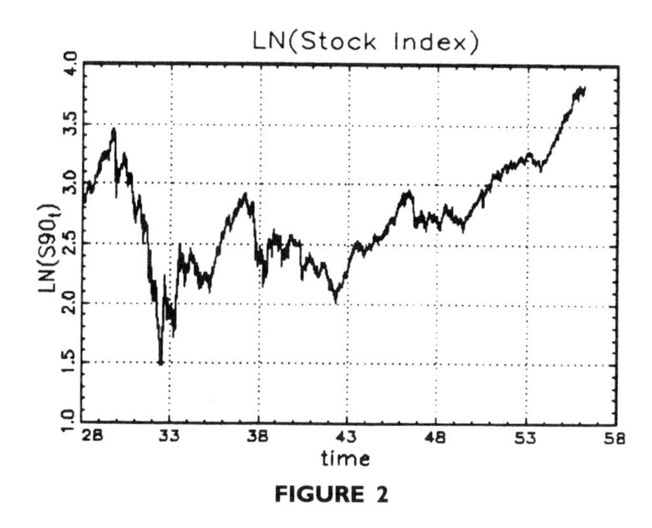

FIGURE 2

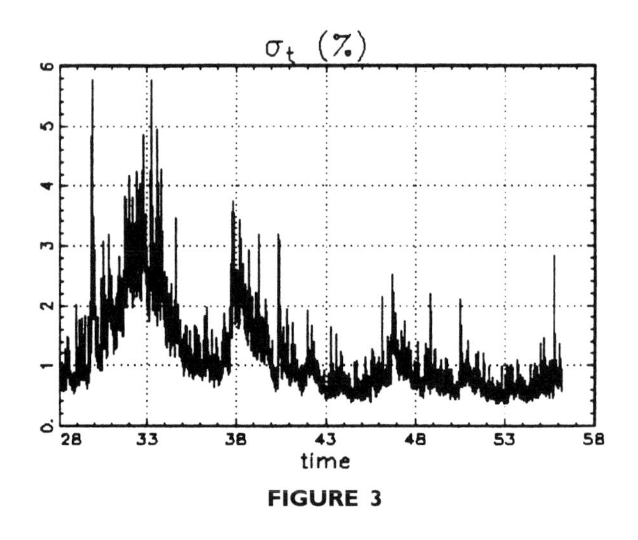

FIGURE 3

returns (the black dots) and the one-day ahead 99% confidence intervals for returns for the last half of 1929. Since $\upsilon \approx 1.29$, $P[|Z_t| < 3] \approx .99$, so the ex ante 99% confidence intervals are approximately $\pm 30\sigma_t$. (The mean term is usually small.)

On October 28, 1929, "Black Monday," the market dropped about 13%. The ex ante 99% confidence interval for that day was about $\pm 7\%$—much higher than a few weeks before, when the one-day ahead confidence interval had been about $\pm 2\%$. Black Monday is the only severe outlier over the period. The large, negative, market moves signalled higher volatility ahead.

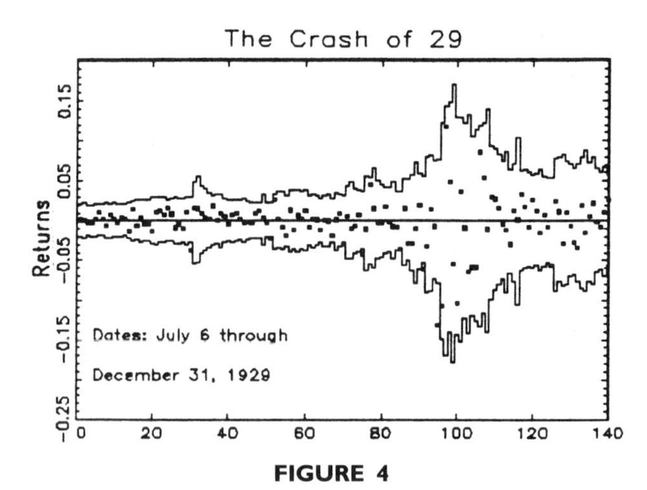

FIGURE 4

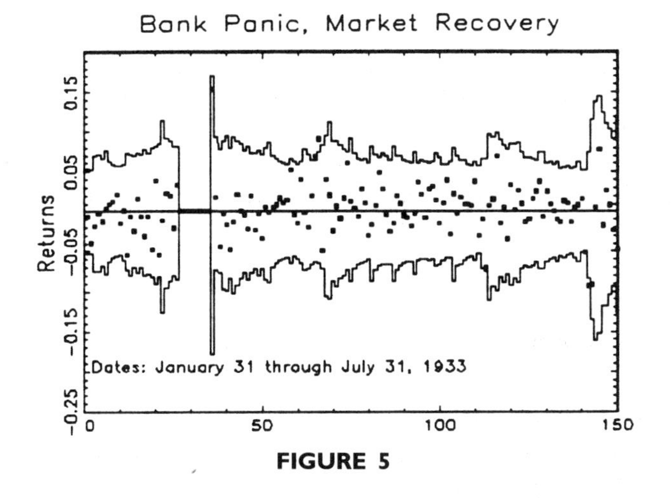

FIGURE 5

Figure 5 depicts returns and confidence intervals in the period surrounding the banking panic and bank holiday of March 1933. The period of zero returns and confidence intervals reflects the 11-day period March 4–14, 1933 when the stock exchanges were closed. The model appears quite successful in forecasting the high volatility of the period: The sharp downward movements of the preceding months (and years!) signalled high volatility, as did the high value of N_t when the market open on March 15. The most severe outliers are given in Table 2.

These sharp market moves took place during periods that were otherwise calm on the market, as evidenced by the low values of σ_t (compare with Figure 3). It's hard to see how any model could predict the market volatility resulting from, say, Eisenhower's heart attack. However Table 2 does give evidence that the GED specification does not permit the tails of

TABLE 2 Most Severe Outliers

Date	R_t(%)	σ_t(%)	Newspaper explanation[a]
9/26/55	−6.85	0.63	Eisenhower heart attack
6/26/50	−5.53	0.79	Korean War begins
11/3/48	−4.72	0.70	Truman defeats Dewey
7/6/55	+3.51	0.50	GM stock split triggers market optimism

[a] These are taken from Cutler *et al.* (1988).

the conditional distribution to be thick enough: The return on September 26, 1955, was almost 11 standard deviations away from zero. Under correct specification, the expected number of outliers this large or larger in a 28-year sample is less than $5 \cdot 10^{-6}$! The contrast, returns on Black Monday and on May 10, 1940, were slightly more than five conditional standard deviations below zero: The expected number of deviations at least this large in the sample period is about 1.9.

3.5. Formal Specification Tests

In essence, we have just performed a series of informal specification tests. More formal specification tests are also available. We can, for example, apply the results of Newey (1985), who developed conditional moment tests for GMM (generalized method of moments) models. This test is based on orthogonality conditions implied by correct specification. Table 3 reports results for a test based on 14 orthogonality conditions. The first

TABLE 3 Specification Test

	Orthogonality conditions $E[R_{j,t}] = 0, j = 1, 14$	$T^{-1} \sum_{t=1}^{T} R_{j,t}$	t Value
1	$E(Z_t) = 0$	−0.024	−1.278
2	$E(Z_t^2) - 1 = 0$	0.011	0.260
3	$E\|Z_t\| - \lambda 2^{1/v}\Gamma(2/v)/\Gamma(1/v) = 0$	−0.001	−0.073
4	$E[Z_t(\|Z_t\| - E\|Z_t\|)] = 0$	−1.103	−3.744
5	$E[(Z_t^2 - 1)(Z_{t-1}^2 - 1)] = 0$	0.079	0.884
6	$E[(Z_t^2 - 1)(Z_{t-2}^2 - 1)] = 0$	−0.007	−0.076
7	$E[(Z_t^2 - 1)(Z_{t-3}^2 - 1)] = 0$	0.067	0.826
8	$E[(Z_t^2 - 1)(Z_{t-4}^2 - 1)] = 0$	−0.019	−0.320
9	$E[(Z_t^2 - 1)(Z_{t-5}^2 - 1)] = 0$	0.047	0.854
10	$E(Z_t Z_{t-1}) = 0$	0.031	1.575
11	$E(Z_t Z_{t-2}) = 0$	−0.044	−3.879
12	$E(Z_t Z_{t-3}) = 0$	0.008	0.688
13	$E(Z_t Z_{t-4}) = 0$	0.021	1.890
14	$E(Z_t Z_{t-5}) = 0$	0.010	0.826

Notes:

χ^2 statistic for conditional moment test using the first nine orthogonality conditions = 30.37. The probability value = $3.8 \cdot 10^{-4}$.

χ^2 statistically for conditional moment test using all 14 orthogonality conditions = 57.13. The probability value = $13.7 \cdot 10^{-7}$.

four conditions compare the unconditional moments of the fitted $\{Z_t\}$ series with the moments of the standardized GED distribution. The next five conditions check for serial correlation in $\{Z_t^2\}$, which, if present, is a sign of conditional heteroskedasticity not captured by the model. The final five conditions check for serial correlation in $\{Z_t\}$, which, if present, is a sign of misspecification of the conditional mean.

The asymptotic t statistics and p values given in the table assume that the regularity conditions in Newey (1985) hold, which is as yet unproven. (Asymptotic distribution theory for ARCH models of all types is seriously underdeveloped.) Assuming that the regularity conditions are satisfied, the conditional moment test has mixed implications for the model. On the positive side, there is little evidence of conditional heteroskedasticity in the residuals (conditions 5–9), and the unconditional variance is approximately correct (i.e., the $\{Z_t\}$ have approximately unit variance, Condition 2). Less positively, the distribution of $\{Z_t\}$ is strongly skewed to the left, so that the symmetry test in Condition 4 has a T statistic of about -3.7. Given the spectacular market crashes of 1929 through 1932, this is not too surprising, but it suggests the need for a more general distribution than the GED for $\{Z_t\}$. There is also evidence of serial correlation in the fitted $\{Z_t\}$. Overall, the model is rejected at any standard significance level in both tests, with probability values of $3.8 \cdot 10^{-4}$ and $3.7 \cdot 10^{-7}$ in the 9 and 14 condition tests, respectively.

4. CONCLUSION

The Exponential ARCH model provides an alternative to standard GARCH models in modelling and forecasting conditionally heteroskedastic processes. Its principle virtue is that it accounts for leverage effects, which appear to be very important in determining stock volatility. It also has the virtue of imposing no inequality constraints on the parameters.

In applying the model to the 1928–1956 stock market, we found that the model was able to forecast many episodes of high volatility, and performed quite well in episodes during which volatility rose sharply for an extended period, such as the crash of 1929. In periods of sudden but short-lived volatility, as occurred during Eisenhower's heart attack, the model does much less well: There seem to be thicker tails in the conditional distribution than can be allowed by the normal or even the GED distribution. There is also evidence of conditional skewness and serially correlated residuals in the model. The model is a step in the right direction, but much work remains to be done in this area.

ACKNOWLEDGMENT

I would like to thank Tim Bollerslev for serving as referee.

REFERENCES

Baillie, R. T., and Bollerslev, T. (1989). "Intra Day and Inter Market Volatility in Foreign Exchange Rates," Mimeo. Northwestern University, Evanston, IL.

Black, F. (1976). Studies of stock market volatility changes. *Proceedings of the American Statistical Association, Business and Economic Statistics Section*, pp. 177-181.

Bollerslev, T. (1986). Generalized autoregressive conditional heteroskedasticity. *Journal of Econometrics* **31**, 307-327.

Christie, A. A. (1982). The stochastic behavior of common stock variances: Value, leverage and interest rate effects. *Journal of Financial Economics*, **10**(4), 407-432.

Cutler, D. M., Poterba, J. M., and Summers, L. H. (1988). What moves stock prices? *NBER Working Paper* No. 2538.

Engle, R. F. (1982). Autoregressive conditional heteroskedasticity with estimates of the variance of United Kingdom inflation. *Econometrica* **50**(4), 987-1008.

Engle. R. F., and Bollerslev, T. (1986). Modeling the persistence of conditional variances. *Econometric Reviews* **5**(1), 1-50.

Fama, E. F. (1965). The behavior of stock market prices. *Journal of Business* **38**, 34-105.

Fama, E. F., and Schwert, G. W. (1977). Asset returns and inflation. *Journal of Financial Economics* **5**, 115-146.

French, K. R., and Roll, R. (1986). Stock return variances: The arrival of information and the reaction of traders. *Journal of Financial Economics* **17**, 5-26.

French, K. R., Schwert, G. W., and Stambaugh, R. F. (1987). Expected stock returns and volatility. *Journal of Financial Economics* **19**(1), 3-30.

Glosten, L. R., Jagannathan, R., and Runkle, D. (1989). "Relationship Between the Expected Value and the Volatility of the Nominal Excess Return on Stocks," Banking Research Center Working Paper No. 166. Northwestern University, Evanston, IL.

Hamilton, J. D. (1989). A new approach to the economic analysis of nonstationary time series and the business cycle. *Econometrica* **57**(2), 357-384.

Hardouvelis, G. (1988). Margin requirements and stock market volatility. *Federal Reserve Bank of New York Quarterly Review*, pp. 80-89.

Hsieh, D. A., and Miller, M. H. (1989). Margin regulation and stock market volatility. *Journal of Finance*.

Mandelbrot, B. (1963). The variation of certain speculative prices. *Journal of Business* **36**, 394-419.

Mulherin, J. H., and Gerety, M. S. (1988). "Trading Volume on the NYSE in the Twentieth Century: A Daily and Hourly Analysis," Mimeo. U.S. Securities Exchange Commission, Washington, D.C.

Nelson, D. B. (1988). The time series behavior of stock market volatility and returns. Doctoral Dissertation, Massachusetts Institute of Technology, Cambridge, MA (unpublished).

Nelson, D. B. (1989). "Conditional Heteroskedasticity in Asset Returns: A New Approach," Working Paper, Series in Economics and Econometrics No. 89-73. University of Chicago, Graduate School of Business, Chicago.

Newey, W. K. (1985). Generalized method of moments specification testing. *Journal of Econometrics* **29**(3), 229-256.

Officer, R. R. (1973). The variability of the market factor of the New York Stock Exchange. *Journal of Business* **46**, 434-453.

Pagan, A. R., and Schwert, G. W. (1989). Alternative models for conditional stock volatility. *Journal of Econometrics*.

Schwarz, G. (1978). Estimating the dimension of a model. *Analysis of Statistics* **6**, 461-464.

Schwert, G. W. (1988). "Business Cycles, Financial Crises and Stock Volatility," Mimeo. University of Rochester, William E. Simon Graduate School of Business, Rochester, NY.

Subbotin, M. T. (1923). On the law of frequency of errors. *Matematicheskii Sbornik*, **31**, 296-301.

2

STATIONARITY AND PERSISTENCE IN THE GARCH(1, 1) MODEL*

DANIEL B. NELSON

Graduate School of Business
University of Chicago,
Chicago, Illinois 60637

1. INTRODUCTION

Since the seminal work of Engle (1982), ARCH (autoregressive conditionally heteroskedastic) models have been widely used to model time-varying volatility and the persistence of shocks to volatility. One member of the family of ARCH processes, GARCH(1, 1), since its introduction by Bollerslev (1986) has been especially popular in econometric modeling. In this chapter we investigate the GARCH(1, 1) model in depth. Analyzing the properties of GARCH(1, 1) not only sheds light on the behavior of this commonly used model, it also provides interesting and natural examples of several important concepts in probability theory, including the distinction between martingales and random walks, between strict and weak stationarity, and between almost sure and L^p convergence.

*From *Econometric Theory* **6** (1990), 318–334, copyright © 1990. Reprinted with permission of Cambridge University Press.

To define the model let

$$\{z_t\}_{t=-\infty,\infty} \sim \text{i.i.d. (independent and identically)},$$

$$z_t^2 \text{ nondegenerate}, P[-\infty < z_t < \infty] = 1, \quad (1.1)$$

$$\sigma_t^2 = \omega + \beta\sigma_{t-1}^2 + \alpha\xi_{t-1}^2, \quad (1.2)$$

with

$$\xi_t = \sigma_t \cdot z_t \quad (1.3)$$

where $\omega \geq 0$, $\beta \geq 0$, and $\alpha > 0$. In most papers using GARCH(1, 1), a further restriction has been placed on $\{z_t\}$, namely that

$$\mathrm{E}[z_t] = 0, \qquad \mathrm{E}[z_t^2] = 1. \quad (1.4)$$

Under (1.4), σ_t^2 is the conditional variance of ξ_t given the history of the system. If we assume $\mathrm{E}[z_t^2] = 1$ but allow $\mathrm{E}[z_t] \neq 0$, then σ_t^2 is the conditional second moment of ξ_t. If we allow the second moment of z_t to be infinite or undefined, then σ_t^2 is a conditional scale parameter. Since the restrictions $\mathrm{E}[z_t] = 0$, $\mathrm{E}[z_t^2] = 1$ play no role in the main results of this chapter, we adopt the less stringent condition (1), along with the replacement

$$\mathrm{E}[\ln(\beta + \alpha z_t^2)] \text{ exists}. \quad (1.5)$$

Note that (1.5) does not require that $\mathrm{E}[\ln(\beta + \alpha z_t^2)]$ be finite, only that the expectations of the positive and negative parts of $\ln(\beta + \alpha z_t^2)$ are not both infinite. For example, (1.5) holds trivially if $\beta > 0$.

Repeatedly substituting for σ_{t-i}^2 in (1.2), we have, for $t \geq 2$,

$$\sigma_t^2 = \sigma_0^2 \prod_{i=1}^{t} (\beta + \alpha z_{t-i}^2) + \omega\left[1 + \sum_{k=1}^{t-1} \prod_{i=1}^{k} (\beta + \alpha z_{t-i}^2)\right]. \quad (1.6)$$

Equation (1.6) holds for $t = 1$ as well, if we take a sum of the form $\Sigma_{k=1,0} X_k$ to equal zero. We can close the system in one of two ways: either by defining a probability measure μ_0 for the starting value σ_0^2 or by assuming that the system extends infinitely far into the past. To start the system at time 0, let (1.7)–(1.9) hold:

$$\mathrm{P}[\sigma_0^2 \in \Gamma] = \mu_0(\Gamma) \qquad \text{for all } \Gamma \in B, \quad (1.7)$$

where B denotes the Borel sets on $[0, \infty)$,

$$\mu_0((0, \infty)) = 1, \quad (1.8)$$

that is, σ_0^2 is strictly positive and finite with probability one, and

$$\sigma_0^2 \text{ and } \{z_t\}_{t=0,\infty} \text{ are independent}. \quad (1.9)$$

We call the model for $\{\sigma_t^2, \xi_t\}_{t=0,\infty}$ defined by (1.1)–(1.3), (1.5), and

(1.6)–(1.9) the *conditional* model. We denote the probability measure for σ_t^2 by μ_t, and the measure for σ_t^2 given $\sigma_0^2 = x$ by $\mu_{t,x}$, that is, for any $\Gamma \in B$, $\mathrm{P}[\sigma_t^2 \in \Gamma] = \mu_t(\Gamma)$ and $\mathrm{P}[\sigma_t^2 \in \Gamma \mid \sigma_0^2 = x] = \mu_{t,x}(\Gamma)$.

To extend the process infinitely far into the past, we define the *unconditional* process $\{{}_u\sigma_t^2, {}_u\xi_t\}_{t=-\infty,\infty}$ by (1.1),

$$ {}_u\sigma_t^2 \equiv \omega\left[1 + \sum_{k=1}^{\infty} \prod_{i=1}^{k} \left(\beta + \alpha z_{t-i}^2\right)\right], \quad \text{and} \tag{1.10}$$

$$ {}_u\xi_t \equiv {}_u\sigma_t \cdot z_t. \tag{1.3$'$}$$

If $\omega = 0$, define ${}_u\sigma_t^2 \equiv {}_u\xi_t \equiv 0$ for all t. If ${}_u\sigma_t^2 = \infty$ and $z_t = 0$, define ${}_u\xi_t = \infty \cdot 0 \equiv 0$.

Without further restrictions, there is no guarantee that ${}_u\sigma_t^2$ and ${}_u\xi_t$ are finite. So for the time being, we define these processes on a subset of the extended real plane—that is, $0 \le {}_u\sigma_t^2 \le \infty$, and $-\infty \le {}_u\xi_t \le \infty$. If ${}_u\sigma_t^2$ has a well-defined probability measure on $[0, \infty)$, we call this measure μ_∞.

In the rest of this chapter we address the following questions: When is μ_∞ well defined? When do $\mu_t \to \mu_\infty$ and $\mu_{t,x} \to \mu_\infty$ as $t \to \infty$?[1] What is the behavior of $\{\sigma_t^2, \xi_t\}_{t=0,\infty}$ when μ_∞ is not well defined? When does μ_∞ admit finite moments, that is, for any real number q, when is $\int \sigma^{2q}\, d\mu_\infty < \infty$, and when does $\int \sigma^{2q}\, d\mu_t \to \int \sigma^{2q}\, d\mu_\infty$ as $t \to \infty$? Under what circumstances do shocks to ${}_u\sigma_t^2$ and σ_t^2 (defined variously in terms of z_k, ${}_u\xi_k$ or ξ_k for some fixed k) decay as $t \to \infty$? Section 2 presents the results. All proofs are in the appendix.

2. MAIN RESULTS

Throughout the chapter, time subscripts t are assumed nonnegative in the conditional model, but may be positive or negative in the unconditional model. All limits are taken as $t \to \infty$, except where otherwise indicated.

The first theorem considers the case $\omega = 0$. Since ${}_u\sigma_t^2 \equiv 0$ for all t in this case, only the conditional model is of interest.

THEOREM 1. *Let* $\omega = 0$. *Then*

$$\sigma_t^2 \to \infty\ a.s.\ (\textit{almost surely})\ \textit{if and only if}\ (\textit{iff})\ \mathrm{E}\left[\ln\left(\beta + \alpha z_t^2\right)\right] > 0, \tag{2.1}$$

$$\sigma_t^2 \to 0\ a.s.\ \textit{iff}\ \mathrm{E}\left[\ln\left(\beta + \alpha z_t^2\right)\right] < 0. \tag{2.2}$$

[1]The convergence is taken to be weak convergence of measures, that is, $\mu_t \to \mu_\infty$ and $\mu_{t,x} \to \mu_\infty$ mean that for any bounded, continuous function f from $[0,\infty)$ into R^1, $\int f\, d\mu_t \to \int f\, d\mu_\infty$ and $\int f\, d\mu_{t,x} \to \int f\, d\mu_\infty$, respectively, as $t \to \infty$.

If $\mathrm{E}[\ln(\beta + \alpha z_t^2)] = 0$, $\ln(\sigma_t^2)$ *is a driftless random walk after time* 0, *with*

$$\limsup_{t\to\infty} \sigma_t^2 = \infty \text{ and } \liminf_{t\to\infty} \sigma_t^2 = 0 \text{ a.s.} \tag{2.3}$$

The next theorem considers the more important case of $\omega > 0$.

THEOREM 2. *Let* $\omega > 0$. *If* $\mathrm{E}[\ln(\beta + \alpha z_t^2)] \geq 0$, *then*

$$\sigma_t^2 \to \infty \text{ a.s. and} \tag{2.4}$$

$$_u\sigma_t^2 = \infty \text{ a.s. for all } t. \tag{2.5}$$

If $\mathrm{E}[\ln(\beta + \alpha z_t^2)] < 0$, *then* (2.6)–(2.10) *hold*:

$$\omega/(1-\beta) \leq {}_u\sigma_t^2 < \infty \qquad \text{for all } t \text{ a.s.}, \tag{2.6}$$

$_u\sigma_t^2$ *is strictly stationary*[2] *and ergodic with a well-defined probability measure*

$$\mu_\infty \text{ on } [\omega/(1-\beta), \infty) \qquad \text{for all } t, \tag{2.7}$$

$$_u\sigma_t^2 - \sigma_t^2 \mapsto 0, \text{ a.s.}, \tag{2.8}$$

$$\mu_t \to \mu_\infty, \qquad \text{and} \tag{2.9}$$

$$\mu_\infty \text{ is nondegenerate.} \tag{2.10}$$

As an application of Theorems 1 and 2, consider the IGARCH(1, 1) model of Engle and Bollerslev (1986a), in which $\mathrm{E}[z_t^2] = 1$ and $\beta + \alpha = 1$. In this model, the conditional expectation of σ_{t+k}^2 $(k \geq 0)$ at time t is given by

$$\mathrm{E}\left(\sigma_{t+k}^2 \mid \sigma_t^2\right) = \sigma_t^2 + \omega \cdot k. \tag{2.11}$$

When $\omega = 0$, σ_t^2 is a martingale. In the behavior of its conditional expectation, σ_t^2 with $\omega > 0$ and $\omega = 0$ is analogous to a random walk with and without drift, respectively. Its behavior in other respects is very different from that of a random walk: for example, Geweke (1986) and Engle and Bollerslev (1986b) showed that the structure of the higher moments of σ_t^2 when $\alpha + \beta = 1$ and $\omega = 0$ implies that the distribution of σ_t^2 becomes more and more concentrated around zero with fatter and fatter tails, which is not the case with a random walk. Theorem 1 strengthens this result: by Jensen's inequality and the strict concavity of $\ln(x)$, we have $\mathrm{E}[\ln(\beta + \alpha z_t^2)] < \ln(\mathrm{E}[\beta + \alpha z_t^2]) = \ln(1) = 0$. In the IGARCH(1, 1) model with no drift $(\omega = 0)$, therefore, $\sigma_t^2 \to 0$ almost surely. In the IGARCH(1, 1) model with $\omega > 0$, $_u\sigma_t^2$ is strictly stationary and ergodic, and $\sigma_t^2 \to {}_u\sigma_t^2$ almost surely. This illustrates the important point that the

[2] Sampson (1988) independently derived the stationarity condition $\mathrm{E}[\ln(\beta + \alpha z_t^2)] < 0$.

behavior of a martingale can differ very sharply from the behavior of a random walk.

For IGARCH(1, 1) with $\omega = 0$, this result could have been obtained by the martingale convergence theorem (Dudley, 1989, Section 10.5). But (2.2) does not just apply to martingales. Consider for example, the model

$$\sigma_t^2 = 3 \cdot z_{t-1}^2 \cdot \sigma_{t-1}^2, \tag{2.12}$$

where $z_t \sim N(0,1)$. In this model, $\mathrm{E}[\sigma_t^2] = 3^t \cdot \mathrm{E}[\sigma_0^2] \to \infty$ as $t \to \infty$. Nevertheless, $\mathrm{E}[\ln(3 \cdot z_{t-1}^2)] < 0$ (see Theorem 6 and Figure 1 below), so $\sigma_t^2 \to 0$ a.s. Theorem 1, therefore, applies not merely when σ_t^2 is a martingale, but often applies when the expectation of σ_t^2 diverges to infinity. If, instead of (2.12), we have

$${}_u\sigma_t^2 = \omega + 3 \cdot z_{t-1}^2 \cdot {}_u\sigma_{t-1}^2, \tag{2.13}$$

for $\omega > 0$ and $z_t \sim N(0,1)$, then the results of Theorem 2 hold, and ${}_u\sigma_t^2$ is strictly stationary.

Next we consider the moments of ${}_u\sigma_t^2$ and σ_t^2. Bollerslev (1986) evaluates the integer moments of σ_t^2, but the finiteness of the noninteger moments is also of interest. For example, $\mathrm{E}[\xi_t^j] = \mathrm{E}[z_t^j] \cdot \mathrm{E}[(\sigma_t^2)^{j/2}]$, that is, the odd integer moments of ξ_t involve fractional moments of σ_t^2. In addition, many limit theorems (see, e.g., White, 1984) impose moment conditions of the form $\mathrm{E}[|Y_t|^{j+\delta}] < \infty$ for $j = 1$ or 2 and $\delta > 0$. To apply these theorems as broadly as possible requires that δ be as small as possible. The next theorem puts upper and lower bounds on the moments of σ_t^2 and ${}_u\sigma_t^2$. Necessary and sufficient conditions for finiteness follow as a corollary.

THEOREM 3. *Define* $\zeta \equiv \mathrm{E}[(\beta + \alpha z_t^2)^p]$. *When* $0 < p \leq 1$,

$$\left[\left(\mathrm{E}(\sigma_0^{2p})\right)^{1/p} \zeta^{t/p} + \omega \sum_{k=0}^{t-1} \zeta^{k/p}\right]^p \leq \mathrm{E}\left[\sigma_t^{2p}\right] \leq \mathrm{E}(\sigma_0^{2p})\zeta^t + \omega^p \sum_{k=0}^{t-1} \zeta^k, \tag{2.14}$$

and

$$\omega^p \left[\sum_{k=0}^{\infty} \zeta^{k/p}\right]^p \leq \mathrm{E}\left[{}_u\sigma_t^{2p}\right] \leq \omega^p \sum_{k=0}^{\infty} \zeta^k. \tag{2.15}$$

When $1 \leq p$, *the inequalities in* (2.14) *and* (2.15) *are reversed.*

COROLLARY. *let* $\omega > 0$, $p > 0$, *and* $\mathrm{E}[\ln(\beta + \alpha z_t^2)] < 0$.

$$\mathrm{E}\left[\sigma_t^{-2p}\right] < \infty, \qquad t \geq 1, \quad \textit{and} \tag{2.16}$$

$$\mathrm{E}\left[{}_u\sigma_t^{-2p}\right] < \infty \qquad \textit{for all } t. \tag{2.17}$$

$$\mathrm{E}\left[\sigma_t^{2p}\right] < \infty \qquad \textit{iff } \mathrm{E}\left[\sigma_0^{2p}\right] < \infty \quad \textit{and} \quad \mathrm{E}\left[\left(\beta + \alpha z_t^2\right)^p\right] < \infty. \tag{2.18}$$

$$\mathrm{E}\left[{}_u\sigma_t^{2p}\right] < \infty \qquad \textit{iff } \mathrm{E}\left[\left(\beta + \alpha z_t^2\right)^p\right] < 1. \tag{2.19}$$

$$\limsup_{t \to \infty} \mathrm{E}\left[\sigma_t^{2p}\right] < \infty \qquad \textit{iff } \mathrm{E}\left[\sigma_0^{2p}\right] < \infty \quad \textit{and} \quad \mathrm{E}\left[\left(\beta + \alpha z_t^2\right)^p\right] < 1. \tag{2.20}$$

If $\mathrm{E}[\sigma_0^{2p}] < \infty$, *then*

$$\lim_{t \to \infty} \mathrm{E}\left[\sigma_t^{2p}\right] = \mathrm{E}\left[{}_u\alpha_t^{2p}\right]. \tag{2.21}$$

The following theorem is useful in verifying the existence of a finite moment of order $j + \delta$, $\delta > 0$:

THEOREM 4. (*a*) *Let* $\omega > 0$ *and* $\mathrm{E}[\ln(\beta + \alpha z_t^2)] < 0$. *If* $\mathrm{E}[|z_t|^{2q}] < \infty$ *for some* $q > 0$, *then there exists a* p, $0 < p < q$, *such that*

$$\mathrm{E}\left[\left(\beta + \alpha z_t^2\right)^p\right] < 1.$$

(*b*) *If, in addition*, $\mathrm{E}[(\beta + \alpha z_t^2)^r] < 1$ *for* $0 < r < q$, *then there exists a* $\delta > 0$ *such that* $\mathrm{E}[(\beta + \alpha z_t^2)^{r+\delta}] < 1$.

Theorem 4(a) says, in essence, that if ${}_u\sigma_t^2$ is strictly stationary and z_t^2 has a finite moment of some (arbitrarily small, possibly fractional) order, then ${}_u\sigma_t^2$ has a finite (possibly fractional) moment as well. The existence of such a finite fractional moment implies, for example, that $\mathrm{E}[\ln({}_u\sigma_t^2)] < \infty$ (see Zucker, 1965, Formula 4.1.30).

Part (b) gives a condition for $\mathrm{E}[{}_u\sigma_t^{2(p+\delta)}] < \infty$ for some $\delta > 0$, given that $\mathrm{E}[{}_u\sigma_t^{2p}] < \infty$. It says, for example, that if $\mathrm{E}[(\beta + \alpha z_t^2)^{1/2}] < 1$ and $\mathrm{E}[|z_t|^{2p}] < \infty$ for some $p > \frac{1}{2}$, then not only is $\mathrm{E}[|{}_u\xi_t|] < \infty$, but there is also a $\delta > 0$ with $\mathrm{E}[|{}_u\xi_t|^{1+\delta}] < \infty$.

In many applications, GARCH processes are used to model the persistence of shocks to the conditional variance (see, e.g., Engle and Bollerslev, 1986a). For example, an increase in the conditional variance of stock market returns may cause the market risk premium to rise. However, if the increase is expected to be short-lived, the term structure of market risk premia may move only at the short end, and the valuation of long-lived assets may be affected only slightly (Poterba and Summers, 1986). Since, in the GARCH(1, 1) model, the changes in σ_t^2 or ${}_u\sigma_t^2$ are driven by past realizations of $\{z_t\}$ (or, equivalently, of $\{\xi_t\}$), it seems desirable to form sensible definitions of "persistence" and develop criteria for the persistence of the effects of ξ_t, ${}_u\xi_t$, or z_t on σ_{t+m}^2 and ${}_u\sigma_{t+m}^2$ for large m.

To define the persistence of z_t in σ_{t+m}^2 and ${}_u\sigma_{t+m}^2$, write the latter two as

$$\sigma_{t+m}^2 = \left[\sigma_0^2 \prod_{i=1}^{t+m} (\beta + \alpha z_{t+m-i}^2) + \omega \left[\sum_{k=m}^{t+m-1} \prod_{i=1}^{k} (\beta + \alpha z_{t+m-i}^2)\right]\right]$$

$$+ \omega\left[1 + \sum_{k=1}^{m-1} \prod_{i=1}^{k} (\beta + \alpha z_{t+m-i}^2)\right]$$

$$\equiv A(\sigma_0^2, z_{t+m-1}, \ldots, z_1, z_0) + B(z_{t+m-1}, \ldots, z_{t+1}) \tag{2.22}$$

and

$${}_u\sigma_{t+m}^2 = \omega\left[\sum_{k=m}^{\infty} \prod_{i=1}^{k} (\beta + \alpha z_{t+m-i}^2)\right] + \omega\left[1 + \sum_{k=1}^{m-1} \prod_{i=1}^{k} (\beta + \alpha z_{t+m-i}^2)\right]$$

$$\equiv C(z_{t+m-1}, \ldots, z_1, z_0, z_{-1}, \ldots) + B(z_{t+m-1}, \ldots, Z_{t+1}). \tag{2.23}$$

By construction, B is independent of z_t, while A and C are not. B, C, and A are all nonnegative a.s.

DEFINITION. We say that z_t is persistent in σ^2 almost surely unless

$$A(\sigma_0^2, z_{t+m-1}, \ldots, z_1, z_0) \to 0 \text{ a.s.} \qquad \text{as } m \to \infty, \tag{2.24}$$

z_t is persistent in σ^2 in probability unless, for all $b > 0$,

$$\mathrm{P}[A(\sigma_0^2, z_{t+m-1}, \ldots, z_1, z_0) > b] \to 0 \qquad \text{as } m \to \infty, \tag{2.25}$$

and z_t is persistent in σ^2 in L^p, $p > 0$, unless

$$\mathrm{E}\big[[A(\sigma_0^2, z_{t+m-1}, \ldots, z_1, z_0)]^p\big] \to 0 \qquad \text{as } m \to \infty. \tag{2.26}$$

Similarly, z_t is persistent in ${}_u\sigma^2$ almost surely unless

$$C(z_{t+m-1}, \ldots, z_1, z_0, z_{-1}, \ldots) \to 0 \text{ a.s.} \qquad \text{as } m \to \infty, \tag{2.27}$$

z_t is persistent in ${}_u\sigma^2$ in probability unless, for all $b > 0$,

$$\mathrm{P}[C(z_{t+m-1}, \ldots, z_1, z_0, z_{-1}, \ldots) > b] \to 0 \qquad \text{as } m \to \infty, \tag{2.28}$$

and z_t is persistent in ${}_u\sigma^2$ in L^p, $p > 0$, unless

$$\mathrm{E}\big[[C(z_{t+m-1}, \ldots, z_1, z_0, z_{-1}, \ldots)]^p\big] \to 0 \qquad \text{as } m \to \infty. \tag{2.29}$$

Alternatively, we can write σ_{t+m}^2 and ${}_u\sigma_{t+m}^2$ as functions of past values of $\{\xi_t\}$ and $\{{}_u\xi_t\}$ and define the persistence of $\{\xi_t\}$ and $\{{}_u\xi_t\}$ in σ^2 and ${}_u\sigma^2$:

$$\sigma_{t+m}^2 = \alpha \cdot \beta^{m-1} \cdot \xi_t^2 + \left[\beta^{m-1} \cdot \omega + \beta^{t+m}\sigma_0^2 + \sum_{i=1, i \neq m}^{t+m} \beta^{i-1}\left(\alpha \cdot \xi_{t+m-i}^2 + \omega\right)\right], \tag{2.30}$$

$${}_u\sigma_{t+m}^2 = \alpha \cdot \beta^{m-1} \cdot {}_u\xi_t^2 + \left[\beta^{m-1} \cdot \omega + \sum_{i=1, i \neq m}^{\infty} \beta^{i-1}\left(\alpha \cdot {}_u\xi_{t+m-i}^2 + \omega\right)\right], \tag{2.31}$$

Equations (2.30)–(2.31) break σ_{t+m}^2 and ${}_u\sigma_{t+m}^2$ into two pieces, the first of which depends directly on ξ_t or ${}_u\xi_t$ and the second does not. While $B(z_{t+m-1}, \ldots, z_{t+1})$ is independent of z_t, it is easy to verify that the bracketed terms in (2.30)–(2.31) are not independent of ξ_t and ${}_u\xi_t$.

DEFINITION. We say that ξ_t is almost surely persistent unless

$$\beta^{m-1} \cdot \xi_t^2 \to 0 \text{ a.s.} \qquad \text{as } m \to \infty, \tag{2.32}$$

ξ_t is persistent in probability unless for every $b > 0$,

$$\mathrm{P}\left[\beta^{m-1} \cdot \xi_t^2 > b\right] \to 0 \qquad \text{as } m \to \infty, \text{ and} \tag{2.33}$$

ξ_t is persistent in L^p, $p > 0$, unless

$$\mathrm{E}\left[\left[\beta^{m-1} \cdot \xi_t^2\right]^p\right] \to 0 \qquad \text{as } m \to \infty. \tag{2.34}$$

We define persistence of ${}_u\xi_t$ similarly, replacing ξ_t with ${}_u\xi_t$ in (2.32)–(2.34). If $\beta = 0$, $m > 1$, and $|{}_u\xi_t| = \infty$, we define $\beta^{m-1} \cdot {}_u\xi_t^2 \equiv 0$.

THEOREM 5. *z_t is persistent in σ^2, both almost surely and in probability*

$$\textit{iff } \mathrm{E}\left[\ln\left(\beta + \alpha z_t^2\right)\right] \geq 0. \tag{2.35}$$

z_t is persistent in ${}_u\sigma^2$, both almost surely and in probability, iff $\omega > 0$ and

$$\mathrm{E}\left[\ln\left(\beta + \alpha z_t^2\right)\right] \geq 0. \tag{2.36}$$

z_t is persistent in L^p in σ^2 iff either $\mathrm{E}[(\beta + \alpha z_t^2)^p] \geq 1$ or

$$\mathrm{E}\left[\sigma_0^{2p}\right] = \infty \qquad (\textit{or both}). \tag{2.37}$$

$$z_t \textit{ is persistent in } L^p \textit{ in } {}_u\sigma^2 \textit{ iff } \mathrm{E}\left[\left(\beta + \alpha z_t^2\right)^p\right] \geq 1. \tag{2.38}$$

$$\xi_t \textit{ is persistent, both almost surely and in probability, iff } \beta \geq 1. \tag{2.39}$$

${}_u\xi_t$ *is persistent, both almost surely and in probability, iff* $\beta > 0$, $\omega > 0$, *and*

$$\mathrm{E}\left[\ln(\beta + \alpha z_t^2)\right] \geq 0. \tag{2.40}$$

ξ_t *is persistent in* L^p *iff* $\beta > 0$ *and either* $\mathrm{E}[(\beta + \alpha z_t^2)^p] = \infty$, $\beta \geq 1$, *or*

$$\mathrm{E}\left[\sigma_0^{2p}\right] = \infty. \tag{2.41}$$

${}_u\xi_t$ *is persistent in* L^p *iff* $\omega > 0$, $\beta > 0$, *and* $\mathrm{E}\left[\left(\beta + \alpha z_t^2\right)^p\right] \geq 1$. (2.42)

From Theorem 5, it is clear that whether a shock "persists" or not depends very much on our definition of persistence. For example, suppose $\omega > 0$, $\mathrm{E}[|z_t|^q] < \infty$ for some $q > 0$, and $\mathrm{E}[\ln(\beta + \alpha z_t^2)] < 0$. Then by Theorem 4, $\mathrm{E}[(\beta + \alpha z_t^2)^p] < 1$ for some $p > 0$, that is, there are L^p norms in which shocks to ${}_u\sigma_t^2$ (i.e., z_t and ${}_u\xi_t$) do not persist. If, however, it is also the case that $\mathrm{P}[(\beta + \alpha z_t^2) > 1] > 0$, then by Hardy *et al.* (1952), Theorem 193, there exists an $r > 0$ such that $\mathrm{E}[(\beta + \alpha z_t^2)^r] > 1$, that is, shocks to ${}_u\sigma_t^2$ persist in some L^p norms (for all sufficiently large p) but not in others. For example, in the IGARCH model with $\omega > 0$ and $\beta > 0$, ${}_u\xi_t$ is persistent (and z_t is persistent in ${}_u\sigma^2$) in L^p, $p \geq 1$, but not almost surely, in probability, or in L^p, $0 < p < 1$.

"Persistence" of shocks also differs in the conditional and unconditional models. Equations (2.39)–(2.40) illustrate this: Suppose that $\mathrm{E}[\ln(\beta + \alpha z_t^2)] \geq 0$, $0 < \beta < 1$, and $\omega > 0$. In this case, ${}_u\sigma_t^2 = \infty$ and $\sigma_t^2 \to \infty$ a.s., so that $\mathrm{P}[\beta^{m-1} \cdot {}_u\xi_t^2 = \infty] > 0$ for all m and t, and ${}_u\xi_t$ is therefore persistent almost surely. Nevertheless, for any fixed t, ξ_t is almost surely not persistent, since the individual ξ_t terms are finite with probability one and $\beta^{m-1} \cdot \xi_t^2 \to 0$ a.s. as $m \to \infty$. In other words, for fixed t, the effect of an individual ξ_t on σ_{t+m}^2 dies out a.s. as $m \to \infty$, but at a progressively slower rate as $t \to \infty$. This is why it is necessary to define persistence of each t rather than uniformly over all t.

When $\mathrm{E}[z_t^2] = 1$, persistence of z_t in L^1 in ${}_u\sigma_t^2$ corresponds to persistence as defined by Engle and Bollerslev (1986a). In this case, we have

$$\mathrm{E}\left[\sigma_{t+m}^2 \mid \sigma_t^2\right] = \sigma_t^2(\beta + \alpha)^m + \omega \cdot \left[\sum_{k=0}^{m-1} (\beta - \alpha)^k\right]. \tag{2.43}$$

Engle and Bollerslev defined shocks to σ_t^2 to be persistent unless the $\sigma_t^2(\beta + \alpha)^m$ term vanishes as $m \to \infty$, which it does if and only if $(\beta + \alpha) < 1$. Note that according to this definition, when $\beta + \alpha \geq 1$, shocks accumulate (and "persist") not necessarily in the sense that $\sigma_{t+m}^2 \to \infty$ as $m \to \infty$, but rather in the sense that $\mathrm{E}[\sigma_{t+m}^2 \mid \sigma_t^2] \to \infty$, so that if $\omega > 0$, $\mathrm{E}[{}_u\sigma_t^2] = \infty$. Our definitions of persistence in L^p all have similar interpretations: Persistence of z_t in L^p in ${}_u\sigma_t^2$ means that shocks accumulate (and persist) in the sense that $E[\sigma_{t+m}^{2p} \mid \sigma_t^2] \to \infty$ as $m \to \infty$, implying (when $\omega > 0$) that $E[{}_u\sigma_t^{2p}] = \infty$. Persistence in L^p therefore corresponds to persistence of shocks in the forecast *moments* of σ_t^2 and ${}_u\sigma_t^2$, while almost surely persistence corresponds to persistence of shocks in the forecast *distributions* of σ_t^2 and ${}_u\sigma_t^2$.

In Theorems 1–5, two moment conditions repeatedly arise, namely, $E[\ln(\beta + \alpha z_t^2)] < 0$ and $E[(\beta + \alpha z_t^2)^p] < 1$. In a number of cases of interest, expressions are available for these expectations in terms of widely tabulated functions. The next theorem evaluates these expectations when z_t is either standard normal or Cauchy. Approximations and algorithms for evaluating the functions in Theorem 6 are found in Abramowitz and Stegun (1965); Lebedev (1972), and Spanier and Oldham (1987).

THEOREM 6. *If* $z \sim N(0,1)$, *and* $\beta > 0$, *then*

$$E\left[\ln(\beta + \alpha z^2)\right] = \ln(2\alpha) + \psi\left(\tfrac{1}{2}\right) + (2\pi\beta/\alpha)^{1/2}\Phi\left(\tfrac{1}{2}, 1.5; \beta/2\alpha\right) - (\beta/\alpha)\,{}_2F_2(1,1;2,1.5;\beta/2\alpha), \qquad (2.44)$$

and

$$E\left[(\beta + \alpha z^2)^p\right] = (2\alpha)^{-1/2}\beta^{p+1/2}\Psi\left(\tfrac{1}{2}, p + 1.5; \beta/2\alpha\right), \qquad (2.45)$$

where $\Phi(\cdot,\cdot,\cdot)$ *is a confluent hypergeometric function* (*Lebedev*, 1972, *Section* 9.9), ${}_2F_2(\cdot,\cdot;\cdot,\cdot;\cdot)$ *is a generalized hypergeometric function* (*Lebedev*, 1972, *Section* 9.14), $\Psi(\cdot,\cdot;\cdot)$ *is a confluent hypergeometric function of the second kind* (*Lebedev*, 1972, *Section* 9.10), *and* $\psi(\cdot)$ *is the Euler psi function, with* $\psi(\frac{1}{2}) \approx -1.96351$ (*Davis*, 1965). *Let* $r(\cdot)$ *be the gamma function. When* $\beta = 0$,

$$E\left[\ln(\alpha z^2)\right] = \ln(2\alpha) + \psi\left(\tfrac{1}{2}\right). \qquad (2.46)$$

$$E\left[(\alpha z^2)^p\right] = \pi^{-1/2}(2\alpha)^p\Gamma\left(p + \tfrac{1}{2}\right). \qquad (2.47)$$

If $z \sim$ *standard Cauchy* (*i.e., the probability density of z is* $f(z) = [\pi(1 + z^2)]^{-1}$, *then*

$$E\left[\ln(\beta + \alpha z^2)\right] = 2 \cdot \ln(\beta^{1/2} + \alpha^{1/2}), \quad \textit{and} \qquad (2.48)$$

$$E\left[(\beta + \alpha z^2)^p\right] = \begin{cases} \dfrac{F\left[-p, \frac{1}{2}; 1 - p; 1 - (\alpha/\beta)\right] \cdot \beta^p \cdot \Gamma\left(\frac{1}{2} - p\right)}{\pi^{-1/2} \cdot \Gamma(1 - p)} & p < \frac{1}{2}, \\ \infty & p \geq \frac{1}{2} \end{cases} \qquad (2.49)$$

where $F(\cdot,\cdot;\cdot;\cdot)$ *is a hypergeometric function* (*Lebedev*, 1972, *Section* 9.1).

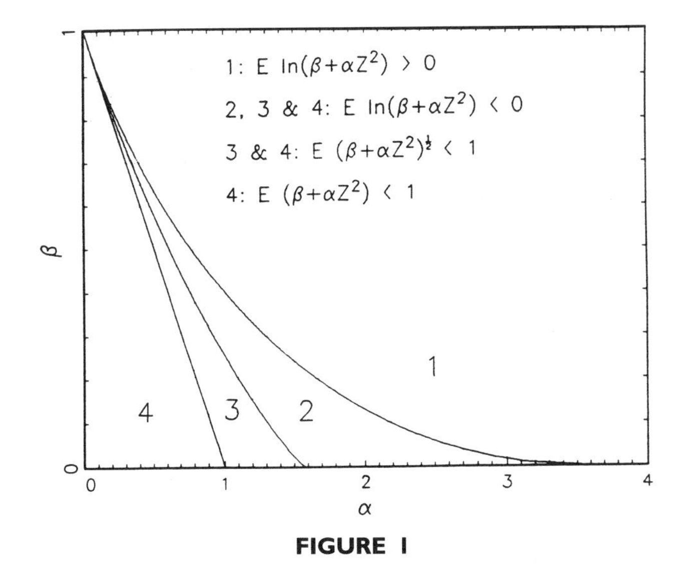

FIGURE 1

When $p < \frac{1}{2}$ *and* $\beta = 0$, (2.49) *simplifies to*

$$\mathrm{E}\left[\left(\alpha z^2\right)^p\right] = \alpha^p/\sin\left[\left(p + \tfrac{1}{2}\right)\pi\right]. \tag{2.50}$$

Figure 1 summarizes some of the information from Theorem 6 for the case in which $z_t \sim N(0,1)$.[3] Assume that $\omega > 0$. In region 1 and on the boundary between regions 1 and 2, ${}_u\sigma_t^2 = \infty$ a.s. and σ_t^2 is explosively nonstationary. In regions 2, 3, and 4, ${}_u\sigma_t^2$ and ${}_u\xi_t$ are strictly stationary and ergodic. In region 2 and on the boundary between regions 2 and 3, however, $\mathrm{E}[{}_u\xi_t]$ is not defined. In region 3, and on the boundary between regions 3 and 4, $\mathrm{E}[{}_u\xi_t] = 0$ but $\mathrm{E}[{}_u\xi_t^2] = \infty$. Region 4 contains the classical, covariance stationary GARCH(1, 1) models with $\mathrm{E}[{}_u\xi_t^2] < \infty$, while in regions 2 and 3, ${}_u\xi_t$ is strictly stationary but not weakly (i.e., covariance) stationary.[4]

Figure 2 is the counterpart of Figure 1 for $z_t \sim$ Cauchy. In this figure there is no counterpart to regions 3 and 4, since $\mathrm{E}[{}_u\xi_t]$ is ill-defined and $\mathrm{E}[{}_u\xi_t^2] = \infty$ for all parameter values as long as $\omega > 0$. In region 2, however, ${}_u\sigma_t^2$ and ${}_u\xi_t$ are strictly stationary and ergodic. Although ${}_u\sigma_t^2$ has no integer moments, Theorem 4(b) implies that for each α, β combination in region 2 there exists some $q > 0$ with $\mathrm{E}[{}_u\sigma_t^{2q}] < \infty$, that is, ${}_u\sigma_t^2$ has *some* finite positive moment.

[3]The graphs were produced using GAUSS 2.0 (1988).
[4]A related paper (Nelson, 1990) derives the distribution of ${}_u\sigma_t^2$ as a continuous time limit is approached.

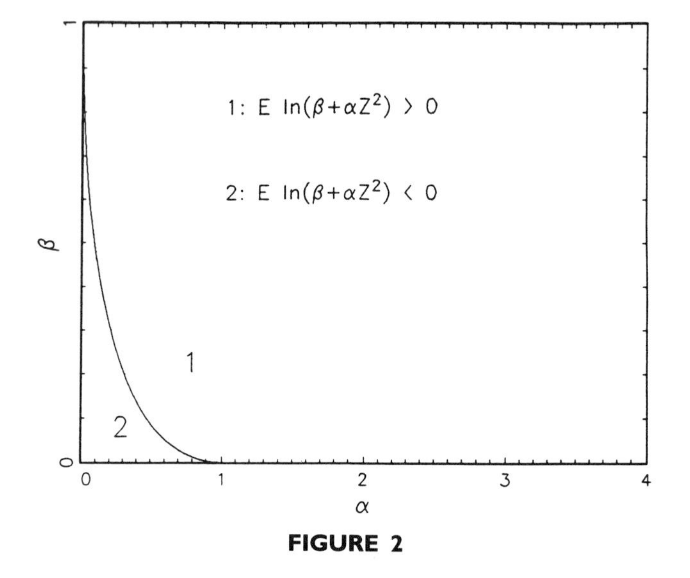

FIGURE 2

As we have seen, in the IGARCH(1, 1) process, ${}_u\xi_t$ is strictly stationary but not covariance stationary, since $\mathrm{E}[{}_u\xi_t{}^2] = \infty$. Since IGARCH(1, 1) has been employed quite often in empirical work (e.g., in Engle and Bollerslev, 1986a), it is of interest to see what the distribution of ${}_u\sigma_t{}^2$ looks like for this model. Figure 3 plots density estimates for $\ln({}_u\sigma_t{}^2)$ generated by five IGARCH(1, 1) models, with α set equal to 1, 0.75, 0.5, 0.25, and 0.1, respectively. ${}_u\sigma_t{}^2$ was simulated by σ_T^2 for a suitably large value of T, using the recursion

$$\sigma_t{}^2 = \omega + (1 - \alpha)\sigma_{t-1}^2 + \alpha\xi_{t-1}^2, \qquad \sigma_0^2 = 1 \tag{2.51}$$

with $z_t \sim$ i.i.d. $N(0, 1)$. ω was set equal to α, giving ${}_u\sigma_t{}^2$ support on $(1, \infty)$ and $\ln({}_u\sigma_t{}^2)$ support on $(0, \infty)$. The technical details on how the density estimates were obtained are in the appendix.

What is most striking about Figure 3 is the extreme values that ${}_u\sigma_t{}^2$ frequently takes. Since ω is only a scale parameter, changing its value would shift the densities of $\ln({}_u\sigma_t{}^2)$ to the right or left, but would leave the shape of the densities unchanged. Note also that the lower α is, the more strongly skewed to the right the density is. This is not surprising, since the larger α is, the more influence a single draw of z_t^2 has in making σ_{t+1}^2 small or large. Since $z_t^2 \sim \chi^2$ with 1 degree of freedom, its density is infinite at $z = 0$ and is strongly skewed to the right. When $\alpha = 1$, this lends a similar shape to the distribution of $\ln({}_u\sigma_t{}^2)$.

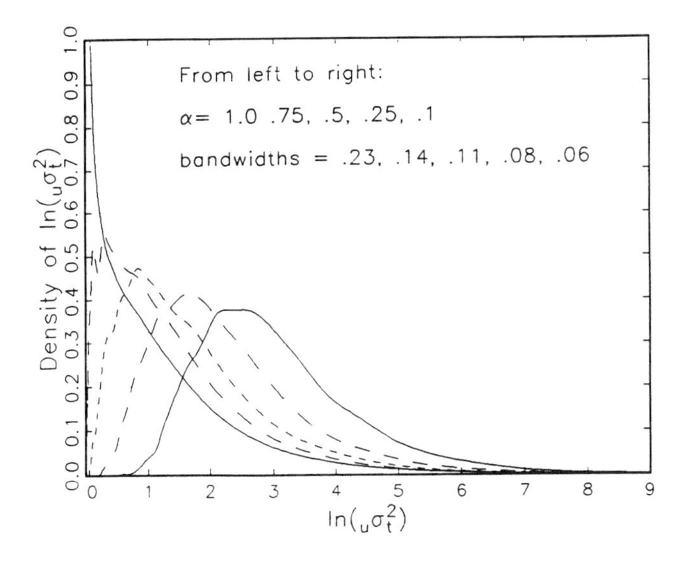

APPENDIX

Proof of Theorem 1. In the conditional model with $\omega = 0$, we have $\ln(\sigma_t^2) = \ln(\sigma_0^2) + \Sigma_{i=0,t-1} \ln(\beta + \alpha z_i^2)$, that is, $\ln(\sigma_t^2)$ is a random walk. Its drift is positive, negative, or zero as $E[\ln(\beta + \alpha z_t^2)]$ is positive, negative, or zero. By (1.8), $|\ln(\sigma_0^2)| < \infty$ a.s. By a direct application of the strong law of large numbers for i.i.d. random variables, $\ln(\sigma_t^2) \to \infty$ if $E[\ln(\beta + \alpha z_t^2)] > 0$ and $\ln(\sigma_t^2) \to -\infty$ if $E[\ln(\beta + \alpha z_t^2)] < 0$ (Stout, 1974, Chapter 6). By Stout (1974) Theorem 6.1.4 and Corollary 6.1.1., when $E[\ln(\beta + \alpha z_t^2)] = 0$ and z_t^2 is nondegenerate (assumed in (1)), $\limsup \ln(\sigma_t^2) = \infty$ and $\liminf \ln(\sigma_t^2) = -\infty$ with probability one.[5]

Proof of Theorem 2. Equation (1.6) implies that

$$\sigma_t^2 \geq \omega\left[\sup_{1 \leq k \leq t-1} \prod_{i=1}^{k} \left(\beta + \alpha z_{t-i}^2\right)\right]. \tag{A.1}$$

By Theorem 1, the term on the right diverges to ∞ as $t \to \infty$ if $E[\ln(\beta + \alpha z_t^2)] \geq 0$, proving (2.4). A similar argument proves (2.5).

[5]A similar argument in a manuscript version of Dudley (1989) inspired the proof of this theorem.

The lower inequality in (2.6) follows trivially from (1.10). To prove the upper inequality, we will show that the terms in the sum

$$\sum_{k=1}^{\infty} \prod_{i=1}^{k} (\beta + \alpha z_{t-i}^2)$$

are $O(\exp(-\lambda k))$ for some $\lambda > 0$ with probability one. Define $\gamma \equiv E[\ln(\beta + \alpha z_t^2)] < 0$. By the strong law of large numbers for i.i.d. random variables (Dudley, 1989, Theorem 8.3.5) with probability one there exists a random positive integer $M < \infty$ such that, for all $k > M$,

$$\left| k^{-1} \sum_{i=1}^{k} \ln(\beta + \alpha z_{t-i}^2) - \gamma \right| < |\gamma|/2 \tag{A.2}$$

so, for all $k > M$,

$$\sum_{i=1}^{k} \ln(\beta + \alpha z_{t-i}^2) < \gamma \cdot k/2, \qquad \text{and} \tag{A.3}$$

$$\prod_{i=1}^{k} (\beta + \alpha t_{t-i}^2) < \exp(\gamma \cdot k/2), \tag{A.4}$$

establishing that, with probability one,

$$\prod_{i=1}^{k} (\beta + \alpha z_{t-i}^2) = O(\exp(-\lambda k))$$

with $\lambda \equiv |\gamma|/2 > 0$, so that the series converges a.s., concluding the proof of (2.6).[6] We next prove (2.7). Consider the function ${}_u\sigma_t^2 \equiv {}_u\sigma^2(z_{t-1}, z_{t-2}, \ldots)$ from R^∞ into the extended real line $[-\infty, \infty]$ given in (1.10). By Stout (1974) Theorem 3.5.8, (2.7) follows if we can show that the function ${}_u\sigma^2(z_{t-1}, z_{t-2}, \ldots)$ is measurable. First consider the sequence of functions given by $f_k = \prod_{t=1,k}(\beta + \alpha z_{t-i}^2)$. Since the product (or sum) of a finite number of measurable functions is measurable (Royden, 1968, Chapter 3, Proposition 5), f_k is measurable for any finite k. By the same reasoning, the functions $r_n \equiv \sum_{k=1,n} \prod_{i=1,k}(\beta + \alpha z_{t-i}^2)$ are measurable for any finite n. Since r_n is increasing in n, we have $\sup_{\{n\}} r_n = r_\infty \equiv \sum_{k=1,\infty} \prod_{i=1,k}(\beta + \alpha z_{t-i}^2) = {}_u\sigma^2(z_{t-1}, z_{t-2}, \ldots)$ which is measurable by Royden, 1968, Chapter 3, Theorem 20. (2.7) therefore holds.

[6]This is one of those rare instances in which the distinction between convergence in probability and almost sure convergence matters, since a weak law of large numbers would not suffice in this proof.

Next we prove (2.8), after which (2.9) follows by Dudley (1989), Proposition 9.3.5. We require

$$\sigma_t^2 - {}_u\sigma_t^2 = \sigma_0^2 \prod_{i=1}^{t} \left(\beta + \alpha z_{t-i}^2 \right) - \omega \left[\sum_{k=t}^{\infty} \prod_{i=1}^{k} \left(\beta + \alpha z_{t-i}^2 \right) \right] \to 0 \quad \text{(A.5)}$$

with probability one. The first term vanishes a.s. by Theorem 1; that is, the log of the first term is a random walk with negative drift and therefore diverges to $-\infty$ a.s. The second term also vanishes a.s., since the individual terms in the summation are $O(\exp(-\lambda k))$ a.s. for some $\lambda > 0$.

Finally, suppose that the distribution of ${}_u\sigma_t^2$ is degenerate. Then by (1.2) and (1.3)′ we have, with probability one for some constant v and for all t,

$$\mathrm{P}\left[v = \omega + \beta \cdot v + \alpha \cdot v \cdot z_t^2 \right] = 1, \quad \text{(A.6)}$$

a contradiction, since $\alpha > 0$, and, by (1), z_t^2 is nondegenerate.

Proof of Theorem 3. The left-hand inequalities in (2.14)–(2.15) follow by Minkowski's integral inequality (Hardy *et al.*, 1952, Theorem 198, p. 146), and the right-hand inequalities follow from a companion to Minkowski's inequality (Hardy *et al.*, 1952, Theorem 199, p. 147). In Hardy *et al.* (1952), these inequalities are originally stated using Riemann integrals, but hold for Stieltjes integrals as well—see Hardy *et al.* (1952, Section 6.17). They hold in (2.14)–(2.15), since the Lebesgue integral $\int \sigma^{2q}\, d\mu_t$ is equivalent to the Stieltjes integral $\int \sigma^{2q}\, dF_t(\sigma^2)$, where the cumulative distribution function $F_t(x) \equiv \mu_t((-\infty, x])$ (Kolmogorov and Fomin, 1970, Sections 36.1 and 36.2). This holds for μ_∞ and $F_\infty(\sigma^2)$ by the same argument.

Proof of the Corollary. Equations (2.16)–(2.17) are immediate, since σ_t^2 and ${}_u\sigma_t^2$ are bounded below by ω and $\omega/(1-\beta)$, respectively. Equations (2.18)–(2.20) follow directly from (2.14) and (2.15). To verify (2.21) in the case $\mathrm{E}[(\beta + \alpha z_t^2)^p] < 1$, define the usual L^p norm by $\|X\|_p \equiv \mathrm{E}[|X|^p]$ for $0 < p \leq 1$ and $\|X\|_p \equiv \mathrm{E}^{1/p}[|X|^p]$ for $p > 1$. By the triangle inequality,

$$\|\sigma_t^2 - {}_u\sigma_t^2\|_p + \|{}_u\sigma_t^2 - 0\|_p \geq \|\sigma_t^2 - 0\|_p, \quad \text{and} \quad \text{(A.7)}$$

$$\|{}_u\sigma_t^2 - \sigma_t^2\|_p + \|\sigma_t^2 - 0\|_p \geq \|{}_u\sigma_t^2 - 0\|_p. \quad \text{(A.8)}$$

Together, (A.7) and (A.8) imply

$$\left| \|{}_u\sigma_t^2\|_p - \|\sigma_t^2\|_p \right| \leq \|\sigma_t^2 - {}_u\sigma_t^2\|_p, \quad \text{so} \quad \text{(A.9)}$$

$$\|\sigma_t^2\|_p - \|{}_u\sigma_t^2\|_p \to 0 \quad \text{and} \quad E\left[\sigma_t^{2p}\right] - \mathrm{E}\left[{}_u\sigma_t^{2p}\right] \to 0$$

$$\text{if } \mathrm{E}\,|{}_u\sigma_t^2 - \sigma_t^2|^p \to 0. \quad \text{(A.10)}$$

Again, define $\zeta \equiv \mathrm{E}[(\beta + \alpha z_t^2)^p]$. When $0 < p \leq 1$ (Hardy *et al.*, 1952) Theorem 199 (again with the generalization in Hardy *et al.*, 1952, Section 6.17) yields

$$\mathrm{E}\,|{}_u\sigma_t^2 - \sigma_t^2|^p \leq \mathrm{E}(\sigma_0^{2p})\zeta^t + \omega^p \sum_{k=t}^{\infty} \zeta^k, \tag{A.11}$$

and when $p \geq 1$, (Hardy *et al.*, 1952) Theorem 198 yields

$$\mathrm{E}|{}_u\sigma_t^2 - \sigma_t^2|^p \leq \left[\mathrm{E}(\sigma_0^{2p})^{1/p}\zeta^{t/p} + \omega \sum_{k=t}^{\infty} \zeta^{k/p}\right]^p. \tag{A.12}$$

Both bounds collapse to zero if $\zeta < 1$ and $\mathrm{E}(\sigma_0^{2p}) < \infty$, proving (2.21) when $\zeta < 1$. The case $\zeta \geq 1$ is immediate by Theorem 3, since $\mathrm{E}[{}_u\sigma_t^{2p}] = \infty$ and $\mathrm{E}[\sigma_t^{2p}] \to \infty$.

Proof of Theorem 4. By Hardy *et al.* (1952), Theorem 194 and the generalization in Section 6.17, if $\mathrm{E}^{1/q}[(\beta + \alpha z^2)^q] < \infty$, then $\mathrm{E}^{1/p}[(\beta + \alpha z^2)^p]$ is a continuous function of p for $0 < p < q$. By Hardy *et al.* (1952), Theorems 198, 199, and their generalizations in Section 6.17, $\mathrm{E}[(\beta + \alpha z^2)^q] < \infty$ if and only if $\mathrm{E}[|z|^{2q}] < \infty$. (b) follows immediately. (a) then follows by Hardy *et al.* (1952), Theorem 187 and Section 6.17.

Proof of Theorem 5. First consider the proofs of (2.35) and (2.36). The case $\omega = 0$ is obvious, so let us take $\omega > 0$. By the reasoning in the proofs of Theorems 1 and 2, if

$$\mathrm{E}\left[\ln(\beta + \alpha z_t^2)\right] < 0, \qquad \text{then} \prod_{i=1,k} \left[(\beta + \alpha z_{t+m-i}^2)\right] = O(\exp(-\lambda k))$$

for some $\lambda > 0$, a.s., guaranteeing that A and C vanish almost surely and in probability as $m \to \infty$. Also by the reasoning of Theorems 1 and 2, A and C diverge a.s. if $\mathrm{E}[\ln(\beta + \alpha z_t^2)] \geq 0$.

To prove (2.38), again define $\zeta \equiv \mathrm{E}[(\beta + \alpha z_t^2)^p]$. When $0 < p \leq 1$, we have, by the reasoning in the proof of Theorem 3,

$$\omega^p\left[\sum_{k=m,\infty} \zeta^{k/p}\right]^p \leq \mathrm{E}[C^p] \leq \omega^p\left[\sum_{k=m,\infty} \zeta^k\right] \tag{A.13}$$

with the inequalities reversed if $1 \leq p$. As $m \to \infty$, both sides collapse to zero if $\zeta < 1$ and diverge to infinity if $\zeta \geq 1$, proving (2.38). The proof of (2.37) is similar, and is left to the reader.

By (1.1) and (1.6)–(1.8) we have, for any t, $|\xi_t| < \infty$ a.s. and $P[\xi_t \neq 0] > 0$. Equation (2.39) follows directly.

Nonpersistence of ${}_u\xi_t$ is obvious if $\beta = 0$ or $\omega = 0$. Suppose $\beta > 0$ and $\omega > 0$ but $\mathrm{E}[\ln(\beta + \alpha z_t^2)] < 0$ (which in turn implies $\beta < 1$). Then by Theorem 2, $|{}_u\xi_t| < \infty$ a.s., so that $\beta^{m-1}{}_u\xi_t^2 \to 0$ a.s. as $m \to \infty$, proving sufficiency in (2.40). Suppose $\beta > 0$ and $\omega > 0$ but $\mathrm{E}[\ln(\beta + \alpha z_t^2)] \geq 0$. In the proof of Theorem 1, we saw that $\Sigma_{i=1,k}\ln(\beta + \alpha z_{t-i}^2)$ is an a.s. nonconvergent random walk if $E[\ln(\beta + \alpha z_t^2)] \geq 0$, which proves necessity in (2.40).

Equation (2.42) is trivial when $\omega = 0$ or $\beta = 0$, so take both to be positive. If $\beta < 1$, $E[(\beta^{m-1}{}_u\xi_t^2)^p] \to 0$ as $m \to \infty$ iff $E[({}_u\xi_t^2)^p] < \infty$. By the corollary to Theorem 3, $E[({}_u\xi_t^2)^p] < \infty$ iff $E[(\beta + \alpha z_t^2)^p] < 1$. $E[(\beta + \alpha z_t^2)^p] < 1$ implies $\beta < 1$, so if $E[(\beta + \alpha z_t^2)^p] < 1$, $E[(\beta^{m-1}{}_u\xi_t^2)^p] \to 0$ as $m \to \infty$. Suppose, on the other hand, that $E[(\beta + \alpha z_t^2)^p] \geq 1$ and $\beta > 0$. Since $E[({}_u\xi_t^2)^p] = \infty$ in this case, $E[(\beta^{m-1}{}_u\xi_t^2)^p] = \infty$ for all m, concluding the proof of (2.42). The proof of (2.41) is similar, except that now, for any $t < \infty$, the necessary and sufficient condition for $E[(\xi_t^2)^p] < \infty$ is $E[\sigma_0^{2p}] < \infty$ and $E[(\beta + \alpha z_t^2)^p] < \infty$.

Proof of Theorem 6. Equations (2.44) and (2.46) are applications of Prudnikov *et al.* (1986), Formula 2.6.23 #4, and Gradshteyn and Ryzhik (1980), Formula 4.352 #1, respectively. Equation (2.45) follows from the integral representation of $\Psi(\cdot,\cdot,\cdot)$ (Lebedev, 1972, Section 9.11). Equations (2.47), (2.48), and (2.49) with $p < \frac{1}{2}$ are applications of Gradshteyn and Ryzhik (1980) Formulas 3.381 #4, 4.295 #7, and 3.227 #1, respectively. When $p \geq \frac{1}{2}$ in (2.49), $E[(\beta + \alpha z^2)^p] \geq E[(\alpha z^2)^p] = \alpha^p \cdot E[|z|^{2p}] = \infty$. Equation (2.50) is an application of Gradshteyn and Ryzhik (1980), Formula 3.222 #2.

Obtaining the Density Estimates in Figure 3. By (1.6) and (1.10), we have, for any $T > 0$,

$$\left({}_u\sigma_T^2 - \sigma_T^2\right) = \left({}_u\sigma_0^2 - \sigma_0^2\right)\prod_{i=1}^{T}\left[(1-\alpha) + \alpha z_{T-i}^2\right]. \tag{A.14}$$

By Theorem 2, ${}_u\sigma_T^2 - \sigma_T^2 \to 0$ a.s. as $T \to \infty$, since the term $\Pi_{i=1,T}[(1-\alpha) + \alpha z_{T-i}^2]$ vanishes at an exponential rate with probability one. After some experimentation, it was found that $T = 5000$ achieved consistently tiny values for $\Pi_{i=1,T}[(1-\alpha) + \alpha z_{T-i}^2]$ (typically, on the order of 10^{-40}) for the selected values of α. σ_{5000}^2 was therefore used to simulate ${}_u\sigma_t^2$.

The z_t's were generated by the normal random number generator in GAUSS. Density estimates for $\ln(\ln(\sigma_T^2))$ (which has support on $(-\infty,\infty)$) were formed using the kernel method with a normal kernel and 7000 draws of σ_T^2. The bandwidth was set equal to $0.9 \cdot A \cdot n^{-1/5}$, where n was the sample size (7000) and A was the minimum of the interquartile range divided by 1.34 and the standard deviation of $\ln(\ln(\sigma_T^2))$. This is the bandwidth selection procedure employed in Spanier and Oldham (1987), Section 3.4.2; see Spanier and Oldham (1987) for details of the statistical properties. By a change of variables, density estimates for $\ln(\sigma_T^2)$ were then obtained. Doubling or halving the bandwidth had little effect on the density estimates.

ACKNOWLEDGMENTS

I thank Patrick Billingsley, Tim Bollerslev, Jeff Wooldridge, three referees and the editor for helpful comments, and the Department of Education and the University of Chicago Graduate School of Business for research support. Any remaining errors are mine alone.

REFERENCES

Abramowitz, M., and Stegun, N., eds. (1965). "Handbook of Mathematical Functions." Dover, New York.

Bollerslev, T. (1986). Generalized autoregressive conditional heteroskedasticity. *Journal of Econometrics* **31**, 307–327.

Davis, P. J. (1965). The gamma function and related functions. *In* "Handbook of Mathematical Functions" (M. Abramowitz and N. Stegun, eds.), Chapter 6, pp. 253–294. Dover, New York.

Dudley, R. M. (1989). "Real Analysis and Probability." Wadsworth & Brooks/Cole, Pacific Grove, CA.

Engle, R. F. (1982). Autoregressive conditional heteroskedasticity with estimates of the variance of United Kingdom inflation. *Econometrica* **50**, 987–1007.

Engle, R. F., and Bollerslev, T. (1986a). Modeling the persistence of conditional variances. *Econometric Reviews* **5**, 1–50.

Engle, R. F., and Bollerslev, T. (1986b). Reply to comments on modeling the persistence of conditional variances. *Econometric Reviews* **5**, 81–88.

GAUSS (1988). "The GAUSS System Version 2.0." Aptech Systems, Kent, WA.

Geweke, J. (1986). Comment on modeling the persistence of conditional variances. *Econometric Reviews* **5**, 57–62.

Gradshteyn,, I. S., and Ryzhik, I. M. (1980). "Table of Integrals, Series and Products." Academic Press, New York.

Hardy, G. H., Littlewood, J. E., and Pólya, G. (1952). "Inequalities," 2nd ed. Cambridge University Press, Cambridge, UK.

Kolmogorov, A. N., and Fomin, S. V. (1968). "Introductory Real Analysis." Dover, New York.

Lebedev, N. N. (1972). "Special Functions and Their Applications." Dover, New York.

Nelson, D. B. (1990). ARCH models as diffusion approximations, *Journal of Econometrics* **45**, 7–38.

Poterba, J. M., and Summers, L. H. (1986). The persistence of volatility and stock market fluctuations. *American Economic Review* **76**, 1142–1151.

Prudnikov, A. P., Brychkov, Y. A., and Marichev, O. I. (1986) "Integrals and Series," Vol. 1. Gordon & Breach, New York.

Royden, H. L. (1970). *Real Analysis*, 2nd ed. Macmillan, New York.

Sampson, M. (1988). "A Stationary Condition for the GARCH(1, 1) Process," Mimeo. Concordia University, Department of Economics, River Forest, IL.

Spanier, J., and Oldham, K. B. (1987). "An Atlas of Functions." Hemisphere Publishing, Washington, DC.
Stout, W. F. (1974). "Almost Sure Convergence." Academic Press, London.
White, H. (1984). "Asymptotic Theory for Econometricians." Academic Press, Orlando, FL.
Zucker, R. (1965). Elementary transcendental functions: Logarithmic; exponential, circular and hyperbolic functions. *In* "Handbook of Mathematical Functions" (M. Abramowitz and N. Stegun, eds.), Chapter 4, pp. 65–226. Dover, New York.

3

CONDITIONAL HETEROSKEDASTICITY IN ASSET RETURNS: A NEW APPROACH*

DANIEL B. NELSON

Graduate School of Business
University of Chicago
Chicago, Illinois 60637

1. INTRODUCTION

After the events of October 1987, few would argue with the proposition that stock market volatility changes randomly over time. Understanding the way in which it changes is crucial to our understanding of many areas in macroeconomics and finance, for example the term structure of interest rates (e.g., Barsky, 1989; Abel, 1988), irreversible investment (e.g., Bernanke, 1983; McDonald and Siegel, 1986), options pricing (e.g., Wiggins, 1987), and dynamic capital asset pricing theory (e.g., Merton, 1973; Cox *et al.*, 1985).

Recent years have also seen a surge of interest in econometric models of changing conditional variance. Probably the most widely used, but by no means the only such models,[1] are the family of ARCH (autoregressive conditionally heteroskedastic) models introduced by Engle (1982). ARCH models make the conditional variance of the time t prediction error a function of time, system parameters, exogenous and lagged endogenous variables, and past prediction errors. For each integer t, let ξ_t be a

* Reprinted from *Econometrica* **59** (1991), 347–370.

[1] See, e.g., Poterba and Summers (1986), French *et al.* (1987), and Nelson (1988, Chapter 1).

model's (scalar) prediction error, b a vector of parameters, x_t a vector of predetermined variables, and σ_t^2 the variance of ξ_t given information at time t. A univariate ARCH model based on Engle (1982) equations 1–5 sets

$$\xi_t = \sigma_t z_t, \tag{1.1}$$

$$z_t \sim \text{i.i.d.} \quad \text{with } \mathrm{E}(z_t) = 0, \quad \mathrm{Var}(Z_t) = 1, \quad \text{and} \tag{1.2}$$

$$\sigma_t^2 = \sigma^2(\xi_{t-1}, \xi_{t-2}, \ldots, t, x_t, b)$$

$$= \sigma^2(\sigma_{t-1}z_{t-1}, \sigma_{t-2}z_{t-2}, \ldots, t, x_t, b). \tag{1.3}$$

The system (1.1)–(1.3) can easily be given a multivariate interpretation, in which case z_t is an n by one vector and σ_t^2 is an n by n matrix. We refer to any model of the form (1.1)–(1.3), whether univariate or multivariate, as an ARCH model.

The most widely used specifications for $\sigma^2(\cdot, \cdot, \ldots, \cdot)$ are the linear ARCH and GARCH models introduced by Engle (1982) and Bollerslev (1986) respectively, which makes σ_t^2 linear in lagged values of $\xi_t^2 = \sigma_t^2 z_t^2$ by defining

$$\sigma_t^2 = \omega + \sum_{j=1}^{p} \alpha_j z_{t-j}^2 \sigma_{t-j}^2, \qquad \text{and} \tag{1.4}$$

$$\sigma_t^2 = \omega + \sum_{i=1}^{q} \beta_i \sigma_{t-i}^2 + \sum_{j=1}^{p} \alpha_j z_{t-j}^2 \sigma_{t-j}^2, \tag{1.5}$$

respectively, where ω, the α_j, and the β_j are nonnegative. Since (1.4) is a special case of (1.5), we refer to both (1.4) and (1.5) as GARCH models, to distinguish them as special cases of (1.3).

The GARCH-M model of Engle and Bollerslev (1986a) adds another equation

$$R_t = a + b\sigma_t^2 + \xi_t, \tag{1.6}$$

in which σ_t^2, the conditional variance of R_t, enters the conditional mean of R_t as well. For example if R_t is the return on a portfolio at time t, its required rate of return may be linear in its risk as measured by σ_t^2.

Researchers have fruitfully applied the new ARCH methodology in asset pricing models: for example, Engle and Bollerslev (1986a) used GARCH(1, 1) to model the risk premium on the foreign exchange market, and Bollerslev *et al.* (1988) extended GARCH(1, 1) to a multivariate context to test a conditional CAPM with time varying covariances of asset returns.

Substituting recursively for the $\beta_i \sigma_{t-i}^2$ terms lets us rewrite (1.5) as[2]

$$\sigma_t^2 = \omega^* + \sum_{k=1}^{\infty} \phi_k z_{t-k}^2 \sigma_{t-k}^2. \tag{1.7}$$

It is readily verified that if ω, the α_j, and the β_i are nonnegative, ω^* and the ϕ_k are also nonnegative. By setting conditional variance equal to a constant plus a weighted average (with positive weights) of past squared residuals, GARCH models elegantly capture the volatility clustering in asset returns first noted by Mandelbrot (1963): "... large changes tend to be followed by large changes—of either sign—and small changes by small changes..." This feature of GARCH models accounts for both their theoretical appeal and their empirical success.

On the other hand, the simple structure of (1.7) imposes important limitations on GARCH models: For example, researchers beginning with Black (1976) have found evidence that stock returns are negatively correlated with changes in returns volatility—i.e., volatility tends to rise in response to "bad news" (excess returns lower than expected) and to fall in response to "good news" (excess returns higher than expected).[3] GARCH models, however, assume that only the magnitude and not the positivity or negativity of unanticipated excess returns determines feature σ_t^2. If the distribution of z_t is symmetric, the change in variance tomorrow is conditionally uncorrelated with excess returns today.[4] In (1.4)–(1.5), σ_t^2 is a function of lagged σ_t^2 and lagged z_t^2, and so is invariant to changes in the algebraic sign of the z_t's—i.e., only the size, not the sign, of lagged residuals determines conditional variance. This suggests that a model in which σ_t^2 responds asymmetrically to positive and negative residuals might be preferable for asset pricing applications.

Another limitation of GARCH models results from the nonnegativity constraints on ω^* and the ϕ_k in (1.7), which are imposed to ensure that σ_t^2 remains nonnegative for all t with probability one. These constraints

[2] The representation (1.7) assumes that $\{\sigma_t^2\}$ is strictly stationary, so that the recursion can be carried into the infinite past.

[3] The economic reasons for this are unclear. As Black (1976) and Christie (1982) note, both financial and operating leverage play a role, but are not able to explain the extent of the asymmetric response of volatility to positive and negative returns shocks. Schwert (1989a,b) presents evidence that stock volatility is higher during recessions and financial crises, but finds only weak relations between stock market volatility and measures of macroeconomic uncertainty. See also Nelson, 1988; Pagan and Hong, 1988.

[4] Adrian Pagan pointed out, however, that in a GARCH-M model σ_t^2 may rise (fall) on average when returns are negative (positive), even though $\sigma_{t+1}^2 - \sigma_t^2$ and R_t are conditionally uncorrelated, since in (1.6) $E[|z_t| \,|\, R_t < 0] > E[|z_t| \,|\, R_t \geq 0]$ if a and b are positive.

imply that increasing z_t^2 in any period increases σ_{t+m}^2 for all $m \geq 1$, ruling out random oscillatory behavior in the σ_t^2 process. Furthermore, these nonnegativity constraints can create difficulties in estimating GARCH models. For example, Engle *et al.* (1987) had to impose a linearly declining structure on the α_j coefficients in (1.4) to prevent some of them from becoming negative.

A third drawback of GARCH modelling concerns the interpretation of the "persistence" of shocks to conditional variance. In many studies of the time series behavior of asset volatility (e.g., Poterba and Summers, 1986; French *et al.*, 1987; Engle and Bollerslev, 1986a), the central question has been how long shocks to conditional variance persist. If volatility shocks persist indefinitely, they may move the whole term structure of risk premia, and are therefore likely to have a significant impact on investment in long-lived capital goods (Poterba and Summers, 1986).

There are many different notions of convergence in the probability literature (almost sure, in probability, in L^p), so whether a shock is transitory or persistent may depend on our definition of convergence. In linear models it typically makes no difference which of the standard definitions we use, since the definitions usually agree. In GARCH models, the situation is more complicated. For example, the IGARCH(1, 1) model of Engle and Bollerslev (1986a) sets

$$\sigma_t^2 = \omega + \sigma_{t-1}^2\left[(1-\alpha) + \alpha z_{t-1}^2\right], \qquad 0 < \alpha \leq 1. \tag{1.8}$$

When $\omega = 0$, σ_t^2 is a martingale. Based on the nature of persistence in linear models, it seems that IGARCH(1, 1) with $\omega > 0$ and $\omega = 0$ are analogous to random walks with and without drift, respectively, and are therefore natural models of "persistent" shocks. This turns out to be misleading, however: in IGARCH(1, 1) with $\omega = 0$, σ_t^2 collapses to zero almost surely, and in IGARCH(1, 1) with $\omega > 0$, σ_t^2 is strictly stationary and ergodic (Geweke, 1986; Nelson, 1990) and therefore does not behave like a random walk, since random walks diverge almost surely.

The reason for this paradox is that in GARCH(1, 1) models, shocks may persist in one norm and die out in another, so that conditional moments of GARCH(1, 1) may explode even when the process itself is strictly stationary and ergodic (Nelson, 1990).

The object of this chapter is to present an alternative to GARCH that meets these objections, and so may be more suitable for modelling conditional variances in asset returns. In section 2, we describe this σ_t^2 process, and develop some of its properties. In section 3 we estimate a simple model of stock market volatility and the risk premium. Section 4 concludes. In appendix 1, we provide formulas for the moments of σ_t^2 and ξ_t in the model presented in section 2. Proofs are in appendix 2.

2. EXPONENTIAL ARCH

If σ_t^2 is to be the conditional variance of ξ_t given information at time t, it clearly must be nonnegative with probability one. GARCH models ensure this by making σ_t^2 a linear combination (with positive weights) of positive random variables. We adopt another natural device for ensuring that σ_t^2 remains nonnegative, by making $\ln(\sigma_t^2)$ linear in some function of time and lagged z_t's. That is, for some suitable function g:

$$\ln(\sigma_t^2) = \alpha_t + \sum_{k=1}^{\infty} \beta_k g(z_{t-k}), \qquad \beta_1 \equiv 1, \tag{2.1}$$

where $\{\alpha_t\}_{t=-\infty,\infty}$ and $\{\beta_k\}_{k=1,\infty}$ are real, nonstochastic, scalar sequences. Pantula (1986) and Geweke (1986) have previously proposed ARCH models of this form; in their log–GARCH models, $g(z_t) = \ln|z_t|^b$ for some $b > 0$.[5]

To accommodate the asymmetric relation between stock returns and volatility changes noted in section 1, the value of $g(z_t)$ must be a function of both the magnitude and the sign of z_t. One choice, that in certain important cases turns out to give σ_t^2 well-behaved moments, is to make $g(z_t)$ a linear combination of z_t and $|z_t|$:

$$g(z_t) \equiv \theta z_t + \gamma\big[|z_t| - \mathrm{E}|z_t|\big]. \tag{2.2}$$

By construction, $\{g(z_t)\}_{t=-\infty,\infty}$ is a zero-mean, i.i.d. random sequence. The two components of $g(z_t)$ are θz_t and $\gamma[|z_t| - \mathrm{E}|z_t|]$, each with mean zero. If the distribution of z_t is symmetric, the two components are orthogonal, though of course they are not independent. Over the range $0 < z_t < \infty$, $g(z_t)$ is linear in z_t with slope $\theta + \gamma$, and over the range $-\infty < z_t \leq 0$, $g(z_t)$ is linear with slope $\theta - \gamma$. Thus, $g(z_t)$ allows the conditional variance process $\{\sigma_t^2\}$ to respond asymmetrically to rises and falls in stock price.

To see that the term $\gamma[|z_t| - \mathrm{E}|z_t|]$ represents a magnitude effect in the spirit of the GARCH models discussed in section 1, assume for the moment that $\gamma > 0$ and $\theta = 0$. The innovation in $\ln(\sigma_{t+1}^2)$ is then positive (negative) when the magnitude of z_t is larger (smaller) than its expected value. Suppose now that $\gamma = 0$ and $\theta < 0$. The innovation in conditional variance is now positive (negative) when returns innovations are negative (positive). Thus the exponential form of ARCH in (2.1)–(2.2) meets the first objection raised to the GARCH models in section 1.

In section 1 we also argued that the dynamics of GARCH models were unduly restrictive (i.e., oscillatory behavior is excluded) and that they

[5] See also Engle and Bollerslev (1986b).

impose inequality constraints that were frequently violated by estimated coefficients. But note that in (2.1)–(2.2) there are no inequality constraints whatever, and that cycling is permitted, since the β_k terms can be negative or positive.

Our final criticism of GARCH models was that it is difficult to evaluate whether shocks to variance "persist" or not. In exponential ARCH, however, $\ln(\sigma_t^2)$ is a linear process, and its stationarity (covariance or strict) and ergodicity are easily checked. If the shocks to $\{\ln(\sigma_t^2)\}$ die out quickly enough, and if we remove the deterministic, possibly time-varying component $\{\alpha_t\}$, then $\{\ln(\sigma_t^2)\}$ is strictly stationary and ergodic. Theorem 2.1 below states conditions for the ergodicity and strict stationarity of $\{\exp(-\alpha_t)\sigma_t^2\}$ and $(\exp(-\alpha_t/2)\xi_t\}$, which are $\{\sigma_t^2\}$ and $\{\xi_t\}$ with the influence of $\{\alpha_t\}$ removed.

THEOREM 2.1. *Define* $\{\sigma_t^2\}, \{\xi_t\}$, *and* $\{z_t\}$ *by* (1.1)–(1.2) *and* (2.1)–(2.2), *and assume that* γ *and* θ *do not both equal zero. Then* $\{\exp(-\alpha_t)\sigma_t^2\}$, $\{\exp(-\alpha_t/2)\xi_t\}$, *and* $\{\ln(\sigma_t^2) - \alpha_t\}$ *are strictly stationary and ergodic and* $\{\ln(\sigma_t^2) - \alpha_t\}$ *is covariance stationary if and only if* $\sum_{k=1}^{\infty} \beta_k^2 < \infty$. *If* $\sum_{k=1}^{\infty} \beta_k^2 = \infty$, *then* $|\ln(\sigma_t^2) - \alpha_t| = \infty$ *almost surely. If* $\sum_{k=1}^{\infty} \beta_k^2 < \infty$, *then for* $k > 0$, $\mathrm{Cov}\{z_{t-k}, \ln(\sigma_t^2)\} = \beta_k[\theta + \gamma \mathrm{E}(z_t|z_t|)]$, *and*

$$\mathrm{Cov}\big[\ln(\sigma_t^2), \ln(\sigma_{t-k}^2)\big] = \mathrm{Var}[g(z_t)] \sum_{j=1}^{\infty} \beta_j \beta_{j+k}.$$

The stationarity and ergodicity criterion in Theorem 2.1 is exactly the same as for a general linear process with finite innovations variance,[6] so if, for example, $\ln(\sigma_t^2)$ follows an AR(1) with AR coefficient Δ, $\ln(\sigma_t^2)$ is strictly stationary and ergodic if and only if $|\Delta| < 1$.

There is often a simpler expression for $\ln(\sigma_t^2)$ than the infinite moving average representation in (2.1). In many applications, an ARMA process provides a parsimonious parameterization:

$$\ln(\sigma_t^2) = \alpha_t + \frac{(1 + \psi_1 L + \cdots + \psi_q L^q)}{(1 - \Delta_1 L - \cdots - \Delta_p L^p)} g(z_{t-1}). \tag{2.3}$$

We assume that $[1 - \Sigma_{i=1,p} \Delta_i y^i]$ and $[1 + \Sigma_{i=1,q} \psi_i y^i]$ have no common roots. By Theorem 2.1, $\{\exp(-\alpha_t)\sigma_t^2)$ and $\{\exp(-\alpha_t/2)\xi_t\}$ are then strictly stationary and ergodic if and only if all the roots of $[1 - \Sigma_{t=1,p} \Delta_i y^i]$ lie

[6] The assumption that z_t has a finite variance can be relaxed. For example, let z_t be i.i.d. Cauchy, and $g(z_t) \equiv \theta z_t + \gamma|z_t|$, with γ and θ not both equal to zero. By the Three Series Theorem (Billingsley, 1986, Theorem 22.8), σ_t^2 is finite almost surely if and only if $|\Sigma_{i\geq 1, \beta_i \neq 0} \gamma\beta_i \cdot \ln(1 + \beta_i^{-2})| < \infty$ and $\Sigma_{i\geq 1} |\beta_i| < \infty$. Then $\{\exp(-\alpha_t)\sigma_t^2\}$ and $\{\exp(-\alpha_t/2)\xi_t\}$ are strictly stationary and ergodic.

outside the unit circle. While an ARMA representation may be suitable for many modelling purposes, Theorem 2.1 and the representation (2.1) also allow "long memory" (Hosking, 1981) processes for $\{\ln(\sigma_t^2)\}$.

Next consider the covariance stationarity of $\{\sigma_t^2\}$ and $\{\xi_t\}$. According to Theorem 2.1, $\Sigma\ \beta_i^2 < \infty$ implies that $\{\exp(-\alpha_t)\sigma_t^2\}$ and $\{\exp(-\alpha_t/2)\xi_t\}$ are strictly stationary and ergodic. This strict stationarity, however, need not imply covariance stationarity, since $\{\exp(-\alpha_t)\sigma_t^2\}$ and $\{\exp(-\alpha_t/2)\xi_t\}$ may fail to have finite unconditional means and variances. For some distributions of $\{z_t\}$ (e.g., the Student's t with finite degrees of freedom), $\{\exp(-\alpha_t)\sigma_t^2\}$ and $\{\exp(-\alpha_t/2)\xi_t\}$ typically have no finite unconditional moments. The results for another commonly used family of distributions, the GED (Generalized Error Distribution) (Harvey, 1981; Box and Tiao, 1973),[7] are more encouraging. The GED includes the normal as a special case, along with many other distributions, some more fat tailed than the normal (e.g., the double exponential), some more thin tailed (e.g., the uniform). If the distribution of $\{z_t\}$ is a member of this family and is not too thick-tailed, and if $\Sigma_{i=1,\infty}\ \beta_i^2 < \infty$, then $\{\sigma_t^2\}$ and $\{\xi_t\}$ have finite unconditional moments of arbitrary order.

The density of a GED random variable normalized to have a mean of zero and a variance of one is given by

$$f(z) = \frac{v \exp\left[-\left(\frac{1}{2}\right)|z/\lambda|^{v}\right]}{\lambda 2^{(1+1/v)}\Gamma(1/v)}, \qquad -\infty < z < \infty, \qquad 0 < v \leq \infty, \quad (2.4)$$

where $\Gamma(\cdot)$ is the gamma function, and

$$\lambda \equiv \left[2^{(-2/v)}\Gamma(1/v)/\Gamma(3/v)\right]^{1/2}. \quad (2.5)$$

v is a tail-thickness parameter. When $v = 2$, z has a standard normal distribution. For $v < 2$, the distribution of z has thicker tails than the normal (e.g., when $v = 1$, z has a double exponential distribution) and for $v > 2$, the distribution of z has thinner tails than the normal (e.g., for $v = \infty$, z is uniformly distributed on the interval $[-3^{1/2}, 3^{1/2}]$).

THEOREM 2.2. *Define* $\{\sigma_t^2, \xi_t\}_{t=-\infty,\infty}$ *by* (1.1)–(1.2) *and* (2.1)–(2.2), *and assume that* γ *and* θ *do not both equal zero. Let* $\{z_t\}_{t=-\infty,\infty}$ *be i.i.d. GED with mean zero, variance one, and tail thickness parameter* $v > 1$, *and let* $\Sigma_{i=1}^{\infty}\ \beta_i^2 < \infty$. *Then* $\{\exp(-\alpha_t)\sigma_t^2\}$ *and* $\{\exp(-\alpha_t/2)\xi_t\}$ *possess finite, time-invariant moments of arbitrary order. Further, if* $0 < p < \infty$, *condition-*

[7] Box and Tiao call the GED the exponential power distribution.

ing information at time 0 *drops out of the forecast pth moment of* $\exp(-\alpha_t)\sigma_t^2$ *and* $\exp(-\alpha_t/2)\xi_t$ *as* $t \to \infty$:

$$p \lim_{t \to \infty} \mathrm{E}\left[\exp(-p\alpha_t)\sigma_t^{2p} \mid z_0, z_{-1}, z_{-2}, \ldots\right] - \mathrm{E}\left[\exp(-p\alpha_t)\sigma_t^{2p}\right] = 0, \tag{2.6}$$

and

$$p \lim_{t \to \infty} \mathrm{E}\left[\exp(-p\alpha_t/2)|\xi_t|^p \mid z_0, z_{-1}, z_{-2}, \ldots\right] - \mathrm{E}\left[\exp(-p\alpha_t)|\xi_t|^p\right] = 0, \tag{2.7}$$

where p lim *denotes the limit in probability.*

That is, if the distribution of the z_t is GED and is thinner-tailed than the double exponential, and if $\Sigma\, \beta_i^2 < \infty$, then $\exp(-\alpha_t)\sigma_t^2$ and $\exp(\alpha_t/2)\xi_t$ are not only strictly stationary and ergodic, but have arbitrary finite moments, which in turn implies that they are covariance stationary.

Since the moments of $\{\exp(-\alpha_t)\sigma_t^2\}$ and $\{\exp(-\alpha_t/2)\xi_t\}$ are of interest for forecasting, Appendix 1 derives the conditional and unconditional moments (including covariances) of $\{\exp(-\alpha_t)\sigma_t^2\}$ and $\{\exp(-\alpha_t/2)\xi_t\}$ under a variety of distributional assumptions for $\{z_t\}$, including Normal, GED, and Student t.

3. A SIMPLE MODEL OF MARKET VOLATILITY

In this section, we estimate and test a simple model of market risk, asset returns, and changing conditional volatility. We use this model to examine several issues previously investigated in the economics and finance literature, namely (i) the relation between the level of market risk and required return, (ii) the asymmetry between positive and negative returns in their effect on conditional variance, (iii) the persistence of shocks to volatility, (iv) "fat tails" in the conditional distribution of returns, and (v) the contribution of nontrading days to volatility.

We use the model developed in section 2 for the conditional variance process, assuming an ARMA representation for $\ln(\sigma_t^2)$. To allow for the possibility of nonnormality in the conditional distribution of returns, we assume that the $\{z_t\}$ are i.i.d. draws from the GED density (2.4). To account for the contribution of nontrading periods to market variance, we assume that each nontrading day contributed as much to variance as some fixed fraction of a trading day, so if, for example, this fraction is one tenth, than σ_t^2 on a typical Monday would be 20% higher than on a typical Tuesday. Other researchers (e.g., Fama, 1965; French and Roll, 1986) have found that nontrading periods contribute much less than do trading periods to market variance, so we expect that $0 < \delta \ll 1$.[8]

Specifically, we model the log of conditional variance as

$$\ln(\sigma_t^2) = \alpha_t + \frac{(1 + \Psi_1 L + \cdots + \Psi_q L^q)}{(1 - \Delta_1 L + \cdots - \Delta_p L^p)} g(z_{t-1}), \qquad (3.1)$$

where z_t is i.i.d. GED with mean zero, variance one, and tail thickness parameter $v > 0$, and $\{\alpha_t\}$ is given by

$$\alpha_t = \alpha + \ln(1 + N_t \delta), \qquad (3.2)$$

where N_t is the number of nontrading days between trading days $t - 1$ and t, and α and δ are parameters. If the unconditional expectation of $\ln(\sigma_t^2)$ exists, then it equals $\alpha + \ln(1 + N_t \delta)$. Together, (3.1)–(3.2) and Theorem 2.1 imply that $\{(1 + N_t \delta)^{-1/2} \xi_t\}$ is strictly stationary and ergodic if and only if all the roots of $(1 - \Delta_1 Y - \cdots - \Delta_p Y^p)$ lie outside the unit circle.[9]

We model excess returns R_t as

$$R_t = a + bR_{t-1} + c\sigma_t^2 + \xi_t, \qquad (3.3)$$

where the conditional mean and variance of ξ_t at time t are 0 and σ_t^2 respectively, and where a, b, and c are parameters. The bR_{t-1} term allows for the autocorrelation induced by discontinuous trading in the stocks making up an index (Scholes and Williams, 1977; Lo and MacKinlay, 1988). The Scholes and Williams model suggests an MA(1) form for index returns, while the Lo and MacKinlay model suggests an AR(1) form, which we adopt. As a practical matter, there is little difference between an AR(1) and an MA(1) when the AR and MA coefficients are small and the autocorrelations at lag one are equal, since the higher-order autocorrelations die out very quickly in the AR model. As Lo and MacKinlay note, however, such simple models do not adequately explain the short-term autocorrelation behavior of the market indices, and no fully satisfactory model yet exists.

The theoretical justification for including the $c\sigma_t^2$ term in (3.3) is meager, since the required excess return on a portfolio is linear in its conditional variance only under very special circumstances. In Merton's (1973) intertemporal CAPM model, for example, the instantaneous expected excess return on the market portfolio is linear in its conditional variance if there is a representative agent with log utility. Merton's conditions (e.g., continuous time, continuous trading, and a true "market" portfolio) do not apply in our model even under the log utility assumption. Backus and Gregory (1989) and Glosten *et al.* (1989) give examples of

[8] French and Roll (1986) and Barclay *et al.* (1990) offer economic interpretations.
[9] Again, we assume that $[1 - \Sigma_{i=1,p} \Delta_i y^i]$ and $[1 + \Sigma_{i=1,q} \Psi_i y^i]$ have no common roots.

equilibrium models in which a regression of returns on σ_t^2 yields a negative coefficient.[10] There is, therefore, no strong theoretical reason to believe that c is positive. Rather, the justification for including $c\sigma_t^2$ is pragmatic: a number of researchers using GARCH models (e.g., French *et al.*, 1987; Chou, 1987) have found a statistically significant positive relation between conditional variance and excess returns on stock market indices, and we therefore adopt the form (3.3).

For a given ARMA(p, q) exponential ARCH model, the $\{z_t\}_{t=1,T}$ and $\{\sigma_t^2\}_{t=1,T}$ sequences can be easily derived recursively given the data $\{R_t\}_{t=1,T}$, and the initial values $\sigma_1^2, \ldots, \sigma_{1+\max\{p,q+1\}}^2$. To close the model, $\ln(\sigma_1^2), \ldots, \ln(\sigma_{1+\max\{p,q+1\}}^2)$ were set equal to their unconditional expectations $(\alpha + \ln(1 + \delta N_1)), \ldots, (\alpha + \ln(1 + \delta N_{1+\max\{p,q+1\}}))$. This allows us to write the log likelihood L_T as

$$L_T = \sum_{t=1}^{T} \ln(v/\lambda) - \left(\tfrac{1}{2}\right)\left|(R_t - a - bR_{t-1} - c\sigma_t^2)/\sigma_t\lambda\right|^v$$
$$- (1 + v^{-1})\ln(2) - \ln[\Gamma(1/v)] - \tfrac{1}{2}\ln(\sigma_t^2), \tag{3.4}$$

where λ is defined in (2.5). Given the parameters and initial states, we can easily compute the likelihood (3.4) recursively, using (3.1) and setting

$$z_t = \sigma_t^{-1}(R_t - a - bR_{t-1} - c\sigma_t^2). \tag{3.5}$$

In light of (3.5), however, it may be that setting the out-of-sample values of $\ln(\sigma_t^2)$ equal to their unconditional expectation is not innocent: using the true parameter values but the wrong σ_t^2 in (3.5) leads to an incorrect fitted value of z_t, in turn leading to an incorrect fitted value for σ_{t+1}^2, and so on. In simulations using parameter estimates similar to those reported below, fitted σ_t^2 generated by (3.4)–(3.5) with an incorrect starting value converge very rapidly to the σ_t^2 generated by (3.4)–(3.5) and a correct starting value. In a continuous time limit, it is possible to prove that this convergence takes place instantaneously (Nelson, 1992).

Under sufficient regularity conditions, the maximum likelihood estimator is consistent and asymptotically normal. Unfortunately, verifying that these conditions hold in ARCH models has proven extremely difficult in both GARCH models and in the exponential ARCH model introduced in this chapter. Weiss (1986) developed a set of sufficient conditions for consistency and asymptotic normality in a variant of the linear ARCH formulation of Engle (1982). These conditions are quite restrictive, and are not satisfied by the coefficient estimates obtained in most studies using this form of ARCH. In the GARCH-M model, in which conditional variance appears in the conditional mean, the asymptotics are even more uncertain, and no sufficient conditions for consistency and asymptotic normality are

[10] See also Gennotte and Marsh (1987).

yet known. The asymptotics of exponential ARCH models are equally difficult, and as with other ARCH models, a satisfactory asymptotic theory for exponential ARCH is as yet unavailable. In the remainder of this chapter we assume (as is the usual practice of researchers using GARCH models) that the maximum likelihood estimator is consistent and asymptotically normal.

For our empirical analysis, we use the daily returns for the value-weighted market index from the CRSP tapes for July 1962–December 1987. An immediate problem in using this data is that we wish to model the excess returns process but do not have access to any adequate daily riskless returns series. As an initial approximation to the riskless rate, we extracted the monthly Treasury Bill returns from the CRSP tapes, assumed that this return was constant for each calendar day within a given month, and computed daily excess returns using this riskless rate series and value-weighted CRSP daily market returns.[11] As a check on whether measurement errors in the riskless rate series are likely to bias the results seriously, we also fit the model using the capital gains series, ignoring both dividends and the riskless interest rate. As shown below, it made virtually no difference in either the estimated parameters or the fitted variances.

To select the order of the ARMA process for $\ln(\sigma_t^2)$, we used the Schwarz Criterion (Schwarz, 1978), which provides consistent order-estimation in the context of linear ARMA models (Hannan, 1980). The asymptotic properties of the Schwarz criterion in the context of ARCH models are unknown.

The maximum likelihood parameter estimates were computed on VAX 8650 and 8550 computers using the IMSL subroutine DUMING. Table 1 lists likelihood values for ARMA models of various orders on the CRSP excess returns series. For both the excess returns and capital gains series, the Schwarz Criterion selected an ARMA(2, 1) model for $\ln(\sigma_t^2)$.[12] Table 2 gives the parameter estimates and estimated standard errors for both ARMA(2, 1) models. The estimated correlation matrix of the parameter estimates in the excess returns model is in Table 3. The asymptotic covariance matrix was computed using the score.

First note that except for the parameter c (the risk premium term in (3.3)), the two sets of coefficient estimates are nearly identical. The fitted values of $\ln(\sigma_t^2)$ from the two models are even more closely related: their means and variances for the 1962–1987 period are nearly equal (-9.9731 and 0.6441, vs. -9.9766 and 0.6415, respectively) and the two series have a sample correlation of 0.9996. In other words, the series are practically

[11] Logarithmic returns are used throughout: i.e., if S_t is the level of the value-weighted market index at time t and d_t are the dividends paid at t, then the value-weighted market return, capital gain, and excess return are computed as $\ln[(S_t + d_t)/S_{t-1}]$, $\ln[S_t/S_{t-1}]$, and $\ln[(S_t + d_t)/S_{t-1}] - RR_t$ respectively, where RR_t is our proxy for the riskless rate.

[12] The AIC (Akaike, 1973) chose the highest-order model estimated.

TABLE 1 Likelihood Values for ARMA Models for CRSP Value-Weighted Excess Returns

Observations = 6408
Deterministic conditional variance model
(i.e., $\gamma = \theta = \Delta = \Psi = 0$), likelihood = 22273.313.
ARMA exponential ARCH model likelihoods:

		AR order				
		0	**1**	**2**	**3**	**4**
MA Order	0	22384.898	22888.052	22891.937	22894.237	22894.902
	1	22429.942	22893.687	22915.454 (SC)	22916.799	22916.894
	2	22473.728	22894.167	22916.762	22917.035	22918.708
	3	22532.768	22894.385	22916.941	22918.853	22922.439
	4	22590.536	22900.990	22917.670	22918.857	22923.752 (AIC)

(SC = Model selected by the criterion of Schwarz (1978).)
(AIC = Model selected by the information criterion of Akaike (1973).)

TABLE 2 Parameter Estimates for the CRSP Excess Returns and Capital Gains Models (Standard Errors in Parentheses)

Parameter	**CRSP excess returns**	**CRSP capital gains**
α	−10.0593	−10.0746
	(0.3462)	(0.3361)
δ	0.1831	0.1676
	(0.0277)	(0.0271)
γ	0.1559	0.1575
	(0.0125)	(0.0126)
Δ_1	1.92938	1.92914
	(0.0145)	(0.0146)
Δ_2	−0.92941	−0.92917
	(0.0145)	(0.0146)
Ψ	−0.9782	−0.9781
	(0.0062)	(0.0063)
θ	−0.1178	−0.1161
	(0.0090)	(0.0090)
a	$3.488 \cdot 10^{-4}$	$3.416 \cdot 10^{-4}$
	$(9.850 \cdot 10^{-5})$	$(9.842 \cdot 10^{-5})$
b	0.2053	0.2082
	(0.0123)	(0.0123)
c	−3.3608	−1.9992
	(2.0261)	(2.0347)
v	1.5763	1.5760
	(0.0320)	(0.0320)

TABLE 3 Estimated Correlation Matrix for Parameter Estimates Excess Returns Model (Only Lower Triangle Reported)

Parameter					
α	1				
δ	−0.0087	1			
γ	−0.0023	−0.1639	1		
Δ_1	0.0139	0.0715	−0.2996	1	
Δ_2	−0.0138	−0.0719	0.2994	−0.999992	1
Ψ	−0.0094	−0.0392	0.1020	−0.8356	0.8352
	1				
θ	−0.0472	0.1909	−0.0591	0.4342	−0.4346
	−0.2211	1			
a	−0.0674	−0.0342	−0.0222	0.0734	−0.0715
	0.0064	0.0251	1		
b	−0.0144	−0.0395	−0.0109	0.0439	−0.0443
	−0.0256	−0.0772	−0.1979	1	
c	−0.0702	0.1111	0.0815	−0.0891	0.0871
	−0.0151	−0.0064	−0.8287	0.1198	1
υ	0.0591	0.1763	0.2853	−0.0182	0.0179
	−0.0414	−0.1554	0.0042	0.0402	0.1240
	1				
	α	δ	γ	Δ_1	Δ_2
	Ψ	θ	a	b	c
	υ				

identical, so ignoring dividends and interest payments appears likely to introduce no important errors in forecasting the volatility of broad market indices.

Next we examine the empirical issues raised earlier in the section:

(i) Market Risk and Expected Return. The estimated risk premium is negatively (though weakly) correlated with conditional variance, with $c \approx -3.361$ with a large standard error of about 2.026. This contrasts with the significant positive relation between returns and conditional variance found by researchers using GARCH-M models (e.g., Chou, 1987; French *et al.*, 1987), but agrees with the findings of other researchers not using GARCH models (e.g., Pagan and Hong, 1988). Given the results of Gennotte and Marsh (1987), Glosten *et al.* (1989), and Backus and Gregory (1989), our findings of a negative (albeit insignificant) coefficient should not be too surprising.

(ii) The asymmetric relation between returns and changes in volatility, as represented by θ, is highly significant. Recall that a negative value of θ indicates that volatility tends to rise (fall) when returns surprises are negative (positive). The estimated value for θ is about -0.118 (with a standard error of about 0.008) which is significantly below zero at any standard level. Figure 1 plots the estimated $g(z)$ function.

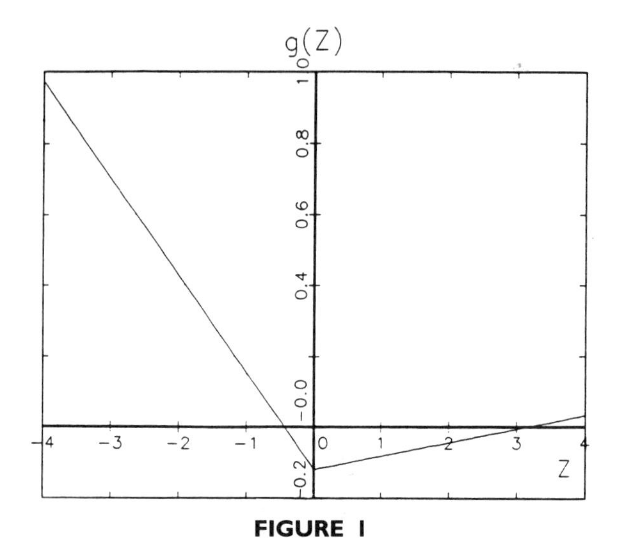

FIGURE 1

Figures 2 and 3 plot σ_t (the daily conditional standard deviation of returns) and the log value of the CRSP value-weighted market index, respectively. The σ_t series is extremely variable, with lows of less than 0.5% and highs over 5%. All the major episodes of high volatility are associated with market drops.

(iii) Persistence of Shocks. The largest estimated AR root is approximately 0.99962 with a standard error of about 0.00086, so the t statistic for

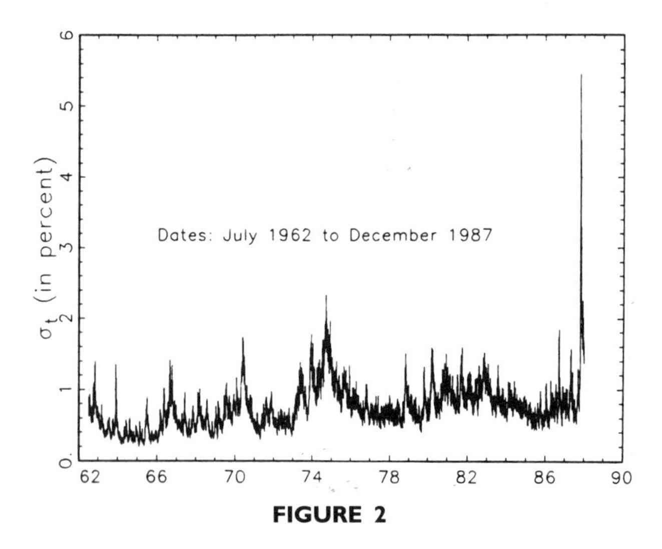

FIGURE 2

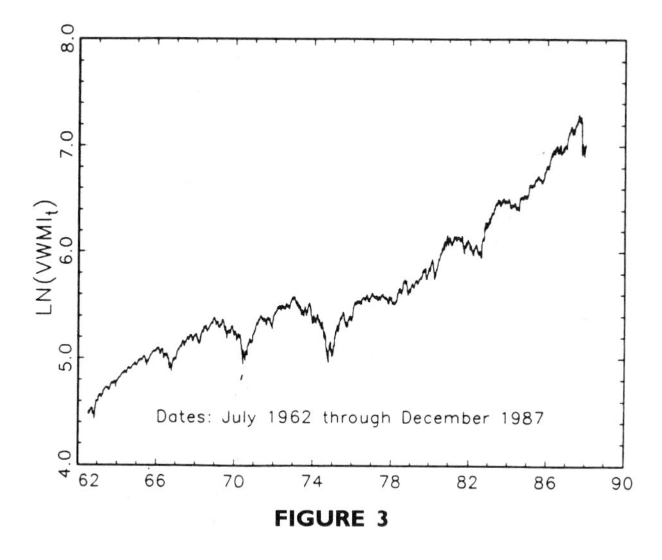

FIGURE 3

a unit root is only about -0.448. To gain intuition about the degree of persistence implied by the largest AR root ρ, it is useful to think of the half-life h of a shock associated with this root, i.e., the number h such that

$$\rho^h = 1/2. \tag{3.6}$$

$\rho = 0.99962$ implies a half-life h of over 1820 trading days, about 7.3 years. In contrast, the half-life implied by the smaller AR root is less than two weeks. While this indicates substantial persistence and perhaps nonstationarity, it is hard to know how seriously to take the point estimates, since we have only about 25 years of data, about four times our estimated half-life for the larger AR root. The usual cautions about interpreting an estimated AR root near the unit circle as evidence of truly infinite persistence also apply[13] (see, e.g., Cochrane, 1988).

(iv) Thick Tails. It is well known that the distribution of stock returns has more weight in the tails than the normal distribution (e.g., Mandelbrot, 1963; Fama 1965), and that a stochastic process is thick tailed if it is conditionally normal with a randomly changing conditional variance (Clark, 1973). Our estimated model generates thick tails with both a randomly changing conditional variance σ_t^2 and a thick-tailed conditional distribu-

[13] It is also unclear what the effect of a unit root in $\ln(\sigma_t^2)$ is on the asymptotic properties of the parameter estimates. The (unverified) regularity conditions for asymptotic normality require that the scoring function and hessian obey a central limit theorem and uniform weak law of large numbers respectively, which may or may not require $\{\sigma_t^2, \xi_t\}$ to have finite moments. It may be that the standard asymptotics are valid even in the presence of a unit root in $\ln(\sigma_t^2)$.

tion for ξ_t. Recall from the discussion of the GED(v) distribution in section 2 that if $v < 2$, the distribution of z_t (and therefore the conditional distribution of ξ_t) has thicker tails than the normal distribution. The estimated v is approximately 1.58 with a standard error of about 0.03, so the distribution of the z_t is significantly thicker-tailed than the normal.

(v) The estimated contribution of nontrading days to conditional variance is roughly consistent with the results of French and Roll (1986). The estimated value of δ is about 0.183, with a standard error of about 0.028, so a nontrading day contributes less than a fifth as much to volatility as a trading day.

The general results just discussed are quite robust to which ARMA model is selected, though of course the parameter estimates change somewhat. The results also appear to be quite robust with respect to the sample period: another paper, Nelson (1989), reports strikingly similar parameter estimates in an exponential ARCH model fit to daily capital gains on the Standard 90 stock index from 1928 to 1956.

3.1. Specification Tests

To test the fit of the model, several conditional moment tests (Newey, 1985) were fit using orthogonality conditions implied by correct specification. Correct specification of the model has implications for the distribution of $\{z_t\}$. For example, $E[z_t] = 0$, $E[z_t^2] = 1$, and $E[g(z_t)] = 0$. Since the GED distribution is symmetric, we also require that $E[z_t \cdot |z_t|] = 0$. The first four orthogonality conditions test these basic properties. Correct specification also requires that $\{\xi_t^2 - \sigma_t^2\}$ and $\{\xi_t\}$ (or equivalently $\{z_t^2 - 1\}$ and $\{z_t\}$) are serially uncorrelated. Accordingly, we test for serial correlation in z_t and z_t^2 at lags one through five.

Table 4 reports first the sample averages for the 14 selected orthogonality conditions and their associated t statistics, and then chi-square statistics and probability values for two combinations of the orthogonality conditions, the first including only conditions relating to correct specification of the conditional variance process σ_t^2 and the second also testing for correct specification of the conditional mean process.

In the first chi-square test, the CRSP model does extremely well, with a probability value of 0.94. Considered individually, none of the first nine orthogonality conditions are significantly different from zero at any standard significance level. When the last five conditions, which test for serial correlation in $\{z_t\}$, are included, the probability value drops to 0.16, which still does not reject at any standard significance level, although statistically significant serial correlation is found at lag two. Overall, the fit of the

TABLE 4 Specification Test Results for CRSP Value-Weighted Excess Returns, ARMA(2, 1) Model[a]

	Orthogonality conditions	Sample averages	t Statistics
1	$E(z_t) = 0$	-0.007	-0.349
2	$E(z_t^2) - 1 = 0$	-0.001	-0.037
3	$E\|z_t\| - \lambda 2^{1/v}\Gamma(2/v)/\Gamma(1/v) = 0$	$2.583 \cdot 10^{-5}$	0.002
4	$E\|z_t\| \cdot (\|z_t\| - E\|z\|) = 0$	-0.027	-1.254
5	$E[(z_t^2 - 1)(z_{t-1}^2 - 1)] = 0$	0.102	1.023
6	$E[(z_t^2 - 1)(z_{t-2}^2 - 1)] = 0$	0.028	0.529
7	$E[(z_t^2 - 1)(z_{t-3}^2 - 1)] = 0$	0.079	0.864
8	$E[(z_t^2 - 1)(z_{t-4}^2 - 1)] = 0$	0.042	0.947
9	$E[(z_t^2 - 1)(z_{t-5}^2 - 1)] = 0$	0.019	0.469
10	$E(z_t \cdot z_{t-1}) = 0$	0.022	0.949
11	$E(z_t \cdot z_{t-2}) = 0$	-0.034	-2.521
12	$E(z_t \cdot z_{t-3}) = 0$	0.018	1.377
13	$E(z_t \cdot z_{t-4}) = 0$	0.015	1.174
14	$E(z_t \cdot z_{t-5}) = 0$	0.020	1.563

[a] χ^2 statistic for conditional moment test using the first 9 orthogonality conditions = 3.46. With 9 degrees of freedom, the probability value = 0.94. χ^2 statistic for conditional moment test using all 14 orthogonality conditions = 19.14. With 14 degrees of freedom, the probability value = 0.16.

CRSP model of the conditional variance process $\{\sigma_t^2\}$ seems remarkably good.[14, 15]

Our conditional moment tests leave many potential sources of misspecification unchecked. It therefore seems desirable to check the forecasting performance of the model during periods of rapidly changing volatility, and to check for large outliers in the data. From Figure 2, two periods stand out as times of high and rapidly changing volatility: the market break of September, 1973–December, 1974, and the last five months of 1987. Figures 4 and 5 plot returns and the one-day-ahead ex ante 99% prediction intervals implied by the estimated model for these periods. The 99% prediction intervals have a width of approximately $5.56 \cdot \sigma_t$.

[14] Engle *et al.* (1987) and Pagan and Sabau (1987) based conditional moment tests on $\{\xi_t\}$ rather than on $\{z_t\}$. Basing tests on $\{z_t\}$ is analogous to a GLS correction and seems likely to increase the power of the specification tests. As a check, chi-square tests were recomputed using $\{\xi_t\}$ instead of $\{z_t\}$. The test statistics were drastically lower in each instance.

[15] The specification tests for the 1928–1956 Standard 90 capital gains data reported in Nelson (1989) were not as favorable: The tests found evidence of negatively skewed returns and serially correlated residuals, rejecting the model at any standard level.

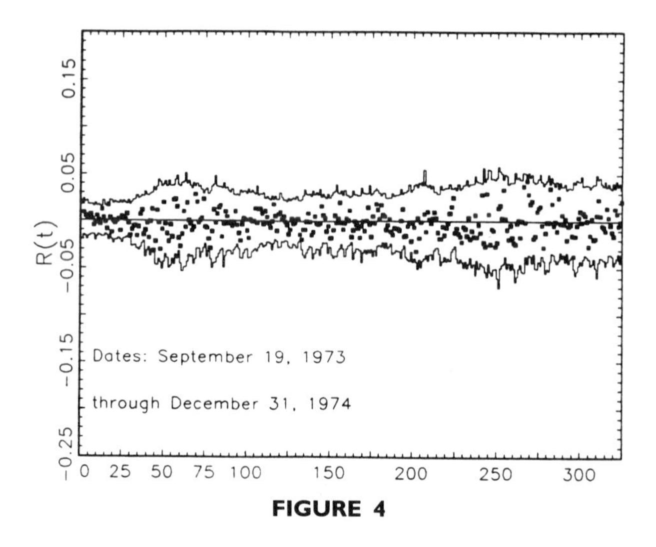

FIGURE 4

In 1973–1974, the model seems to track the change in volatility quite closely, with no serious outliers (Figure 4). The model also succeeds quite well in picking up the volatility of the period *after* October 19, 1987, with no serious outliers (Figure 5). On October 19, 1987, however, the model's performance is mixed at best: the ex ante prediction intervals for the day are approximately $\pm 7\%$, the widest in the data set up to that time, brought on by the sharp drops in the market during the preceding week. Unfortunately, the drop in the index that day was approximately 20.25%,

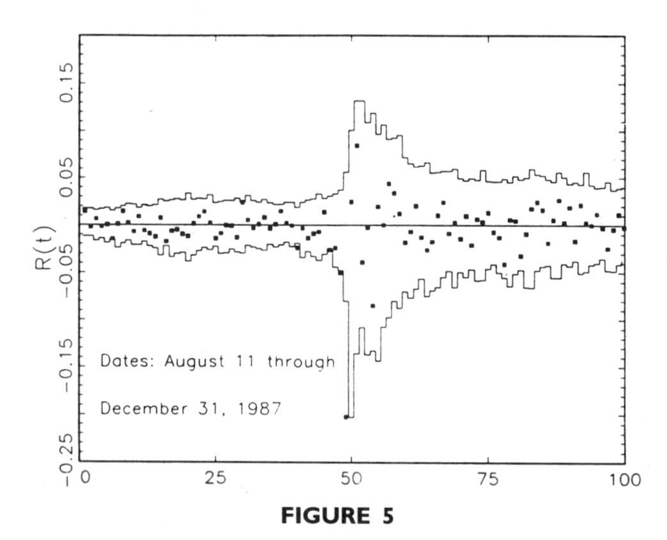

FIGURE 5

TABLE 5 Largest Outliers in the Sample

Date	$R_t(\%)$	$\sigma_t(\%)$	z_t	Expected frequency[a]
10/19/87	−20.25	2.44	−7.78	$1.11 \cdot 10^{-5}$
9/11/87	−4.58	0.84	−5.38	0.05
8/17/82	4.36	0.87	4.95	0.20
1/8/86	−2.42	0.57	−4.81	0.31
8/2/78	2.23	0.48	4.53	0.67
8/16/71	3.50	0.82	4.33	1.21
10/9/79	−3.37	0.73	−4.29	1.36
7/7/86	−3.02	0.72	−4.12	2.20

[a] The expected number of $|z_t|$ values of this size or greater in a $25\frac{1}{2}$ year sample (6408 observations).

about $7.78 \cdot \sigma_t$, a serious outlier. If the estimated model were literally true, the expected number of outliers of this size or greater in a $25\frac{1}{2}$ year period (the length of the data set) is only about $1 \cdot 10^{-5}$—i.e., the probability of observing such an outlier in a data set of this length is extremely low.

Although October 19, 1987, is the most extreme outlier in the sample, it is not the only large outlier. Table 5 lists the largest, ranked in order of the implied value for $|z_t|$. The standardized GED has only one parameter, v, to control the shape of the conditional distribution, and this may well not be flexible enough—i.e., there are two many "large" $|z_t|$ values. Nonparametric methods (as in Engle and González-Rivera, 1989), or more flexible parametric families of distributions, would probably improve the model.

4. CONCLUSION

This chapter has presented a new class of ARCH models that do not suffer from some of the drawbacks of GARCH models. Ideally, we would like ARCH models that allow the same degree of simplicity and flexibility in representing conditional variances as ARIMA and related models have allowed in representing conditional means. While this chapter has made a contribution to this end, the goal is far from accomplished: it remains to develop a multivariate version of exponential ARCH, and a satisfactory asymptotic theory for the maximum likelihood parameter estimates. These tasks await further research.

APPENDIX 1. THE MOMENTS OF σ_t^2 AND ξ_t

By (2.1) and the independence of the $\{z_t\}$, the joint moments and conditional moments of $\{\sigma_t^2\}$ and $\{\xi_t\}$ take either the form

$$E\left[\exp\left(a_t + \sum_{i=1}^{\infty} b_i g(z_{t-i})\right)\right] = \exp(a_t) \prod_{i=1}^{\infty} E\left[\exp(b_i g(z_{t-i}))\right] \quad \text{(A1.1)}$$

or

$$E\left[z_{t-k}^p z_{t-j}^q \exp\left(a_t + \sum_{i=1}^{\infty} b_i g(z_{t-i})\right)\right] = \left[\prod_{i \geq 1, i \neq j, k} E\left[\exp(b_i g(z_{t-i}))\right]\right]$$
$$\times \exp(a_t) \cdot E\left[z_{t-k}^p \exp(b_k g(z_{t-k}))\right] E\left[z_{t-j}^q \exp(b_j g(z_{t-j}))\right] \quad \text{(A1.2)}$$

for nonnegative integers p, q, j, and $k \neq j$. For example, to get the unconditional expectation of σ_t^2, we set $\{a_t\} = \{\alpha_t\}$ and $\{b_t\} = \{\beta_t\}$ in (A1.1). For the conditional expectation of σ_t^2 given $z_{t-k}, z_{t-k-1}, \ldots$, set

$$a_t = \alpha_t + \sum_{i=k}^{\infty} \beta_i g(z_{t-i}), \; b_i = \beta_i \text{ for } 1 \leq i \leq k-1, \text{ and } b_i = 0 \text{ for } i \geq k.$$

To obtain the moments of ξ_t and $\xi_{t-k}\sigma_t^2$, we proceed similarly, using (A1.2).

To evaluate the expectations in (A1.1)–(A1.2), we must make a further distributional assumption about $\{z_t\}$. When $\{z_t\} \sim$ i.i.d. $N(0,1)$, the following result, combined with (A1.1)–(A1.2), gives the joint conditional and unconditional moments of the σ_t^2 and ξ_t processes:

THEOREM A1.1. *Let $z \sim N(0,1)$. For any finite, real scalar b and positive integer p,*

$$E\left[\exp(g(z)b)\right] = \left\{\Phi(\gamma b + \theta b)\exp\left[b^2(\theta - \gamma)^2/2\right] + \Phi(\gamma b - \theta b)\right.$$
$$\left. \times \exp\left[b^2(\gamma - \theta)^2/2\right]\right\} \exp\left[-b\gamma(2/\pi)^{1/2}\right] < \infty, \quad \text{(A1.3)}$$

and

$$E\left[z^p \exp(g(z)b)\right] = \exp\left[-b\gamma(2/\pi)^{1/2}\right] \cdot \Gamma(p+1) \cdot (2\pi)^{-1/2}$$
$$\cdot \left\{\exp\left[b^2(\theta + \gamma)^2/4\right] \cdot D_{-(p+1)}\left[-b(\gamma + \theta)\right]\right.$$
$$\left. + (-1)^p \exp\left[b^2(\gamma - \theta)^2/4\right] \cdot D_{-(p+1)}\left[-b(\gamma - \theta)\right]\right\} < \infty, \quad \text{(A1.4)}$$

where $\Phi(\cdot)$ is the standard normal cumulative distribution function, $\Gamma[\cdot]$ is the Gamma function, and $D_q[\cdot]$ is the parabolic cylinder function (Gradshteyn and Ryzhik, 1980).

Proof of Theorem A1.1. (A1.3) follows by straightforward but tedious calculus. (A1.4) is easily proven with the help of Gradshteyn and Ryzhik (1980) Formula 3.461 #1. The finiteness of the expression in (A1.4) follows as a special case in the proof of Theorem A1.2 below. Q.E.D.

Theorem A1.2 deals with the more general GED case.

THEOREM A1.2. *Let p be a nonnegative integer, and let* $z \sim GED(v)$ *with* $\mathrm{E}(z) = 0$, $\mathrm{Var}(z) = 1$, *and* $v > 1$. *Then*

$$\mathrm{E}[z^p \exp(g(z)b)]$$
$$= \exp\left[-b\gamma\Gamma(2/v)\lambda 2^{1/v}/\Gamma(1/v)\right] \cdot 2^{p/v} \cdot \lambda^p$$
$$\cdot \sum_{k=0}^{\infty} (2^{1/v}\lambda b)^k \left[(\gamma + \theta)^k + (-1)^p(\lambda - \theta)^k\right]$$
$$\times \frac{\Gamma[(p + k + 1)/v]}{2\Gamma(1/v)\Gamma(k + 1)} < \infty. \tag{A1.5}$$

If $z \sim GED(v)$ *with* $v < 1$, *or* $z \sim$ *Student's t with d degrees of freedom* $(d > 2)$, *and z is normalized to satisfy* $\mathrm{E}(z) = 0$, $\mathrm{Var}(z) = 1$, *then* $\mathrm{E}[\exp(g(z)b)]$ *and* $\mathrm{E}[z^p \exp(g(z)b)]$ *are finite if and only if*

$$b\gamma + |b\theta| \leq 0. \tag{A1.6}$$

If $z \sim GED(1)$, *then* $\mathrm{E}[\exp(g(z)b)]$ *and* $\mathrm{E}[z^p \exp(g(z)b)]$ *are finite if and only if*

$$b\gamma + |b\theta| < 2^{1/2}. \tag{A1.7}$$

The restriction (A1.6) is rarely satisfied in practice. In computing the unconditional expectation of σ_t^2, b is one of the moving average coefficients $\{\beta_i\}$, at least some of which are positive, since $\beta_0 \equiv 1$. If $b > 0$, (A1.6) implies either $\gamma < 0$ or $\gamma = \theta = 0$. If $\gamma < 0$, residuals larger than expected decrease conditional variance, which goes against the intuition developed in Section 1. In the author's experience in fitting exponential ARCH models, the estimated value of γ is always positive.

Proof of Theorem A1.2. The density of z given in (2.4) and Gradshteyn and Ryzhik (1980) Formula 3.381 #4 yield

$$\mathrm{E}(|z|) = \lambda 2^{1/v}\Gamma(2/v)/\Gamma(1/v). \tag{A1.8}$$

Straightforward calculus then yields

$$E[z^p \cdot \exp(g(z)b)] = \frac{\lambda^p 2^{p/v}}{2\Gamma(1/v)} \exp[-b\gamma\lambda 2^{1/v}\Gamma(2/v)/\Gamma(1/v)]$$

$$\cdot \int_0^\infty y^{(-1+(p+1)/v)} e^{-y} \left[e^{b\gamma(\theta+\gamma)(2y)^{1/v}} + (-1)^p e^{b\lambda(\gamma-\theta)(2y)^{1/v}} \right] dy. \quad \text{(A1.9)}$$

Expanding the part of the integrand in square brackets in a Taylor series,

$$E[z^p \cdot \exp(g(z)b)]$$

$$= \frac{2^{p/v}\lambda^p}{2\Gamma(1/v)} \exp[-b\lambda 2^{1/v}\Gamma(2/v)/\Gamma(1/v)]$$

$$\cdot \int_0^\infty \sum_{k=0}^{\infty} \left[[(\theta+\gamma)2^{1/v}b\lambda]^k + (-1)^p [(\gamma-\theta)2^{1/v}b\lambda]^k \right]$$

$$\times \frac{e^{-y} y^{-1+(p+1+k)/v}}{\Gamma(k+1)} dy. \quad \text{(A1.10)}$$

If we can interchange the order of summation and integration in (A1.10), then Gradshteyn and Ryzhik (1980) Formula 3.381 #4 yields (A1.5). First, consider the related expression

$$\int_0^\infty \sum_{k=0}^{\infty} 2\Delta^k \frac{e^{-y} y^{-1+(p+1+k)/v}}{\Gamma(k+1)} dy, \quad \text{where} \quad \text{(A1.11)}$$

$$\Delta \equiv \max\{|(\theta+\gamma)b\lambda 2^{1/v}|, |(\gamma-\theta)b\lambda 2^{1/v}|\}. \quad \text{(A1.12)}$$

The terms in (A1.11) are nonnegative, so by monotone convergence (Rudin, 1976, Theorem 11.28), the order of integration and summation in (A1.11) can be reversed. If the integral in (A1.11) is finite, then by dominated convergence (Rudin, 1976, Theorem 11.32), we can interchange the order of summation and integration in (A1.10) and the integral in (A1.10) is finite. To prove (A1.5), therefore, it remains only to show that (A1.11) is finite if $v > 1$. By Gradshteyn and Ryzhik (1980) Formula 3.381 #4,

$$\int_0^\infty \sum_{k=0}^{\infty} 2\Delta^k \frac{e^{-y} y^{-1+(p+1+k)/v}}{\Gamma(k+1)} dy = \sum_{k=0}^{\infty} \frac{2\Delta^k \Gamma[(p+1+k)/v]}{\Gamma(k+1)}.$$

$$\text{(A1.13)}$$

By construction, $\Delta \geq 0$. When $\Delta = 0$, the convergence of (A1.13) is trivial. When $\Delta > 0$, then by the root test (Rudin, 1976, Theorem 3.33), the sum in (A1.13) converges if

$$\limsup_{k\to\infty} \ln(\Delta) + k^{-1}\ln(\Gamma[(k+1+p)/v]) - k^{-1}\ln(\Gamma(k+1)) < 0. \tag{A1.14}$$

By the asymptotic expansion for $\ln(\Gamma(x))$ in Davis (1965, Equation 6.1.41),

$$\begin{aligned}
&\limsup_{k\to\infty} \ln(\Delta) + k^{-1}\ln(\Gamma[(k+1+p)/v]) - k^{-1}\ln(\Gamma(k+1)) \\
&\quad = \limsup_{k\to\infty} \ln(\Delta) + k^{-1}[(k+1+p)/v - 1/2]\cdot\ln(k+1+p) \\
&\quad\quad +(k+1)/k - k^{-1}[(k+1+p)/v - 1/2]\cdot\ln(v) \\
&\quad\quad - k^{-1}[k+1/2]\cdot\ln(k+1) - (k+p+1)/kv + O(k^{-1}).
\end{aligned} \tag{A1.15}$$

Expanding $\ln(k+1+p)$ in a Taylor series around $p = 0$ and substituting into (A1.15),

$$\begin{aligned}
&\limsup_{k\to\infty} \ln(\Delta) + k^{-1}\ln(\Gamma[(k+1+p)/v]) - k^{-1}\ln(\Gamma(k+1)) \\
&\quad = \limsup_{k\to\infty} \ln(\Delta) - k^{-1}[(k+1+p)/v - 1/2] \\
&\quad\quad \cdot\ln(v) - (k+p+1)/kv + 1 + O(k^{-1}) \\
&\quad\quad -\left[(k+1+p)/kv - 1 - k^{-1}\right]\cdot\ln(k+1).
\end{aligned} \tag{A1.16}$$

The last term on the right-hand side of (A1.16) asymptotically dominates every other term if $v \neq 1$, diverging to $-\infty$ if $v > 1$, so the series converges when $v > 1$ and $\Delta < \infty$.

If $z \sim$ Student t, or GED with $v > 1$, it is easy to verify that $\mathrm{E}[\exp(b\cdot g(z))] = \infty$ unless $(\gamma+\theta)b \leq 0$ and $(\gamma-\theta)b \leq 0$, which holds if and only if $b\gamma + |\theta b| \leq 0$. If this inequality is satisfied, $\mathrm{E}[\exp(g(z)b] < \infty$. The GED case with $v = 1$ is similar. Q.E.D.

APPENDIX 2. PROOFS

Proof of Theorem 2.1. That $|\ln(\sigma_t^2 - \alpha_t|$ is finite almost surely when $\sum_{k=1}^{\infty} \beta_k^2 < \infty$ follows immediately from the independence and finite variance of the $g(z_t)$ terms in (2.1) and from Billingsley (1986, Theorem 22.6).

Since $|\ln(\sigma_t^2) - \alpha_t|$ is finite almost surely, so are $\exp(-\alpha_t)\sigma_t^2$ and $\exp(-\alpha_t/2)\xi_t$. This, combined with Stout (1974, Theorem 3.5.8) and the representation in (1.1)–(1.2) and (2.1)–(2.2) implies that these series are strictly stationary and ergodic. For all t, the expectation of $(\ln(\sigma_t^2) - \alpha_t) = 0$, and the variance of $(\ln(\sigma_t^2) - \alpha_t)$ is $\text{Var}(g(z_t))\Sigma_{k=1}^{\infty} \beta_k^2$. Since $\text{Var}(g(z_t))$ is finite and the distribution of $\ln(\sigma_t^2) - \alpha_t$ is independent of t, the first two moments of $(\ln(\sigma_t^2) - \alpha_t)$ are finite and time invariant, so $\{\ln(\sigma_t^2) - \alpha_t\}$ is covariance stationary.

If $\Sigma_{k=1}^{\infty} \beta_k^2 = \infty$, then $|\ln(\sigma_t^2 - \alpha_t| = \infty$ almost surely by Billingsley (1986, Theorems 22.3 and 22.8). Q.E.D.

Proof of Theorem 2.2. By Theorem 2.1, the distributions of $\exp(-\alpha_t)\sigma_t^2$ and $\exp(-\alpha_t/2)\xi_t$ and any existing moments are time invariant. We will show that $\exp(-\alpha_t)\sigma_t^2$ and $\exp(-\alpha_t/2)\xi_t$ have finite moments of arbitrary positive order. As shown in Appendix 1, the conditional, unconditional, and cross moments of $\exp(-\alpha_t)\sigma_t^2$ and $\exp(-\alpha_t/2)\xi_t$ have the form (A1.1)–(A1.2). By Hölder's Inequality, if σ_t^2 and ξ_t have arbitrary finite moments, the cross moments are also finite. By the independence of the z_t, $\text{E}[|(\exp(-\alpha_t/2)\xi_t)|^d] = \text{E}|z|^d\text{E}(\exp(-\alpha_t/2)\sigma_t)^d$. Since z has arbitrary finite moments, we need only show that $\text{E}[(\exp(-\alpha_t/2)\sigma_t)^d]$ is finite for all $d > 0$ if $\{\beta_i\}$ is square-summable. This expectation is given by

$$\text{E}\left[(\exp(-\alpha_t/2)\sigma_t)^d\right] = \prod_{i=1}^{\infty} \text{E}\exp\left[\tfrac{1}{2}d\beta_i g(z_{t-i})\right], \tag{A2.1}$$

where the individual expectation terms in (A2.1) are obtained by setting $b = \beta_i d/2$ in Theorem A1.1.

A sufficient condition for an infinite product $\Pi_{t=1,\infty}\,\alpha_i$ to converge to a finite, nonzero number is that the series $\Sigma_{i=1,\infty}|a_i - 1|$ converge (Gradshteyn and Ryzhik, 1980, Section 0.25). Let a_i equal the ith term in (A2.1). Define

$$S(\beta) \equiv \exp\left[\tfrac{1}{2}d\beta\gamma\lambda\Gamma(2/v)2^{1/v}/\Gamma(1/v)\right]\cdot \text{E}\left(\exp\left[g(z)\beta\tfrac{1}{2}d\right]\right). \tag{A2.2}$$

We then have

$$S(0) = 1, \qquad S'(0) = \tfrac{1}{2}d\gamma\lambda\Gamma(2/v)2^{1/v}/\Gamma(1/v), \qquad \text{and}$$
$$S''(\beta) = O(1) \quad \text{as} \quad \beta \to 0. \tag{A2.3}$$

Expanding $S(\beta)$ and $\exp[-\tfrac{1}{2}d\beta\gamma\lambda\Gamma(2/v)2^{1/v}/\Gamma(1/v)]$ in Taylor series around $\beta = 0$ and substituting into (A1.5), we have

$$\begin{aligned} a_i - 1 &= \mathrm{E}\big(\exp[g(z)\beta_i \tfrac{1}{2} d]\big) - 1 \\ &= \big[1 - \tfrac{1}{2} d\beta_i \gamma\lambda\Gamma(2/v)2^{1/v}/\Gamma(1/v) + O(\beta_i^2)\big] \\ &\quad \cdot\big[1 + \tfrac{1}{2} d\beta_i \gamma\lambda\Gamma(2/v)2^{1/v}/\Gamma(1/v) + O(\beta_i^2)\big] - 1 \\ &= O(\beta_i^2) \quad \text{as} \quad \beta_i \to 0. \end{aligned} \tag{A2.4}$$

$O(\beta_i^2)$ in (A2.4) means that for some $\varepsilon > 0$, there exists a finite M independent of i such that

$$\sup_{|\beta_i| < \varepsilon,\, \beta_i \neq 0} \beta_i^{-2}\, |O(\beta_i^2)| < M. \tag{A2.5}$$

By (A2.4)–(A2.5), $\Sigma\, \beta_i^2 < \infty$ implies $\Sigma_{i=1,\infty} |a_i - 1| < \infty$ and thus $\Pi_{i=1,\infty}\, a_i < \infty$.

Finally, we must prove (2.6)–(2.7). The proofs of (2.6) and (2.7) are substantially identical, so we prove only (2.6). By Theorem A1.1,

$$\begin{aligned} &\mathrm{E}\big[\exp(-p\alpha_t)\sigma_t^{2p}\, |z_0, z_{-1}, z_{-2}, \cdots\big] - \mathrm{E}\big[\exp(-p\alpha_t)\sigma_t^{2p}\big] \\ &\quad = \left[\exp\left(p \sum_{j=t}^{\infty} \beta_j g(z_{t-j})\right) - \prod_{j=t}^{\infty} \mathrm{E}\big[\exp(p\beta_j g(z_{t-j}))\big]\right] \\ &\qquad \cdot \prod_{i=1}^{t-1} \mathrm{E}\big[\exp(p\beta_i g(z_{t-i}))\big]. \end{aligned} \tag{A2.6}$$

The last term on the right-hand side of (A2.6) is finite by Theorem A1.2. (2.6) will therefore be proven if we can show that

$$p \lim_{t\to\infty} \left[\exp\left(p \sum_{j=t}^{\infty} \beta_j g(z_{t-j})\right) - \prod_{j=t}^{\infty} \mathrm{E}\big[\exp(p\beta_j g(z_{t-j}))\big]\right] = 0. \tag{A2.7}$$

First, consider the unconditional variance of the log of the first term on the left-hand side of (A2.7). We have

$$\mathrm{Var}\left[\sum_{j=t}^{\infty} p\beta_j g(z_{t-j})\right] = p^2\, \mathrm{Var}(g(z_t)) \sum_{j=t}^{\infty} \beta_j^2 \to 0 \qquad \text{as} \quad t \to \infty. \tag{A2.8}$$

Since convergence in L^2 implies convergence in probability,

$$\exp\left[\sum_{j=t}^{\infty} p\beta_j g(z_{t-j})\right] \xrightarrow{P} 1. \tag{A2.9}$$

Finally,

$$\lim_{t\to\infty}\prod_{j=t}^{\infty}\mathrm{E}\big[\exp(p\beta_j g(z_{t-j}))\big] = \lim_{t\to\infty}\exp\left[\sum_{j=t}^{\infty}\ln\big(\mathrm{E}\big[\exp(p\beta_j g(z_{t-j}))\big]\big)\right]$$

$$= \exp\left[\sum_{j=t}^{\infty}\ln\big[1 + O(\beta_j^2)\big]\right]$$

$$= \exp\left[\sum_{j=t}^{\infty}O(\beta_j^2)\right] \to 1 \qquad \text{as } t\to\infty \tag{A2.10}$$

by (A2.4)–(A2.5) and the square summability of $\{\beta_j\}$. Q.E.D.

ACKNOWLEDGMENTS

This paper is a revision of part of Chapter III of my Ph.D. dissertation (Nelson, 1988). The Department of Education and the University of Chicago Graduate School of Business provided financial support. I am indebted to George Constantinides, John Cox, Nancy Hammond, Daniel McFadden, Mark Nelson, Adrian Pagan, James Poterba, G. William Schwert, Stephen Taylor, Jeffrey Wooldridge, Arnold Zellner, two referees, and a co-editor for helpful comments. Seminar participants at the 1987 Econometric Society Winter meetings, at the 1989 A.S.A. meetings, at Berkeley, Chicago, M.I.T., Northwestern, Princeton, the Research Triangle Econometrics Workshop, Rochester, Stanford, Wharton, and Yale also made helpful suggestions. Jiahong Shi provided able research assistance. Remaining errors are mine alone.

REFERENCES

Abel, A. B. (1988). Stock prices under time-varying dividend risk: An exact solution in an infinite-horizon general equilibrium model. *Journal of Monetary Economics* **22**, 375–393.

Akaike, H. (1973). Information theory and an extension of the maximum likelihood principle. *In* "Second International Symposium on Information Theory" (B. N. Petrov and F. Csáki, eds.), pp. 267–281. Akadémiai Kiadó, Budapest.

Backus, D., and Gregory, A. (1989). "Theoretical Relations Between Risk Premiums and Conditional Variances," Mimeo. Federal Reserve Bank of Minneapolis and Queen's University.

Barclay, M. J., Litzenberger, R. H., and Warner, J. (1990). Private information, trading volume, and stock return variances. *Review of Financial Studies* **3**, 233–254.

Barsky, R. B. (1989). Why don't the prices of stocks and bonds move together? *American Economic Review* **79**, 1132–1145.

Bernanke, B. S. (1983). Irreversibility, uncertainty and cyclical investment. *Quarterly Journal of Economics* **98**, 85–106.

Billingsley, P. (1986). "Probability and Measure," 2nd ed. Wiley, New York.

Black, F. (1976). Studies of stock market volatility changes. *Proceedings of the American Statistical Association, Business and Economic Statistics Section*, pp. 177–181.

Bollerslev, T. (1986). Generalized autoregressive conditional heteroskedasticity. *Journal of Econometrics* **31**, 307–327.

Bollerslev, T., Engle, R. F., and Wooldridge, J. M. (1988). A capital asset pricing model with time varying covariances. *Journal of Political Economy* **96**, 116–131.

Box, G. E. P., and Tiao, G. C. (1973). "Bayesian Inference in Statistical Analysis." Addison-Wesley, Reading, MA.

Chou, R. Y. (1987). Volatility persistence and stock returns—Some empirical evidence using GARCH. *Journal of Applied Econometrics* **3**, 279–294.

Christie, A. A. (1982). The stochastic behavior of common stock variances: Value, leverage and interest rate effects. *Journal of Financial Economics* **10**, 407–432.

Clark, P. K. (1973). A subordinated stochastic process model with finite variance for speculative prices. *Econometrica* **41**, 135–155.

Cochrane, J. (1988). How big is the random walk in GNP? *Journal of Political Economy* **96**, 893–920.

Cox, J., Ingersoll, J., and Ross, S. (1985). An intertemporal general equilibrium model of asset prices. *Econometrica* **53**, 363–384.

Davis, P. J. (1965). The Gamma function and related functions. *In* "Handbook of Mathematical Functions" (M. Abramowitz and I. A. Stegun, eds.), 253–294. Dover, New York.

Engle, R. F. (1982). Autoregressive conditional heteroskedasticity with estimates of the variance of United Kingdom inflation. *Econometrica* **50**, 987–1008.

Engle, R. F., and Bollerslev, T. (1986a). Modelling the persistence of conditional variances. *Econometric Reviews* **5**, 1–50.

Engle, R. F., and Bollerslev, T. (1986b). Modelling the persistence of conditional variances: A reply. *Econometric Reviews* **5**, 81–87.

Engle, R. F., and González-Rivera, G. (1989). "Semiparametric ARCH Models," Discuss. Pap. No. 89–17. University of California at San Diego, Economics Department.

Engle, R. F., Lilien, D. M., and Robins, R. P. (1987). Estimating time varying risk premia in the term structure: The ARCH-M model. *Econometrica* **55**, 391-408.

Fama, E. F. (1965). The behavior of stock market prices. *Journal of Business* **38**, 34–105.

French, K. R., and Roll, R. (1986). Stock return variances: The arrival of information and the reaction of traders. *Journal of Financial Economics* **17**, 5–26.

French, K. R., Schwert, G. W., and Stambaugh, R. F. (1987). Expected stock returns and volatility. *Journal of Financial Economics* **19**, 3–29.

Gennotte, G., and Marsh, T. A. (1987). "Variations in Economic Uncertainty and Risk Premiums on Capital Assets," Mimeo. University of California at Berkeley, Graduate School of Business.

Geweke, J. (1986). Modelling the persistence of conditional variances: A comment. *Econometric Reviews* **5**, 57–61.

Glosten, L. R., Jagannathan, R., and Runkle, D. (1989). "Relationship Between the Expected Value and the Volatility of the Nominal Excess Return on Stocks," Banking Research Center Working Paper, No. 166. Northwestern University, Evanston, IL.

Gradshteyn, I. S., and Ryzhik, I. M. (1980). "Table of Integrals, Series and Products." Academic Press, New York.

Hannan, E. J. (1980). The estimation of the order of an ARMA process. *Annals of Statistics* **8**, 1071–1081.

Harvey, A. C. (1981). "The Econometric Analysis of Time Series." Philip Allan, Oxford.

Hosking, J. R. M. (1981). Fractional differencing. *Biometrika* **68**, 165–176.

Lo, A., and MacKinlay, C. (1988). Stock market prices do not follow random walks: Evidence from a simple specification test. *Review of Financial Studies* **1**, 41–66.

Mandelbrot, B. (1963). The variation of certain speculative prices. *Journal of Business* **36**, 394–419.

McDonald, R. M., and Siegel, D. (1986). The value of waiting to invest. *Quarterly Journal of Economics* **101**, 707–728.

Merton, R. C. (1973). An intertemporal capital asset pricing model. *Econometrica* **41**, 867–888.

Nelson, D. B. (1988). The time series behavior of stock market volatility and returns. Doctoral Dissertation, Massachusetts Institute of Technology, Economics Department, Cambridge, MA (unpublished).

Nelson, D. B. (1989). Modelling stock market volatility changes. *Proceedings of the American Statistical Association, Business and Economic Statistics Section*, pp. 93–98.

Nelson, D. B. (1990). Stationarity and persistence in the GARCH(1, 1) Model. *Econometric Theory* **6**, 318–334.

Nelson, D. B. (1992). Filtering and forecasting with misspecified ARCH Models I: Getting the right variance with the wrong model. *Journal of Econometrics* **52**, 61–90.

Newey, W. K. (1985). Generalized method of moments specification testing. *Journal of Econometrics* **29**, 229–256.

Pagan, A. R., and Hong, Y. (1988). Non-parametric estimation and the risk premium. *In* "Nonparametric and Semiparametric Methods in Econometrics" (W. Barnett, J. Powell, and G. Tauchen, eds.). Cambridge University Press, Cambridge, UK.

Pagan, A. R., and Sabau, H. C. L. (1987). "Consistency Tests for Heteroskedastic and Risk Models," Mimeo. University of Rochester Economics Department, Rochester, NY.

Pantula, S. G. (1986). Modelling the persistence of conditional variances: A comment. *Econometric Reviews* **5**, 71–73.

Poterba, J. M., and Summers, L. H. (1986). The persistence of volatility and stock market fluctuations. *American Economic Review* **76**, 1142–1151.

Rudin, W. (1976). "Principles of Mathematical Analysis." McGraw-Hill, New York.

Scholes, M., and Williams, J. (1977). Estimating betas from nonsynchronous data. *Journal of Financial Economics* **5**, 309–327.

Schwarz, G. (1978). Estimating the dimension of a model. *Annals of Statistics* **6**, 461–464.

Schwert, G. W. (1989a). Business cycles, financial crises and stock volatility. *Carnegie-Rochester Conference Series on Public Policy* **31**, 83–126.

Schwert, G. W. (1989b). Why does stock market volatility change over time? *Journal of Finance* **44**, 1115–1154.

Stout, W. F. (1974). "Almost Sure Convergence." Academic Press, London.

Weiss, A. A. (1986). Asymptotic theory for ARCH models: Estimation and testing. *Econometric Theory* **2**, 107–131.

Wiggins, J. B. (1987). Option values under stochastic volatility: Theory and empirical estimates. *Journal of Financial Economics* **19**, 351–372.

4

GOOD NEWS, BAD NEWS, VOLATILITY, AND BETAS

PHILLIP A. BRAUN*
DANIEL B. NELSON[†]
ALAIN M. SUNIER[‡]

**J. L. Kellogg Graduate School of Management*
Northwestern University
Evanston, Illinois 60208
[†]*National Bureau of Economic Research and*
Graduate School of Business
University of Chicago
Chicago, Illinois, 60637
[‡]*Goldman, Sachs & Co.*
New York, New York 10004

Many researchers have documented that stock return volatility tends to rise following good and bad news. This phenomenon, which we call "predictive asymmetry of second moments," has been noted both for individual stocks and for market indices.[1] Given this evidence, there is also good reason to expect such an effect to exist in conditional betas as well. We provide a method for estimating time-varying conditional betas based on a bivariate version of the exponential ARCH (EGARCH) model of Nelson (1991), allowing for the possibility that positive and negative returns affect β's differently.

[1] For example, see Black (1976), Christie (1982), French *et al.* (1987), Nelson (1989, 1991), Schwert (1989), Ng (1991), Gallant *et al.* (1992), Glosten *et al.* (1993), and Engle and Ng (1993). See Bollerslev *et al.* (1994) for a complete review of this literature.

The literature has focused on two classes of explanations for predictive asymmetry of second moments: The first, and most obvious, highlights the role of financial and operating leverage—e.g., if the value of a leveraged firm drops, its equity will, in general, become more leveraged, causing the volatility on equity's rate of return to rise. As Black (1976), Christie (1982), and Schwert (1989) show, however, financial and operating leverage cannot fully account for predictive asymmetry of second moments. A second set of explanations focuses on the role of volatility in determining the market risk premium: If the market risk premium is an increasing function of market volatility, and if increases in market risk premia due to increased volatility are not offset by decreases in riskless rates,[2] then increases in market volatility should lead to drops in the market, contributing to the predictive asymmetry of return variances.[3] If shocks to volatility persist for long periods, then the changes in asset prices due to volatility movements can be large (Poterba and Summers, 1986). Critics of this second explanation have argued that the theoretical link between the market risk premium and market volatility may be weak and that shocks to market volatility are not sufficiently persistent to account for predictive asymmetry of return variances. Whether the link is weak or not depends on whether the market risk premium is constant and how it moves through time. The evidence in Campbell (1987) and Harvey (1988) suggest that the link is not weak due to these facts.

There is also good reason to expect asymmetric responses of conditional betas to good and bad news. First, an exogenous shock to the value of a firm's assets which raises (lowers) the firm's financial leverage will raise (lower) the beta of the firm's equity. Thus the unexpected component of stock returns should be negatively correlated with changes in conditional beta.[4] Second, a persistent shock to the riskiness or conditional beta of a firm's equity, ceteris paribus, will manifest itself in a change in the price of equity. Again, an unexpected increase (decrease) in the equity beta will be associated with negative (positive) unexpected returns. While the direction of causality differs in these two arguments, in both cases the result is the same: negative correlation between movements in betas and unexpected returns.[5]

Time variation in return volatilities and betas is deeply connected to how asset prices are determined in equilibrium. Better knowledge of the

[2] As they sometimes are, for example, in the model of Barsky (1989).

[3] See, among others, Malkiel (1979), Pindyck (1984), Poterba and Summers (1986), French *et al.* (1987), and Campbell and Hentschel (1992).

[4] See Hamada (1972).

[5] Note that the second of the arguments for correlation between stock price movements and changes in beta relies crucially on two assumptions: (1) that changes in beta are persistent and (2) that shocks to beta are not also negatively correlated with shocks to the market risk premium.

time series properties of conditional second moments of asset returns, whether it be of volatilities or betas, therefore is crucial to advance our understanding of asset pricing. Hence, our model can also be used to advantage in investigations of empirical anomalies in asset payoffs. We focus on one challenge to asset pricing theory in particular: the finding by De Bondt and Thaler (1989) that stocks that have recently experienced sharp price declines ("losers") tend to subsequently outperform stocks that have recently experienced sharp price increases ("winners"). De Bondt and Thaler interpret this as evidence of investor overreaction. Chan (1988) and Ball and Kothari (1989), however, argue that the time-series behavior of conditional betas and the market risk premium can explain the returns performance of winners and losers. Both Chan and Ball and Kothari find evidence that betas of individual stocks rise (fall) in response to negative (positive) abnormal returns—i.e., they find predictive asymmetry in conditional betas. Ball and Kothari argue that the asymmetric response of conditional betas to good and bad news is sufficient to account for the relative returns performance of winners versus losers.

We offer an alternative method of estimating time-varying conditional betas. In contrast to other multivariate ARCH models,[6] EGARCH allows for the possibility that good and bad news may affect covariances differently. The bivariate extension we propose is a natural way to capture asymmetries in conditional betas. An alternative approach, which also captures asymmetric effects, is the model of Glosten *et al.* (1993). Their model allows a quadratic response of volatility to news, with different responses permitted for good and bad news, but maintains the assertion that the minimum volatility will result when there is no news.

The econometric evidence on whether conditional betas are significantly variable is mixed (see, for example, the discussion and references in Ferson, 1989, Section IV). Other methods for estimating time-varying betas include the rolling regression method of Fama and MacBeth (1973), ARCH betas [implicit in Bollerslev *et al.* (1988) and Ng (1991)], as well as betas conditional on a set of information variables (Shanken, 1990; Ferson and Harvey, 1993).

In this study, we compare the bivariate EGARCH betas to rolling regression betas. To anticipate the empirical results developed below, we (nor surprisingly) find very strong evidence of conditional heteroskedasticity in both the market and nonmarket components of returns and weaker evidence of time-varying conditional betas. What is surprising is that while predictive asymmetry is very strong in both the market component and the firm-specific component of volatility, it appears to be entirely absent in conditional betas for both industry and decile portfolios.

[6] For example, Engle *et al.* (1990) and Bollerslev *et al.* (1988).

The structure of the chapter is as follows. We present the bivariate EGARCH model in section 1. Section 2 presents empirical results using monthly equity portfolio data. Section 3 provides a detailed specification analysis. Section 4 concludes.

1. THE BIVARIATE EGARCH MODEL

Let $R_{m,t}$ and $R_{p,t}$ denote the time t excess returns of a market index and a second portfolio, respectively. Our strategy is to model the conditional covariance matrix of $R_{m,t}$ and $R_{p,t}$ by splitting it into three pieces: the conditional variance of the market $\sigma_{m,t}^2$; the conditional beta of portfolio p with respect to the market index, $\beta_{p,t}$, and the variance of the nonmarket component of $R_{p,t}$, $\sigma_{p.t}^2$. Specifically, we assume that $R_{m,t}$ and $R_{p,t}$ can be written as

$$R_{m,t} = \mu_{m,t} + \sigma_{m,t} \cdot z_{m,t}, \tag{1.1}$$

and

$$\begin{aligned} R_{p,t} &= \mu_{p,t} + \beta_{p,t} \cdot R_{m,t} + \sigma_{p,t} \cdot z_{p,t} \\ &= \mu_{p,t} + \beta_{p,t} \cdot \mu_{m,t} + \beta_{p,t} \cdot \sigma_{m,t} \cdot z_{m,t} + \sigma_{p,t} \cdot z_{p,t}, \end{aligned} \tag{1.2}$$

where $\{z_{m,t}\}_{t=-\infty}^{\infty}$ and $\{z_{p,t}\}_{t=-\infty}^{\infty}$ are contemporaneously uncorrelated, i.i.d. standardized residual processes with zero means and unit variances, and $\mu_{m,t}$, $\mu_{p,t}$ are, respectively, the conditional means of $R_{m,t}$ and $R_{p,t}$. The conditional beta of $R_{p,t}$ with respect to $R_{m,t}$ is given by

$$\beta_{p,t} = \frac{\mathrm{E}_{t-1}\left[(R_{p,t} - \mu_{p,t}) \cdot (R_{m,t} - \mu_{m,t})\right]}{\mathrm{E}_{t-1}\left[(R_{m,t} - \mu_{m,t})^2\right]}, \tag{1.3}$$

where $\mathrm{E}_{t-1}[\cdot]$ denotes expectation at time $t-1$. $\mu_{m,t}$, $\mu_{p,t}$, $\sigma_{m,t}^2$, $\sigma_{p,t}^2$, and $\beta_{p,t}$ are taken to be measurable with respect to information at time $t-1$.

Equations (1.1) and (1.2) split up portfolio returns in a readily interpretable way: $\beta_{p,t} \cdot \sigma_{m,t} \cdot z_{m,t}$ is the market factor in $R_{p,t}$ with conditional variance $\beta_{p,t}^2 \sigma_{m,t}^2$ while $\sigma_{p,t} \cdot z_{p,t}$ is the portfolio-specific component of risk, with conditional variance $\sigma_{p,t}^2$. By construction, these two compounds of returns are uncorrelated.

We assume that the market conditional variance follows a univariate EGARCH process (Nelson, 1991); i.e.,

$$\ln(\sigma_{m,t}^2) = \alpha_{m,t} + \sum_{k=1}^{\infty} \phi_{m,k} g_m(z_{m,t-k}), \qquad \phi_{m,1} \equiv 1 \tag{1.4}$$

where $\{\alpha_{m,t}\}_{t=-\infty}^{\infty}$ and $\{\phi_{m,k}\}_{k=1}^{\infty}$ are real, nonstochastic, scalar sequences, and where we define the function $g_m(\cdot)$ by

$$g_m(z_{m,t-k}) \equiv \theta_m z_{m,t-k} + \gamma_m\big[|z_{m,t-k}| - \mathrm{E}\,|z_m|\big]. \qquad (1.5)$$

By construction, $g_m(z_{m,t})$ is the innovation in $\ln(\sigma_{m,t+1}^2)$. The $\theta_m z_m$ term in (1.5) allows for leverage effects; recall that the surprise component of returns has the same sign as z_m, so when $\theta_m < 0$, $\ln(\sigma_{m,t}^2)$ tends to rise (fall) following market drops (rises). When $\gamma_m > 0$, the $\gamma_m[|z_{m,t-k}| - \mathrm{E}|z_m|]$ term raises (lowers) $\ln(\sigma_{m,t}^2)$ when the magnitude of market movements is large (small). Taken together, the $\theta_m z_m$ and $\gamma_m[|z_m| - \mathrm{E}|z_m|]$ terms allow the market's conditional variance to respond asymmetrically to positive and negative returns.

For the portfolio-specific conditional variance $\sigma_{p,t}^2$, we modify the univariate EGARCH model:

$$\ln\big(\sigma_{p,t}^2\big) = \alpha_{p,t} + \sum_{k=1}^{\infty} \phi_{p,m,k}\, g_{p,m}(z_{m,t-k}) + \sum_{k=1}^{\infty} \phi_{p,k}\, g_p(z_{p,t-k}),$$

$$\phi_{p,m,1} \equiv \phi_{p,1} \equiv 1 \qquad (1.6)$$

where, as before, $\{\alpha_{p,t}\}_{t=-\infty}^{\infty}$, $\{\phi_{p,m,k}\}_{k=1}^{\infty}$, and $\{\phi_{p,k}\}_{k=1}^{\infty}$ are nonstochastic, and $g_{p,m}(\cdot)$ and $g_p(\cdot)$ are functions of the form (1.5), with $\theta_{p,m}$ and $\gamma_{p,m}$ (and θ_p and γ_p) replacing θ_m and γ_m in (1.5). The intuition for the functional form (1.6) is similar to that for (1.4): If $\theta_p < 0$ and $\gamma_p > 0$, then the portfolio-specific conditional variance rises (falls) in response to negative portfolio-specific shocks and portfolio-specific shocks are large (small).

The $g_{p,m}(\cdot)$ function allows contemporaneous correlation between $z_{m,t-1}$ and the innovations in $\ln(\sigma_{p,t}^2)$ and $\ln(\sigma_{m,t}^2)$. This is in line with the work of Black (1976), who found that volatilities of individual stocks tend to change in the same direction. Recall that, by (1.2), the conditional variance of $R_{p,t}$ is $\beta_{p,t}^2\sigma_{m,t}^2 + \sigma_{p,t}^2$. Our model allows two channels through which the volatilities of different portfolios can move together. First, if $\sigma_{m,t}^2$ rises, individual portfolio volatilities with nonzero conditional betas rise because of the $\beta_{p,t}^2 \cdot \sigma_{m,t}^2$ terms. Second, the $g_{p,m}$ terms allow contemporaneous correlation between changes in $\ln(\sigma_{m,t}^2)$ and $\ln(\sigma_{p,t}^2)$—i.e., shocks to the market may feed directly into the non-market component of volatility.

We model the conditional beta, $\beta_{p,t}$, as

$$\beta_{p,t} = \alpha_{\beta,t} + \sum_{k=1}^{\infty} \phi_{\beta,k}\big[\lambda_{p,m} \cdot z_{m,t-k} \cdot z_{p,t-k} + \lambda m \cdot z_{m,t-k}\big], \qquad \phi_{\beta,1} = 1. \qquad (1.7)$$

The $\lambda_m \cdot z_m$ and $\lambda_p \cdot z_p$ terms allow for leverage effects in the conditional betas of the sort envisioned by Chan (1988) and Ball and Kothari (1989): If

λ_m is negative, the conditional beta rises in response to negative market returns and drops in response to positive market returns. Similarly, if λ_p is negative, the conditional beta rises in response to negative nonmarket (i.e., idiosyncratic) returns and drops in response to positive nonmarket returns. Since a weighted average of the betas must equal one, however, we would not expect λ_m to be negative for all portfolios. The roles of financial and operating leverage, however, lead us to expect that λ_p should be negative for all portfolios. Since $z_{m,t}$ and $z_{p,t}$ are uncorrelated, the conditional expectation (at time $t-1$) of $\lambda_{p,m} \cdot z_{m,t} \cdot z_{p,t}$ equals zero because $z_{m,t} \cdot z_{p,t}$ is positive (negative) and $R_{m,t}$ and $R_{p,t}$ moved together more (less) than expected at time $t-1$. If $\lambda_{p,m} > 0$, the model responds by increasing (decreasing) the conditional covariance by increasing (decreasing) the next period's conditional beta $\beta_{p,t+1}$.

Positive values of $\lambda_{p,m}$, γ_m, and γ_p make the conditional covariance matrix estimates produced by the model robust to model misspecification. To see this, suppose that the current value of $\sigma_{m,t}^2$ is too low—i.e., below the "true" conditional volatility for the market. In estimation we form the standardized residual $\hat{z}_{m,t}$ by

$$\hat{z}_{m,t} = \frac{R_{m,t} - \hat{\mu}_{m,t}}{\hat{\sigma}_{m,t}}. \tag{1.8}$$

When $\hat{\sigma}_{m,t} < \sigma_{m,t}$ then $\mathrm{E}_{t-1}|\hat{z}_{m,t}| - \mathrm{E}_{t-1}|z_{m,t}| > 0$ so when $\gamma_p > 0$, $\mathrm{E}_{t-1} g_m(\hat{z}_{m,t}) > 0$. Similarly, when $\hat{\sigma}_{m,t} > \sigma_{m,t}$ $\mathrm{E}_{t-1} g_m(\hat{z}_{m,t}) < 0$. Therefore, $\sigma_{m,t}^2$ tends to rise when it is too low (i.e., below the true $\sigma_{m,t}^2$) and fall when it is too high. $\lambda_{p,m}$ and γ_p, when they are positive, play a similar role. Accordingly, we impose the conditions $\lambda_{p,m} \geq 0$, $\gamma_p \geq 0$, and $\gamma_m \geq 0$ in the empirical implementation reported below. For detailed discussion of this robustness property of ARCH models, see Nelson (1992).[7]

Conditions for strict stationarity and ergodicity of the model are easily derived: Since the $\beta_{p,t}$, $\ln(\sigma_{p,t}^2)$, and the $\ln(\sigma_{m,t}^2)$ are linear with i.i.d. errors, the first requirement for strict stationarity is the familiar condition that the moving average coefficients are square summable—i.e., that $\Sigma_{k=1}^{\infty} \phi_{\beta,k}^2 < \infty$, $\Sigma_{k=1}^{\infty} \phi_{p,k}^2 < \infty$, and $\Sigma_{k=1}^{\infty} \phi_{m,k}^2 < \infty$. Second, the intercept terms $\alpha_{\beta,t}$, $\alpha_{p,t}$ and $\alpha_{m,t}$ must be time invariant. Third, we require that

[7] The basic intuition underlying Nelson's results is straightforward. Suppose that the data are generated in continuous time by a (heteroskedastic) diffusion process, which the econometrician observes at discrete time intervals of length h. Suppose a conditional covariance is estimated for each point in time using a misspecified bivariate EGARCH model. As $h \downarrow 0$, more and more information is available about the instantaneous covariance of the underlying diffusion [see Merton (1980) or Nelson (1992) for intuition on why this is so]. So much information arrives as $h \downarrow 0$, in fact, that even a misspecified bivariate EGARCH can extract the true underlying conditional covariance of the limit diffusion. Many ARCH models have similar continuous time consistency properties.

$\sum_{k=1}^{\infty} \phi_{\beta,k} \cdot z_{m,t} \cdot z_{p,t} < \infty$ almost surely.[8] The proof of strict stationarity then proceeds along the same lines as the proof in the univariate EGARCH case in Nelson (1991).

There are several limitations to the bivariate EGARCH model. First, while $\ln(\sigma_{m,t}^2)$, $\ln(\sigma_{p,t}^2)$, and $\beta_{p,t}$ each follow linear processes, they are linked only by their innovations terms, so feedback is not allowed, as it would be if $\ln(\sigma_{m,t}^2)$, $\ln(\sigma_{p,t}^2)$, and $\beta_{p,t}$ followed, say, a vector ARMA process. Second, although leverage effects enter the model, the way they enter is fairly ad hoc—i.e., it follows no economic theory explaining leverage effects. As is true of ARCH models in general, bivariate EGARCH is more a statistical model than an economic model. An interesting future project is to give the bivariate EGARCH more economic content by imbedding it into an explicit model of firm capital structure to directly account for the effects of financial and operating leverage. Third, as is the case with ARCH models in general, asymptotic normality for the MLE estimates is as yet unproven. Fourth, we consider portfolios one at a time, rather than estimating the joint covariance of a whole set of portfolios at once. In the same vein, we also ignore other possible determinants of conditional covariances, such as interest rates (Ferson, 1989) and the list of macroeconomic variables found by Schwert (1989) to influence volatility. There are benefits and costs to doing this: The benefit is parsimony and (relative) ease of computation. The costs are that certain parameter restrictions cannot be used (i.e., the weighted average of the conditional betas equals one, and $\sigma_{m,t}^2$ should not depend on any portfolio p with which it is paired in the bivariate system) and that we lose some information that may help predict volatility (e.g., interest rates).

The asymmetric model of Glosten, Jagannathan, and Runkle (GJR, 1993) circumvents the above limitations of the EGARCH model. Specifically, the GJR model is linked up more directly with theory. Their inclusion of an exogenous variable in the conditional variance equation permits the incorporation into the model of information that may be relevant for predicting conditional variance.

Despite these limitations, the model has several appealing properties. First, as indicated above, it allows both the size and sign of $R_{m,t}$ and $R_{p,t}$ (as well as their comovement) to enter the model in a straightforward way. In particular, it allows for leverage effects. Second, it captures, in a natural way, Black's (1976) observation that volatilities on different portfolios tend

[8] In allowing market volatility movements to drive the conditional covariance of asset returns, our model is related to the single index model of Schwert and Seguin (1990) in which the conditional covariance of returns on any two assets is linear conditional variance. If we take $\beta_{p,t}$ and $\sigma_{p,t}$ to be constant, our model is a special case of Schwert and Seguin equation (1). Schwert and Seguin do not require conditional betas to be constant, but the mechanism they propose for changes in conditional betas is different from ours.

to move in the same direction. We allow for this both through the $\beta_{p,t} R_{m,t}$ term in $R_{p,t}$, and through the $g_{m,p}$ function. While *allowing* the market to drive all movement in conditional covariances, we do not force it to—portfolio-specific shocks may also affect conditional covariances through the $\lambda_{p,m}$, λ_p, and g_p terms. Third, under some mild regularity conditions, the model has a desirable robustness property: As argued above, when γ_m, γ_p, and $\lambda_{p,m}$ are all positive, the model produces reasonable conditional covariance estimates even if it is misspecified. As we show below, our model produces conditional covariance estimates strikingly similar to (and by some measures better than) the covariance estimates produced by a rolling regression.

2. EMPIRICAL APPLICATIONS

2.1. Data

For this empirical analysis we use monthly returns from CRSP for the period July 1926–December 1990. For the market return, $R_{m,t}$, we use the CRSP equally weighted market index. For portfolio returns we use two different sets. For the first set, we created 12 broad-based industry portfolios of NYSE stocks according to two-digit SIC codes. The classification of these industries follows exactly from the SIC groupings used by Breeden *et al.* (1989) and Ferson and Harvey (1991a,b). These industry portfolios are equally weighted. For the second set of portfolios we used sized-based equally weighted decile portfolios. All portfolios were converted to excess returns using the one-month T-Bill rate from the CRSP tapes.

There are two main reasons why we present the results for the equally weighted rather than value-weighted portfolios, both related to the fact that value-weighted portfolios give less weight to "losers" than equally weighted portfolios.[9] First, each of the industry portfolios is a nontrivial component of the total market. When bad news hits an industry, say construction, its portfolio becomes a smaller component of the value-weighted market index. Since it is a smaller component of the market index, the impact of "construction news" on the market index returns may drop, causing the covariance of market returns and the construction portfolio returns to drop. This works in the opposite direction from the leverage effect. Second, there is wide cross-sectional variation in the returns of the stocks making up any given industry portfolio—i.e., even in a "loser" industry, some individual firms will be relative "winners." If the leverage effect holds for conditional betas, then value weighting the

[9] The first condition implies the third if $\mathrm{Var}[z_{m,t} \cdot z_{p,t}] < \infty$.

industry portfolios gives more weight to relative "winner" firms than to relative "loser" firms. Again, this works in the opposite direction of the leverage effect—i.e., when bad news hits the portfolio, the leverage effect story says that portfolio beta should rise. But in computing the portfolio value weights, we give the most weight to the relative winners, whose betas rose the *least*, or even may have fallen. When we use equal weights, we avoid these issues.

Although the bivariate EGARCH model, equations (1.1) through (1.7), allows for intercept terms in the mean equations, we de-meaned each series by its unconditional mean. This greatly simplifies the estimation. The rationale is that for monthly stock returns the conditional variance–covariance matrix seems likely to have much larger elements than the outer product of the conditional means. This is clearly true of the *unconditional* means and covariances. Moreover, the unconditional covariance matrix is very large relative to the outer product of the unconditional means. If expected returns are constant and the conditional covariance matrix is the same order of magnitude as the unconditional covariance, using the unconditional means should have a minor effect in estimating the second moment matrix. Researchers have also found that omitting relatively slowly varying components of expected returns, for example riskless interest rates and dividends, has a very minor effect on estimated conditional variances (see, e.g., Schwert, 1990; Nelson, 1991).[10]

Although, some researchers (for example, Ferson and Harvey, 1991a,b) have found evidence that expected returns are highly variable and that a substantial part of the unconditional variance of returns is due to time-varying expected returns, we effectively treat all this time variation in expected returns as noise. Using monthly returns data from 1964 to 1986, Ferson and Harvey (1991a, Figure 1) estimate conditional prices of the market beta ranging from -3% to 14% *per month* in January and -4% to 6% *per month* during other months of the year. Regressing portfolio returns on their generated expected returns series, yields adjusted R^2's of up to 19.6%, with an R^2 of about 10% for many portfolios (Ferson and Harvey, 1991b, Table II). To the extent that the Ferson–Harvey results accurately represent ex ante expected returns rather than market inefficiency, data mining, or sampling error, we potentially overstate the conditional variance of returns by as much as 20% in some portfolios.

[10] This was suggested by Philip Dybvig and Christopher Lamoureux. We also conducted estimations using value-weighted industry and decile portfolios. There was no qualitative difference in the results; hence, we present only those results using the equally weighted portfolios. Furthermore, since the decile portfolios are so highly correlated with each other we report the results only for deciles 1, 5, and 10. *All* of the value-weighted and equally weighted results are reported in an earlier working paper version of this chapter.

2.2. Estimation Procedure

We fit the model using a conditional normal likelihood function. The rationale for assuming conditional normality is predominantly ease of computation. However, as shown by Bollerslev and Wooldridge (1992), quasi-maximum-likelihood estimators using conditional normality of the error terms yield consistent and asymptotically normal parameter estimates as long as the conditional means and variances are correctly specified, even when the errors are not conditionally normal. We also note that although the normality assumption directly determines the $E[|z|]$ in equations (1.4) and (1.6), all parameters of our model except α_m and α_p are numerically invariant to the assumption of no unit or explosive autoregressive roots in the variance equations.

We employ Bollerslev and Wooldridge's (1992) robust variance–covariance estimator for computing asymptotic standard errors as well as Wald statistics.[11] Bollerslev and Wooldridge show that, as long as the first two conditional moments are correctly specified, a Wald statistic, W_T, has an asymptotic χ^2 distribution under the null hypothesis whether or not the conditional normality assumption holds. Note that the robust Wald statistic has an asymptotic χ^2 distribution if the alternative hypothesis has an unrestricted form; e.g., the parameter vector is not equal to zero. In our implementation of the EGARCH model, however, we have imposed the three inequality restrictions: $\lambda_{p,m} \geq 0$, $\gamma_m \geq 0$, and $\gamma_p \geq 0$. Wolak (1989a) shows that the presence of inequality restrictions such as these in either the null or alternative hypothesis changes the asymptotic null distribution of the Wald statistic to a weighted sum of χ^2's. Design of an exact size α test generally requires the numerical solution of a highly nonlinear equation (see Wolak, 1989b). Alternatively, upper and lower bounds on the critical value for a size α test are available from Kodde and Palm (1986). We adopt this simpler approach of Kodde and Palm. Whenever W_T is greater than the appropriate upper bound critical value, $c_u(\alpha)$, one may reject the null hypothesis at the α level of significance; whenever W_T is less than the appropriate lower bound critical value, $c_l(\alpha)$, one is unable to reject the null hypothesis at the α level of significance. For values of W_T between c_l and c_u, the test is inconclusive.

To implement the model, we reduce the infinite-order moving averages in (4), (6), and (7) to AR(1) processes with constant intercepts:

$$\ln(\sigma_{m,t}^2) = \alpha_m + \delta_m \cdot \left[\ln(\sigma_{m,t-1}^2) - \alpha_m\right] + g_m(z_{m,t-1}), \tag{2.1}$$

[11] Nelson (1992) has shown that, when passing to continuous time, misspecification in the conditional means does not affect the estimated conditional covariances. In addition, if the Sharpe–Linter CAPM holds conditionally each period, then $\mu_{p,t} = 0$ and can be ignored (see equation (1.2)). Of course, this does not by itself justify ignoring $\mu_{m,t}$.

$$\ln\left(\sigma_{p,t}^2\right) = \alpha_p + \delta_p \cdot \left[\ln\left(\sigma_{p,t-1}^2\right) - \alpha_p\right] + g_{p,m}(z_{m,t-1}) + g_p(z_{p,t-1}), \tag{2.2}$$

$$\beta_{p,t} = \alpha_\beta + \delta_\beta \cdot \left[\, \beta_{p,t-1} - \alpha_\beta \right] + \lambda_{p,m} \cdot z_{m,t-1} \cdot z_{p,t-1} + \lambda_m \cdot z_{m,t-1} + \lambda_p \cdot z_{p,t-1}. \tag{2.3}$$

If $|\delta_\beta| < 1$, then the unconditional expectation of $\beta_{p,t}$ exists and equals α_β. Similarly, α_p and α_m are, respectively, the unconditional means of $\ln \sigma_{p,t}^2$ and $\ln \sigma_{m,t}^2$ if $|\delta_p| < 1$ and $|\delta_m| < 1$, respectively.

For a given multivariate EGARCH(p, q), the $\{z_{m,t}\}$, $\{z_{p,t}\}$, $\{\sigma_{m,t}^2\}$, $\{\sigma_{m,t}^2\}$, and $\{\beta_{p,t}\}$ sequences can be easily derived recursively given startup values and the data sequences $\{R_{m,t}\}$ and $\{R_{p,t}\}$. To get initial values for these sequences, the variance and the beta series were set to their unconditional expectations. Given parameter values and these initial states we computed the quasi-likelihood function recursively.

2.3. Results

Conditional Betas: Persistence, Variability, and Leverage Effects

The quasi-maximum-likelihood parameter estimates and their robust standard errors for the industry portfolios are reported in Table 1 and the fitted $\beta_{p,t}$ in Figure 1. Looking at the parameter estimates for the $\beta_{p,t}$ equation across the different portfolios, we see that, for all of the portfolios, shocks to the $\beta_{p,t}$ processes exhibit strong persistence as indicated by δ_β always being greater than 0.9. To gain intuition about the degree of persistence these autoregressive parameters imply, it is useful to think about the half-life of a shock associated with this parameter; i.e., the number h such that $\delta(h/\beta) = \frac{1}{2}$. The largest δ_β estimated (ignoring the δ_β greater than 1 for the construction portfolio) was .998 for the consumer durables portfolio. This estimate implies a half-life of 29 years. In contrast, the half-life implied by the smallest estimated δ_β, .946 for the finance portfolio, is only one year.

The market leverage term from the $\beta_{p,t}$ equation, λ_m, is statistically insignificant at the .05 level for all but two industry portfolios, construction and consumer durables, and decile portfolios 1 and 5. The portfolio leverage term from the $\beta_{p,t}$ equation, λ_p, is insignificant for all industry and decile portfolios. The cross term, $\lambda_{p,m}$, is positive and significant for all but four industry portfolios and decile 1. Another way of looking at these results is to consider the decomposition of the time-series variation in β into its constituent parts.[12] The most important source of variation in

[12] Bollerslev and Wooldridge show how to consistently estimate the sample hessian matrix using only first derivatives. This involves substituting a consistent estimate of the one-step-ahead conditional expectation of the hessian for its analytic counterpart.

TABLE 1 Bivariate EGARCH(1, 0) Maximum Likelihood Parameter Estimates

Portfolio	Bivariate EGARCH parameter estimates														
	α_β	δ_β	$\lambda_{p,m}$	λ_m	λ_p	α_m	δ_m	θ_m	γ_m	α_p	δ_p	θ_p	γ_p	$\theta_{p,m}$	$\gamma_{p,m}$
Basic industries	1.028	0.989	0.012	0.003	−0.001	−5.320	0.979	−0.058	0.182	−7.900	0.991	0.035	0.076	−0.015	0.040
	(0.055)	(0.008)	(0.003)	(0.003)	(0.003)	(0.318)	(0.012)	(0.022)	(0.033)	(0.359)	(0.004)	(0.012)	(0.027)	(0.017)	(0.035)
Capital goods	1.091	0.960	0.004	0.003	0.001	−5.318	0.981	−0.061	0.169	−8.114	0.994	0.020	0.096	−0.028	0.041
	(0.012)	(0.042)	(0.003)	(0.003)	(0.003)	(0.308)	(0.010)	(0.021)	(0.032)	(0.391)	(0.004)	(0.016)	(0.028)	(0.016)	(0.036)
Construction	1.036	1.007	0.000	−0.007	0.001	−5.335	0.981	−0.074	0.159	−7.204	0.991	−0.038	0.087	−0.019	−0.012
	(0.018)	(0.005)	(0.003)	(0.003)	(0.002)	(0.315)	(0.010)	(0.020)	(0.031)	(0.250)	(0.007)	(0.017)	(0.033)	(0.017)	(0.035)
Consumer durables	1.196	0.998	0.005	0.004	−0.002	−5.392	0.973	−0.067	0.177	−8.273	0.924	0.047	0.266	−0.028	0.175
	(0.060)	(0.004)	(0.002)	(0.002)	(0.002)	(0.289)	(0.013)	(0.021)	(0.036)	(0.160)	(0.028)	(0.037)	(0.057)	(0.031)	(0.054)
Finance/real estate	0.909	0.946	0.019	−0.008	0.003	−5.327	0.979	−0.056	0.186	−7.979	0.980	−0.019	0.105	−0.024	0.093
	(0.026)	(0.035)	(0.004)	(0.005)	(0.004)	(0.320)	(0.011)	(0.021)	(0.033)	(0.239)	(0.009)	(0.015)	(0.036)	(0.017)	(0.030)
Food	0.786	0.947	0.010	−0.004	0.001	−5.331	0.981	−0.067	0.159	−8.000	0.972	0.047	0.111	−0.053	0.073
	(0.023)	(0.038)	(0.006)	(0.004)	(0.006)	(0.310)	(0.009)	(0.020)	(0.031)	(0.194)	(0.010)	(0.017)	(0.039)	(0.021)	(0.034)
Leisure	1.015	0.980	0.010	−0.006	0.006	−4.993	0.987	−0.057	0.174	−6.950	0.996	−0.006	0.000	−0.029	0.055
	(0.037)	(0.013)	(0.005)	(0.005)	(0.004)	(0.318)	(0.007)	(0.020)	(0.029)	(0.231)	(0.004)	(0.009)	(0.033)	(0.018)	(0.036)
Petroleum	0.875	0.976	0.010	0.009	−0.004	−5.348	0.980	−0.067	0.173	−6.393	0.981	0.016	0.147	−0.033	0.045
	(0.039)	(0.019)	(0.007)	(0.007)	(0.007)	(0.324)	(0.011)	(0.021)	(0.033)	(0.322)	(0.009)	(0.024)	(0.035)	(0.026)	(0.031)
Services	0.827	0.988	0.028	0.002	0.006	−5.320	0.981	−0.063	0.167	−5.663	1.000	0.026	0.084	−0.035	0.021
	(0.100)	(0.008)	(0.006)	(0.004)	(0.004)	(0.310)	(0.010)	(0.021)	(0.032)	(0.373)	(0.002)	(0.013)	(0.024)	(0.014)	(0.033)
Textiles/trade	0.984	0.955	0.018	0.006	−0.002	−5.313	0.979	−0.061	0.182	−7.558	0.989	−0.010	0.053	−0.014	0.061
	(0.032)	(0.024)	(0.004)	(0.006)	(0.005)	(0.320)	(0.011)	(0.021)	(0.033)	(0.234)	(0.006)	(0.011)	(0.022)	(0.016)	(0.026)
Transportation	1.120	0.981	0.020	−0.007	0.006	−5.380	0.986	−0.056	0.142	−6.887	0.996	−0.031	0.000	−0.056	0.044
	(0.077)	(0.024)	(0.005)	(0.007)	(0.006)	(0.294)	(0.008)	(0.020)	(0.029)	(0.196)	(0.003)	(0.013)	(0.018)	(0.016)	(0.028)

Utilities	0.588	0.992	0.008	−0.005	0.001	−5.359	0.977	−0.071	0.175	−6.887	0.976	−0.008	0.277	−0.075	0.020
	(0.057)	(0.011)	(0.004)	(0.005)	(0.003)	(0.307)	(0.012)	(0.022)	(0.034)	(0.359)	(0.016)	(0.032)	(0.064)	(0.030)	(0.042)
Decile 1	1.186	0.996	0.006	−0.009	0.003	−5.344	0.980	−0.067	0.153	−6.269	0.998	0.010	0.055	−0.079	0.008
	(0.067)	(0.006)	(0.004)	(0.004)	(0.004)	(0.301)	(0.010)	(0.020)	(0.031)	(0.359)	(0.003)	(0.018)	(0.020)	(0.015)	(0.026)
Decile 5	1.020	0.989	0.004	0.007	0.000	−5.220	0.980	−0.059	0.195	−8.537	0.993	0.009	0.050	−0.045	0.077
	(0.033)	(0.015)	(0.002)	(0.002)	(0.003)	(0.290)	(0.011)	(0.023)	(0.034)	(0.341)	(0.004)	(0.020)	(0.023)	(0.013)	(0.039)
Decile 10	0.685	0.982	0.006	0.003	0.007	−5.364	0.987	−0.070	0.160	−7.847	0.994	0.033	0.000	−0.083	0.003
	(0.029)	(0.009)	(0.003)	(0.004)	(0.004)	(0.292)	(0.011)	(0.020)	(0.032)	(0.145)	(0.004)	(0.008)	(0.020)	(0.016)	(0.013)

Note:

Bivariate EGARCH estimates using equally weighted CRSP industry and decile portfolios with the equally weighted CRSP market return for the time period of July 1926 through December 1990 for a total of 780 observations. The parameters are estimated by maximum likelihood from the following discrete-time system of equations:

$$\beta_{p,t} = \alpha_\beta + \delta_\beta \cdot \left[\, \beta_{p,t-1} - \alpha_\beta \right] + \lambda_{p,m} \cdot z_{m,t-1} \cdot z_{p,t-1} + \lambda_m \cdot z_{m,t-1} + \lambda_p \cdot z_{p,t-1}$$

$$\ln\left(\sigma_{m,t}^2\right) = \alpha_m + \delta_m \cdot \left[\ln\left(\sigma_{m,t-1}^2\right) - \alpha_m\right] + \theta_m z_{m,t} + \gamma_m\left[|z_{m,t-1}| - \mathrm{E}|z_m|\right]$$

$$\ln\left(\sigma_{p,t}^2\right) = \alpha_p + \delta_p \cdot \left[\ln\left(\sigma_{p,t-1}^2\right) - \alpha_p\right] + \theta_p z_{p,t-1} + \gamma_p\left[|z_{p,t-1}| - \mathrm{E}|z_p|\right] + \theta_{p,m} z_{m,t-1} + \gamma_{p,m}\left[|z_{m,t-1}| - \mathrm{E}|z_m|\right].$$

β_p is the estimated conditional beta for the industry (or decile) portfolio; $\ln(\sigma_m^2)$ is the log of the conditional variance of the market return and $\ln(\sigma_p^2)$ is the log of the conditional industry (or decile) portfolio return. z_m is the standardized residual for the market portfolio and is calculated as $z_{m,t} = r_{m,t}/\sigma_{m,t}$, where r_m is the de-meaned return on the market portfolio. z_p is the standardized residual for the industry (or decile) portfolio and is calculated as $z_{p,t} = (r_{p,t} - \beta_{p,t} \cdot r_{m,t})/\sigma_{p,t}$, where r_p is the de-meaned return on the industry (or decile) portfolio. Standard errors are estimated using the robust variance–covariance estimator of Bollerslev and Wooldridge (1992) and appear below the coefficient estimates in parentheses.

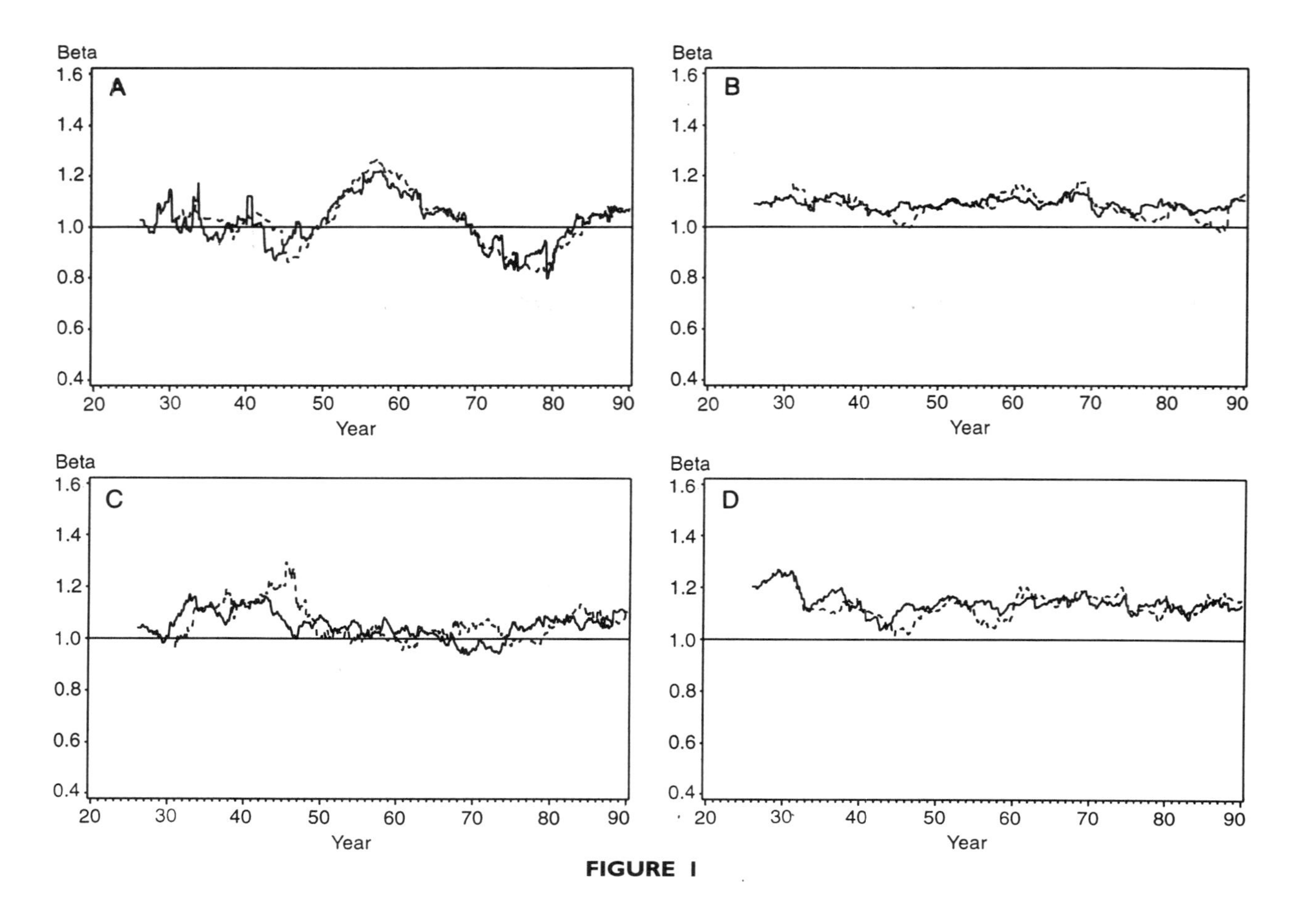

FIGURE 1

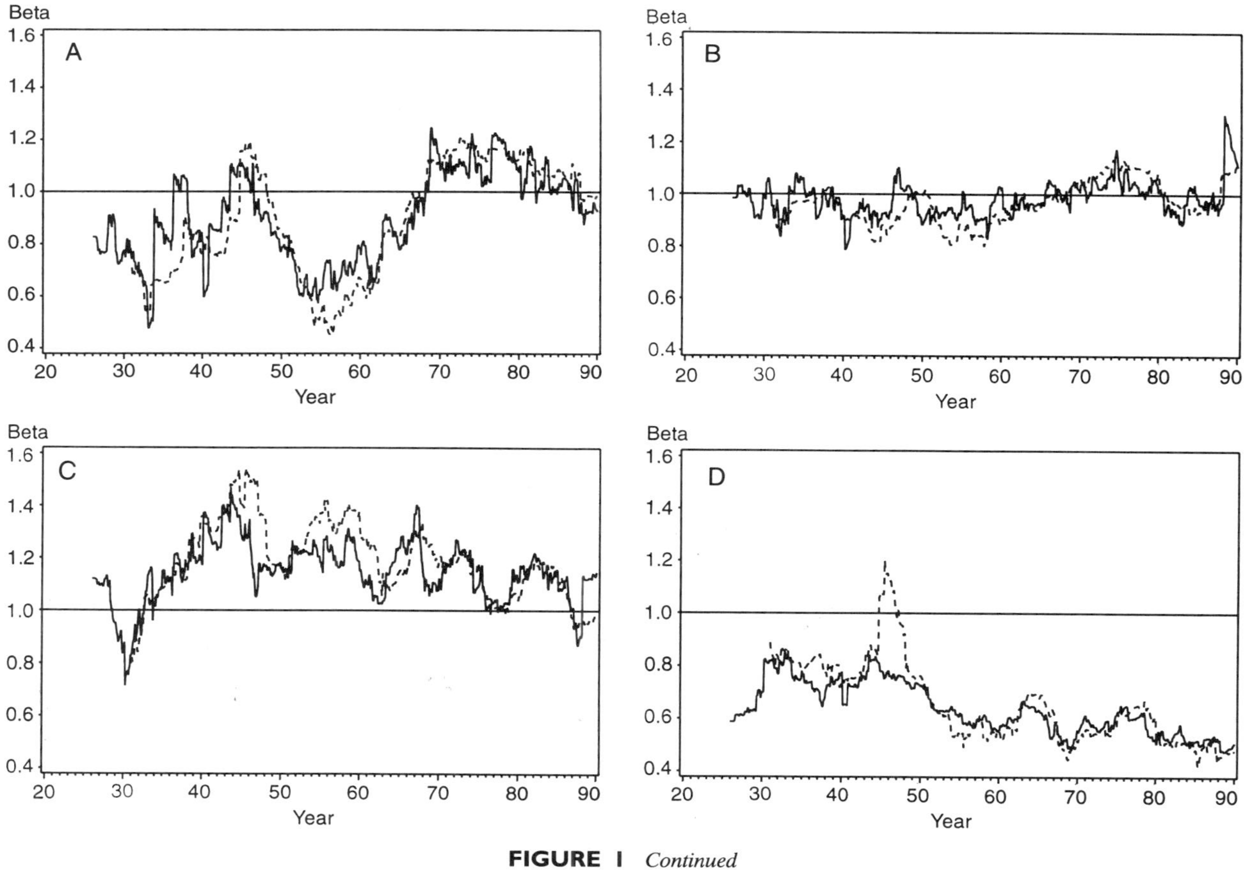

FIGURE 1 *Continued*

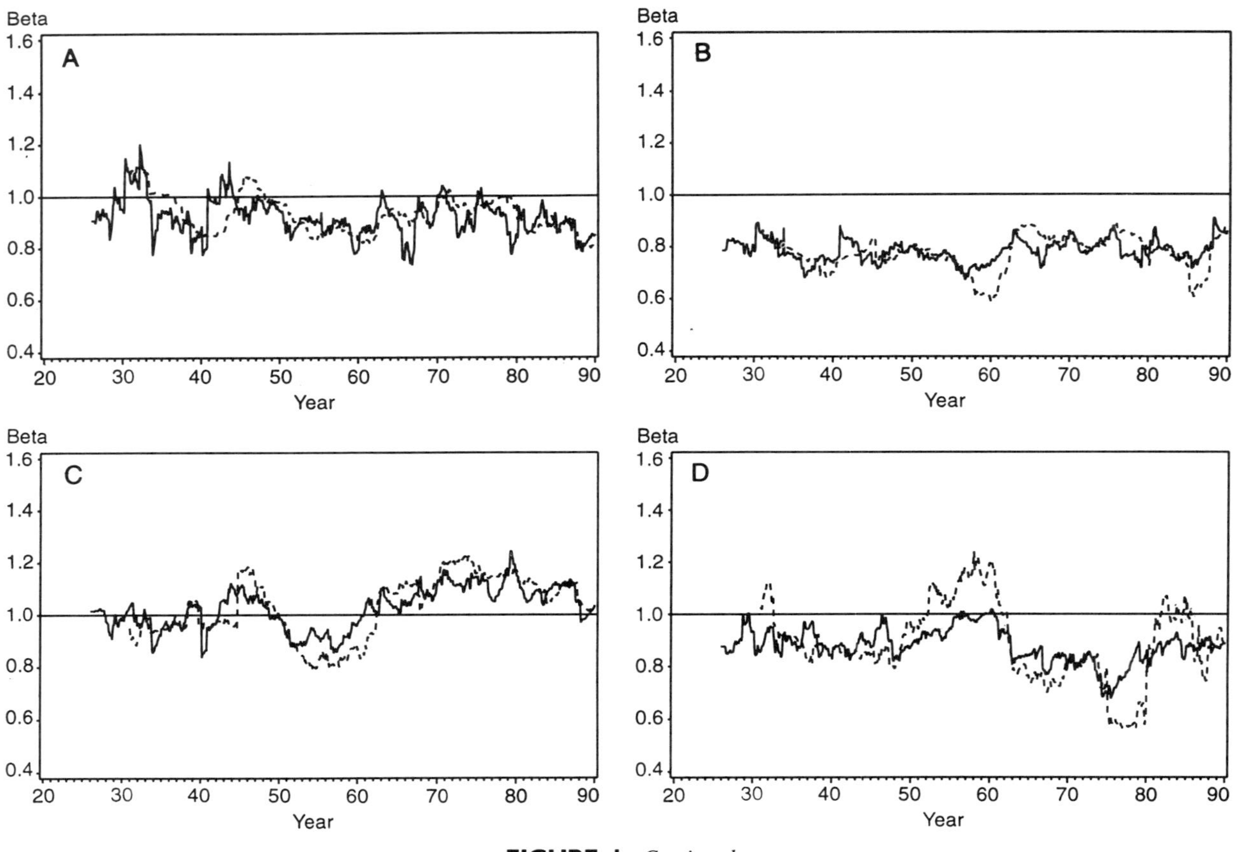

FIGURE 1 *Continued*

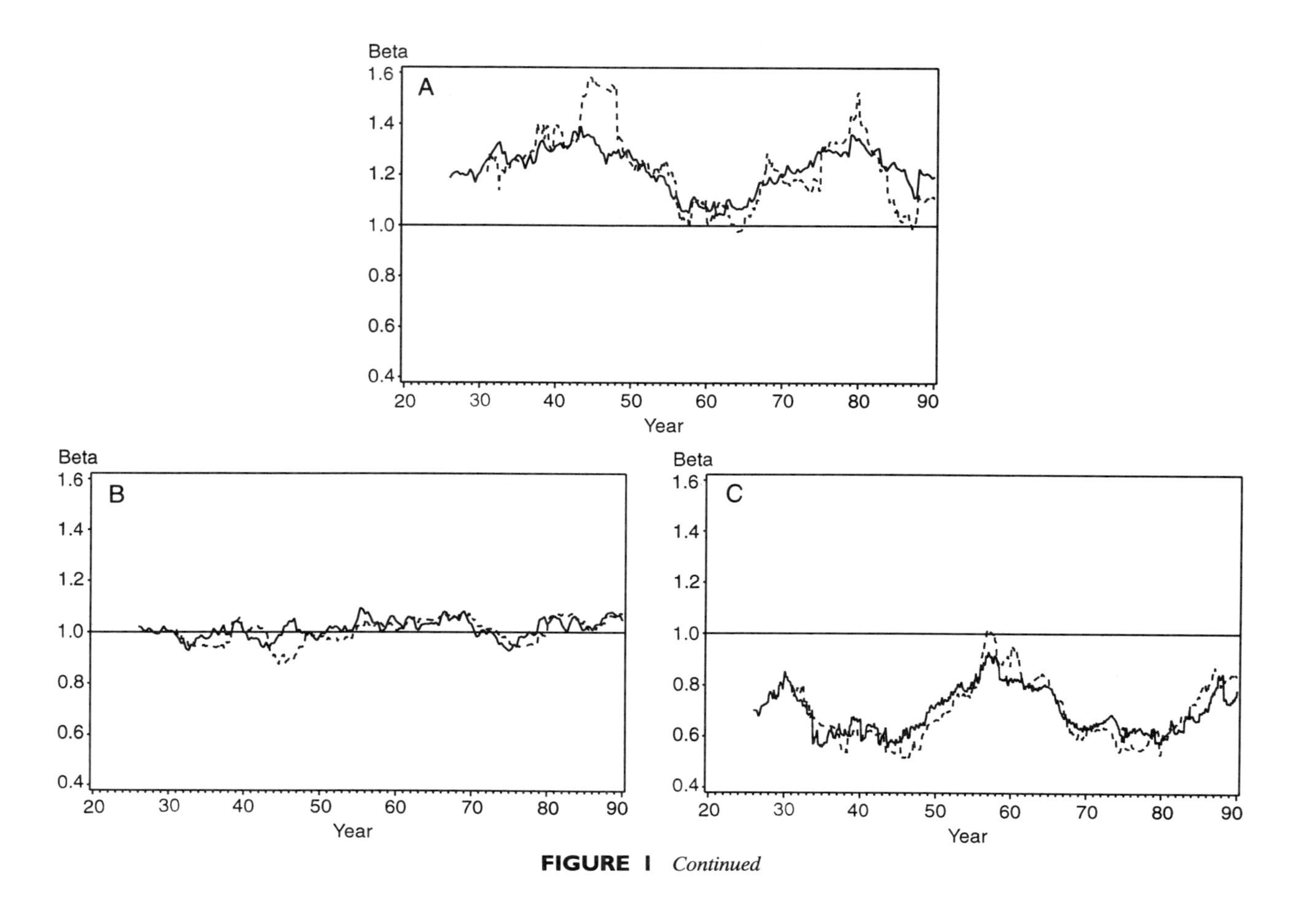

FIGURE 1 *Continued*

the industry betas is the cross-term, which accounts for at least half of the total variance in almost every case. For the deciles the cross-term accounts for a somewhat smaller portion of beta's total variation, with the other terms accounting equally for the difference. We find these results interesting because, within our bivariate EGARCH model, conditional betas do not exhibit leverage effects in the direction expected given the work of Chan (1988) and Ball and Kothari (1989).

To test the joint hypothesis that both leverage terms and the cross-term are jointly equal to zero, we computed a Wald statistic. We present these results in column one of Table 2. This test, while testing the joint significance of $\lambda_{p,m}$, λ_m, and λ_p, equivalently tests for whether the $\beta_{p,t}$ processes are constant across time against the alternative hypothesis that $\lambda_{p,m} > 0$ and $\lambda_m, : \lambda_p \neq 0$. Note that all of the test statistics presented in this table are correlated and hence one must draw an overall conclusion from this table with caution. We cannot reject the null hypothesis of constant betas in 3 of the 12 industry portfolios and all three decile portfolios at .05 critical value. Although this evidence for time-varying betas is somewhat strong, as we show below, this evidence for time-varying betas is weaker than the evidence for time-varying volatility.

Market Volatility

We present the variance equation parameters for the market equation, $\ln(\sigma_{m,t}^2)$, in columns six through nine of Table 1 for the industry and decile portfolios. As with previous studies, we find evidence of strong persistence in the market's conditional variance as indicated by $\delta_m > .95$ regardless of the portfolio with which the market is paired in estimation. The estimated half-lives for shocks to $\sigma_{m,t}^2$ range between 24 and 52 months.

The leverage and magnitude effect results for the market variance equation also confirm previous studies as seen from the $\theta_m < 0$ and $\gamma_m > 0$ with small errors. Recall that a negative value of θ_m indicates that volatility tends to rise when return surprises are negative and vice versa, while a positive value of γ_m implies that $\ln(\sigma_{m,t}^2)$ rise when the magnitude of market movements is larger than expected.[13]

Finally, and also as expected, the market variance parameters are virtually invariant to the portfolio with which we pair the market in estimation as seen from the similarity of the parameter estimates in columns six through nine of Table 1. Furthermore, we conducted a specification check of the fitted values of the market's volatility from these bivariate models with a standard univariate EGARCH for the market and found the parameter estimates and the fitted values to be very similar.

[13] It can be shown that the contribution to total variance for a particular shock is $\lambda_i^2/(\lambda_{p,m}^2 + \lambda_m^2 + \lambda_p^2)$, where i is equal to either (p, m), m or p.

TABLE 2 Bivariate EGARCH(1, 0) Wald Statistics

	H_0: $\lambda_p = \lambda_m = \lambda_{p,m} = 0$	H_0: $\theta_p = \lambda_p = 0$	H_0: $\theta_{m,p} = \gamma_{m,p} = 0$	H_0: $\theta_p = \gamma_p = \theta_{m,p} = \gamma_{m,p} = 0$
Critical values				
($\alpha = .05$)				
χ^2	7.815	5.991	5.991	9.488
c_u	7.045	5.138	5.138	8.761
c_l	5.138	2.706	5.138	7.045
Portfolio				
Basic industries	17.61	15.90	5.99	20.41
Capital goods	3.34	17.75	7.22	21.76
Construction	7.92	11.91	1.04	12.84
Consumer durables	6.31	22.36	33.91	33.29
Finance/real estate	27.71	9.36	28.94	23.88
Food	8.42	15.16	12.43	41.67
Leisure	5.89	0.46	20.96	21.96
Petroleum	3.48	17.33	3.37	20.06
Services	29.99	21.44	5.73	36.42
Textiles/trade	22.13	6.30	16.77	16.93
Transportation	18.80	5.68	20.73	60.37
Utilities	5.05	20.84	2.08	29.92
Decile 1	8.20	7.79	16.98	44.77
Decile 5	9.79	5.36	44.70	36.11
Decile 10	10.46	19.66	19.30	56.77

Note:

Wald tests of parameter restrictions for the bivariate EGARCH. The EGARCH parameters are estimated using the equally weighted CRSP industry and decile portfolios with the equally weighted CRSP market return for the time period July 1926 through December 1990 (780 observations) and are estimated by maximum likelihood from the following discrete-time system of equations:

$$\beta_{p,t} = \alpha_\beta + \delta_\beta \cdot [\beta_{p,t-1} - \alpha_\beta] + \lambda_{p,m} \cdot z_{m,t-1} \cdot z_{p,t-1} + \lambda_m \cdot z_{m,t-1} + \lambda_p \cdot z_{p,t-1}$$

$$\ln(\sigma_{m,t}^2) = \alpha_m + \delta_m \cdot [\ln(\sigma_{m,t-1}^2) - \alpha_m] + \theta_m z_{m,t-1} + \gamma_m[|z_{m,t-1}| - \mathrm{E}|z_m|]$$

$$\ln(\sigma_{p,t}^2) = \alpha_p + \delta_p \cdot [\ln(\sigma_{p,t-1}^2) - \alpha_p] + \theta_p z_{p,t-1} + \gamma_p[|z_{p,t-1}| - \mathrm{E}|z_p|]$$
$$+ \theta_{p,m} z_{m,t-1} + \gamma_{p,m}[|z_{m,t-1}| - \mathrm{E}|z_m|].$$

β_p is the estimated conditional beta for the industry (or decile) portfolio; $\ln(\sigma_m^2)$ is the log of the conditional variance of the market return and $\ln(\sigma_p^2)$ is the log of the conditional industry (or decile) portfolio return. z_m is the standardized residual for the market portfolio and is calculated as $z_{m,t} = r_{m,t}/\sigma_{m,t}$, where r_m is the de-meaned return on the market portfolio. z_p is the standardized residual for the industry (or decile) portfolio and is calculated as $z_{p,t} = (r_{p,t} - \beta_{p,t} \cdot r_{m,t})/\sigma_{p,t}$, where r_p is the de-meaned return on the industry (or decile) portfolio. These Wald statistics are computed using the robust variance–covariance estimator of Bollerslev and Wooldridge (1992). The c_u and c_l are upper and lower bounds on the critical value as reported in Kodde and Palm (1986). These values take into account the effect of inequality constraints on the distribution of the Wald statistic. If the Wald statistic is greater than c_u, the null can be rejected at the .05 level. If the Wald statistic is less than c_l, one cannot reject at the .05 level. The test is inconclusive if the Wald statistic falls between these bounds.

Portfolio-Specific Volatility

We present the parameter estimates for the $\ln(\sigma_{p,t}^2)$ equation in columns 10 through 15 of Table 1 for the industry and decile portfolios. As with the market's equation, all of the portfolios show strong persistence as indicated by all of the δ_p's except one being greater than .97, with half-lives ranging from nine months for the consumer durables portfolio to over 14 years for decile 1 (ignoring the unit root in the services portfolio).

The portfolio-specific leverage and magnitude terms, θ_p and γ_p (columns 12 and 13), respectively, differ from their counterparts in the market equation. For 2 of the 12 industry portfolios the estimated θ_p is significantly negative and 3 times significantly positive. One of the decile portfolios, decile 10, has a significantly positive θ_p. Conversely, for both industry and decile portfolios, all but three of the γ_p's are significant at the .05 level and all are positive. For the industry portfolios the market-specific leverage and magnitude terms, $\theta_{p,m}$ and $\gamma_{p,m}$, are usually insignificant. The $\theta_{p,m}$'s are significant at the .05 level for four industry portfolios and the $\gamma_{p,m}$'s for four portfolios—in only one case (the food portfolio) are both terms simultaneously significant. For the decile portfolios the $\theta_{p,m}$ is always significantly negative and the $\gamma_{p,m}$ is always insignificant.

To examine the role of leverage and magnitude effects in the above estimations in more detail we estimated three more Wald statistics. The first one, presented in column two of Table 2, tests the null hypothesis that all movements in the portfolio-specific conditional variance are caused by only the market return; i.e., $\theta_p = \gamma_p = 0$.[14] Again, note that all of the test statistics presented in this table are correlated and therefore overall conclusions must be drawn with care. We can reject this null hypothesis at the .05 level for all portfolios except the leisure portfolio. The second null hypothesis is that the market return does not enter the $\sigma_{p,t}^2$ equation; i.e., $\theta_{p,m}; = \gamma_{p,m} = 0$.[15] We present these results in column three of Table 2. We can reject this null hypothesis for all but three of the portfolios: construction, petroleum, and the utilities.

We also estimated Wald statistics for the null hypothesis of constant portfolio-specific conditional variance; i.e., $\theta_p = \gamma_p = \theta_{p,m} = \gamma_{p,m} = 0$. These results are presented in column four of Table 2. As was the case for the market equation, for all portfolios we reject the null hypothesis of homoskedasticity at the .01 level. We additionally tested the joint hypothesis of constant conditional variances in both the market and portfolio

[14] We tested for constant conditional variance in the market by testing the joint hypothesis that θ_m and γ_m are equal to zero against the alternative that $\theta_m \neq 0$ and $\gamma_m \geq 0$. As expected for all 12 industry portfolios and all of the deciles, we overwhelmingly reject the null hypothesis of constant market conditional variances. For sake of brevity, we do not present these statistics.

[15] The alternative hypothesis is $\theta_p \neq 0$ and $\gamma_p \geq 0$.

equations. Again, as expected given the individual results for the two equations, we reject the null hypothesis overwhelmingly in all cases and hence to not present the results.

The insignificance or varying signs of the market leverage terms in the portfolio variance equation, (2.2), and in the beta equation, (2.3), indicate that, overall, leverage effects affect portfolio variance only through the market variance equation, (2.1). Although one or the other of the portfolio-specific leverage terms in a majority of the portfolio conditional variance equations are significant for both the industry and decile portfolios, in only one case are they both simultaneously significant. This evidence is not consistent with the financial and operating leverage explanations of the "leverage effect" in conditional variances. It is difficult to understand why leverage effects in portfolio-specific variances are not more evident and consistent across portfolios if these explanations are important. Furthermore, these results suggest that the strong leverage effects in market conditional variances may be better explained by the action of macro forcing variables that have little or no effect on nonsystematic risk. It seems that, on an industry or decile portfolio level, the leverage effects are small with respect to portfolio-specific return movements, but when these return shocks are aggregated across portfolios in the market return, there is a marketwide leverage component.

3. SPECIFICATION TESTS

To check the adequacy of the model, we subject it to an array of specification tests, both formal and informal. Our formal specification tests examine the predictions of the model regarding conditional second moments, in particular, the prediction of the model that for each t, $z_{m,t}^2 - 1$, $z_{m,t} \cdot z_{p,t}$, and $z_{p,t}^2 - 1$ have a conditional mean of zero—i.e., are not predictable given past information. We test these restrictions using the regression-based tests developed in Wooldridge (1990). As we see below, the results of these tests are supportive of the model. Our informal specification tests consist of cross-validating our model's performance to a simple and intuitively appealing benchmark, five-year rolling regressions along the lines of Fama and MacBeth (1973). Again, the model performs well.

3.1. Conditional Moment Tests

An extensive literature has developed on testing conditional moment restrictions such as these (see, e.g., Newey, 1985; Tauchen, 1985; White, 1987). In the bivariate EGARCH model, implementing these tests is complicated by both the relative complexity of the model and by the use of quasi-maximum-likelihood estimation. However, Wooldridge (1990) has shown how to compute conditional moment tests that are both simple to compute and valid for quasi-maximum-likelihood estimators.

To compute the Wooldridge conditional moment tests, we first define an $L \times 1$ vector process ψ_t with conditional expectation zero under the null hypothesis of correct specification. Our model specifies the form of the conditional covariance matrix of $[R_{m,t}, R_{p,t}]$. In particular, correct specification implies that $(R_{m,t}^2 - \sigma_{m,t}^2)$, $(R_{m,t} \cdot R_{p,t} - \beta_{p,t} \cdot \sigma_{m,t}^2)$, and $((R_{p,t}^2 - \beta_{m,t}^2 \cdot \sigma_{m,t}^2 - \sigma_{p,t}^2)$ have conditional means of zero or, equivalently, that $z_{m,t}^2 - 1$, $z_{m,t} \cdot z_{p,t}$ and $z_{p,t}^2 - 1$ have conditional means of zero. We base our conditional moment tests on the latter set of restrictions, since they correct for conditional heteroskedasticity and thereby increase the power of the specification tests. For our applications, ψ_t will be set equal to either $z_{m,t}^2 - 1$, $z_{m,t} \cdot z_{p,t}$, or $z_{p,t}^2 - 1$ individually, or to a 3×1 vector containing all three.

Next, if p is the number of parameters in the bivariate EGARCH system, define the $L \times p$ process $\Psi_t = \mathrm{E}_t[\nabla_\vartheta : \psi_t]$, where E_t is the time $t - 1$ conditional expectation and ∇_ϑ denotes differentiation with respect to the system's parameters. Further define Λ_t to be an $L \times q$ matrix of predetermined variables, which will be used to try to detect predictability in ψ_t. In the specification tests reported below, each row of Λ_t will consist of a constant, the current one-month Treasury Bill rate r_t, and one lag each of $z_{m,t}^2 : z_{p,t}^2 : z_{m,t} \cdot z_{p,t} : z_{m,t}$ and $z_{p,t}$. For each portfolio, we compute four conditional moment specification tests; the first test, reported in the first column of Table 3, sets ψ_t equal to $z_{m,t}^2 - 1$, the second and third columns of Table 3 set ψ_t equal to $z_{m,t} \cdot z_{p,t}$ and $z_{p,t}^2 - 1$, respectively. The specification test reported in the fourth column of Table 3 sets Λ_t equal to a 3×6 matrix with each row equal to

$$\left[1,: r_t,: z_{m,t-1}^2,: z_{p,t-1}^2, z_{m,t-1} \cdot z_{p,t-1},: z_{m,t-1},: z_{p,t-1}\right]$$

and sets

$$\psi_t \equiv \left[z_{m,t}^2 - 1,: z_{m,t} \cdot z_{p,t},: z_{p,t}^2 - 1\right]'.$$

All of these tests are distributed asymptotically χ^2 with seven degrees of freedom (i.e., the number of columns in Λ_t).

Our choice of one lag is somewhat arbitrary, since theory gives us little guidance on this matter.[16] Adding "too many" variables to the conditioning information set reduces power by adding degrees of freedom from the χ^2 test, while including too few reduces power by omitting useful conditioning information. We included the current one-month Treasury-bill rate since

[16] The alternative hypothesis is $\theta_{p,m} \neq 0$ and $\gamma_{p,m} \neq 0$. Note that this is not equivalent to a null hypothesis that market returns do not enter the conditional variance of $R_{p,t}$.

TABLE 3 Bivariate EGARCH(1, 0) Conditional Moment Specification Tests

Portfolio	Moment restriction			
	$E[z_m^2 - 1] = 0$	$E[z_m z_p] = 0$	$E[z_p^2 - 1] = 0$	$E\begin{bmatrix} z_m^2 - 1 \\ z_m z_p \\ z_p^2 - 1 \end{bmatrix} = 0$
Basic industries	13.55 (0.060)	9.75 (0.203)	15.32 (0.032)	6.88 (0.441)
Capital goods	14.73 (0.040)	6.37 (0.497)	9.13 (0.244)	8.93 (0.258)
Construction	15.00 (0.036)	18.18 (0.011)	3.85 (0.797)	7.55 (0.374)
Consumer durables	10.99 (0.139)	12.44 (0.087)	3.77 (0.806)	14.84 (0.038)
Finance/real estate	7.64 (0.366)	4.00 (0.780)	10.51 (0.161)	4.86 (0.678)
Food	11.93 (0.103)	8.75 (0.271)	6.57 (0.475)	12.02 (0.100)
Leisure	8.43 (0.296)	1.79 (0.971)	6.40 (0.494)	6.27 (0.509)
Petroleum	13.51 (0.061)	16.43 (0.021)	9.74 (0.204)	7.88 (0.343)
Services	8.41 (0.298)	7.36 (0.393)	9.42 (0.224)	6.59 (0.472)
Textiles/trade	14.08 (0.050)	4.87 (0.676)	14.60 (0.042)	5.43 (0.608)
Transportation	13.25 (0.066)	14.83 (0.038)	13.47 (0.061)	13.22 (0.067)
Utilities	7.88 (0.343)	5.60 (0.587)	4.63 (0.705)	11.48 (0.119)
Decile 1	16.05 (0.025)	19.74 (0.006)	3.79 (0.804)	2.78 (0.905)
Decile 5	8.57 (0.285)	9.73 (0.204)	9.64 (0.210)	11.19 (0.131)
Decile 10	19.98 (0.006)	13.62 (0.058)	12.32 (0.090)	21.29 (0.003)

Note:
These are regression-based conditional moment tests computed by the method of Wooldridge (1990). They check for the failure of conditional moment restrictions using a constant, the comtemporaneous T-Bill rate, and one lag of

$$\left[z_{m,t}^2, z_{p,t}^2, z_{m,t} \cdot z_{p,t}, z_{m,t}, z_{p,t} \right].$$

z_m is the standardized residual for the market portfolio and is calculated as $z_{m,t} = r_{m,t}/\sigma_{m,t}$, where r_m is the de-meaned return on the market portfolio and σ_m is the conditional standard deviation of the market return calculated from the bivariate EGARCH. z_p is the standardized residual for the industry (or decile) portfolio and is calculated as $z_{p,t} = (r_{p,t} - \beta_{p,t} \cdot r_{m,t})/\sigma_{p,t}$, where r_p is the de-meaned return on the industry (or decile) portfolio, β_p is the bivariate EGARCH estimated conditional beta for the industry (or decile) portfolio and σ_p is the conditional standard deviation of the industry (or decile) portfolio return estimated from the bivariate EGARCH. The time series are for the period July 1926 through December 1990 for a total of 780 observations. See Table 1 for the corresponding parameter estimates. The asymptotic $\chi^2(6)$ statistics are reported above the p values.

earlier research (e.g., Ferson, 1989; Glosten *et al.* (1993) have found that nominal interest rates have power to forecast stock volatility.

Overall, the models perform well, especially the industry portfolio models. Of the 48 test statistics reported for the industry portfolios, 8 reject at the .05 level. Three of the rejections find evidence of misspecification in the conditional betas—i.e., of predictability in $z_{m,t} \cdot z_{p,t}$. None of the rejections for the industry portfolios is particularly strong, however, with only one rejection at the .01 level. Results are not as favorable for the decile portfolios: Of the 12 test statistics calculated, 4 reject at the .05 level and 3 at the .01 level. These results imply that there is evidence that returns on decile portfolios have power to predict market variances. An anomalous aspect of the results we present in Table 3 is that, for some cases, e.g., decile 1, where we reject the null hypothesis for some subset of moment conditions (columns one through three), we fail to reject when we use the complete set of moment restrictions (column four). We attribute these reversals to a combination of (i) the relative variance of a particular moment restriction to the other moment restrictions and (ii) large correlations between some of the moment restrictions.

3.2. Comparison to Rolling Regression

As an informal specification test we compare the EGARCH model to the Fama–MacBeth (1973) or rolling regression approach to estimating betas.[17] The latter method relies on ordinary least squares regressions to obtain betas. The rolling regression model was chosen as a benchmark for comparison because of its simplicity and wide usage. It is also attractive as a benchmark because it allows us to estimate a series of time-varying betas without imposing any particular structure on the way in which conditional covariances change. In contrast, the EGARCH model imposes a considerable degree of structure on the data generating process. The rolling regression approach provides a useful reality check on the EGARCH structure and in this sense we can view the comparisons that follow as informal specification tests. We use these informal specification checks as a means of *cross-validating* our model against a standard benchmark.

To estimate the rolling regression betas we employ a lagged window of 60 months. Thus, we estimate the following model to obtain a beta for date τ:

$$R_{p,t} = \alpha_\tau + \beta_{p,\tau} R_{m,t} + \epsilon_{p,t}, \qquad t = \tau - 61, \ldots, \tau - 1 \qquad (3.1)$$

[17] We repeated the same conditional moment tests using a larger information set: each row of Λ_t consisting of a constant, the current one-month Treasury-Bill rate r_t, and six lags each of $z^2_{m,t}$, :: $z^2_{p,t}$, :: $z_{m,t} \cdot z_{p,t}$, :: $z_{m,t}$ and $z_{p,t}$. The results from these six lag tests almost always failed to reject the null hypothesis and hence we do not report the results for sake of brevity.

where $\{\epsilon_{p,t}\}$ is a white noise process. This is the conventional way of forming beta estimates; see, for example, Merrill Lynch's beta book (1990). We estimated both the EGARCH and rolling regression models for each portfolio over the entire data set from January 1926 to December 1990. Because the rolling regression method requires 60 months of data, we had 720 observations for comparison.

In Table 4, we gauge the in-sample performance of the two models with four statistics. The first two statistics, mean square error (MSE), in columns one and two, and mean absolute error (MAE), in columns three and four, indicate how well each model's betas can explain the in-sample variation of portfolio returns. Note that we could have employed alternative metrics to cross-validate our model with rolling regressions; however, we selected MSE and MAE again because of their simplicity. Because each model's ability to match in-sample variation of portfolio returns depends on the quality of its beta estimates, these statistics provide evidence of how well each model matches the true underlying covariation of portfolio and market returns.

The mean square error and mean absolute error of the EGARCH model is consistently lower that the mean square error and mean absolute error of the rolling regression model. The differences in MAE between the two models are somewhat smaller than the differences in MSE. Because the MAE statistic puts less weight on large residuals than the MSE statistic, this indicates that, at least in periods of low market volatility when most residuals are small, the rolling regression model may perform nearer to the EGARCH model. In periods of high market volatility, however, the EGARCH model performs better.

The second set of columns in Table 4 indicate how well the two models are able to capture changes in market and portfolio volatility. Columns five and six report the mean absolute error of each model in predicting the square of the de-meaned *market* return over the whole sample. Columns seven and eight report the same mean absolute error comparison for the square of *individual* industry or decile portfolio returns.

The EGARCH model is consistently better than the rolling regression model in matching these second moments of market returns. The improvement in mean absolute error is in the range of 5% to 15%. The EGARCH model also does consistently better than the rolling regression model in the second moments of the individual industry or decile portfolios. The percentage improvement ranges between 0.5 to 20%.

The collection of plots in Figure 1 provide a useful look at how EGARCH and rolling regression betas compare. The solid line in the charts represents the betas from the EGARCH model and the dashed line represents betas from rolling regressions. The betas from the two models track each other broadly. There are significant differences between portfo-

TABLE 4 EGARCH(1, 0) versus Rolling Regression In-Sample Performance Comparison

	MSE $\times 10^{-4}$		MAE $\times 10^{-2}$		MAE(r_m^2) $\times 10^{-2}$		MAE(r_p^2) $\times 10^{-2}$	
Portfolio	EGARCH	Rolling regression	EGARCH	Rolling regression	EGARCH	Rolling regression	EGARCH	Rolling regression
Basic industries	2.42	2.51	1.11	1.14	5.47	6.22	5.78	6.65
Capital goods	1.97	2.12	1.01	1.03	5.46	6.22	6.66	7.52
Construction	6.54	6.88	1.83	1.88	5.46	6.22	7.09	8.24
Consumer durables	2.52	2.69	1.18	1.21	5.29	6.22	7.12	8.07
Finance/real estate	2.76	3.19	1.24	1.27	5.46	6.22	5.13	6.07
Food	3.38	3.59	1.33	1.37	5.43	6.22	3.79	4.13
Leisure	6.54	6.93	1.85	1.92	5.90	6.22	6.81	7.01
Petroleum	13.64	14.64	2.72	2.79	5.45	6.22	5.72	6.39
Services	13.72	14.60	2.51	2.56	5.46	6.22	6.00	6.00
Textiles/trade	4.05	4.31	1.49	1.52	5.49	6.22	5.86	6.33
Transportation	13.51	14.22	2.52	2.64	5.37	6.22	8.45	9.92
Utilities	10.63	11.15	2.23	2.28	5.40	6.22	3.71	4.71
Decile 1	12.36	12.90	2.29	2.36	5.38	6.22	9.87	11.70
Decile 5	1.49	1.64	0.83	0.87	5.65	6.22	5.78	6.14
Decile 10	5.16	5.42	1.60	1.61	5.34	6.22	2.94	3.20

Note:

In-sample comparison of the fitted conditional means and variances from the bivariate EGARCH and rolling regressions. The statistics used for comparison are

Statistic	EGARCH	Rolling regression
MSE	$\frac{1}{T}\sum_{t=1}^{T}(R_{p,t}-\hat{\beta}_{p,t}R_{m,t})^2$	$\frac{1}{T}\sum_{t=1}^{T}(R_{p,t}-\hat{\alpha}_{p,t}-\hat{\beta}_{p,t}R_{m,t})^2$
MAE	$\frac{1}{T}\sum_{t=1}^{T}\lvert R_{p,t}-\hat{\beta}_{p,t}R_{m,t}\rvert$	$\frac{1}{T}\sum_{t=1}^{T}\lvert R_{p,t}-\hat{\alpha}_{p,t}-\hat{\beta}_{p,t}R_{m,t}\rvert$
MAE(r_m^2)	$\frac{1}{T}\sum_{t=1}^{T}\lvert r_{m,t}^2-\hat{\sigma}_{m,t}^2\rvert$	$\frac{1}{T}\sum_{t=1}^{T}\lvert r_{m,t}^2-\hat{\sigma}_{m,t}^2\rvert$
MAE(r_p^2)	$\frac{1}{T}\sum_{t=1}^{T}\lvert r_{p,t}^2-\hat{\beta}_{p,t}^2\hat{\sigma}_{m,t}^2-\hat{\sigma}_{p,t}^2\rvert$	$\frac{1}{T}\sum_{t=1}^{T}\lvert r_{p,t}^2-\hat{\beta}_{p,t}^2\hat{\sigma}_{m,t}^2-\hat{\sigma}_{p,t}^2\rvert$

R_p is the return on the equally weighted CRSP industry (or decile) portfolio; R_m is the return on the CRSP equally weighted market return; β_p is the bivariate EGARCH or rolling regression estimated conditional beta for the industry (or decile) portfolio; α_p is the estimated intercept term from the rolling regressions; r_m is the de-meaned return on the market portfolio; r_p is the de-meaned return on the industry (or decile) portfolio; σ_m^2 is the conditional variance of the market return calculated from the bivariate EGARCH or the rolling regression; σ_p^2 is the conditional variance of the industry (or decile) portfolio return estimated from the bivariate EGARCH or rolling regression; T is the sample size, 780 observations.

lios in these plots. Unfortunately, differences across industry portfolios are difficult to reconcile with a casual knowledge about corresponding differences in financial and operating leverage. The EGARCH model generally produces higher frequency movements in beta estimates than the rolling regression method.[18] Plots of portfolio-specific volatilities given by the two models are similar to Figure 1. That is, the EGARCH model produces more variable estimates of conditional volatility, which, as indicated by the MAE statistics, produce a better fit to the actual pattern of squared return residuals.

To sum up, the EGARCH model compares quite well to the rolling regression model. It is consistently better than the rolling regression model in explaining in-sample variation of portfolio returns and in fitting second moments of returns. We find these results encouraging and supportive of the thesis that the EGARCH model can build better estimates of beta and volatility than rolling regressions. Furthermore, we conclude that our bivariate EGARCH model is cross-validated with respect to rolling regressions given the MSE and MAE metrics we employ.

4. CONCLUSION

The empirical finance literature has amply documented the asymmetric response of volatility to good news and bad news. In this chapter, we shed additional light on the issues that surround the evidence for predictive asymmetry of second moments of returns. In particular, it appears that predictive asymmetry occurs mainly at the level of the market—i.e., market volatility tends to rise strongly in response to bad news and fall in response to good news. Surprisingly, however, our evidence indicates that predictive asymmetry is entirely absent (or of the "wrong" sign) in conditional betas and weak in idiosyncratic (e.g., industry-specific) sources of risk for industry portfolios. Predictive asymmetry *does* show up in the volatility of all portfolios, but it appears most consistently through the market factor rather than through the other terms—i.e., predictive asymmetry appears in portfolio conditional variance $\beta_{p,t}^2\sigma_{m,t}^2 + \sigma_{p,t}^2$ through the $\sigma_{m,t}^2$ term, not through the $\beta_{p,t}^2$ or the $\sigma_{p,t}^2$ terms.

We find this quite surprising, especially given the role of financial and operating leverage in the determination of beta. While it has long been known that leverage per se could not explain all of the leverage effect, the link may be even weaker than had been realized. Our results thus lend indirect support to the explanations of a leverage effect based on macro

[18] Of course, Fama and MacBeth's motivation when they developed the rolling regression procedure is different from our current motivation. Consequently, some of the details of our implementation will differ.

forcing variables (e.g., Malkiel, 1979; French *et al.*, 1987; Glosten *et al.*, 1993) rather than financial structure.

At first glance, our results also appear to contradict the assertions of Chan (1988) and Ball and Kothari (1989) that shifts in conditional betas can explain mean reversion in asset prices. We find evidence for variation in betas at the industry and decile levels, but the variation in betas that we find appears unrelated to the leverage effects postulated by Chan and Ball and Kothari. This offers some support for De Bondt and Thaler's (1989) claim that betas were not responsive enough to account for the differing return performances of "winners" and "losers."

Still, Chan and Ball and Kothari examined individual securities and portfolios of winners and losers, which we have not done. Perhaps because our test design aggregates to the industry and decile level, we lose much of the cross-sectional variation in betas and in particular miss an asymmetric response of the conditional betas of individual firms to good and bad news. In other words, the predictive asymmetry in betas may take place at the firm level rather than the industry and decile levels. This suggests that it may be desirable to model conditional covariances at the individual firm level, carefully separating returns into market-, industry-, and firm-specific components and checking the effects of each on conditional betas, perhaps in the context of a model of financial structure. Still, it is surprising to find virtually *no* evidence that betas rise (fall) in response to bad (good) news at the industry level. Since we have not modelled expected returns, our results say nothing about Chan's (1988) claim that the market risk premium is positively correlated with loser betas. Nevertheless, our results are more supportive of the conclusions of Chopra *et al.* (1992), that a leverage effect in betas is not sufficient to explain the market's overreaction to winners and losers, the conclusions of Chan and Ball and Kothari.

It is also possible that an existing leverage effect in betas is obscured by our use of an EGARCH rather than a stochastic volatility model. Stochastic volatility models allow for an independent error term in the conditional variance equation. This inclusion of the error term permits realistic estimation of some standard continuous time problems.[19] Although Nelson (1994) finds that as one approaches a continuous time limit, multivariate ARCH models asymptotically approach stochastic volatility models. It is possible that a one-month interval is not sufficient to approach this limit. Therefore, the recent work of Jacquier *et al.* (1995) may be pertinent in interpreting the results of this chapter. In particular, Jacquier and colleagues find that negative correlations between the mean and variance equation errors for stock returns produce a leverage effect.

[19] Of course, these movements could be picked up by the rolling regressions by choosing a narrower window than 60 months. Please also see, for example, Melino and Turnbull (1990) and the discussion in Jacquier *et al.* (1994).

However, they show that even very high negative correlations could not produce as large asymmetries as those implied by common EGARCH parameter estimates. A full resolution of these issues would require the estimation of stochastic volatility models with time-varying betas.

Our results also indicate that the bivariate EGARCH model may be useful in other contexts, which we have not explored. For example, volatility forecasting at the portfolio level can be improved by using information from *both* past portfolio and market returns. Such forecasts may be valuable for option pricing and hedging purposes. The model is also able to generate improved beta forecasts, which could be useful in dynamic hedging applications. Finally, while we have focused on a one-factor ARCH model, the modelling approach can be extended to allow for additional factors besides the market that drives expected portfolio returns. Such a multi-factor extension of our model is a potential tool for investigating a wide range of practical issues in asset pricing.

ACKNOWLEDGMENTS

We thank Torben G. Andersen, Tim Bollerslev, John Campbell, Phillip Dybvig, Gene Fama, Wayne Ferson, Campbell Harvey, Eric Jacquier, Christopher Lamoureux, Anna Paulson, Peter Rossi, G. William Schwert, Robert Stambaugh, Jeff Wooldridge, participants at the 1990 NBER Summer Institute, workshops at Chicago, Columbia, Harvard, Michigan, and Vanderbilt, and an anonymous referee for helpful comments. The first author thanks the Center for Research in Security Prices for research support.

REFERENCES

Ball, R., and Kothari, S. P. (1989). Nonstationary expected returns: Implications for tests of market efficiency and serial correlation in returns. *Journal of Financial Economics* **25**, 51–74.

Barsky, R. B. (1989). Why don't the prices of stocks and bonds move together? *American Economic Review* **79**, 1132–1145.

Black, F. (1976). Studies of stock market volatility changes. *Proceedings of the American Statistical Association, Business and Economic Statistics Section*, pp. 177–181.

Bollerslev, T., and Wooldridge, J. M. (1992). Quasi-maximum likelihood estimation of dynamic models with time varying covariances. *Econometric Reviews* **11**, 143–172.

Bollerslev, T., Engle, R. F., and Wooldridge, J. M. (1988). A capital asset pricing model with time varying covariance. *Journal of Political Economy* **96**, 116–131.

Bollerslev, T., Engle, R. F., and Nelson, D. (1994). ARCH models. *In* "Handbook of Econometrics" (R. F. Engle and D. L. McFadden, eds.), Vol. 4, Chapter 49, pp. 2959–3038. North-Holland Publ., Amsterdam.

Breeden, D., Gibbons, M. R., and Litzenberger, R. H. (1989). Empirical tests of the consumption-oriented CAPM. *Journal of Finance* **44**, 231–262.

Campbell, J., and Hentschel, L. (1992). No news is good news: An asymmetric model of changing volatility in stock returns. *Journal of Financial Economics* **31**, 281–318.

Chamberlain, G. (1983). Finds, factors, and diversification in arbitrage pricing models. *Econometrica* **51**, 1305–1323.

Chamberlain, G., and Rothschild, M. (1983). Arbitrage, factor structure, and mean-variance analysis of large asset markets. *Econometrica* **51**, 1281–1304.

Chan, K. C. (1988). On the contrarian investment strategy. *Journal of Business* **61**, 147–163.

Christie, A. A. (1982). The stochastic behavior of common stock variances: Value, leverage and interest rate effects. *Journal of Financial Economics* **10**, 407–432.

Cox, J. C., Ingersoll, J. E., and Ross, S. A. (1985). An intertemporal general equilibrium model of asset prices. *Econometrica* **51**, 363–384.

De Bondt, W. F. M., and Thaler, R. H. (1989). Anomalies: A mean-reverting walk down Wall Street. *Journal of Economic Perspectives* **3**, 189–202.

Engle, R. F., and Ng, V. M. (1993). Measuring and testing the impact of news on volatility. *Journal of Finance* **48**, 1749–1778.

Engle, R. F., Ng, V. M., and Rothschild, M. (1990). Asset pricing with a factor-ARCH structure: Empirical estimates for treasury bills. *Journal of Econometrics* **45**, 213–237.

Engle, R. F., Hong, C. H., Kane, A., and Noh, J. (1993). Arbitrage valuation of variance forecasts with simulated options. *In* "Advances in Futures and Options Research" (D. M. Chance and R. R. Trippi, eds.). JAI Press, Greenwich, CT.

Fama, E. F., and MacBeth, J. D. (1973). Risk, return, and equilibrium: Empirical tests. *Journal of Political Economy* **81**, 607–636.

Ferson, W. E. (1989). Changes in expected security returns, risk, and the level of interest rates. *Journal of Finance* **44**, 1191–1218.

Ferson, W. E., and Harvey, C. R. (1991a). The variation of economic risk premiums. *Journal of Political Economy* **99**, 385–415.

Ferson, W. E., and Harvey, C. R. (1991b). Sources of predictability in portfolio returns. *Financial Analysts Journal* **47**(3), 49–56.

Ferson, W. E., and Harvey, C. R. (1993). The risk and predictability of international equity returns. *Review of Financial Studies* **6**, 527–566.

French, K. R., Schwert, G. W., and Stambaugh, R. F. (1987). Expected stock returns and volatility. *Journal of Financial Economics* **19**, 3–29.

Gallant, A. R., Rossi, P. E., and Tauchen, G. (1992). Stock prices and volume. *Review of Financial Studies* **5**, 199–242.

Glosten, L. R., Jagannathan, R., and Runkle, D. (1993). On the relation between the expected value and the volatility of the nominal excess return on stocks. *Journal of Finance* **48**, 1779–1801.

Hamada, R., (1972). The effect of the capital structure on the systematic risk of common stocks. *Journal of Finance* **27**, 435–452.

Hansen, L. P., and Richard, S. (1987). The role of conditioning information in deducing testable restrictions implied by dynamic asset pricing models. *Econometrica* **55**, 587–613.

Jacquier, E., Polson, N. G., and Rossi, P. E. (1994). Bayesian analysis of stochastic volatility models. *Journal of Business & Economic Statistics* **12**, 371–389.

Jacquier, E., Polson, N. G., and Rossi, P. E. (1995). "Priors and Models for Multivariate Stochastic Volatility," Working Paper. Cornell University, Ithaca, NY.

Kodde, D. E., and Palm, F. C. (1986). Wald criteria for jointly testing equality and inequality restrictions. *Econometrica* **54**, 1243–1248.

Malkiel, B. G. (1979). The capital formation problem in the United States. *Journal of Finance* **40**, 677–87.

Melino, A., and Turnbull, S. (1990). Pricing foreign currency options with stochastic volatility. *Journal of Econometrics* **45**, 239–266.

Merrill Lynch, Pierce, Fenner, and Smith, Inc. (1990). "Security Risk Evaluation." Merrill Lynch, New York.

Merton, R. C. (1980). On estimating the expected return on the market. *Journal of Financial Economics* **8**, 323–361.
Nelson, D. B. (1989). Modeling stock market volatility changes. *Proceedings of the American Statistical Association, Business and Economic Statistics Section*, pp. 93–98.
Nelson, D. B. (1991). Conditional heteroskedasticity in asset returns: A new approach. *Econometrica* **59**, 347–370.
Nelson, D. B. (1992). Filtering and forecasting with misspecified ARCH models I: Getting the right variance with the wrong model. *Journal of Econometrics* **52**, 61–90.
Newey, W. K. (1985). Generalized method of moments specification testing. *Journal of Econometrics* **29**, 229–256.
Ng, L. (1991). Tests of the CAPM with time-varying covariances: A multivariate GARCH approach. *Journal of Finance* **46**, 1507–1521.
Pagan, A. R., and Schwert, G. W. (1990). Alternative models for conditional stock volatility. *Journal of Econometrics* **45**, 267–290.
Pindyck, R. S. (1984). Risk, inflation and the stock market. *American Economic Review* **74**, 335–351.
Poterba, T. M., and Summers, L. H. (1986). The persistence of volatility and stock market fluctuations. *American Economic Review* **76**, 1142–1151.
Schwert, G. W. (1989). Why does stock market volatility change over time? *Journal of Finance* **44**, 1115–1154.
Schwert, G. W. (1990). Indexes of U.S. stock prices. *Journal of Business* **63**, 399–426.
Schwert, G. W., and Seguin, P. J. (1990). Heteroskedasticity in stock returns. *Journal of Finance* **45**, 1129–1156.
Shanken, J. (1990). Intertemporal asset pricing: An empirical investigation. *Journal of Econometrics* **45**, 99–120.
Tauchen, G. (1985). Diagnostic testing and evaluation of maximum likelihood models. *Journal of Econometrics* **30**, 415–443.
White, H. (1987). Specification testing in dynamic models. *In* "Advances in Econometrics—Fifth World Congress" (T. Bewley, ed.), Vol. 1. Cambridge University Press, New York.
Wolak, F. (1989a). Local and global testing of linear and nonlinear inequality constraints in nonlinear econometric models. *Econometric Theory* **5**, 1–35.
Wolak, F. (1989b). Testing inequality constraints in linear econometric models. *Journal of Econometrics* **41**, 205–235.
Wooldridge, J. M. (1990). A unified approach to robust, regression-based specification tests. *Econometric Theory* **6**, 17–43.

PART II

CONTINUOUS TIME LIMITS AND OPTIMAL FILTERING FOR ARCH MODELS

5

ARCH MODELS AS DIFFUSION APPROXIMATIONS*

DANIEL B. NELSON

Graduate School of Business
University of Chicago
Chicago, Illinois 60637

1. INTRODUCTION

Many econometric studies (e.g., Officer, 1973; Black, 1976; Engle and Bollerslev, 1986; French *et al.*, 1987) have documented that financial time series tend to be highly heteroskedastic. This has important implications for many areas of macroeconomics and finance, including the term structure of interest rates (e.g., Barsky, 1989; Abel, 1988, irreversible investment (e.g., Bernanke, 1983; McDonald and Siegel, 1986, options pricing (e.g., Wiggins, 1987), and dynamic capital asset pricing theory (e.g., Merton, 1973; Cox *et al.*, 1985). Many of these theoretical models have made extensive use of the Ito stochastic calculus (see, e.g., Liptser and Shiryayev, 1977), which provides an elegant and relatively straightforward means to analyze the properties of many diffusion processes.[1]

Econometricians have also been very active in developing models of conditional heteroskedasticity. The most widely used models of dynamic

* Reprinted with permission from Nelson, D. "ARCH Models as Diffusion Approximations" *Journal of Econometrics* **45** (1990), Elsevier Science SA, Lausanne, Switzerland.

[1] For formal definitions of diffusions and Ito processes, see, e.g., Lipster and Shiryayev (1977) and Arnold (1974).

conditional variance have been the ARCH models first introduced by Engle (1982). In its most general form (see Engle, 1982, eqs. 1–5), a univariate ARCH model makes conditional variance at time t a function of exogenous and lagged endogenous variables, time, parameters, and past residuals. Formally, let e_t be a sequence of (orthogonal) prediction errors, b a vector of parameters, x_t a vector of exogenous and lagged endogenous variables, and σ_t^2 the variance of e_t given information at time t:

$$e_t = \sigma_t Z_t, \tag{1.1}$$

$$Z_t \sim \text{i.i.d. with } E(Z_t) = 0, \quad \text{var}(Z_t) = 1, \tag{1.2}$$

$$\begin{aligned}\sigma_t^2 &= \sigma^2(e_{t-1}, e_{t-2}, \ldots, x_t, t, b) \\ &= \sigma^2(\sigma_{t-1}Z_{t-1}, \sigma_{t-2}Z_{t-2}, \ldots, x_t, t, b).\end{aligned} \tag{1.3}$$

Many different parameterizations for the function $\sigma^2(\cdot, \ldots, \cdot)$ have been used in the literature, including the original ARCH(p) specification of Engle (1982), the GARCH and GARCH-M models of Bollerslev (1986) and Engle and Bollerslev (1986), respectively, the Log–GARCH models of Pantula (1986) and Geweke (1986), and the exponential ARCH model of Nelson (1991).

In contrast to the stochastic differential equation models so frequently found in the theoretical finance literature, ARCH models are discrete time stochastic difference equation systems. It is clear why empiricists have favored the discrete time approach of ARCH: Virtually all economic time series data are recorded only at discrete intervals, and a discrete time ARCH likelihood function is usually easy to compute and maximize. By contrast, the likelihood of a nonlinear stochastic differential equation system observed at discrete intervals can be very difficult to derive, especially when there are unobservable state variables (e.g., conditional variance) in the system.[2]

To date, relatively little work has been done on the relation between the continuous time nonlinear stochastic differential systems, used in so much of the theoretical literature, and the ARCH stochastic difference equation systems, favored by empiricists. Indeed, the two literatures have developed quite independently, with little attempt to reconcile the discrete and continuous models. In this chapter we partially bridge the gap by developing conditions under which ARCH stochastic difference equations systems converge in distribution to Ito processes as the length of the discrete time intervals goes to zero.

[2] But see Lo (1988) and Duffie and Singleton (1989) for some progress in estimating such models.

What do we hope to gain from such an enterprise? First, as we've indicated above, it may be much easier to do estimation and forecasting with a discrete time ARCH model than with a diffusion model observed at discrete intervals. We may therefore, as the title of the paper suggests, want to use ARCH models as diffusion approximations.

Second, in some cases we may find the distributional results are available for the diffusion limit of a sequence of ARCH processes that are not available for the discrete time ARCH processes themselves. In such cases, we may be able to use diffusion processes as ARCH approximations. For example, consider the much used GARCH(1, 1) model of Bollerslev (1986): though Nelson (1990) and Sampson (1988) have recently developed necessary and sufficient conditions for the strict stationarity of the conditional variance process, relatively little is known about the stationary distribution. In continuous time, however, we will see later in the chapter that the stationary distribution for the GARCH(1, 1) conditional variance process is an inverted gamma. This implies that in a conditionally normal GARCH(1, 1) process observed at short time intervals, the innovations process is approximately distributed as a Student's t. In the exponential ARCH model of Nelson (1991), on the other hand, the conditional variance in continuous time is lognormal, so that in the discrete time model when time intervals are short, the stationary distribution of the innovations is approximately a normal–lognormal mixture, as in the model of Clark (1973).

In section 2, we first present conditions developed by Stroock and Varadhan (1979) for a sequence of stochastic difference equations to converge weakly to an Ito process, and then present an alternate, somewhat simpler, set of conditions. We apply these results to find and analyze the diffusion limit of the GARCH(1, 1) process. In section 3, we introduce a class of ARCH diffusion approximations, and show that they can approximate quite a wide variety of Ito processes. Two examples are provided, the first based on the AR(1) exponential ARCH model of Nelson (1989) and the second a model with the same diffusion limit as GARCH(1, 1). Section 4 is a brief conclusion. Appendix A summarizes some regularity conditions needed for the convergence results in section 2, and all proofs are in appendix B.

2. CONVERGENCE OF STOCHASTIC DIFFERENCE EQUATIONS TO STOCHASTIC DIFFERENTIAL EQUATIONS

In this section we present general conditions for a sequence of finite-dimensional discrete time Markov processes $\{_hX_t\}_{h\downarrow 0}$ to converge weakly to an Ito process. As we noted in the Introduction, these are drawn largely from Stroock and Varadhan (1979). Kushner (1984) and Ethier and Kurtz

(1986) have extended these convergence theorems to jump–diffusion processes, but it would be beyond the scope of this paper to deal with this more general case.

2.1. The Main Convergence Result

Our formal setup is as follows: Let $D([0, \infty), R^n)$ be the space of mappings from $[0, \infty)$ into R^n that are continuous from the right with finite left limits, and let $B(R^n)$ denote the Borel sets on R^n. D is a metric space when endowed with the Skorohod metric (Billingsley, 1968). For each $h > 0$, let M_{kh} be the σ–algebra generated by $kh, {}_hX_0, {}_hX_h, {}_hX_{2h}, \ldots, X_{kh}$, and let ν_h be a probability measure on $(R^n, B(R^n))$. For each $h > 0$ and each $k = 0, 1, 2, \ldots,$ let $\Pi_{h,kh}(x, \cdot)$ be a transition function on R^n, i.e.,

(a) $\Pi_{h,kh}(x, \cdot)$ is a probability measure on $(R^n, B(R^n))$ for all $x \in R^n$,
(b) $\Pi_{h,kh}(\cdot, \Gamma)$ is $B(R^n)$ measurable for all $\Gamma \in B(R^n)$.

For each $h > 0$, let P_h be the probability measure on $D([0, \infty), R^n)$ such that

$$\mathrm{P}_h[{}_hX_0 \in \Gamma] = \nu_h(\Gamma) \quad \text{for any} \quad T \in B(R^n), \tag{2.1}$$

$$\mathrm{P}_h[{}_hX_t = {}_hX_{kh}, kh \le t < (k+1)h] = 1, \tag{2.2}$$

$$\mathrm{P}_h[{}_hX_{(k+1)h} \in \Gamma \mid M_{kh}] = \Pi_{h,kh}({}_hX_{kh}, \Gamma)$$

$$\text{almost surely under } \mathrm{P}_h \quad \text{for all } k \ge 0 \quad \text{and} \quad \Gamma \in B(R^n). \tag{2.3}$$

For each $h > 0$, (2.1) specifies the distribution of the random starting point and (2.3) the transition probabilities of the n-dimensional discrete time Markov process ${}_hX_{kh}$. We form the continuous time process ${}_hX_t$ from the discrete time process ${}_hX_{kh}$ by (2.2), making ${}_hX_t$ a step function with jumps at times h, $2h$, $3h$, and so on. In this somewhat complicated setup, our notation must keep track of three distinct kinds of processes:

(a) The sequence of discrete time processes $\{{}_hX_{kh}\}$ that depend both on h and on the (discrete) time index kh, $k = 0, 1, 2, \ldots,$
(b) The sequence of continuous time processes $\{{}_hX_t\}$ formed as step functions from the discrete time process in (a) using (2.2) (this process also depends on h and on a (continuous) time index $t, t \ge 0$),
(c) A limiting diffusion process X_t to which, under conditions given below, the sequence of processes $\{{}_hX_t\}_{h \downarrow 0}$ weakly converges.

To accommodate these different processes, we indicate dependence on h to the lower left of X and dependence on the time index to the lower right.

Next, let A' denote the transpose of the matrix A and define the vector/matrix form $\|A\|$ as

$$\|A\| = \begin{cases} [A'A]^{1/2} & \text{when } A \text{ is a column vector,} \\ [\text{trace}(A'A)]^{1/2} & \text{when } A \text{ is a matrix.} \end{cases} \tag{2.4}$$

For each $h > 0$ and each $\varepsilon > 0$, define

$$a_h(x,t) \equiv h^{-1} \int_{\|y-x\| \le 1} (y-x)(y-x)' \Pi_{h,h[t/h]}(x, \mathrm{d}y), \tag{2.5}$$

$$b_h(x,t) \equiv h^{-1} \int_{\|y-x\| \le 1} (y-x) \Pi_{h,h[t/h]}(x, \mathrm{d}y), \tag{2.6}$$

$$\Delta_{h,\varepsilon}(x,t) \equiv h^{-1} \int_{\|y-x\| \ge \varepsilon} \Pi_{h,h[t/h]}(x, \mathrm{d}y), \tag{2.7}$$

where $[t/h]$ is the integer part of t/h, i.e., the largest integer $k \le t/h$.

The integration in (2.5)–(2.6) is taken over $\|y - x\| \le 1$ rather than over R^n because the usual conditional moments may not be finite.[3] $a_h(x,t)$ and $b_h(x,t)$ are measures of the (truncated) second moment and drift per unit of time, respectively. The convergence results we present below will require that $a_h(x,t)$ and $b_h(x,t)$ converge to finite limits, and that $\Delta_{h,\varepsilon}(x,t)$ goes to zero for all $\varepsilon > 0$. $\Delta_{h,\varepsilon}(x,t)$ is a measure of the probability per unit of time of a jump of size ε or greater. Since diffusion processes have sample paths that are continuous with probability one, it should not be surprising that the probability of discrete jumps of any fixed size ε ($\varepsilon > 0$) or greater must go to zero.

The first convergence results of this section will require the following assumptions:

ASSUMPTION 1. *There exists a continuous, measurable mapping $a(x,t)$ from $R^n \times [0,\infty)$ into the space of $n \times n$ nonnegative definite symmetric matrices and a continuous, measurable mapping $b(x,t)$ from $R^n \times [0,\infty)$ into R^n such that for all $R > 0$, $T > 0$, and $\varepsilon > 0$*

$$\lim_{h \downarrow 0} \sup_{\|x\| \le R, 0 \le t \le T} \|a_h(x,t) - a(x,t)\| = 0, \tag{2.8}$$

$$\lim_{h \downarrow 0} \sup_{\|x\| \le R, 0 \le t \le T} \|b_h(x,t) - b(x,t)\| = 0, \tag{2.9}$$

$$\lim_{h \downarrow 0} \sup_{\|x\| \le R, 0 \le t \le T} \Delta_{h,\varepsilon}(x,t) = 0. \tag{2.10}$$

[3] For example, $X_t = \exp[\exp(W_t)]$, with W_t a Brownian motion, is a diffusion but has no moments of any order over any time interval of positive length.

Equations (2.8) and (2.9) require that the second moment and drift per unit of time converge uniformly on compact sets to well-behaved functions of time and the state variables x. (2.10) requires that the probability per unit of time of a jump of size ε or greater vanishes uniformly on compacts for all $\varepsilon > 0$, so the sample paths of the limit process are continuous with probability one.

ASSUMPTION 2. *There exists a continuous, measurable mapping $\sigma(x,t)$ from $R^n \times [0,\infty)$ into the space of $n \times n$ matrices, such that for all $x \in R^n$ and all $t > 0$,*

$$a(x,t) = \sigma(x,t)\sigma(x,t)'. \tag{2.11}$$

Assumption 2 requires that the function $a(x,t)$, the second moment per unit of time of the limit process, has a well-behaved matrix square root $\sigma(x,t)$. The next assumption requires the probability measures ν_h of the random starting points ${}_hX_0$ to converge to a limit measure ν_0 as $h \downarrow 0$:

ASSUMPTION 3. *As $h \downarrow 0$, ${}_hX_0$ converges in distribution to a random variable X_0 with probability measure ν_0 on $(R^n, B(R^n))$.*

We have specified a probability measure ν_0 for the initial value of the limit process X_t, an instantaneous drift function $b(x,t)$, an instantaneous covariance[4] matrix $a(x,t)$, and have guaranteed that the limit process, if it exists, will have sample paths that are continuous with probability one. At this point, there are two things that can go wrong: first, a limit process may not exist, because when taken together ν_0, $a(x,t)$, and $b(x,t)$ may imply that the process explodes (with positive probability) to infinity in finite time. Second, ν_0, $a(x,t)$, and $b(x,t)$ may not *uniquely* define a limit process. For example, we may need additional information in the form of boundary conditions.[5] There is an extensive literature on conditions under which ν_0, $a(x,t)$, and $b(x,t)$ uniquely define a limiting diffusion (e.g., see Stroock and Varadhan, 1979). Appendix A summarizes a few of these conditions.

ASSUMPTION 4. *ν_0, $a(x,t)$, and $b(x,t)$ uniquely specify the distribution of a diffusion process X_t, with initial distribution ν_0, diffusion matrix $a(x,t)$, and drift vector $b(x,t)$.*

[4] As $h \downarrow 0$, the difference between the conditional covariance matrix and the conditional second moment matrix vanishes, so that in the continuous time limit the two are identical.

[5] For examples of explosion and nonuniqueness, see Stroock and Varadhan (1979, sect. 10.0 and problem 6.7.7).

THEOREM 2.1. *Under Assumptions* 1 *through* 4, *the sequence of* ${}_hX_t$ *processes defined by* (2.1)–(2.3) *converges weakly* (*i.e., in distribution*) *as* $h \downarrow 0$ *to the* X_t *process defined by the stochastic integral equation*

$$X_t = X_0 + \int_0^t b(X_s, s)\, ds + \int_0^t \sigma(X_s, s)\, dW_{n,s}, \tag{2.12}$$

where $W_{n,t}$ *is an n-dimensional standard Brownian motion,*[6] *independent of* X_0, *and where for any* $\Gamma \in B(R^n)$, $\mathrm{P}(X_0 \in \Gamma) = \nu_0(\Gamma)$. *Such an* X_t *process exists and is distributionally unique. This distribution does not depend on the choice of* $\sigma(\cdot,\cdot)$ *made in Assumption* 2. *Finally,* X_t *remains finite in finite time intervals almost surely, i.e., for all* $T > 0$,

$$\mathrm{P}\left[\sup_{0 \le t \le T} \|X_t\| < \infty\right] = 1. \tag{2.13}$$

The convergence in distribution in Theorem 2.1 is not merely a convergence in distribution of $\{{}_hX_t\}$ for a fixed value of t: rather for any T, $0 \le T < \infty$, the probability laws generating the *entire sample paths* $\{{}_hX_t\}$, $0 \le t \le T$, converge to the probability law generating the sample path of X_t, $0 \le t \le T$.[7] From this point on, we will denote this type of convergence in distribution by the symbol $\Rightarrow$, while $\xrightarrow{\text{d}}$ will denote convergence in distribution of random variables in R^1 or R^n.

2.2. An Alternate Set of Conditions

Assumption 1 is cumbersome to verify, since the integrals are taken over $\|y - x\| \le 1$ and $\|y - x\| \ge \varepsilon$. This somewhat limits the usefulness of Theorem 2.1. We therefore introduce a simpler, though somewhat less general, alternative to Assumption 1. First, for each $i, i = 1, \ldots, n$, each $\delta > 0$, and each $h > 0$, define

$$c_{h,i,\delta}(x,t) \equiv h^{-1} \int_{R^n} |(y - x)_i|^{2+\delta}\, \Pi_{h,h[t/h]}(x, dy), \tag{2.14}$$

[6] That is, $W_{n,t}$ is an n-dimensional Brownian motion with, for all t, $E[W_{n,t}] = W_{n,0} = 0_n$ and $dW_{n,t}\, dW'_{n,t} = I_{n,n}\, dt$, where $I_{n,n}$ is an $n \times n$ identity matrix.

[7] Weak convergence implies, for example, that given any times $t_1, t_2, \ldots, t_n > 0$, the joint distributions of $\{{}_hX_{t_1}, {}_hX_{t_2}, \ldots, {}_hX_{t_n}\}$ converge to the distribution of $(X_{t_1}, X_{t_2}, \ldots, X_{t_n})$ as $h \downarrow 0$. More generally, weak convergence implies that if $f(\cdot)$ is a continuous functional of the sample path of ${}_hX_t$, then $f({}_hX_t)$ converges in distribution to $f(X_t)$ as $h \downarrow 0$. See Billingsley (1968).

where $(y - x)_i$ is the ith element of the vector $(y - x)$. If for some $\delta > 0$, and all $i, i = 1, \ldots, n$, $c_{h,i,\delta}(x,t)$ is finite, then the following integrals will be well-defined and finite:

$$a_h^*(x,t) \equiv h^{-1} \int_{R^n} (y-x)(y-x)' \Pi_{h,h[t/h]}(x, dy), \tag{2.15}$$

$$b_h^*(x,t) \equiv h^{-1} \int_{R^n} (y-x) \Pi_{h,h[t/h]}(x, dy). \tag{2.16}$$

Like $a(\cdot,\cdot)$ and $b(\cdot,\cdot)$, $a^*(\cdot,\cdot)$ and $b^*(\cdot,\cdot)$ are measures of the second moment and drift per unit of time, with the difference that the integrals are taken over R^n instead of $\|y - x\| \leq 1$.

ASSUMPTION 5. *There exists a $\delta > 0$ such that for each $R > 0$, each $T > 0$, and each $i = 1, \ldots, n$,*

$$\lim_{h \downarrow 0} \sup_{\|x\| \leq R, 0 \leq t \leq T} c_{h,i,\delta}(x,t) = 0. \tag{2.17}$$

Further, there exists a continuous mapping $a(x,t)$ from $R^n \times [0,\infty)$ into the space of $n \times n$ nonnegative definite symmetric matrices and a continuous mapping $b(x,t)$ from $R^n \times [0,\infty)$ into R^n such that

$$\lim_{h \downarrow 0} \sup_{\|x\| \leq R, 0 \leq t \leq T} \|a_h^*(x,t) - a(x,t)\| = 0, \tag{2.18}$$

$$\lim_{h \downarrow 0} \sup_{\|x\| \leq R, 0 \leq t \leq T} \|b_h^*(x,t) - b(x,t)\| = 0. \tag{2.19}$$

THEOREM 2.2. *Under Assumptions 2 through 5, the conclusions of Theorem 2.1 hold.*[8]

Theorems 2.1 and 2.2 provide a relatively simple way to prove convergence to an Ito process limit. In essence, Theorem 2.2 says that if the drift and second moment of ${}_hX_{kh}$ per unit of time converge to well-behaved limits, and if the first differences of ${}_hX_{kh}$ have an absolute moment higher than two that collapses to zero at an appropriate rate as $h \downarrow 0$, then the ${}_hX_t$ processes converge in distribution to the solution of the stochastic integral equation system (2.12).

[8] This theorem was inspired by a standard result for diffusions (as opposed to a sequence of discrete time Markov processes). See Arnold (1974, p. 40).

2.3. An Example: GARCH(1, 1)-M[9]

In discrete time, the GAUCH(1, 1)-M process of Engle and Bollerslev (1986) for the log of cumulative excess returns Y_t on a portfolio is

$$Y_t = Y_{t-1} + c\sigma_t^2 + \sigma_t Z_t, \tag{2.20}$$

$$\times \sigma_{t+1}^2 + \omega + \sigma_t^2[\beta + \alpha Z_t^2], \tag{2.21}$$

where $\{Z_t\} \sim$ i.i.d. $N(0, 1)$.

Now we consider the properties of the stochastic difference equation system (2.20)–(2.21) as we partition time more and more finely. We allow the parameters of the system α, β, and ω to depend on h, and make both the drift term in (2.20) and the variance of Z_t proportional to h:

$${}_hY_{kh} = {}_hY_{(k-1)h} + h \cdot c \cdot {}_h\sigma_{kh}^2 + {}_h\sigma_{kh} \cdot {}_hZ_{kh}, \tag{2.22}$$

$${}_h\sigma_{(k+1)}^2 = \omega_h + {}_h\sigma_{kh}^2[\beta_h + h^{-1}\alpha_h \cdot {}_hZ_{kh}^2], \tag{2.23}$$

and

$$\mathrm{P}[({}_hY_{0,h}\sigma_0^2) \in \Gamma] = \nu_h(\Gamma) \quad \text{for any} \quad \Gamma \in B(R^2), \tag{2.24}$$

where $\{_hZ_{kh}\} \sim$ i.i.d N(0, h). We also assume that the sequence of measures $\{\nu_h\}_{h \downarrow 0}$ satisfies Assumption 3, and that for each

$$h \geq 0,\ \nu_h((Y_0, \sigma_0^2)\colon \sigma_0^2 > 0) = 1.$$

Finally, we create the continuous time processes ${}_hY_t$ and ${}_h\sigma_t^2$ by

$${}_hY_t \equiv {}_hT_{kh} \quad \text{and} \quad {}_h\sigma_t^2 \equiv {}_h\sigma_{kh}^2 \quad \text{for} \quad kh \leq t < (k+1)h. \tag{2.25}$$

We allow ω, α, and β to depend on h because our object is to discover which sequences $\{\omega_h, \alpha_h, \beta_h\}$ make the $\{_h\sigma_t^2, {}_hY_t\}$ process (in which ${}_h\sigma_t^2$ now represents variance per unit of time and $c \cdot {}_h\sigma_t^2$ the risk premium per unit of time) converge in distribution to an Ito process limit as h goes to zero. For all h, however, we require that ω_h, α_h, and β_h be nonnegative, which ensures that the σ_t^2 process remains positive with probability one (Bollerslev, 1986).

The discrete time process (2.22)–(2.24) is clearly Markovian, and the drift per unit of time [conditioned on information at time $(k-1)h$] is

[9] Nelson (1988a, ch. 3) noted that no conditionally heteroskedastic diffusion limit had previously been found for GARCH processes, and boldly conjectured that none was possible. Unfortunately, this bold conjecture turned out to be wrong, and this chapter corrects the error. As Sandburg (1969) said: "I can eat crow, but I don't hanker after it."

given by

$$\mathrm{E}\left[h^{-1}\left({}_hY_{kh} - {}_hY_{(k-1)h}\right) \mid M_{kh}\right] = c \cdot {}_h\sigma_{kh}^2, \tag{2.26}$$

$$\mathrm{E}\left[h^{-1}\left({}_h\sigma_{(k+1)h}^2 - {}_h\sigma_{kh}^2\right) \mid M_{kh}\right] = h^{-1}\omega_h + h^{-1}(\beta_h + \alpha_h - 1)\,{}_h\sigma_{kh}^2, \tag{2.27}$$

where M_{kh} is the σ-algebra generated by kh, ${}_hY_0, \ldots, {}_hY_{(k-1)h}$, and ${}_h\sigma_0^2, \ldots, {}_h\sigma_{kh}^2$. For the drift per unit of time to converge as required by Assumption 5, the limits

$$\lim_{h\downarrow 0} h^{-1}\omega_h = \omega \geq 0, \tag{2.28}$$

$$\lim_{h\downarrow 0} h^{-1}(1 - \beta_h - \alpha_h) = \theta, \tag{2.29}$$

must exist and be finite. Note that ω is required to be nonnegative, whereas α can be of either sign.

The second moment per unit of time is given by

$$\mathrm{E}\left[\left({}_h\sigma_{(k+1)h}^2 - {}_h\sigma_{kh}^2\right)^2/h \mid M_{kh}\right] = h^{-1}\omega_h^2 + 2h^{-1}\omega_h(\beta_h + \alpha_h - 1)\,{}_h\sigma_{kh}^2$$
$$+ h^{-1}(\beta_h + \alpha_h - 1)^2\,{}_h\sigma_{kh}^4 + 2h^{-1}\alpha_h^2\,{}_h\sigma_{kh}^4, \tag{2.30}$$

$$\mathrm{E}\left[h^{-1}\left({}_hY_{kh} - {}_hY_{(k-1)h}\right)^2 \mid M_{kh}\right] = hc^2\,{}_h\sigma_{kh}^4 + {}_h\sigma_{kh}^2, \tag{2.31}$$

$$\mathrm{E}\left[h^{-1}\left({}_hY_{kh} - {}_hY_{(k-1)h}\right)\left({}_h\sigma_{(k+1)h}^2 - {}_h\sigma_{kh}^2\right) \mid M_{kh}\right]$$
$$= c \cdot {}_h\sigma_{kh}^2 \cdot \omega_h + c \cdot {}_h\sigma_{kh}^4(\beta_h + \alpha_h - 1). \tag{2.32}$$

Substituting from (2.28)–(2.29) into (2.30)–(2.32) and assuming that

$$\lim_{h\downarrow 0} 2h^{-1}\alpha_h^2 = \alpha^2 > 0, \tag{2.33}$$

exists and is finite, we have

$$\mathrm{E}\left[h^{-1}\left({}_h\sigma_{(k+1)h}^2 - {}_h\sigma_{kh}^2\right)^2 \mid M_{kh}\right] = \alpha_h^2\sigma_{kh}^4 + o(1), \tag{2.34}$$

$$\mathrm{E}\left[h^{-1}\left({}_hY_{kh} - {}_hY_{(k-1)h}\right)^2 \mid M_{kh}\right] = {}_h\sigma_{kh}^2 + o(1), \tag{2.35}$$

$$\mathrm{E}\left[h^{-1}\left({}_hY_{kh} - {}_hY_{(k-1)h}\right)\left({}_h\sigma_{(k+1)h}^2 - {}_h\sigma_{kh}^2\right) \mid M_{kh}\right] = o(1), \tag{2.36}$$

where the $o(1)$ terms vanish uniformly on compact sets.

It is not difficult to find $\{\omega_h, \alpha_h, \beta_h\}$ sequences satisfying (2.28), (2.29), and (2.33). For example, set $\omega_h = \omega h$, $\beta_h = 1 - \alpha(h/2)^{1/2} - \theta h$, and $\alpha_h = \alpha(h/2)^{1/2}$. It is also straightforward though tedious to verify that the limits of

$$\mathrm{E}\Big[h^{-1}\big({}_hY_{kh} - {}_hY_{(k-1)/h}\big)^4 \mid M_{kh}\Big] \quad \text{and} \quad \mathrm{E}\Big[h^{-1}\big({}_h\sigma^2_{(k+1)h} - {}_h\sigma^2_{kh}\big)^4 \mid M_{kh}\Big]$$

exist and converge to zero, so that with $\delta = 2$,

$$b(Y, \sigma^2) \equiv \begin{bmatrix} c \cdot \sigma^2 \\ \omega - \theta\sigma^2 \end{bmatrix}, \tag{2.37}$$

$$a(Y, \sigma^2) \equiv \begin{bmatrix} \sigma^2 & 0 \\ 0 & \alpha^2\sigma^4 \end{bmatrix}, \tag{2.38}$$

and with α_h, β_h, and ω_h satisfying (2.28), (2.29), and (2.33), Assumption 5 holds. Setting $\sigma(\cdot,\cdot)$ in Assumption 2 equal to the element-by-element square root of $a(Y, \sigma^2)$, Assumption 2 holds as well. (2.37)–(2.38) suggest a limit diffusion of the form

$$dY_t = c \cdot \sigma_t^2 \, dt = \sigma_t \, dW_{1,t}, \tag{2.39}$$

$$d\sigma_t^2 = \big(\omega - \theta\sigma_t^2\big)\, dt + \alpha\sigma_t^2 \, dW_{2,t}, \tag{2.40}$$

$$\mathrm{P}\big[\big(Y_0, \sigma_0^2\big) \in \Gamma\big] = \nu_0(\Gamma) \quad \text{for any} \quad \Gamma \in B(R^2), \tag{2.41}$$

where $W_{1,t}$ and $W_{2,t}$ are independent standard Brownian motions, independent of the initial values (Y_0, σ_0^2).

Unfortunately, none of the Conditions A–D in appendix A directly apply to this diffusion. To verify distributional uniqueness, it helps to define $V_t = \ln(\sigma_t^2)$ and apply Ito's Lemma to (2.39)–(2.41):

$$dY_t = c \cdot \exp(V_t)\, dt + \exp(V_t/2)\, dW_{1,t}, \tag{2.39$'$}$$

$$dV_t = \big[\omega \cdot \exp(-V_t) - \theta - \alpha^2/2\big]\, dt + \alpha \, dW_{2,t}, \tag{2.40$'$}$$

$$P\big[\big(Y_0, \exp(V_0)\big) \in \Gamma\big] = \nu_0(\Gamma) \quad \text{for any} \quad \Gamma \in B(R^2). \tag{2.41$'$}$$

It is easy to check that Condition B and the nonexplosion condition in appendix A hold for (2.39′)–(2.41′).[10] Since distributional uniqueness holds for (2.39′)–(2.41′), it must by the Continuous Mapping Theorem (Billingsley, 1968) holds for (2.39)–(2.41) as well.[11] We now apply Theorem 2.2 to conclude that $\{{}_hY_{t}, {}_h\sigma_t^2\} \Rightarrow (Y_t, \sigma_t^2)$ as $h \downarrow 0$.

[10] For the Liapunov function $\varphi(Y, V)$ in the nonexplosion condition, use $\varphi(Y, V) \equiv K + f(Y) \cdot |Y| + f(V) \cdot \exp(|V|)$, where $f(x) \equiv \exp(-1/|x|)$ if $x \neq 0$ and $\equiv 0$ otherwise. $\varphi(\cdot,\cdot)$ is arbitrarily differentiable, despite the presence of the $|V|$ and $|Y|$ terms. To verify the inequality (A.7), use the fact that for large Y and V, $\partial\varphi(Y, V)/\partial Y \approx \operatorname{sign}(Y)$, $\partial^2\varphi(Y, V)/\partial Y^2 = 0$, $\partial\varphi(Y, V)/\partial V \approx \operatorname{sign}(V) \cdot \exp(|V|)$, and $\partial^2\varphi(Y, V)/\partial V^2 \approx \exp(|V|)$.

[11] That is, as long as $\nu_h((Y_0, \sigma_0^2)\colon \sigma_0^2 > 0) = 1$, which we have already assumed. Using a simple Taylor series argument, it is also easy to check that Assumption 1 holds for the transformed variables ${}_hV_{kh} \equiv \ln({}_h\sigma_{kh}^2)$ and ${}_hY_{kh}$: i.e., the jump sizes go to zero and the local drifts and second moments converge to the limits implied by (2.39′)–(2.41′).

While no closed form exists for the stationary distribution of GARCH(1, 1) in discrete time, an application of the results of Wong (1964) allows us to solve for the stationary distribution of σ_t^2 implied by (2.40):

THEOREM 2.3. *For each $t \geq 0$, define the conditional precision process $\lambda_t \equiv \sigma_t^{-2}$, where σ_t^2 is generated by the stochastic differential equation system* (2.39)–(2.41). *Then if $2\theta/\alpha^2 > -1$ and $\omega > 0$,*

$$\lambda_t \xrightarrow{d} \Gamma(1 + 2\theta/\alpha^2, 2\omega/\alpha^2) \quad \text{as} \quad t \to \infty, \tag{2.42}$$

where $\Gamma(\cdot, \cdot)$ is the Gamma distribution.[12,13] *If $\lambda_0 \sim \Gamma(1 + 2\theta/\alpha^2, 2\omega/\alpha^2)$, then the σ_t^2 and λ_t processes generated by* (2.39)–(2.41) *are strictly stationary, and for all $t \geq 0$,*

$$\lambda_t \sim \Gamma(1 + 2\theta/\alpha^2, 2\omega/\alpha^2). \tag{2.43}$$

If (a) the distribution of ${}_h\lambda_0 \equiv {}_h\sigma_0^{-2}$ converges to a $\Gamma(1 + 2\theta/\alpha^2, 2\omega/\alpha^2)$ as $h \downarrow 0$, (b) the sequence $\{\omega_h, \alpha_h, \beta_h\}_{h \downarrow 0}$ satisfies (2.28), (2.29), *and* (2.31), *and (c) $2\theta/\alpha^2 > -1$, and $\omega > 0$, then in the discrete time system* (2.22)–(2.24),

$$ {}_h\sigma_{kh}^{-2} \xrightarrow{d} \Gamma(1 + 2\theta/\alpha^2, 2\omega/\alpha^2), \tag{2.44}$$

$$h^{-1/2}{}_hZ_{kh} \cdot {}_h\sigma_{kh} \cdot \left[(2\theta + \alpha^2)/2\omega\right]^{1/2} \xrightarrow{d} t(2 + 4\theta/\alpha^2), \tag{2.45}$$

for any constant value of kh as $h \downarrow 0$, where $t(2 + 4\theta/\alpha^2)$ is the Student t distribution with $2 + 4\theta/\alpha^2$ degrees of freedom.

Finally, if (b) and (c) are satisfied and in addition there exists a $d > 0$ such that $\limsup E[{}_h\sigma_0^{2d}] < \infty$, *then*

$$ {}_h\sigma_{kh}^{-2} \xrightarrow{d} \Gamma(1 + 2\theta/\alpha^2, 2\omega/\alpha^2), \tag{2.46}$$

$$h^{-1/2}{}_hZ_{kh} \cdot {}_h\sigma_{kh} \cdot \left[(2\theta + \alpha^2)/2\omega\right]^{1/2} \xrightarrow{d} t(2 + 4\theta/\alpha^2), \tag{2.47}$$

as $h \downarrow 0$ and $kh \to \infty$.[14]

[12] $x \sim \Gamma(r, s)$ means that the probability density function for x is given by $f(x) = s^r \cdot x^{(r-1)} \cdot \exp(-sx)/\Gamma(r)$ for $x > 0$, where $\Gamma(\cdot)$ is the gamma function.

[13] When $\omega = 0$, we have $d[\ln(\sigma_t^2)] = -(\alpha^2/2 + \theta)\,dt + \alpha\,dW_t$, so that $\sigma_t^2 \to 0$, $\sigma_t^2 \to \infty$, or $\ln(\sigma_t^2)$ is a driftless Brownian motion as $(2\theta/\alpha^2 + 1) > 0$, < 0, or $= 0$, respectively. This is the continuous time analog to the results of Nelson (1988b) for GARCH(1, 1) in discrete time. The stationarity condition $2\theta/\alpha^2 > -1$ is also the continuous time analog of the discrete time stationarity condition in Nelson (1990).

[14] Unfortunately, Theorems 2.1 and 2.2 do not allow us to simultaneously take $t \to \infty$ and $h \downarrow 0$, so that (2.46) and (2.47) do not follow directly from Theorem 2.1. However, Kushner (1984, ch. 6) gives conditions for the steady state (i.e., $t = \infty$) distributions of the ${}_hX_t$ processes to converge to the steady state distribution of X_t. In appendix B, Kushner's results are employed to prove (2.46) and (2.47).

Theorem 2.3 tells us that although the innovations process ${}_hZ_{kh} \cdot {}_h\sigma_{kh}$ is conditionally normal, its unconditional distribution is approximately Student t when h is small and kh is large. The degrees of freedom term $(2 + 4\theta/\alpha^2)$ is also instructive: the Student t distribution has a finite variance when it has more than two degrees of freedom, which it does in the GARCH(1, 1) limit case only when $\theta > 0$. The case $\theta = 0$ corresponds to the diffusion limit of the IGARCH(1, 1) model of Engle and Bollerslev (1986), which separates the covariance stationary GARCH (1, 1) case ($\theta > 0$) from the strictly stationary but not covariance stationary GARCH(1, 1) model of Nelson (1990) and Sampson (1988) ($\theta < 0$). In discrete time, the condition for the strict stationarity of the conditionally normal GARCH(1, 1) is the $\mathrm{E}[\ln(\beta_h + \alpha_h Z^2)] < 0$, $Z \sim N(0, 1)$ (Nelson, 1990), whereas in continuous time this requirement is that $(2 + 4\theta/\alpha^2) > 0$ (i.e., the t distribution has strictly positive, though not necessarily integer, degrees of freedom).

3. A CLASS OF UNIVARIATE ITO PROCESS APPROXIMATIONS

In this section we develop a class of ARCH processes based on the exponential ARCH model of Nelson (1991), and use Theorem 2.2 to show how these processes can approximate a wide variety of stochastic differential equations.

3.1. Exponential ARCH

In a conditionally normal exponential ARCH model, we have an innovations process ε_t and its conditional variance σ_t^2 given by

$$\varepsilon_t = \sigma_t Z_t, \tag{3.1}$$

$$Z_t \sim \text{i.i.d. N}(0, 1), \tag{3.2}$$

$$\ln(\sigma_t^2) = \alpha_t + \sum_{k=1}^{\infty} \beta_k g(Z_{t-k}), \qquad \beta_1 \equiv 1, \tag{3.3}$$

where $\{\alpha_t\}_{t=-\infty,\infty}$ and $\{\beta_k\}_{k=1,\infty}$ are real, nonstochastic scalar sequences, and

$$g(Z_t) \equiv \theta Z_t + \gamma\left[|Z_t| - (2/\pi)^{1/2}\right], \tag{3.4}$$

i.e., the log of σ_t^2 is linear in lagged Z_t's, $|Z_t|$'s, and some function of time. $g(Z_t)$ is a zero mean, i.i.d. innovations term, with two zero mean, serially and contemporaneously uncorrelated components, θZ_t and $\gamma[|Z_t| - (2/\pi)^{1/2}]$. The two components allow any desired degree of conditional correlation between Z_t and the change in $\ln(\sigma_t^2)$. In GARCH models, this conditional correlation is always fixed at zero. Black (1976) noted that changes in the level of the stock market are negatively correlated with changes in its volatility. The two terms in $g(\cdot)$ allow exponential ARCH to capture this correlation in a way that GARCH cannot, since in GARCH models the change in σ_t^2 is driven by Z_t^2, which is uncorrelated with Z_t.

3.2. ARCH Diffusion Approximations

The class of diffusion approximations that we now propose is a variant of the exponential ARCH model (3.1)–(3.4). Like exponential ARCH, this class of models uses Z_t and $|Z_t|$ as the driving noise processes. Unlike exponential ARCH, however, the log–linearity assumption for σ_t^2 is relaxed, while a finite-order Markov representation for the system is imposed, ruling out, for example, a fractionally differenced (Hosking, 1981) representation for $\ln(\sigma_t^2)$.

First, define the stochastic differential equation system

$$dS_t = f(S_t, Y_t, t)\, dt + g(S_t, Y_t, t)\, dW_{1,t}, \tag{3.5}$$

$$dY_t = F(S_t, Y_t, t)\, dt + G(S_t, Y_t, t)\, dW_{2,t}, \tag{3.6}$$

$$\begin{bmatrix} dW_{1,t} \\ dW_{2,t} \end{bmatrix} [dW_{1,t}\; dW_{2,t}'] = \begin{bmatrix} 1 & \Omega_{1,2} \\ \Omega_{2,1} & \Omega_{2,2} \end{bmatrix} dt \equiv \Omega\, dt, \tag{3.7}$$

where Ω is an $(n+1) \times (n+1)$ positive semi-definite matrix of rank two or less, Y is an n-dimensional vector of (unobservable) state variables, S is an (observable) scalar process, W_1 is a one-dimensional standard Brownian motion, and W_2 is an n-dimensional Brownian motion. $f(\cdot,\cdot,\cdot)$ and $g(\cdot,\cdot,\cdot)$ are real-valued, continuous scalar functions, and $F(\cdot,\cdot,\cdot)$ and $G(\cdot,\cdot,\cdot)$ are real, continuous $n \times 1$ and $n \times n$ valued functions, respectively. We also assume that (S_0, Y_0) are random variables with joint probability measure ν_0, and are independent of the W_1 and W_2 processes. Define the vector and matrix functions b and a by

$$b(s, y, t) \equiv [f(s, y, t) F(s, y, t)']', \tag{3.8}$$

$$a(s, y, t) = \begin{bmatrix} g^2 & g\Omega_{1,2}G' \\ G\Omega_{2,1}g & G\Omega_{2,2}G' \end{bmatrix}. \tag{3.9}$$

$b(\cdot,\cdot,\cdot)$ is an $(n+1) \times 1$ vector and $a(\cdot,\cdot,\cdot)$ is an $(n+1) \times (n+1)$ matrix.

Next we propose a sequence of approximating processes that converge to (3.5)–(3.7) in distribution as $h \downarrow 0$. We model ${}_hS_t$ and ${}_hY_t$ as step functions with jumps at times h, $2h$, $3h$, and so on. The jumps are given by

$$ {}_hS_{(k+1)h} = {}_hS_{kh} + f({}_hS_{kh}, {}_hY_{kh}, kh)h + g({}_hS_{kh}, {}_hY_{kh}, kh)\,{}_hZ_{kh}, \tag{3.10} $$

$$ {}_hY_{(k+1)h} = {}_hY_{kh} + F({}_hS_{kh}, {}_hY_{kh}, kh)h + G({}_hS_{kh}, {}_hY_{kh}, kh)\,{}_hZ^*_{kh}, \tag{3.11} $$

where

$$ {}_hZ_{kh} \sim \text{i.i.d. N}(0, h), \tag{3.12} $$

$$ {}_hZ^*_{kh} \equiv \begin{bmatrix} \theta_1 \cdot {}_hZ_{kh} + \gamma_1\left[|{}_hZ_{kh}| - (2h/\pi)^{1/2}\right] \\ \cdots \\ \theta_n \cdot {}_hZ_{kh} + \gamma_n\left[|{}_hZ_{kh}| - (2h/\pi)^{1/2}\right] \end{bmatrix}, \tag{3.13} $$

and $\{\theta_1, \gamma_1, \ldots, \theta_n, \gamma_n\}$ are selected so that

$$ E\begin{bmatrix} {}_hZ_{kh} \\ {}_hZ^*_{kh} \end{bmatrix}[{}_hZ_{kh}\ {}_hZ^{*\prime}_{kh}] = \Omega h. \tag{3.14} $$

We convert the discrete time process $[{}_hS_{kh}, {}_hY'_{kh}]$ into a continuous time process by defining

$$ {}_hS_t \equiv {}_hS_{kh}, \qquad {}_hY_t \equiv {}_hY_{kh} \quad \text{for} \quad kh \le t < (k+1)h. \tag{3.15} $$

THEOREM 3.1. *If $b(s, y, t)$ and $a(s, y, t)$ satisfy Assumption* 4, *with $x \equiv [s, y']'$, and if the joint probability measures ν_h of the starting values $({}_hS_0, {}_hY'_0)$ converge to the measure ν_0 as $h \downarrow 0$, then $({}_hS_t, {}_hY'_t) \Rightarrow (S_t, Y'_t)$ as $h \downarrow 0$.*

The proof of Theorem 3.1 is a straightforward application of Theorem 2.2. The details are in appendix B. Intuitively,

$$ ({}_hS_{(k+1)h} - {}_hS_{kh}),\ ({}_hY_{(k+1)h} - {}_hY_{kh}), $$

and h are the discrete time counterparts of dS, dY, and dt, respectively. It is a little bit harder to see why ${}_hZ_{kh}$ and ${}_hZ^*_{kh}$ are the discrete time counterparts of $dW_{1,t}$ and $dW_{2,t}$. To see that they are, the following result may be helpful:

THEOREM 3.2. *Let W_t be a standard Brownian motion on* $[0, 1]$. *For each $k = 1, 2, \ldots$ and each $t \in [0, 1]$ define $Q_{k,t}$ by*

$$ Q_{k,t} \equiv (1 - 2/\pi)^{-1/2} \sum_{j=1}^{[kt]} \left\{ |W_{(j+1)/k} - W_{j/k}| - (2/\pi k)^{1/2} \right\}, \tag{3.16} $$

where $[kt]$ *is the integer part of* kt. *Then as* $k \to \infty$,

$$[Q_{k,t}, W_t] \Rightarrow W_t^{**}, \tag{3.17}$$

where W_t^{**} *is a two-dimensional standard Brownian motion on* $[0, 1]$.[15]

Although $Q_{k,t}$ is a function of the path of W_t, it converges in distribution to a Brownian motion independent of W_t as $k \to \infty$. To manufacture a sequence of processes that converges to a Brownian motion that is imperfectly correlated with W_t, we take a linear combination of W_t and $Q_{k,t}$. Heuristically, this is why ${}_hZ_{kh}$ and ${}_hZ_{kh}^*$ can serve as the discrete time counterparts of $dW_{1,t}$ and $dW_{2,t}$, respectively. Though ${}_hZ_{kh}$ and $|{}_hZ_{kh}| - E\,|{}_hZ_{kh}|$ are uncorrelated but not independent for each $h > 0$, their partial sums are independent in the limit as $h \downarrow 0$. Thus, although there is only one observable random process $({}_hS_{kh})$, we are able to approximate continuous time systems driven by more than one Brownian motion.

One note of caution is in order on the use of the diffusion approximation just developed: Suppose that the true data-generating process in some model is a diffusion of the form (3.5)–(3.7) over a time span $[0, T]$, and that we observe the diffusion only at discrete intervals of length h, and wish to estimate the diffusion's coefficients. Theorem 3.1 takes θ, γ, f, g, F, and G as nonrandom, and does *not* tell us that we can fit a stochastic difference equation model of the form (3.10)–(3.14) to the observed data and consistently estimate the parameters of the diffusion as $h \downarrow 0$ and $T \to \infty$. There may be general conditions under which this procedure yields consistent parameter estimates, but these conditions are as yet unknown.

For the sake of simplicity, we have restricted the rank of Ω to be less than or equal to two. This allowed us to have only two driving noise terms, ${}_hZ_{kh}$ and $|{}_hZ_{kh}|$. We could easily allow Ω to have a higher rank by considering arbitrary piecewise linear functions of ${}_hZ_{kh}$.

Another easy generalization of Theorem 3.1 is to relax the requirement that the ${}_hZ_{kh}$ are $N(0, h)$. We could instead allow $h^{-1/2}{}_hZ_{kh}$ to be i.i.d. from any symmetric distribution (the same for each h) with mean zero, variance one, and a finite absolute moment of order higher than two. We then substitute $E\,|h^{-1/2}{}_hZ_{kh}|$ for $(2/\pi)^{1/2}$ in (3.13).

The approximation scheme (3.10)–(3.15) is closely related to the Euler approximation for stochastic differential equations discussed, for example, in Duffie and Singleton (1989) and Pardoux and Talay (1985). There are two differences between the Euler approximation and the approximation

[15] Richard Dudley pointed out that $Q_{k,t}$ does not converge in probability to any Brownian motion, since as $k \to \infty$, $[Q_{k,t}, Q_{k^2,t}, W_t] \Rightarrow W_t^{***}$, a three-dimensional standard Brownian motion. In other words, the Q function corresponding to a fine partition of time is asymptotically independent of the Q function corresponding to a partition that is much finer still.

(3.10)–(3.15): First, the Euler approximation does not include the $|_h Z_{kh}|$ terms in (3.10)–(3.15).[16] Second, the Euler approximation assumes global Lipschitz continuity of f, F, g, and G, while Theorem 3.1 does not. In the Euler approximation, however, the convergence result analogous to Theorem 3.1 is stronger in that the approximating sequence converges in probability rather than in distribution. The assumption of Lipschitz continuity, however, rules out several interesting cases, including the next example.

3.3. AR(1) Exponential ARCH

Next, consider a simple model of changing volatility on a portfolio, in which the log of the conditional variance follows a continuous time AR(1) [i.e., an Ornstein–Uhlenbeck process (Arnold, 1974)]. Variations on the basic model are found in Wiggins (1987), Nelson (1991), and Gennotte and Marsh (1987):

$$d[\ln(S_t)] = \theta\sigma_t^2\, dt + \sigma_t\, dW_{1,t}, \tag{3.18}$$

$$d[\ln(\sigma_t^2)] = -\beta[\ln(\sigma_t^2) - \alpha]\, dt + dW_{2,t}, \tag{3.19}$$

$$\mathrm{P}[(\ln(S_0), \ln(\sigma_0^2)) \in \Gamma] = \nu_0(\Gamma) \quad \text{for any} \quad T \in B(R^2), \tag{3.20}$$

where S_t is the value of the portfolio at time t, $W_{1,t}$, and $W_{2,t}$ are Brownian motions with

$$\begin{bmatrix} dW_{1,t} \\ dW_{2,t} \end{bmatrix} [dW_{1,t}\, dW_{2,t}] = \begin{bmatrix} 1 & C_{12} \\ C_{12} & C_{22} \end{bmatrix} dt \equiv C\, dt, \tag{3.21}$$

and $C_{22} \geq C_{12}^2 \cdot \sigma_t^2$ is the instantaneous variance per unit of time and $\theta\sigma_t^2$ is the instantaneous risk premium.[17]

The diffusion (3.18)–(3.21) does not satisfy global Lipschitz conditions and cannot be transformed to satisfy them. Hence, the standard theorems on convergence of the Euler approximation do not apply. Theorem 3.1 can be applied, however, to find a sequence of ARCH models that converge weakly (rather than path by path) to (3.18)–(3.21). Since $\ln(\sigma_t^2)$ follows a continuous time AR(1) process in (3.19), our discrete time models will also make $\{\ln({}_h\sigma_{kh}^2)\}_{k=0,\infty}$ an AR(1) process. Specifically, we assume that ${}_h\sigma_{kh}^2$ follows an AR(1) exponential ARCH process (Nelson, 1991). For each h

[16] The referee pointed out that in a simulation context, inclusion of the $|_h Z_{kh}|$ terms can be interpreted as a trick to reduce the number of psuedo-random numbers needed to carry out the simulation.

[17] The conditions under which this is the equilibrium risk premium are very restrictive, however—see Gennotte and Marsh (1987).

we have

$$\ln[{}_hS_{kh}] = \ln[{}_hS_{(k-1)h}] + h\theta_h\sigma_{kh}^2 + {}_h\sigma_{kh}\cdot{}_hZ_{kh}, \tag{3.22}$$

$$\ln[{}_h\sigma_{(k+1)h}^2] = \ln({}_h\sigma_{kh}^2) - \beta[\ln({}_h\sigma_{kh}^2) - \alpha]h + C_{12}\cdot{}_hZ_{kh} + \gamma\left[|{}_hZ_{kh}| - (2h/\pi)^{1/2}\right], \tag{3.23}$$

$$\mathrm{P}[(\ln({}_hS_0), \ln({}_h\sigma_0^2)) \in \Gamma] = \nu_h(\Gamma) \quad \text{for any} \quad \Gamma \in B(R^2), \tag{3.24}$$

where $\gamma \equiv [(C_{22} - C_{12}^2)/(1 - 2/\pi)]^{1/2}$ and ${}_hZ_{kh} \sim$ i.i.d. N(0, h). It is easy to check that

$$\mathrm{E}\begin{bmatrix} {}_hZ_{kh} \\ C_{12}\cdot{}_hZ_{kh} + \gamma\left[|{}_hZ_{kh}| - (2h/\pi)^{1/2}\right] \end{bmatrix} \times \left[{}_hZ_{kh} \quad C_{12}\cdot{}_hZ_{kh} + \gamma\left[|{}_hZ_{kh}| - (2h/\pi)^{1/2}\right]\right] = h\begin{bmatrix} 1 & C_{12} \\ C_{12} & C_{22} \end{bmatrix}, \tag{3.25}$$

which is the discrete time analog of (3.21). Once again, we create the continuous time step function processes ${}_hS_t$ and ${}_h\sigma_t^2$. As before, $\ln({}_hS_{kh}) - \ln({}_hS_{(k-1)h})$, $\ln[{}_h\sigma_{(k+1)h}^2] - \ln({}_h\sigma_{kh}^2)$, ${}_hZ_{kh}$, $(C_{12}\cdot{}_hZ_{kh} + \gamma[|{}_hZ_{kh}| - (2h/\pi)^{1/2}])$, and h are the discrete time counterparts of d(ln(S)), d(ln(σ^2)), $dW_{1,t}$, $dW_{2,t}$, and dt, respectively.

THEOREM 3.3. *If* $\nu_h \to \nu_0$, *then* $\{{}_hS_t, {}_h\sigma_t^2\} \Rightarrow \{S_t, \sigma_t^2\}$.

Recall from section 2 that the continuous time limit of the GARCH(1, 1) process has an inverted Gamma stationary distribution, so that although the normalized discrete time innovations process $h^{-1/2}{}_hZ_{kh}\cdot{}_h\sigma_{kh}$ is normally distributed given ${}_h\sigma_{kh}$, its unconditional distribution is approximately Student t when h is small and kh is large. In the exponential ARCH case, if $\ln(\sigma_0^2)$ is normally distributed, then the $\ln(\sigma_t^2)$ process is Gaussian, and even if $\ln(\sigma_0^2)$ is not normally distributed, then a Gaussian stationary limit distribution for $\ln(\sigma_t^2)$ exists as long as $\beta > 0$ (Arnold, 1974, sect. 8.3)].

THEOREM 3.4. *For a fixed initial condition* ${}_h\sigma_0^2 \equiv \sigma_0^2$ *for all* h, *and for any finite* β,

$$h^{-1/2}{}_hZ_{kh}\cdot{}_h\sigma_{kh} \xrightarrow{\mathrm{d}} Z\cdot\sigma_{kh}, \tag{3.26}$$

for fixed kh *and* $h \downarrow 0$, *where* $Z \sim$ N(0, 1) *independent of* σ_{kh}, *and*

$$\ln(\sigma_{kh}) \sim \mathrm{N}(e^{-\beta kh}(\ln(\sigma_0) - \alpha/2) + \alpha/2, C_{22}(1 - e^{-2\beta kh})/8\beta). \tag{3.27}$$

If $\beta > 0$, *then in the discrete time model* (3.22)–(3.25),

$$h^{-1/2}{}_hZ_{kh} \cdot {}_h\sigma_{kh} \xrightarrow{d} Z \cdot \sigma, \tag{3.28}$$

as $h \downarrow 0$ *and* $kh \to \infty$, *where* Z *and* σ *are independent random variables with*

$$Z \sim \mathrm{N}(0,1), \qquad \ln(\sigma) \sim \mathrm{N}(\alpha/2, C_{22}/8\beta). \tag{3.29}$$

In contrast to the GARCH(1, 1) model, in which the innovations terms are approximately Student when the time interval between observations is short, the exponential ARCH model produces innovations that are approximately a normal–lognormal mixture. The normal–lognormal mixture has been used in modelling stock returns by a number of researchers (e.g., Clark, 1973). In Clark's model, however, the conditional variance process was assumed to be i.i.d., whereas it is serially correlated in the exponential ARCH case.

It is a simple exercise to apply the results of this section to show that exponential ARCH processes, in which $\ln(\sigma^2)$ is either a finite-order AR or else is the sum of a finite number of finite-order AR processes, have well-defined continuous time limits. For a discussion of continuous time AR processes and their relation to discrete time AR processes, see Priestley (1981, ch. 3, sects. 3.7.4, 3.7.5). Passage to continuous time of a sequence of exponential ARCH models in which $\ln(\sigma_t^2)$ follows an AR(k) is directly analogous to the case discussed in Priestley.

3.4. GARCH(1, 1) Revisited

As a final example, consider again the Ito process limit of the GARCH(1, 1) model discussed in section 2. As before, define $V_t \equiv \ln(\sigma_t^2)$. Applying Ito's Lemma, the limiting diffusion becomes (2.39′)–(2.41′). Our discrete time system is now

$${}_hY_{kh} = {}_hY_{(k-1)h} + h \cdot c \cdot \exp({}_hV_{kh}) + \exp({}_hV_{kh}/2) \cdot {}_hZ_{kh}, \tag{3.30}$$

$$\begin{aligned}{}_hV_{(k+1)h} &= {}_hV_{kh} + h\left[\omega \cdot \exp(-{}_hV_{kh}) - \theta - \alpha^2/2\right] \\ &\quad + \alpha\left[|{}_hZ_{kh}| - (2h/\pi)^{1/2}\right][1 - 2/\pi]^{-1/2},\end{aligned} \tag{3.31}$$

$$\mathrm{P}\left[({}_hY_0, \exp({}_hV_0)) \in \Gamma\right] = \nu_h(\Gamma) \quad \text{for any} \quad \Gamma \in B(R^2), \tag{3.32}$$

where $\{{}_hZ_{kh}\} \sim$ i.i.d N(0, h). As in earlier examples, we create the continuous time step function process ${}_hY_t$ and ${}_hV_t$ from the discrete time processes ${}_hY_{kh}$ and ${}_hV_{kh}$.

THEOREM 3.5. *If* $\nu_h \to \nu_0$ *as* $h \downarrow 0$, *then* $[{}_hY_t, {}_hV_t] \Rightarrow [Y_t, V_t]$.

4. CONCLUSION

In this chapter, we have presented some basic tools for investigating the relationship between stochastic difference equations and Ito processes, and have made two concrete applications: First, we have derived the approximate distributions of GARCH(1, 1) and AR(1) exponential ARCH for small sampling intervals. Second, we have introduced a class of ARCH models (in which exponential ARCH is a special case) which can approximate a wide variety of stochastic differential equations.

In recent years, researchers in many disciplines have developed and applied methods of approximating Ito processes with stochastic difference equations (and vice versa). For example, see Kushner (1984) for applications to signal processing and Ethier and Kurtz (1986, ch. 10) for applications to genetics. Apart from the options pricing literature (e.g., Cox *et al.*, 1979), however, economists have made relatively little use of these tools, though they can be very useful, especially in understanding nonlinear time series models such as ARCH.

While the results in this chapter contribute to our understanding of the relation between stochastic difference and differential equations, several limitations have yet to be overcome. For example, it would be useful to extend the results of the chapter to jump–diffusion processes. Results on jump–diffusion approximation have been developed, e.g., by Kushner (1984) and Ethier and Kurtz (1986). It would be interesting to see how these could be integrated into the ARCH framework. A second generalization would be to extend the results of section 3 to multivariate systems. Perhaps most useful would be to find circumstances under which we can use the diffusion approximations of section 3 to estimate consistently the parameters of discretely sampled diffusions. These tasks await future research.

APPENDIX A. DISTRIBUTIONAL UNIQUENESS

In this appendix, we state conditions under which ν_0, $a(x, t)$, and $b(x, t)$ uniquely specify the distribution of a stochastic integral equation of the form given in Theorem 2.1. The conditions are of two types: first, those that ensure distributional uniqueness of the process on compact sets, and second, a condition that ensures that the limit process doesn't fly off to infinity in finite time. The assumptions made in Theorem A.1 below to ensure distributional uniqueness are considerably less restrictive than the Lipschitz and Growth conditions that are often assumed (e.g., Liptser and

Shiryayev, 1977, Theorem 4.6, Corollary)] to ensure strong (pathwise) existence and uniqueness of solutions to stochastic differential equations. Our conditions are less restrictive largely because we are interested only in distributional existence and uniqueness. The source of these results is Stroock and Varadhan (1979) (henceforth S & V). For a much fuller treatment of these and related conditions, see S & V and Ethier and Kurtz (1986) (henceforth E & K).

First, we present several conditions, any one of which ensures weak existence and uniqueness of the process X_t on compact sets:

CONDITION A. *Let $a(x,t)$ and $b(x,t)$ be continuous in both x and t with two continuous partial derivatives with respect to x.*

CONDITION B. *Let $a(x,t)$ and $b(x,t)$ be locally bounded [i.e., bounded on bounded (x,t) sets] and measurable, and let $a(x,t)$ be continuous in x. Let I_n be the $n \times n$ identity matrix. For every $R > 0$, and every $T > 0$, let there be a number $\Lambda_{T,R} > 0$ such that for all (x,t) satisfying $0 \le t \le T$ and $\|x\| \le R$, $a(x,t) - \Lambda_{T,R} \cdot I_n$ is positive definite.*

CONDITION C. *Let $a(x,t)$ and $b(x,t)$ be locally bounded and measurable, and let $\sigma(x,t)$ be a locally bounded, measurable function satisfying $a(x,t) = \sigma(x,t)'\sigma(x,t)$. For every $R > 0$ and every $T > 0$, let there be a number $\Lambda_{T,R} > 0$ such that*

$$\sup_{0 \le t \le T, \|x\| \le R, \|y\| \le R} \|\sigma(y,t) - \sigma(x,t)\| + \|b(y,t) - b(x,t)\| - \Lambda_{T,R}\|y - x\| \le 0. \tag{A.1}$$

CONDITION D. *Let x be a scalar (i.e., $n = 1$), let $a(x,t)$ and $b(x,t)$ be locally bounded and measurable, and let $\sigma(x,t) \equiv a(x,t)^{1/2}$. Let there be an increasing, nonnegative function $\rho(u)$ from $[0,\infty)$ into $[0,\infty)$ such that*

$$\rho(u) > 0 \quad \text{for} \quad u > 0, \tag{A.2}$$

$$\lim_{\varepsilon \downarrow 0} \int_\varepsilon^1 [\rho(u)]^{-2}\, du = \infty. \tag{A.3}$$

For every $R > 0$ and every $T > 0$, let there be a number $\Lambda_{T,R} > 0$ such that

$$\sup_{|x| \le R, |y| \le R, 0 \le t \le T} |\sigma(x,t) - \sigma(y,t)| - \Lambda_{T,R} \cdot \rho(|x - y|) \le 0, \tag{A.4}$$

$$\sup_{|x| < R, |y| < R, 0 \le t \le T} |b(x,t) - b(y,t)| - \Lambda_{T,R} \cdot |x - y| \le 0. \tag{A.5}$$

Next, we state a condition that ensures that the limit process does not run off to infinity in finite time. S & V Theorem 10.2.3 provides a second (somewhat more complicated) such condition.

NONEXPLOSION CONDITION. *There exists a nonnegative function* $\varphi(x,t)$ *that is differentiable with respect to* t *and twice differentiable with respect to* x *such that for each* $T > 0$,

$$\lim_{\|x\|\to\infty} \inf_{0\le t\le T} \varphi(x,t) = \infty, \tag{A.6}$$

and there exists a positive, locally bounded function $\lambda(T)$ *such that for each* $T > 0$, *all* $x \in R^n$ *and all* t, $0 \le t \le T$,

$$\sum_{i=1}^{n} b_i(x,t)\frac{\partial\varphi(x,t)}{\partial x_i} + \frac{1}{2}\sum_{i=1}^{n}\sum_{j=1}^{n} a_{ij}(x,t)\frac{\partial^2\varphi(x,t)}{\partial x_i \partial x_j} + \frac{\varphi(x,t)}{\partial t} \le \lambda(T)\varphi(x,t). \tag{A.7}$$

The left-hand side of (A.7) is the instantaneous drift of $\varphi(X_t, t)$ when $X_t = x$. The right-hand side of (A.7) bounds this drift to grow at most linearly with $\varphi(X_t, t)$, which can be used to ensure that $\varphi(X_t, t)$ does not explode in finite time (see the proof of S & V Theorem 10.2.1). (A.6) guarantees that if $\varphi(X_t, t)$ does not explode, neither will X_t. Since the inequality (A.7) holds uniformly for $x \in R^n$, S & V Theorem 10.2.1 easily extends to a random initial condition.

THEOREM A.1. *Let the nonexplosion condition be satisfied, and let one or more of Conditions A, B, C, and D be satisfied. Then a solution to the stochastic integral equation*

$$X_t = X_0 + \int_0^t b(X_s, s)\, ds + \int_0^t \sigma(X_s, s)\, dW_{n,s}, \tag{A.8}$$

where $\{W_{n,t}\}$ *is an n-dimensional standard Brownian motion independent of* X_0, *and where for any* $\Gamma \in B(R^n)$, $\mathrm{P}(X_0 \in \Gamma) = \nu_0(\Gamma)$, *has a unique weak-sense solution (Liptser and Shiryayev, 1977, Ethier and Kurtz, 1986), in that all solutions to (A.8) have the same probability law. This distribution does not depend on the choice of* $\sigma(\cdot,\cdot)$. *Finally,* X_t *remains finite in finite time intervals almost surely.*

APPENDIX B. PROOFS

Proof of Theorem 2.1. For the case when Condition (2.2) is replaced by

$$\mathrm{P}_h\Big[{}_hX_t = \big\{h^{-1}((k+1)h - t)_h X_{kh} + h^{-1}(t - kh)_h X_{(k+1)h}\big\}, kh \le t < (k+1)h\Big] = 1. \tag{2.2'}$$

S & V Theorem 11.2.3 proved that when 1) $a(\cdot, \cdot)$, $b(\cdot, \cdot)$, and a fixed starting point X_0 uniquely specify the distribution of a diffusion, 2) ${}_hX_0 = X_0$ for all h, and 3). Assumption 1 holds, then ${}_hX_t$ converges weakly to X_t, the diffusion generated by $a(\cdot, \cdot)$, $b(\cdot, \cdot)$, and X_0. That this still holds when (2.2′) is replaced by (2.2), and when the assumption of a fixed initial state is weakened to Assumption 3, is proven in E & K, chapter 7, Corollary 4.2 [Note: In E & K and S & V, time appeared in the transition functions as an element of x, not as a separate argument. It is a simple exercise to show that the change of notation is made without loss of generality. See S & V, p. 166, Problem 6.7.2, and p. 266.]

Under Assumption 2, X_t has the stochastic integral representation (2.12) by E & K, chapter 5, Theorem 3.3 and Corollary 3.4. Finally, since the distribution of X_t is characterized only by ν_0, $a(\cdot, \cdot)$, and $b(\cdot, \cdot)$, $\sigma(\cdot, \cdot)$ only enters through the $a(\cdot, \cdot)$ function. Therefore the distribution of X_t does not depend on our choice of $\sigma(\cdot, \cdot)$ as long as $\sigma(\cdot, \cdot)\sigma(\cdot, \cdot)' = a(\cdot, \cdot)$.

Proof of Theorem 2.2. To prove Theorem 2.2 it is sufficient to show that Assumption 5 implies Assumption 1. The Theorem then follows immediately by Theorem 2.1. To show that Assumption 5 implies Assumption 1, it is sufficient to show that (2.17)–(2.19) imply that for every $i, i = 1, \ldots, n$, every $R, 0 < R < \infty$, every $T, 0 < T < \infty$, and every $\varepsilon > 0$,

$$\lim_{h \downarrow 0} \sup_{\|x\| \le R, 0 \le t \le T} h^{-1} \int_{\|y-x\| \ge 1} (y - x)_i^2 \Pi_{h, h[t/h]}(x, \mathrm{d}y) = 0, \quad \text{(B.1)}$$

$$\lim_{h \downarrow 0} \sup_{\|x\| \le R, 0 \le t \le T} h^{-1} \int_{\|y-x\| \ge 1} |y - x|_i \Pi_{h, h[t/h]}(x, \mathrm{d}y) = 0, \quad \text{(B.2)}$$

$$\lim_{h \downarrow 0} \sup_{\|x\| \le R, 0 \le t \le T} \Delta_{h, \varepsilon}(x, t) = 0. \quad \text{(B.3)}$$

By Markov's Inequality,

$$\Delta_{h, \varepsilon}(x, t) \le h^{-1} \varepsilon^{-(2+\delta)} \int_{R^n} \|y - x\|^{2+\delta} \Pi_{h, h[t/h]}(x, dy). \quad \text{(B.4)}$$

By Minkowski's Integral Inequality,

$$\le \varepsilon^{-(2+\delta)} \left[\sum_{i=1}^{n} \left[c_{h,i,\delta}(x, t)\right]^{2/(2+\delta)} \right]^{(2+\delta)/2}. \quad \text{(B.5)}$$

$$h^{-1} \varepsilon^{-(2+\delta)} \int_{R^n} \|y - x\|^{2+\delta} \Pi_{h, h[t/h]}(x, dy)$$

By (2.17), there is some $\delta > 0$ such that for all $R, T > 0$, this expression vanishes for every $\varepsilon > 0$ as $h \downarrow 0$ uniformly on $\|x\| \le R$, $0 \le t \le T$, proving (B.3).

Next, we prove (B.2). By Hölder's Integral Inequality,

$$h^{-1}\int_{\|y-x\|\geq 1}|y-x|_i\Pi_{h,h[t/h]}(x,dy)$$
$$\leq\left[c_{h,i,\delta}(x,t)\right]^{1/(2+\delta)}\left[\Delta_{h,1}(x,t)\right]^{(1+\delta)/(2+\delta)}, \tag{B.6}$$

which vanishes in the same manner (i.e., uniformly on compacts) as (B.5). Finally, we must prove (B.1). Again by Hölder's Integral Inequality,

$$h^{-1}\int_{\|y-x\|\geq 1}(y-x)_i^2\Pi_{h,h[t/h]}(x,dy)$$
$$\leq\left[c_{h,i,\delta}(x,t)\right]^{2/(2+\delta)}\cdot\left[\Delta_{h,1}(x,t)\right]^{\delta/(2+\delta)}, \tag{B.7}$$

which vanishes in the same manner as (B.5) and (B.6).

Proof of Theorem 2.3. Let $A(x)$ and $B(x)$ be a real-valued scalar function on R^1, with $B(x)\geq 0$ for all $x\in R^1$. Let x_t be the stochastic process given by the stochastic differential equation

$$dx_t = A(x_t)\,dt + \left[2\cdot B(x_t)\right]^{1/2}dW_t, \tag{B.8}$$

where W_t is a standard Brownian motion. The corresponding Fokker–Planck equation (Arnold, 1974) is

$$\partial^2[B(x)p]/\partial x^2 - \partial[A(x)p]/\partial x = \partial p/\partial t, \tag{B.9}$$

where $p = p(x\mid x_0,t)$, $0\leq t<\infty$, is the probability density of x_t given x_0. Wong (1964) showed that if there is a stationary density for x_t, i.e., a density $p^*(x)$ such that

$$p^*(x) = \lim_{t\to\infty}p(x\mid x_0,t), \tag{B.10}$$

then it satisfies the differential equation

$$d[p^*(x)\cdot B(x)]/dx = A(x)\cdot p^*(x). \tag{B.11}$$

Further, if a density $p^*(x)$ satisfying (B.11) is a member of the Pearson family of distributions (Johnson and Kotz, 1970), then it is the stationary density of the x_t process. For the GARCH(1, 1) diffusion limit (2.40), we have

$$A(\sigma^2) = \omega - \theta\sigma^2, \tag{B.12}$$
$$B(\sigma^2) = \sigma^4\alpha^2/2. \tag{B.13}$$

The results of Wong (1964, sect. 2.F) suggest that we try a density $p^*(\sigma^2)$ of the form $p^*(\sigma^2)\propto\sigma^{2a}\cdot\exp(b/\sigma^2)$ for some a and b. This guess turns out to be correct, and the result is an inverted Gamma density. A change of variables $\lambda \equiv \sigma^{-2}$ then yields the stationary density $f(\lambda)$:

$$f(\lambda) = \Gamma(1 + 2\theta/\alpha^2, 2\omega/\alpha^2), \tag{B.14}$$

where $\Gamma(\cdot,\cdot)$ is the Gamma density. (2.42)–(2.43) follow immediately. Because ${}_h\sigma_{kh}^2 \xrightarrow{d} \sigma_t^2$ for $kh = t$ as $h \downarrow 0$, (2.44) immediately follows. Since $h^{-1/2}{}_hZ_{kh} \sim \mathrm{N}(0,1)$ and is independent of λ, we integrate λ out to obtain (2.45).

By Kushner (1984, ch. 6, theorem 6), (2.46) and (2.47) will follow if we can show that the sequence of random functions $\{{}_h\sigma_t^2\}_{h\downarrow 0}$ is tight in $D[0,\infty)$. By Kushner (1984, ch. 6, theorem 3), (2.46)–(2.47) will hold if we can find a twice-continuously differentiable Liapunov function $\varphi(\cdot)$ defined on $[0,\infty)$ and strictly positive numbers h^*, Δ, and η such that

$$\min_{0 \le \sigma^2} \varphi(\sigma^2) = 0, \tag{B.15}$$

$$\lim_{\sigma^2 \to \infty} \varphi(\sigma^2) = \infty, \tag{B.16}$$

$$\sup_{0 < h \le h^*} \mathrm{E}\left[\varphi\left({}_h\sigma_0^2\right)\right] < \infty, \tag{B.17}$$

and for all ${}_h\sigma_{kh}^2 \ge 0$ and all h, $0 < h < h^*$,

$$h^{-1}\mathrm{E}\left[\varphi\left({}_h\sigma_{(k+1)h}^2\right) - \varphi\left({}_h\sigma_{kh}^2\right) \mid M_{kh}\right] < \Delta - \eta \cdot \varphi\left({}_h\sigma_{kh}^2\right). \tag{B.18}$$

Define, for any d, $0 < d < 1$,

$$\varphi_d(\sigma^2) \equiv \begin{cases} 0 & \text{if } \sigma^2 = 0, \\ \sigma^{2d} \cdot \exp(-1/\sigma^2) & \text{otherwise}. \end{cases} \tag{B.19}$$

For each $d > 0$, $\varphi_d(\sigma^2)$ and its derivatives of arbitrary order are continuous on $[0,\infty)$. $\varphi_d(\sigma^2)$ also satisfies (B.15)–(B.16). (B.18) was already assumed in the statement of Theorem 2.3. All that remains therefore is to verify that (B.18) holds for some $d > 0$.

To simplify notation, we write $\varphi(\sigma^2)$, σ_+^2, and σ^2 instead of $\varphi_d(\sigma^2)$, ${}_h\sigma_{(k+1)h}^2$, and ${}_h\sigma_{kh}^2$. By S & V Lemma 11.2.1,

$$h^{-1}\mathrm{E}\left[\varphi(\sigma_+^2) - \varphi(\sigma^2) \mid M_{kh}\right] - (\alpha^2\sigma^4/2) \cdot \varphi''(\sigma^2)$$
$$-(\omega - \theta\sigma^2) \cdot \varphi'(\sigma^2) \to 0, \tag{B.20}$$

uniformly on compacts for fixed d. Therefore, if we confine σ^2 to any compact set $\Theta \subset [0,\infty)$, (B.20) reduces to

$$(\alpha^2\sigma^4/2) \cdot \varphi'' + (\omega - \theta\sigma^2) \cdot \varphi' + o(1) < \Delta - \eta \cdot \varphi. \tag{B.18$'$}$$

Since $\varphi(\cdot)$ and its derivatives are locally bounded, it is clear that for any compact Θ, a finite Δ can be found such that (B.20) holds uniformly on $\sigma^2 \in \Theta$. Now consider (B.18) for large values of σ^2: expanding

$h^{-1}[\varphi(\sigma_+^2) - \varphi(\sigma^2)]$ in a four-term Taylor series and taking expectations, (B.18) is satisfied if the inequality

$$\begin{aligned} h^{-1}\mathrm{E}\left[\varphi(\sigma_+^2) - \varphi(\sigma^2) \mid M_{kh}\right] &= -d\theta\sigma^{2d} + O(\sigma^{-2}) + d(d-1)(\alpha^2/2)\sigma^{2d} \\ &\quad + \sigma^{2d}\left[O(h^{1/2}) + O(\sigma^{-2}) \cdot O(h)\right] \\ &\quad + h^{-1}d(d-1)(d-2)(d-3) \\ &\qquad \cdot \mathrm{E}\left[\left(\varphi(\sigma_+^2) - \varphi(\sigma^2)\right)^4 \cdot \sup_{0 \le \delta \le 1}\left(\delta\sigma^2 + (1-\delta)\sigma_+^2\right)^d \mid M_{kh}\right] \\ &< \Delta - \eta \cdot \sigma^{2d}, \end{aligned} \tag{B.21}$$

is satisfied. The last term on the left-hand side of (B.21) is nonpositive as long as $0 < d < 1$. Clearly also, a finite Δ can be chosen that dominates the $O(\cdot)$ and $o(\cdot)$ terms, (B.18) will therefore hold if we can find a d, $0 < d < 1$, and a $\eta > 0$ such that

$$(\eta - \theta) \cdot \sigma^{2d} + (d-1)(\alpha^2/2) \cdot \sigma^{2d} < 0, \tag{B.22}$$

which we can always do if $2\theta/\alpha^2 > -1$.

Proof of Theorem 3.1. We need only verify Assumptions 2 and 5. (3.5)–(3.7) factor $a(s, y, t)$ into $\sigma(s, y, t)\sigma(s, y, t)'$, satisfying Assumption 2. To verify Assumption 5, we need to check that $b_h^*(\cdot, \cdot, \cdot)$ and $a_h^*(\cdot, \cdot, \cdot)$ converge, and that $c_{h,i,\delta}(\cdot, \cdot, \cdot) \to 0$ uniformly on compacts,

$$b_h^*(s, y, t) = \begin{bmatrix} f(s, y, t) \\ F(s, y, t) \end{bmatrix}. \tag{B.23}$$

Since $b_h^*(\cdot, \cdot, \cdot) = b(\cdot, \cdot, \cdot)$, (2.19) is satisfied. Similarly, we have (dropping the function arguments for the sake of notational clarity):

$$a_h^*(s, y, t) = \begin{bmatrix} hf^2 + g^2 & hfF' + G\Omega_{1,2}g \\ hFf + g\Omega_{2,1}G' & G\Omega_{2,2}G' \end{bmatrix}. \tag{B.24}$$

Since f, F, g, and G are locally bounded, it is clear that

$$\lim_{h \downarrow 0} \sup_{\|x\| R, 0 \le t \le T} \left\| a_h^*(s, y, t) - a(s, y, t) \right\| = 0. \tag{B.25}$$

Finally, choose $\delta = 1$. Stacking the elements of $c_{h,i,1}(\cdot, \cdot, \cdot)$ in a vector, we have

$$c_{h,1}({}_hS_{kh}, {}_hY_{kh}, t) = h^{-1}\mathrm{E}\begin{bmatrix} |hf + g_h Z_{kh}|^3 \\ |hF + G_h Z_{kh}^*|^3 \end{bmatrix} = O(h^{1/2}), \tag{B.26}$$

uniformly on compacts, so that (2.17) is satisfied.

Proof of Theorem 3.2. Let $\{Z_j\}_{j=1,\infty}$ be an i.i.d N(0, 1) sequence, and note that

$$Q_{k,t} \sim (1 - 2/\pi)^{-1/2}[kt]^{-1/2} \sum_{j=1}^{[kt]} \big[|Z_j| - \mathrm{E}\,|Z_j|\big], \qquad \text{(B.27)}$$

where $\sim$ denotes equality in distribution. Since $\mathrm{E}\,|Z_j| = (2/\pi)^{1/2}$, the Theorem follows immediately by Donsker's Theorem (Billingsley, 1968).

Proof of Theorem 3.3. Define $V_t \equiv \ln(\sigma_t^{\,2})$ and $S^* \equiv \ln(S_t)$ and substitute into (3.19). Our equation system becomes

$$dS^* = \theta \mathrm{e}^V\, dt + \mathrm{e}^{V/2}\, dW_1, \qquad \text{(B.28)}$$

$$dV = -\beta(V - \alpha)\, dt + dW_2, \qquad \text{(B.29)}$$

with

$$b = \big[\theta \mathrm{e}^V, -\beta(V - \alpha)\big]', \qquad \text{(B.30)}$$

$$a = \begin{bmatrix} \mathrm{e}^V & \mathrm{e}^{V/2}C_{12} \\ \mathrm{e}^{V/2}C_{12} & C_{22} \end{bmatrix}. \qquad \text{(B.31)}$$

It is readily verified that Condition B of appendix A holds. To verify the nonexplosion condition, define for $K > 0$,

$$\varphi(V, S^*) \equiv K + f(S^*) \cdot |S^*| + f(V) \cdot \exp(|V|), \qquad \text{(B.32)}$$

where $f(x) = \exp(-1/|x|)$ if $x \neq 0$ and $\equiv 0$ otherwise. $\varphi(\cdot,\cdot)$ is nonnegative, arbitrarily differentiable, and satisfies (A.6). Its derivatives are locally bounded, so that positive K and λ can be chosen to satisfy (A.7) on any compact set. For large values of S^* and V, $\varphi_V \approx \mathrm{sign}(V) \cdot \exp(|V|)$, $\varphi_{VV} \approx \exp(|V|)$, $\varphi_{S^*} \approx \mathrm{sign}(S^*)$, $\varphi_{S^*S^*} \approx 0$, and $\varphi_{VS^*} = 0$, so that with $\lambda > 1 + \beta\alpha + C_{22}/2 + |\theta|$, a finite K can be found to satisfy (A.7). The result then follows by Theorem 2.2.

Proof of Theorem 3.4. Since $h^{-1/2}{}_hZ_{kh} \sim \mathrm{N}(0, 1)$ and since ${}_h\sigma_{kh}$ and ${}_hZ_{kh}$ are independent, (3.26)–(3.27) follow immediately from Theorem 3.3 and from Arnold (1974, sect. 8.3). The proof of (3.28)–(3.29) proceeds along the same lines as the proof of (2.46)–(2.47) in Theorem 2.3: Label $V_+ \equiv \ln({}_h\sigma^2_{(k+1)h})$ and $V \equiv \ln({}_h\sigma^2_{kh})$, and define the Liapunov function $\varphi(V) \equiv [V - \alpha]^2$. We then have

$$h^{-1}\mathrm{E}[\varphi(V_+) - \varphi(V) \mid M_{kh}] = -2 \cdot \beta \cdot \varphi(V) \pm h\beta^2 \cdot \varphi^{1/2}(V) + C_{22}. \qquad \text{(B.33)}$$

(B.18) then requires that there exist positive numbers Δ, η, and h^* such that

$$\sup_{0<h<h^*, V\in R^1} (\eta - 2\cdot\beta)\cdot\varphi(V) \pm h\beta^2\cdot\varphi^{1/2}(V) + C_{22} - \Delta < 0, \tag{B.34}$$

which is clearly possible as long as $\beta > 0$.

Proof of Theorem 3.5. We proved nonexplosion and Condition B in section 2. The result then follows by a direct application of Theorem 3.1.

Proof of Theorem A.1. For a given ν_0, the conclusion of the theorem is equivalent to the "martingale problem" of S & V being well-posed (S & V, chapter 6, and E & K, chapters 4 and 5), or, equivalently, to a unique weak sense solution to the stochastic integral equation (2.12) (Ethier and Kurtz, 1986, ch. 6, corollary 3.4). S & V assume that ν_0 puts all its mass at a given point, and E & K make the extension to a random initial condition. E & K, chapter 4, Problem 49 and the nonexplosion condition allow us to conclude that, if the martingale problem as defined in S & V is well-posed, then the martingale problem with random initial condition is well-posed for any ν_0 on R^n.

The Theorem using Condition A follows by S & V, Corollary 6.3.3 and Theorem 10.2.1. Using Condition B, it follows by S & V, Theorems 7.2.1 and 10.2.1. Using Condition C, it follows by S & V, Theorems 6.3.4 and 10.2.1, and with Condition D, it follows by S & V, Theorems 8.2.1 and 10.2.1.

ACKNOWLEDGMENTS

This chapter is a revision of parts of chapter III of Nelson (1988a). The Department of Education and the University of Chicago Graduate School of Business provided research support. I am indebted to John Cox, Richard Dudley, Ricardo Garcia-Badarcco, Mark Nelson, Richard Parkus, James Poterba, Peter Rossi, G. William Schwert, Daniel Stroock, S. R. S. Varadhan, Jeffrey Wooldridge, Arnold Zellner, an editor (Angelo Melino), and a referee for helpful comments. Seminar participants at the 1987 Econometric Society Winter Meetings, at the NBER Conference on Econometric Methods and Financial Time Series, and at workshops at M.I.T., Chicago, Princeton, Wharton, Stanford, Berkeley, Northwestern, Yale, and Rochester also made useful suggestions. Any remaining errors are mine.

REFERENCES

Abel, A. B. (1988). Stock prices under time-varying dividend risk: An exact solution in an infinite-horizon general equilibrium model. *Journal of Monetary Economics* **22**, 375–393.

Arnold, L. (1974). "Stochastic Differential Equations: Theory and Applications." Wiley, New York.

Barsky, R. B. (1989). Why don't the prices of stocks and bonds move together? *American Economic Review* **79**, 1132–1146.

Bernanke, B. S. (1983). Irreversibility, uncertainty and cyclical investment. *Quarterly Journal of Economics* **98**, 85–106.

Billingsley, P. (1968). "Convergence of Probability Measures." Wiley, New York.

Black, F. (1976). Studies of stock market volatility changes. *Proceedings of the American Statistical Association, Business and Economic Statistics Section*, pp. 177–181.

Bollerslev, T. (1986). Generalized autoregressive conditional heteroskedasticity. *Journal of Econometrics* **31**, 307–327.

Clark, P. K. (1973). A subordinated stochastic process model with finite variance for speculative prices. *Econometrica* **41**, 135–155.

Cox, J., Ross, S., and Rubinstein, M. (1979). Options pricing: A simplified approach, *Journal of Financial Economics* **7**, 229–263.

Cox, J., Ingersoll, J., and Ross, S. (1985). An intertemporal general equilibrium model of asset prices. *Econometrica* **53**, 363–384.

Duffie, D., and Singleton, K. J. (1989). "Simulated Moments Estimation of Markov Models of Asset Prices," Mimeo. Stanford University, Graduate School of Business, Stanford, CA.

Engle, R. F. (1982). Autoregressive conditional heteroskedasticity with estimates of the variance of United Kingdom inflation. *Econometrica* **50**, 987–1008.

Engle, R. F., and Bollerslev, T. (1986). Modelling the persistence of conditional variances, *Econometric Reviews* **5**, 1–50.

Ethier, S. N., and Kurtz, T. G. (1986). "Markov Processes: Characterization and Convergence." Wiley, New York.

French, K. R., Schwert, G. W., and Stambaugh, R. F. (1987). Expected stock returns and volatility. *Journal of Financial Economics* **19**, 3–30.

Gennotte, G., and Marsh, T. A. (1987). "Variations in Economic Uncertainty and Risk Premiums on Capital Assets," Mimeo. University of California, Berkeley.

Geweke, J. (1986). Modelling the persistence of conditional variances: A comment. *Econometric Reviews* **5**, 57–61.

Hosking, J. R. M. (1981). Fractional differencing. *Biometrika* **68**, 165–176.

Johnson, N. L., and Kotz, S. (1970). "Distribution in Statistics: Continuous Univariate Distributions," Vol. 1. Wiley, New York.

Kushner, H. J. (1984). "Approximation and Weak Convergence Methods for Random Processes, with Applications to Stochastic Systems Theory. MIT Press, Cambridge, MA.

Liptser, R. S. and Shiryayev, A. N. (1977). "Statistics of Random Processes," Vol. 1. Springer-Verlag, New York.

Lo, A. (1988). Maximum likelihood estimation of generalized Ito processes with discretely sampled data. *Econometric Theory* **4**, 231–247.

McDonald, R. M., and Siegel, D. (1986). The value of waiting to invest. *Quarterly Journal of Economics* **101**, 707–728.

Merton, R. C. (1973). An intertemporal capital asset pricing model. *Econometrica* **41**, 867–888.

Nelson, D. B. (1988a). The time series behavior of stock market volatility and returns. Doctoral Dissertation, Massachusetts Institute of Technology, Cambridge, MA (unpublished).

Nelson, D. B. (1990). Stationarity and persistence in the GARCH(1, 1) model, *Econometric Theory* **6**, 318–334.

Nelson, D. B. (1991). Conditional heteroskedasticity in asset returns: A new approach. *Econometrica* **59**, 347–370.

Officer, R. R. (1973). The variability of the market factor of the New York Stock Exchange. *Journal of Business* **46**, 434–453.

Pantula, S. G. (1986). Modelling the persistence of conditional variances: A comment. *Econometric Reviews* **5**, 71–73.
Pardoux, E., and Talay, D. (1985). Discretization and simulation of stochastic differential equations. *Acta Applicandae Mathematica* **3**, 23–47.
Priestley, M. B. (1981). "Spectral Analysis and Time Series." Academic Press, London.
Sampson, M. (1988). "A Stationarity Condition for the GARCH(1, 1) Model," Mimeo. Concordia University, Department of Economics, Montreal.
Sandburg, C. (1969). "The People, Yes, The Complete Poems of Carl Sandburg." Harcourt Brace Jovanovich, San Diego, CA.
Stroock, D. W., and Varadhan, S. R. S. (1979). "Multidimensional Diffusion Processes." Springer-Verlag, Berlin.
Wiggins, J. B. (1987). Option values under stochastic volatility: Theory and empirical estimates. *Journal of Financial Economics* **19**, 351–372.
Wong, E. (1964). The construction of a class of stationary Markov processes. *In* "Sixteenth Symposia in Applied Mathematics—Stochastic Processes in Mathematical Physics and Engineering" (R. Bellman, ed.), pp. 264–276. American Mathematical Society, Providence, RI.

6

FILTERING AND FORECASTING WITH MISSPECIFIED ARCH MODELS I: GETTING THE RIGHT VARIANCE WITH THE WRONG MODEL*

DANIEL B. NELSON

Graduate School of Business
University of Chicago
Chicago, Illinois 60637

1. INTRODUCTION

Most theories of asset pricing, for example the CAPM of Sharpe (1964) and Lintner (1965), the option pricing formula of Black and Scholes (1973), and the arbitrage pricing theory of Ross (1976), relate required returns on assets to their variances and covariances. An enormous literature in empirical finance has explored the nature of this relation between risk and return. This literature has made it clear that the variability of returns and the degree of co-movement between assets change stochastically over time. Practical experience (as in the October stock market crashes of 1929, 1987, and 1989) points to the same conclusion. This realization has lead many researchers to recast asset pricing theory in terms of the *conditional*

* Reprinted with permission from Nelson, D. "Filtering and Forecasting with Misspecified ARCH Models I" *Journal of Econometrics* **52** (1992), Elsevier Science 5A, Lausanne, Switzerland.

variances and covariances of returns (see for example, Engle *et al.*, 1990; Harvey, 1989; Hull and White, 1987; Merton, 1973; Shanken, 1990; Wiggins, 1987).

Researchers have examined the time-series behavior of stock market volatility for other reasons as well: for example, Schwert (1989) investigates the connection between macroeconomic variables and market volatility. The nature of market volatility changes has also entered into public policy debates on financial market regulation, margin requirements, transactions taxes, etc. (see, e.g., Hardouvelis, 1990; Hsieh and Miller, 1990; Summers and Summers 1989; and the Summer 1988 issue of the *Journal of Economic Perspectives*).

Since the seminal work of Engle (1982), ARCH models have become one of the most widely used means of modelling changing conditional variances.[1] As are all statistical models, ARCH models are at best a rough approximation to reality, and in applied work a misspecified model will inevitably be chosen. If we are concerned with accurately measuring conditional variances and covariances, it is important to ask how well a *misspecified* ARCH model can approximate the true structure of conditional heteroskedasticity in a time series.

This chapter examines the ability of a misspecified ARCH model to estimate the conditional covariance matrix of a stochastic process. To anticipate the results below, ARCH models are remarkably robust to certain types of misspecification. In particular, *if the process is well approximated by a diffusion*, broad classes of ARCH models provide consistent (in a sense developed below) estimates of the conditional covariances. The ARCH models may seriously misspecify both the conditional mean of the process and the dynamic behavior of the conditional variances. In fact, the misspecification can be so severe that the ARCH models would make no sense as data-generating processes and may make terrible medium and long-term forecasts—without affecting the consistency of the one-step-ahead conditional covariance estimates. As we will see, however, this robustness does not extend to all ARCH models, and it breaks down altogether to the extent that the process is not well approximated by a diffusion.

1.1. The Initial Setup

Define the $n \times 1$ diffusion process $\{X_t\}$ by the stochastic integral equation

$$X_t = X_0 + \int_0^t \mu(X_s)\, ds + \int_0^t \Omega^{1/2}(X_s)\, dW_s, \tag{1.1}$$

[1] The survey of applications of ARCH in finance by Bollerslev *et al.* (1992) lists several hundred papers on ARCH in finance, most dated 1989 and after.

where $\{W_t\}$ is an $n \times 1$ standard Brownian motion, and $\mu(\cdot)$ and $\Omega^{1/2}(\cdot)$ are continuous functions from R^n into R^n and into the space of $n \times n$ matrices, respectively. Under mild moment conditions (see Arnold, 1973) $\Omega(X_t) \equiv \Omega^{1/2}(X_t) \cdot \Omega^{1/2}(X_t)'$ and $\mu(X_t)$ are, respectively, the instantaneous conditional covariance matrix and instantaneous conditional mean (per unit of time) of increments in the $\{X_t\}$ process. X_0 is assumed random with probability measure υ_0 independent of $\{W_t\}_{0 \le t < \infty}$.

In applications, $\{X_t\}$ might include asset prices and other state variables describing an economy. Some elements of $\{X_t\}$ may never be directly observable. Others, we assume, are observable, but only at discrete time intervals of length h. For example, our model may define the stock price process in continuous time, while we observe only stock prices at discrete intervals of length h. Formally, we partition X_t as $[X'_{1:q,t} X'_{q+1:n,t}]'$, where $X_{1:q,t}$ consists of the first q elements of X_t (observable at intervals of length h) and $X_{q+1:n,t}$ consists of the last $n - q$ elements (which are never observable.[2] We partition μ and Ω accordingly.

For each t, our goal is to estimate $\Omega_{1:q,1:q}(X_t)$ (the instantaneous covariance of the increments in the observable variables $X_{1:q,t}$ given t and the history of time t of the $\{X_t\}$ process) given the smaller information set $\{t, X_{1:q,0}, X_{1:q,h}, X_{1:q,2h}, \ldots, X_{1:q,h[t/h]}\}$ (i.e., the time index and the past observed values of $X_{1:q,t}$). Under these assumptions, $\Omega_{1:q,1:q}(X_t)$ is unobservable since $X_{q+1:n,t}$ is unobservable, and unless t is an integer multiple of h, $X_{1:q,t}$ is as well—i.e., $\Omega_{1:q,1:q}(X_t)$ is a conditional covariance matrix, but is conditional on a larger information set than is possessed by the econometrician.

1.2. Consistent Filters: A Definition

In our examples, we generate conditional covariance estimates $\{{}_h\hat{\Omega}_{1:q,1:q,t}\}$ with a sequence of ARCH models, whose coefficients may depend on h. We say that a given sequence of ARCH conditional covariance estimates $\{{}_h\hat{\Omega}_{1:q,1:q,t}\}_{h \downarrow 0}$ provides a *consistent filter* for the $\{\Omega_{1:q,1:q,t}\}$ if for each $t > 0$,

$$ {}_h\hat{\Omega}_{1:q,1:q,t} - \Omega_{1:q,1:q,t} \to 0_{q \times q} \tag{1.2} $$

(a $q \times q$ matrix of zeros) in probability for all $t > 0$ as $h \downarrow 0$. Thus, the ARCH model consistently estimates (pointwise in t) the true underlying conditional covariance as h, the time between observations, goes to zero. Note that our use of the term "estimate" corresponds to its use in the filtering literature rather than the statistics literature; that is, an ARCH

[2] Similarly, in our notation, $A_{i,j,t}$ is the $(i-j)$th element of the matrix function A_t at time t, and $b_{i,t}$ is the ith element of the vector function b_t at time t. Where A_t can be written as $A(X_t)$, we use the notations interchangeably.

model with (given) fixed parameters produces "estimates" of the true underlying conditional covariance matrix at each point in time in the same sense that a Kalman filter produces "estimates" of unobserved state variables in a linear system.[3]

1.3. Some Intuition on Why Misspecified ARCH Models Can Provide Consistent Filters

In (1.1), we assumed that $\{X_t\}$ follows a diffusion process and that $\Omega(x)$ is continuous. The sample path of $\{\Omega_t\}$ is therefore continuous with probability one. Suppose we are trying to estimate the value of the $\Omega_{1,1,T}$, the time T instantaneous variance of increments in $\{X_{1,t}\}$, the first element of X_t, using information in the sample path of X_1 up to time T. Since the path of $\{\Omega_t\}$ is continuous almost surely, for every $\varepsilon > 0$ and every $T > 0$ there exists, with probability one, a random $\Delta(T) > 0$ such that

$$\sup_{T-\Delta(T) \le s \le T} |\Omega_{1,1,s} - \Omega_{1,1,T}| < \varepsilon. \tag{1.3}$$

That is, over a suitably small time interval, the change in $\Omega_{1,1}$ can be made arbitrarily small, as can the change in μ_1, the drift in X_1. Now take a small interval $[T - \Delta, T]$, chop it into M equal pieces, and estimate $\Omega_{1,1,T}$ by

$$\hat{\Omega}_{1,1,T}(\Delta, M) \equiv \Delta^{-1} \sum_{j=1}^{M} \left[X_{1,T-(j-1)\Delta/M} - X_{1,T-j\Delta/M}\right]^2. \tag{1.4}$$

For small Δ, μ_1 and $\Omega_{1,1}$ are effectively constant. Given $\mu_{1,T-\Delta}$ and $\Omega_{1,1,T-\Delta}$ the increments $X_{1,T-(j-1)\Delta/M} - X_{1,T-j\Delta/M}$ are approximately i.i.d. $N(M^{-1}\Delta\mu_{1,T-\Delta}, M^{-1}\Delta\Omega_{1,1,T-\Delta})$. Under suitable moment conditions, a law of large numbers yields $[\hat{\Omega}_{1,1,T}(\Delta, M) - \Omega_{1,1,T}] \to 0$ in probability as $\Delta \downarrow 0$ and $M \to \infty$. Note that failing to correct for the nonzero drift in X_1 does not interfere with consistency, since its effect on $\hat{\Omega}_{1,1,T}(\Delta, M)$ vanishes as $M \to \infty$ and $\Delta \downarrow 0$.

Like $\hat{\Omega}_{1,1,T}(\Delta, M)$, the conditional covariance estimates produced by GARCH(1, 1) are long distributed lags of squared residuals. We can create a sequence of GARCH(1, 1) models that, like $\hat{\Omega}_{1,1,T}(\Delta, M)$, form estimates of $\Omega_{1,1,T}$ by averaging increasing numbers of squared residuals from the increasing recent past. In this way, a sequence of GARCH(1, 1) models can consistently estimate $\Omega_{1,1,T}$, despite misspecifying the drift in X_1, ignoring the other state variables $X_{2:n,T}$, and misspecifying the dynamics of $\Omega_{1,1,T}$, whose random behavior may be very different from that of a GARCH(1, 1).

[3] See, e.g., the use of the term in Anderson and Moore (1979, ch. 2) or Arnold (1973, ch. 12).

To see that this consistency is not limited to GARCH(1, 1), consider the following estimation procedure inspired by the methods in Taylor (1986) and Schwert (1989): Instead of estimating $\Omega_{1,1,T}$ by a sum of lagged squared residuals, estimate $\Omega^{1/2}_{1,1,T}$ by a sum of lagged absolute residuals:

$$\hat{\Omega}^{1/2}_{1,1,T}(\Delta, M) \equiv (2\,\Delta M/\pi)^{-1/2} \sum_{j=1}^{M} |X_{1,T-(j-1)\Delta/M} - X_{1,T-j\Delta/M}|, \tag{1.5}$$

which again is consistent under sufficient moment conditions as $\Delta \downarrow 0$ and $M \to \infty$.

Thus far, we have presented heuristics: we have not specified regularity conditions on the diffusion (1.1) and on the sequence of ARCH models to guarantee consistency. In addition, there are other ARCH models, such as the exponential ARCH (EGARCH) model of Nelson (1991), in which the conditional variance estimate is not a simple moving average of transformed residuals. To make our heuristics rigorous and broadly applicable requires more work. In section 2, we present basic results, developed by Ethier and Nagylaki (1988), on diffusion approximations. In section 3, we use these results to prove consistency theorems for three ARCH models: the multivariate GARCH(1, 1) process of Bollerslev *et al.* (1988), the AR(1) exponential ARCH model of Nelson (1991), and a variant of GARCH(1, 1) inspired by Taylor (1986) and Schwert (1989). Section 4 discusses some circumstances in which consistency breaks down, while section 5 concludes the chapter. Proofs are gathered in the appendix.

2. WEAK CONVERGENCE OF MARKOV PROCESSES TO DIFFUSIONS

In this section we present results, drawn largely from Stroock and Varadhan (1979) and Ethier and Nagylaki (1988), on the weak convergence of a sequence of Markov processes to a diffusion. We begin by defining our sequence of processes, which we divide into two pieces: $\{_hX_t\}$ and $\{_hY_t\}$, $n \times 1$ and $m \times 1$ vector processes, respectively.[4] $\{_hX_t\}$ and $\{_hY_t\}$ will be random step functions taking jumps at times h, $2h$, $3h$, and so on. In the convergence results below, we present sufficient conditions for $\{_hX_t\} \Rightarrow \{X_t\}$, where $\{X_t\}$ is a solution of (1.1) and $\Rightarrow$ denotes weak convergence. In the applications in section 3, $\{_hX_t\}$ represents the underlying stochastic system generating the data, while $\{_hY_t\}$ represents the difference between the true conditional covariance matrix of $\{_hX_t\}$ and the estimate generated

[4] In our notation, $\{_hX_t\}$ refers to $_hX_t$ as a random function of time—i.e., it refers to the $_hX_t$ *process*. On the other hand, $_hX_t$ (without the curly brackets) refers to the value that the $\{_hX_t\}$ process takes at a particular time t.

by an ARCH model. The conditions below will be used in section 3 to guarantee that for every $t > 0$, ${}_hY_t$ converges in probability to an $m \times 1$ vector of zeros as $h \downarrow 0$—i.e., that for each $t > 0$, our conditional covariance estimator is consistent as $h \downarrow 0$.

Formally, let $D([0,\infty), R^n \times R^m)$ be the space of functions from $[0,\infty)$ into $R^n \times R^m$ that are right continuous with finite left limits (RCLL). D is a metric space when endowed with the Skorohod metric (see Ethier and Kurtz, 1986, ch. 3, for formal definitions). For each $h > 0$, let ${}_h\mathscr{L}_{kh}$ be the information set at time kh—i.e., the σ-algebra generated by ${}_hX_0, {}_hX_h, {}_hX_{2h,\dots,}{}_hX_{kh}$ and ${}_hY_0, {}_hY_h, {}_hY_{2h,\dots,}{}_hY_{kh}$. Let $B(R^n \times R^m)$ denote the Borel sets of $R^n \times R^m$, and let ν_h be a probability measure on $(R^n, B(R^n \times R^m))$. For each $h > 0$, let $\Pi_h(x, y, \cdot)$ be a transition function on $R^n \times R^m$—i.e.,

(a) $\Pi_h(x, y, \cdot)$ is a probability measure on $(R^n \times R^m, B(R^n \times R^m))$ for all $(x, y) \in R^n \times R^m$.
(b) $\Pi_h(\cdot, \cdot, \Gamma)$ is $B(R^n \times R^m)$ measurable for all $\Gamma \in B(R^n \times R^m)$.

Let P_h be the probability measure on $D([0,\infty), R^n \times R^m)$ such that

$$\mathrm{P}_h\left[({}_hX_0, {}_hY_0) \in \Gamma\right] = \nu_h(\Gamma) \quad \text{for any} \quad \Gamma \in B(R^n \times R^m), \tag{2.1}$$

$$\mathrm{P}_h\left[({}_hX_t, {}_hY_t) = ({}_hX_{kh}, {}_hY_{kh}), \qquad kh \le t < (k+1)h\right] = 1, \tag{2.2}$$

$$\mathrm{P}_h\left[\left({}_hX_{(k+1)h}, {}_hY_{(k+1)h}\right) \in \Gamma \mid {}_h\mathscr{L}_{kh}\right] = \Pi_h({}_hX_{kh}, {}_hY_{kh}, \Gamma) \tag{2.3}$$

almost surely under P_h for all $k \ge 0$ and all $\Gamma \in B(R^n \times R^m)$.

It is possible to make the transition probability depend on time by making the time index kh an element of ${}_hX_{kh}$.

It is convenient to impose the following condition ruling out feedback from $\{{}_hY_{kh}\}$ to $\{{}_hX_{kh}\}$: for every Borel subset Γ_x of R^n and for all $h > 0$,

$$\Pi_h(x, y, \Gamma_x \times R^m) \quad \text{is independent of } y. \tag{2.4}$$

For each $h > 0$, the $(n + m)$-dimension discrete time Markov process $\{{}_hX_{kh}, {}_hY_{kh}\}$ has transition probabilities given by (2.3) and a random starting point $({}_hX_0, {}_hY_0)$ given by (2.1). We form the continuous time process $\{{}_hX_t, {}_hY_t\}$ from $\{{}_hX_{kh}, {}_hY_{kh}\}$ by (2.2), making $\{{}_hX_t, {}_hY_t\}$ a step function with jumps at times h, $2h$, $3h$, and so on.

Our notation must keep track of three distinct kinds of processes:

(a) The sequence of discrete time processes $\{{}_hX_{kh}\}$ and $\{{}_hY_{kh}\}$ that depend both on h and on the (discrete) time index kh, $k = 0, 1, 2, \dots,$
(b) The sequence of continuous time processes $\{{}_hX_t\}$ and $\{{}_hY_t\}$ formed as step functions from the discrete time process in (a) using (2.2), and
(c) A limit diffusion process $\{X_t\}$, with $\{{}_hX_t\} \Rightarrow \{X_t\}$ as $h \downarrow 0$.

To understand the weak convergence results below, it is useful to think of $\{X_t\}$, a solution to (1.1), as a random variable taking a value in $D([0,\infty), R^n)$. Under regularity conditions explored in Stroock and Varadhan (1979) and Ethier and Kurtz (1986), four characteristics uniquely define the distribution of $\{X_t\}$:

(a) Starting point X_0 (or its distribution if X_0 is random),
(b) Drift function $\mu(x)$,
(c) Covariance function $\Omega(x)$, and
(d) Almost sure continuity of $\{X_t\}$ as a function of t.

We achieve $\{{}_hX_t\} \Rightarrow \{X_t\}$ by making $\{{}_hX_t\}$ match (a)–(d) in the limit as $h \downarrow 0$. To match (a), we require Assumption 1.

ASSUMPTION 1. *$({}_hX_0, {}_hY_0) \Rightarrow (X_0, Y_0)$ as $h \downarrow 0$, where (X_0, Y_0) has probability measure ν_0.*

Next, we require that the discrete analogs of μ and Ω, μ_h and Ω_h, are well-defined for all $(x, y) \in R^{n+m}$:

$$\mu_h(x) \equiv h^{-1}\mathrm{E}\big[{}_hX_{(k+1)h} - {}_hX_{kh} \mid {}_hX_{kh} = x, {}_hY_{kh} = y\big], \tag{2.5}$$

$$\Omega_h(x) \equiv h^{-1}\mathrm{E}\big[\big({}_hX_{(k+1)h} - {}_hX_{kh}\big) \times \big({}_hX_{(k+1)h} - {}_hX_{kh}\big)' \mid {}_hX_{kh} = x, {}_hY_{kh} = y\big], \tag{2.6}$$

where the expectations in (2.5)–(2.6) are taken under P_h.[5] μ_h and Ω_h are independent of y by (2.4). Next, define the norm of a $q \times r$ real matrix A as

$$\|A\| \equiv \left[\sum_{i=1,q} \sum_{j=1,r} A_{i,j}^2\right]^{1/2}. \tag{2.7}$$

To match (b)–(c) in the limit as $h \downarrow 0$, we require

ASSUMPTION 2. *For every $\eta > 0$,*

$$\lim_{h \downarrow 0} \sup_{\|x\| \le \eta} \|\mu_h(x) - \mu(x)\| = 0, \tag{2.8}$$

$$\lim_{h \downarrow 0} \sup_{\|x\| \le \eta} \|\Omega_h(x) - \Omega(x)\| = 0, \tag{2.9}$$

where $\mu(\cdot)$ and $\Omega(\cdot)$ are continuous.

[5] Equation (2.6) defines $\Omega_h(x)$ as an uncentered second moment. None of our results would be altered by using the centered moments instead [i.e., by replacing $({}_hX_{h(k+1)} - {}_hX_{hk})$ with $({}_hX_{h(k+1)} - {}_hX_{hk} - h\mu_h({}_hX_{hk}))$ in 2.6], since the difference between the centered and uncentered second moments vanishes as $h \downarrow 0$.

Assumption 2 requires μ_h and Ω_h to converge uniformly on every bounded subset of R^n, which is much less restrictive than uniform convergence on R^n.

The following fourth moment condition guarantees sample path continuity of the limit process [see Arnold (1973, p.40) and Nelson (1990a, theorem 2.2)], and so matches (d) in the limit as $h \downarrow 0$.

ASSUMPTION 3. *For every* $\eta > 0$ *and all* $i = 1, \ldots, n$,

$$\lim_{h \downarrow 0} \sup_{\|x\| \leq \eta} h^{-1} \mathrm{E}\left[\left({}_hX_{i,(k+1)h} - {}_hX_{i,kh}\right)^4 \mid {}_hX_{kh} = x\right] = 0. \quad (2.10)$$

Finally, (a)–(d) must completely characterize the distribution of $\{X_t\}$.

ASSUMPTION 4. *There is a distributionally unique solution to* (1.1).[6]

THEOREM 2.1 (Stroock–Varadhan). *Under Assumptions* 1–4, $\{{}_hX_t\} \Rightarrow \{X_t\}$ *as* $h \downarrow 0$.

We next turn to $\{{}_hY_t\}$. First, for every $\delta > 0$, $h > 0$, and $(x, y) \in R^{n+m}$, let the following conditional expectations be well-defined:

$$c_{h,\delta}(x, y) \equiv h^{-\delta} \mathrm{E}\left[{}_hY_{(k+1)h} - {}_hY_{kh} \mid {}_hX_{kh} = x, {}_hY_{kh} = y\right], \quad (2.11)$$

$$d_{h,\delta}(x, y) \equiv h^{-\delta} \mathrm{E}\left[\left({}_hY_{(k+1)h} - {}_hY_{kh}\right) \times \left({}_hY_{(k+1)h} - {}_hY_{kh}\right)' \mid {}_hX_{kh} = x, {}_hY_{kh} = y\right]. \quad (2.12)$$

ASSUMPTION 5. *For some* δ, $0 < \delta < 1$, *and for every* $\eta > 0$,

$$\lim_{h \downarrow 0} \sup_{\|(x', y')\| \leq \eta} \|c_{h,\delta}(x, y) - c(x, y)\| = 0, \quad (2.13)$$

where for all $x \in R^n$, $c(x, 0) = 0$, *and*

$$\lim_{h \downarrow 0} \sup_{\|(x', y')\| \leq \eta} \|d_{h,\delta}(x, y)\| = 0. \quad (2.14)$$

While the drift and conditional second moment of the increments in $\{{}_hX_{kh}\}$ are $O(h)$, Assumption 5 guarantees that the drift and (uncentered) second moment of the increments in $\{{}_hY_{kh}\}$ are $O(h^\delta)$ and $o(h^\delta)$, respectively. This has two important implications: First, $\{{}_hY_t\}$ operates on a faster time scale than $\{{}_hX_t\}$, since the drift (and possibly the variance) per unit of time of $\{{}_hY_t\}$ grow at a faster rate (as $h \downarrow 0$) than the drift and variance per unit of time of $\{{}_hX_t\}$. This implies that if $\{{}_hY_t\}$ mean-reverts to a vector of zeros, it does so with increasing speed as $h \downarrow 0$. Second, as $h \downarrow 0$ the drift of

[6] See Ethier and Kurtz (1986, pp. 290–291) for formal definitions. Several sets of sufficient conditions for Assumption 4 are summarized in appendix A of Nelson (1990a).

$\{_hY_t\}$ [which is $O(h^\delta)$] dominates the variance of $\{_hY_t\}$ [which is $o(h^\delta)$]. This allows us to approximate the behavior of $\{_hY_t\}$ by a *deterministic* differential equation. The next assumption assures that this differential equation is well-behaved, pulling $\{_hY_t\}$ back to a vector of zeros.

ASSUMPTION 6. *For each* $x \in R^n$, $y \in R^m$, *define the differential equation*

$$dY(t,x,y)/dt = c(x, Y(t,x,y)), \tag{2.15}$$

with initial condition

$$Y(0,x,y) = y. \tag{2.16}$$

Then $0_{m\times 1}$ *is a globally asymptotically stable solution of* (2.15)–(2.16) *for bounded values of* x, y*—i.e., for every* $\eta > 0$,

$$\lim_{t\to\infty} \sup_{\|(x',y')\|\le\eta} \|Y(t,x,y)\| = 0.\text{[7]} \tag{2.17}$$

Finally, we require a Lyapunov condition which guarantees that $\{_hY_t\}$ does not diverge to infinity in finite time.[8]

ASSUMPTION 7. *The exists a nonnegative function* $\rho(x,y,h)$, *twice differentiable in* x *and* y, *and a positive function* $\lambda(\eta,h)$ *such that*

$$\lim_{\eta\to\infty} \liminf_{h\downarrow 0} \inf_{\|(x',y')\|\ge\eta} \rho(x,y,h) = \infty, \tag{2.18}$$

$$\limsup_{\eta\to\infty} \limsup_{h\downarrow 0} \lambda(\eta,h) < \infty, \tag{2.19}$$

$$\limsup_{h\downarrow 0} \mathrm{E}[\rho(_hX_0, {}_hY_0, h)] < \infty, \tag{2.20}$$

and for every $\eta > 0$,

$$\limsup_{h\downarrow 0} \sup_{\|(x',y')\|\le\eta} h^{-1}\mathrm{E}\big[\rho(_hX_{(k+1)h}, {}_hY_{(k+1)h}, h) - \rho(x,y,h) \mid {}_hX_{kh} = x, {}_hY_{kh} = y\big] - \lambda(\eta,h)\rho(x,y,h) \le 0. \tag{2.21}$$

THEOREM 2.2 (Ethier–Nagylaki). *Let Assumptions* 1–7 *hold. Then*

$$_hY_t \to 0_{m\times 1} \quad \textit{in probability as } h\downarrow 0 \quad \textit{for every} \quad t > 0. \tag{2.22}$$

[7] It is often possible to verify Assumption 6 without solving (2.15)–(2.16). For example, the following Lyapunov condition suffices: There is a nonnegative $V(x,y)$, bounded on compact (x,y) sets and differentiable in y functions $\kappa(\eta)$ and $g(x)$, and a $\Delta > 0$ such that for each $N > 0$, $\inf_{0\le\eta\le N}\kappa(\eta) > 0$, $\inf_{0\le\|x\|\le\eta} g(x) > 0$, $\|y\|^\Delta g(x) \le V(x,y)$, and for each $\eta > 0$ and all $y \in R^m$, $\sup_{\|x\|\le\eta}\kappa(\eta)V(x,y) + (\partial V(x,y)/\partial y)' \cdot c(x,y) \le 0$.

[8] This condition also guarantees that $\{X_t\}$ does not explode in finite time, but this was already implied by Assumption 4 (see Stroock and Varadhan 1979, ch. 10).

3. EXAMPLES OF CONSISTENT ARCH FILTERS

In this section, we prove filter consistency for three ARCH models [multivariate GARCH(1, 1), AR(1) EGARCH, and the Taylor–Schwert model] when (1.1) generates $\{_hX_t\}$—i.e., we let $\{X_t\}$ be a solution to (1.1), and then define for all $t \geq 0$, $_hX_t \equiv X_{[t/h]\cdot h}$ where $[t/h]$ is the integer part of t/h. We then expand these results to allow $\{_hX_t\}$ to be generated by a stochastic *difference* equation.

3.1. Multivariate GARCH(1, 1)

In the multivariate GARCH(1, 1) model of Bollerslev *et al.* (1988), conditional covariance estimates are formed by the recursion

$$\begin{aligned} \operatorname{vech}\left[{}_h\hat{\Omega}_{1:q,1:q,(k+1)h}\right] &= W_h + B_h \operatorname{vech}\left[{}_h\hat{\Omega}_{1:q,1:q,kh}\right] \\ &\quad + h^{-1}A_h \operatorname{vech}\left[{}_h\hat{\xi}_{1:q,kh}\,{}_h\hat{\xi}'_{1:q,kh}\right], \end{aligned} \tag{3.1}$$

where vech(·) stacks the columns of the lower portion of a symmetric matrix. W_h, B_h, and A_h are $\frac{1}{2}(q+1)q \times \frac{1}{2}(q+1)q$ matrices. The $\{_h\hat{\xi}_{1:q,kh}\}$ in (3.1) are fitted residuals obtained using the (possibly misspecified) drift $\hat{\mu}_h({}_hX_{kh}, {}_h\hat{\Omega}_{1:q,1:q,kh})$:

$${}_h\hat{\xi}_{1:q,kh} \equiv {}_hX_{1:q,(k+1)h} - {}_hX_{1:q,kh} - h \cdot \hat{\mu}_h\left({}_hX_{kh}, {}_h\hat{\Omega}_{1:q,1:q,kh}\right). \tag{3.2}$$

For some δ, $0 < \delta < 1$, let W_h, A_h, B_h, and $\hat{\mu}_h(\cdot,\cdot)$ satisfy

$$\hat{\mu}_h = o(h^{-1/2}),^{9} \tag{3.3}$$

$$W_h = o(h^{\delta}), \tag{3.4}$$

$$I - B_h - A_h = o(h^{\delta}), \tag{3.5}$$

$$A_h = h^{\delta}A + o(h^{\delta}), \tag{3.6}$$

where I is a $\frac{1}{2}(q+1)q \times \frac{1}{2}(q+1)q$ identity matrix and A is a $\frac{1}{2}(q+1)q \times \frac{1}{2}(q+1)q$ matrix independent of h. Defining

$${}_hY_t \equiv \operatorname{vech}\left[{}_h\hat{\Omega}_{1:q,1:q,t} - \Omega_{1:q,1:q,t}\right], \tag{3.7}$$

we have under regularity conditions given below,

$$h^{-1}\mathrm{E}\left[{}_hY_{(k+1)h} - {}_hY_{kh} \mid {}_hX_{kh} = x, {}_hY_{kh} = y\right] = -h^{\delta-1}A \cdot y + O(1). \tag{3.8}$$

[9] $f(x, y, h) = O(h^{\gamma})$ means that for every positive finite η, there is a $\lambda \geq 0$ such that

$$\limsup_{h \downarrow 0} \sup_{\|(x', y')\| < \eta} h^{-\gamma}\|f(x, y, h)\| \leq \lambda.$$

That is, $h^{-\gamma}f$ is uniformly bounded in h for bounded (x, y). If $\lambda = 0$, then we say that $f(x, y, h) = o(h^{\gamma})$.

${}_hY_t$ is the time t estimation error–i.e., the vech of the difference between the estimated and true condition covariances. For each $h > 0$ $\{{}_hY_t\}$ fluctuates randomly, but mean-reverts to a vector of zeros if A is well-behaved. As $h \downarrow 0$, the speed of this mean-reversion goes to infinity, implying under regularity conditions that for each $t > 0$, ${}_hY_t$ converges in probability to a vector of zeros as $h \downarrow 0$. All of the filters presented in this section achieve consistency in this way—i.e., by increasing rapid mean-reversion in $\{{}_hY_t\}$ as $h \downarrow 0$.

We next give regularity conditions on the diffusion (1.1), on $\hat{\mu}_h$, W_h, A_h, and B_h, and on the starting value ${}_h\hat{\Omega}_{1:q,1:q,0}$ guaranteeing that Assumptions 1–7 in section 2 hold. We are then able to apply Theorem 2.2 to prove that the measurement error ${}_hY_t$ converges to zero in probability for each $t > 0$ as $h \downarrow 0$. We emphasize that the conditions given below, while *sufficient* for filter consistency, may or may not be *necessary*. We use Conditions 3.1–3.3 repeatedly in this section. Conditions 3.4–3.5 are specific to the GARCH(1, 1) case. While these conditions impose conditional moment and nonexplosion conditions on the diffusion (1.1), they do not impose stationarity or ergodicity.

Condition 3.1. For each $h > 0$ (1.1) generates $\{{}_hX_t\}$ and satisfies Assumptions 1, 3, and 4 of section 2.

Condition 3.2. For some $\varepsilon > 0$, $\limsup_{h \downarrow 0} \mathrm{E}\,\|{}_hY_0\|^{2+\varepsilon} < \infty$.

Condition 3.3. There is a twice differentiable, nonnegative $\omega(x)$ and a $\theta > 0$ such that for every $\eta > 0$,

$$\limsup_{\|x\| \to \infty} \omega(x) = \infty, \tag{3.9}$$

$$\lim_{h \downarrow 0} \mathrm{E}[\,\omega({}_hX_0)] < \infty, \tag{3.10}$$

$$\lim_{h \downarrow 0} \sup_{\|x\| \le \eta} h^{-1}\mathrm{E}\Big(\Big[\big|\,\omega\big({}_hX_{(k+1)h}\big) - \omega(x)\big|^{1+\theta}\,|\,{}_hX_{kh} = x\Big]\Big) < \infty, \tag{3.11}$$

and there is a $\lambda > 0$ for all $x \in R^n$,

$$\sum_{i=1}^{n} \mu_i(x)\frac{\partial\omega(x)}{\partial x_i} + \frac{1}{2}\sum_{i=1}^{n}\sum_{j=1}^{n} \Omega_{i,j}(x)\frac{\partial^2\omega(x)}{\partial x_i\,\partial x_j} \le \lambda\omega(x). \tag{3.12}$$

We use Conditions 3.2–3.3 to verify Assumption 7. Condition 3.2 is trivially satisfied if ${}_h\Omega_{1:q,1:q,0}$ and ${}_h\hat{\Omega}_{1:q,1:q,0}$ are nonrandom and independent of h. Though Condition 3.3 looks formidable, it is often easy to verify, since verifying Assumption 4 usually involves a condition close to Condition 3.3 (see the examples and discussion in Nelson, 1990a, esp. app. A).

Condition 3.4. For every $\eta > 0$, there is an $\varepsilon > 0$ such that for every $i, j, 1 \leq i, j \leq q$,

$$\lim_{h \downarrow 0} \sup_{\|x\| \leq \eta} h^{-1} \mathrm{E}\left[\left| \Omega_{i,j}({}_hX_{(k+1)h}) - \Omega_{i,j}(x) \right|^{2+\varepsilon} \mid {}_hX_{kh} = x \right] = 0, \quad (3.13)$$

$$\lim_{h \downarrow 0} \sup_{\|x\| \leq \eta} h^{-1} \mathrm{E}\left[|{}_hX_{i,(k+1)h} - x_i|^{4+\varepsilon} \mid {}_hX_{kh} = x \right] = 0. \quad (3.14)$$

Condition 3.5. W_h, A_h, and B_h satisfy (3.3)–(3.5). All the eigenvalues of A have strictly positive real parts.

The differential equation of Assumption 6 is now

$$dY/dt = -A \cdot Y, \quad (3.15)$$

which has the unique solution $Y_t = \exp(-At)Y_0$ where $\exp(\cdot)$ is the matrix exponential. Condition 3.5 assures that Y satisfies Assumption 6 (Hochstadt, 1975).

THEOREM 3.1. *Let Conditions* 3.1–3.5 *hold. Then for each* $t > 0$, $\|{}_hY_t\| \to 0$ *in probability as* $h \downarrow 0$.

The GARCH model is misspecified, since it is not the true data-generating process. Nevertheless, it achieves filter consistency as $h \downarrow 0$. The GARCH filter need not be heavily parameterized, regardless of how high q (the dimension of $X_{1:q,t}$) is. For example, let α be a positive number, set $\hat{\mu}_h \equiv 0_{q \times 1}$, $A_h = h^{1/2}\alpha I$, $B_h = (1 - h^{1/2}\alpha)I$, and $W_h = 0$, where "I" and "0" are $\frac{1}{2}(q+1)q \times \frac{1}{2}(q+1)q$ identity and zero matrices, respectively. Under Conditions 3.1–3.5, this defines a consistent filter for $\{\Omega_{1:q,1:q,t}\}$.

Under (3.3)–(3.5), $A_h + B_h$ converge to an identity matrix as $h \downarrow 0$. This also holds for GARCH models considered as data-generating processes converging to a diffusion limit (Nelson, 1990a) and is consistent with findings in the empirical ARCH literature. For example, Baillie and Bollerslev (1989) estimated univariate GARCH(1, 1) models on exchange rate data for several currencies using various sampling intervals. The estimated values of $A_h + B_h$ using daily data were very close to one, less so using weekly data. Using monthly data, $A_h + B_h$ were much less than one.[10] In interpreting results from ARCH models fit to high-frequency data, it is important to remember that values of $A_h + B_h$ close to one do

[10] However, Baillie and Bollerslev (1990) find that when even finer time increments are used (i.e., hourly) MLE estimates of $A_h + B_h$ begin to move away from one. It may be that lumpy information and/or market microstructure effects destroy the validity of the diffusion approximation at extremely high frequencies. On the other hand, Chan and Chung (1993) estimate higher-order GARCH models on intra-day data and find $\Sigma_j A_{j,h} + B_{j,h}$ close to one again.

not necessarily indicate nonstationarity—i.e., even for a sequence of covariance-stationary GARCH(1, 1) models converging to a covariance-stationary diffusion limit, $A_h + B_h$ converge to an identity matrix as $h \downarrow 0$.[11]

Caution is required in applying our results when the parameters of the ARCH model (i.e., $\hat{\mu}_h$, W_h, A_h, and B_h) are random, as they are when they are estimated, for example by quasi-maximum likelihood. There are reasons to believe that our results apply when W_h, A_h, and B_h are estimated and the data are generated by a diffusion: first, because the class of consistent filters in Theorem 3.1 is very broad, and second, because better conditional covariance estimates should produce higher quasi-likelihoods. This heuristic is not a proof, however.

For a given GARCH model to make sense as a data-generating process, the conditional covariance matrix ${}_h\hat{\Omega}_{kh}$ must be nonnegative definite for all (h, k) with probability one. This need not be true of a consistent filter. It is easy to construct examples of sequences of GARCH(1, 1) models which are consistent filters, but in which "variance" becomes *negative* with probability one *when the model is considered as a data-generating process*. For example, let a univariate IGARCH(1, 1) model (Bollerslev, 1986) generate $\{{}_h\sigma_{kh}^2\}$:

$$\alpha_h = h^{1/2}\alpha, \tag{3.16}$$

$$\beta_h = 1 - h^{1/2}\alpha, \qquad 0 < \alpha < \infty, \qquad 0 < h < 1/\alpha^2, \tag{3.17}$$

$$\omega_h = h \cdot \omega, \tag{3.18}$$

$${}_h\sigma_{(k+1)h}^2 = \omega_h + \beta_h \cdot {}_h\sigma_{kh}^2 + h^{-1}\alpha_h \cdot {}_h\sigma_{kh}^2 \cdot {}_h z_{kh}^2, \tag{3.19}$$

$${}_h z_{kh} \sim \text{i.i.d. N}(0, h). \tag{3.20}$$

When $\omega \geq 0$, ${}_h\sigma_{kh}^2$ remains nonnegative with probability one for all (h, k), and the process has a well-defined diffusion limit (Nelson, 1990a). For each h and k we can define ${}_h\xi_{kh} \equiv {}_h z_{kh} \cdot {}_h\sigma_{kh}$ and write

$$h^{-1}\operatorname{var}[{}_h\xi_{kh} \mid {}_h\mathscr{L}_{kh}] = {}_h\sigma_{kh}^2. \tag{3.21}$$

When $\omega < 0$, however, (3.16)–(3.20) imply that for every h, $\{{}_h\sigma_{kh}^2\}$ eventually becomes negative (for large k) with probability one and remains negative forever (Nelson, 1990b). Clearly such a model yields unacceptable long-term forecasts if we insist on interpreting $\{{}_h\sigma_{kh}^2\}$ as a conditional variance! Nevertheless, (3.16)–(3.19) define a consistent filter, even when $\omega < 0$.

The misspecification in the conditional mean and covariance processes permitted by Theorem 3.1 is even wider. For example, while the true drift

[11] The assumption of a diffusion limit here is critical—it may well be that convergence to, say, a jump–diffusion limit can occur without $A_h + B_h$ converging to an identity matrix.

per unit of time $\mu(x)$ is bounded for bounded values of x, (3.3) permits the drift per unit of time $\hat{\mu}_h$ assumed in the ARCH model to explode to infinity as $h \downarrow 0$. Next, we have ignored the state variables $X_{q+1:n,t}$. As a final example of the ways in which the ARCH model may misspecify the data-generating mechanism yet still provide consistent filtering, suppose we replace the vech operators in (3.1) with ordinary vec operators: we could then create ${}_h\hat{\Omega}_{1:q,1:q,t}$ matrices that are *asymmetric* with probability one for all (h, t), and yet are consistent filters.[12]

3.2. AR(1) EGARCH

As a discrete time data-generating process, the AR(1) exponential ARCH model is given by

$$ {}_hX_{1,kh} = {}_hX_{1,(k-1)h} + h \cdot \mu_h\left({}_hX_{1,(k-1)h}, {}_h\sigma_{kh}^2, kh\right) + {}_hz_{kh} \cdot {}_h\sigma_{kh}, \quad (3.22) $$

$$ \ln\left({}_h\sigma_{(k+1)h}^2\right) = \ln\left({}_h\sigma_{kh}^2\right) - \beta_h \cdot h \cdot \left[\ln\left({}_h\sigma_{kh}^2\right) - \alpha_h\right] $$

$$ + \theta_h \cdot {}_hz_{kh} + \gamma_h\left[|{}_hz_{kh}| - m \cdot h^{1/2}\right], \quad (3.23) $$

$$ {}_hz_{kh} \sim \text{i.i.d. N}(0, h), \quad (3.24) $$

$$ m \equiv \mathrm{E}\left[|h^{-1/2}{}_hz_{kh}|\right] = (2/\pi)^{1/2}. \quad (3.25) $$

Nelson (1990a) shows that if $\mu_h \to \mu$ uniformly on compacts,

$$ \beta_h = \beta + o(1), \qquad \alpha_h = \alpha + o(1), $$

$$ \theta_h = \theta + o(1), \qquad \gamma_h = \gamma + o(1), \quad (3.26) $$

if the starting values ${}_hX_0$ and $\ln({}_h\sigma_0^2)$ converge in distribution, and if μ satisfies fairly mild regularity conditions, then as $h \downarrow 0$, the sequence of step function processes $\{{}_hX_{1,t}, {}_h\sigma_t^2\}_{h \downarrow 0}$ generated by (3.22)–(3.26) and (2.2) converge weakly to the diffusion[13]

$$ dX_{1,t} = \mu\left(X_t, \sigma_t^2, t\right) dt + \sigma_t \, dW_{1,t}, \quad (3.27) $$

$$ d\ln\left(\sigma_t^2\right) = -\beta\left(\ln\left(\sigma_t^2\right) - \alpha\right) dt + dW_{2,t}, \quad (3.28) $$

[12] This distinction between the behavior of a model as a filter and its behavior as a data-generating process is found in linear models as well. For example, a stationary series that has been exponentially smoothed remains stationary. As a data-generating process, however, an exponential smoothing model corresponds to an ARIMA(0, 1, 1) model, which is nonstationary (Granger and Newbold, 1986).

[13] This diffusion is employed in the options pricing models of Hull and White (1987) and Wiggins (1987).

where $W_{1,t}$ and $W_{2,t}$ are one-dimensional Brownian motions with no drift and with

$$\begin{bmatrix} dW_{1,t} \\ dW_{2,t} \end{bmatrix} [dW_{1,t} \quad dW_{2,t}] = \begin{bmatrix} 1 & \theta \\ \theta & \theta^2 + (1 - m^2)\gamma^2 \end{bmatrix} dt. \quad (3.29)$$

But suppose, once again, that the data are generated by (1.1) and that a misspecified EGARCH model is used to produce an estimate of the true underlying conditional variance process $\{\Omega_{1,1}(X_t)\}$. We generate fitted $\{_h\hat{\sigma}_{kh}^2\}$ recursively by (3.23) with some $_h\hat{\sigma}_0^2$, and generate $\{_h\hat{z}_{kh}\}$ by

$$_h\hat{z}_{kh} \equiv \left[_hX_{1,kh} - {_hX_{1,(k-1)h}} - h \cdot \hat{\mu}_h\left(_hX_{(k-1)h}, {_h\hat{\sigma}_{kh}^2}, kh\right)\right] / {_h\hat{\sigma}_{kh}}. \quad (3.30)$$

When the model is misspecified, $\{_h\hat{z}_{kh}\}$ is neither i.i.d. nor $N(0, h)$. The requirements we place on $\hat{\mu}_h$, β_h, α_h, θ_h, and γ_h to achieve consistent filtering are much weaker than (3.26): for some δ, $0 < \delta < 1$, we require

$$\hat{\mu}_h = o(h^{-1/2}) \quad (3.31)$$

$$\beta_h = o(h^{\delta-1}), \quad (3.32)$$

$$\alpha_h \beta_h = o(h^{\delta-1}), \quad (3.33)$$

$$\theta_h = o(h^{(\delta-1)/2}), \quad (3.34)$$

$$\gamma_h = \gamma \cdot h^{\delta-1/2} + o(h^{\delta-1/2}) \quad \text{where} \quad \gamma > 0. \quad (3.35)$$

Defining the measurement error

$$_hY_t \equiv \left[\ln\left(_h\hat{\sigma}_t^2\right) - \ln\left(_h\Omega_{1,1,t}\right)\right], \quad (3.36)$$

we have

$$\begin{aligned} c(x, y) &\equiv \lim_{h \downarrow 0} h^{-\delta} \mathrm{E}\left[_hY_{(k+1)h} - {_hY_{kh}} \mid {_hX_{kh}} = x, {_hY_{kh}} = y\right] \\ &= \gamma \cdot m \cdot [\exp(-y/2) - 1]. \end{aligned} \quad (3.37)$$

The drift per unit of time in $\{_hY_{kh}\}$ is approximately $h^{\delta-1} \cdot c(x, y)$. By (3.37), this drift explodes to $+\infty$ $(-\infty)$ whenever $_hY_{kh}$ is below (above) zero, causing the measurement error to mean-revert to zero with increasing speed and vanish in the limit as $h \downarrow 0$. The differential equation of Assumption 6 is

$$dY/dt = \gamma \cdot m \cdot [\exp(-Y/2) - 1], \quad (3.38)$$

which is globally asymptotically stable when $\gamma > 0$. Condition 3.6 replaces Condition 3.4.

Condition 3.6. For every $\eta > 0$, there is an $\varepsilon > 0$ such that

$$\lim_{h \downarrow 0} \sup_{\|x\| \leq \eta} h^{-1} \mathrm{E}\left[\left| \ln\left(\Omega_{1,1}({}_{h}X_{(k+1)h})\right) - \ln(\Omega_{1,1}(x)) \right|^{2+\varepsilon} \Big| \, {}_{h}X_{kh} = x \right] = 0. \tag{3.39}$$

THEOREM 3.2. *Let* (3.31)–(3.36) *and Conditions* 3.1–3.3 *and* 3.6 *hold. Then* ${}_{h}Y_{t} \to 0$ *in probability for every* $t > 0$ *as* $h \downarrow 0$.

As was the case with GARCH(1, 1), the effect of misspecification in the conditional mean washes out of the conditional variance as $h \downarrow 0$. This is in accord with experience from estimated ARCH models. For example, Nelson (1991) fit an EGARCH model to daily stock index returns computed two different ways: first, using capital gains, and second, using including dividends and a (crude) adjustment for the riskless interest rate. Since dividends and riskless interest rates are almost entirely predictable one day ahead, their inclusion in the returns series amounts to a perturbation in the conditional mean of the capital gains series. The inclusion or exclusion of dividends and riskless rates made virtually no difference in the fitted conditional variances of the two series, which had nearly identical in-sample means and variances and a correlation of approximately 0.9996. Considering our results, this is not surprising.[14] Similarly, Schwert (1990) found that neglecting dividends made little difference in stock volatility estimates even in monthly data. Note also that (3.23) and (3.32) imply an approach to a unit root in $\{\ln({}_{h}\sigma_{kh}^{2})\}$ as $h \downarrow 0$. As the GARCH(1, 1) case, this holds for the model both as a filter and as a data-generating process and does not necessarily imply nonstationarity.[15]

3.3. The Taylor – Schwert Variant

Univariate GARCH(1, 1) creates its estimate of the conditional variance as a distributed lag of squared residuals. An equally natural procedure is to estimate the conditional standard deviation using a distributed lag of absolute residuals. Taylor (1986) employed this method to estimate conditional variances for several financial time series. Schwert (1989) used a closely related procedure, generating conditional standard deviations with a 12th-order moving average of absolute residuals. We follow Taylor

[14] Merton (1980) demonstrated the near-irrelevance of the mean as $h \downarrow 0$ when the variance is constant.

[15] Once again, this is in accord with the empirical literature—compare, for example, the AR parameters generated in Nelson (1991) using daily data and in Glosten *et al.* (1989) using monthly data. See also Priestley (1981, section 3.7.2) for a related discussion of passage to continuous time in the linear AR(1) model. Sims (1984) uses a related result to explain the martingale-like behavior of asset prices and interest rates.

(1986) by using an AR(1). As a data-generating process, the system is

$$ {}_hX_{1,kh} = {}_hX_{1,(k-1)h} + h \cdot \mu_h({}_hX_{1,(k-1)h}, {}_h\sigma_{kh}, kh) + {}_hz_{kh} \cdot {}_h\sigma_{kh}, \quad (3.40) $$

$$ {}_h\sigma_{(k+1)h} = \omega_h + \beta_h \cdot {}_h\sigma_{kh} + h^{-1/2}\alpha_h \, |{}_hz_{kh}| \, {}_h\sigma_{kh}, \quad (3.41) $$

$$ {}_hz_{kh} \sim \text{i.i.d. N}(0, h) \quad (3.42) $$

where ω_h, β_h, and α_h are nonnegative, and distribution $\nu_h(x, \sigma)$ of $\{{}_hX_{1,0}, {}_h\sigma_0\}$ converges to a limit $\nu_0(x, \sigma)$ as $h \downarrow 0$. The $\{{}_hX_{1,t}, {}_h\sigma_t\}$ are formed using (2.2).[16]

The properties of the procedure as a filter are similar to GARCH(1, 1) Let ${}_hX_{1,t}$ consist of the first element of ${}_hX_t$: As in the EGARCH case, define an initial value ${}_h\hat{\sigma}_0$ and define $\{{}_h\hat{z}_{kh}\}$ and $\{{}_h\hat{\sigma}_{kh}\}$ by (3.30) and (3.41). Again, define $m \equiv (2/\pi)^{1/2}$. for some δ with $0 < \delta < 1$, let

$$ \omega_h = o(h^\delta), \quad (3.43) $$

$$ \alpha_h = h^\delta\alpha + o(h^\delta) \quad \text{with } \alpha > 0, \quad (3.44) $$

$$ \beta_h = 1 - h^\delta\alpha \cdot m + o(h^\delta), \quad (3.45) $$

$$ \hat{\mu}_h = o(h^{-1.2}). \quad (3.46) $$

Defining

$$ {}_hY_t \equiv ({}_h\hat{\sigma}_t - {}_h\Omega_{1,1,t}^{1/2}), \quad (3.47) $$

the differential equation (2.15) is

$$ dY = -\alpha \cdot m \cdot Y dt, \quad (3.48) $$

which satisfies Assumption 6.[17]

Condition 3.7. For every $\eta > 0$, there is an $\varepsilon > 0$ such that

$$ \lim_{h \downarrow 0} \sup_{\|x\| \le \eta} h^{-1}\mathrm{E}\left[\left|\Omega_{1,1}^{1/2}({}_hX_{(k+1)h}) - \Omega_{1,1}^{1/2}(x)\right|^{2+\varepsilon} \big| \, {}_hX_{kh} = x\right] = 0. \quad (3.49) $$

[16] Higgins and Bera (1989) nest the Taylor–Schwert model and GARCH in a class of "NARCH" (nonlinear ARCH) models. They make $\sigma_t^{2\delta}$ a distributed tag of absolute residuals each raised to the 2δ power. The Taylor–Schwert model corresponds to $\delta = \frac{1}{2}$, while univariate GARCH sets $\delta = 1$.

[17] Considered as a data-generating process, the model's diffusion limit is also similar to GARCH(1, 1)'s. An application of Theorem 2.1 shows that if $\nu_h \to \nu_0$ as $h \downarrow 0$, $\omega_h = h\omega$, $\alpha_h = h^{1/2}\alpha$, $\beta_h = 1 - h^{1/2}\alpha \cdot m - \theta h$, with ω and α nonnegative, and if $\mu(\cdot)$ satisfies minimal regularity conditions, $\{{}_hX_{1,t}, {}_h\sigma_t\}$ converges weakly to the solution of

$$ dX_{1,t} = \mu(X_{1,t}, \sigma_t, t)\, dt + \sigma_t \, dW_{1,t}, $$

$$ d\sigma_t = (\omega - \theta\sigma_t)\, dt + \alpha(1 - m^2)^{1/2} \sigma_t \, dW_{2,t}, $$

as $h \downarrow 0$, where $\{W_{1,t}\}$ and $\{W_{2,t}\}$ are independent Brownian motions.

THEOREM 3.3. *Under* (3.43)–(3.46) *and Conditions* 3.1–3.3 *and* 3.7, ${}_hY_t \to 0$ *in probability as* $h \downarrow 0$ *for every* $t > 0$.

3.4. A Generalization: From Diffusions to Near-Diffusions

Thus far we have generated $\{{}_hX_t\}$ by the stochastic integral equation (1.1) observed at intervals of length h. This can be generalized: it is not necessary that $\{{}_hX_t\}$ is generated by a diffusion, though it is essential that the sequence of processes $\{{}_hX_t\}_{h \downarrow 0}$ converges to the diffusion (1.1) as $h \downarrow 0$. We call a process $\{{}_hX_t\}$ embedded in such a sequence a "near-diffusion."

For example $\{{}_hX_t\}$ may be generated by a stochastic volatility model (e.g., Melino and Turnbull, 1990) or by an ARCH model. This requires minor modifications to Theorems 3.1–3.3:

Condition 3.1′. The sequence of processes $\{{}_hX_t\}_{h \downarrow 0}$ is generated as in section 2 and satisfies Assumptions 1–4.

Condition 3.3′. There exists a twice differentiable, nonnegative function $\omega(x)$ such that

$$\lim_{\|x\| \to \infty} \omega(x) = \infty, \tag{3.50}$$

$$\limsup_{h \downarrow 0} \mathrm{E}\left[\omega({}_hX_0)\right] < \infty, \tag{3.51}$$

and there exists a $\lambda > 0$ such that for all $x \in R^n$

$$\limsup_{h \downarrow 0} h^{-1}\mathrm{E}\left[\omega({}_hX_{(k+1)h}) - \omega(x) \mid {}_hX_{kh} = x\right] - \lambda\omega(x) \leq 0. \tag{3.52}$$

THEOREM 3.4. *Let Conditions* 3.1′ *and* 3.3′ *replace Conditions* 3.1 *and* 3.3 *in the statement of Theorem* 3.1. *Then the conclusions of Theorems* 3.1 *hold. Next, let the function*

$$m(h,x) = \mathrm{E}\left[h^{-1/2}{}_h\Omega_{1,1}(x)^{-1/2}|{}_hX_{1,(k+1)h} - x_1| \mid {}_hX_{kh} = x\right] \tag{3.53}$$

be well-defined for all $x \in R^n$, *and let*

$$m(h,x) = m + o(1), \tag{3.54}$$

where m *is a constant,*[18] *and the* $o(1)$ *term in* (3.54) *vanishes uniformly on bounded* x *sets as* $h \downarrow 0$. *Let Conditions* 3.1′ *and* 3.3′ *replace Conditions* 3.1 *and* 3.3 *and* (3.54) *replace* (3.25) *in the statements of Theorems* 3.2–3.3. *Then the conclusions of Theorems* 3.2–3.3 *hold.*

[18] When the increments in $\{X_{1,t}\}$ are conditionally normal, $m = (2/\pi)^{1/2}$. In general, however, it is not. When $\{{}_hX_t\}$ is generated by (1.1), the increments in $\{{}_hX_t\}$ are approximately normal as $h \downarrow 0$, so $m = (2/\pi)^{1/2}$.

4. INCONSISTENT ARCH FILTERS

These consistency results do not, alas, apply to all ARCH models and all data-generating processes. There are at least three important cases in which consistency breaks down.

4.1. Not Using Enough Lagged Residuals

All of the consistent filters discussed in the chapter achieved consistency by (in effect) employing a law of large numbers to extract covariance estimates from an infinite number of lagged residuals. This suggests that consistent filtering will break down if the number of lagged residuals used to form the covariance estimates remain bounded as $h \downarrow 0$. For example, the ARCH(p) model of Engle (1982) and the semi-nonparametric (SNP) model of Gallant *et al.* (1990) employ only a finite number of residuals, p, in forming the conditional densities. If $p \to \infty$ at a suitable rate as $h \downarrow 0$, then consistency could no doubt be achieved. This may be difficult to prove formally, however: The convergence theorems of section 2 allowed only a finite number of state variables. Both for consistency filtering and for parsimony therefore, GARCH(p, q) may be preferable to ARCH(p) in applications where the data are observed at high frequencies and are generated by a near-diffusion. Similarly, it may be useful to extend the SNP model to allow infinite (geometrically declining) lags of residuals to appear in the conditional distribution of the observable variables.

4.2. Restrictions on the Functional Form of ${}_h\hat{\Omega}_{1:q,1:q}$

The Factor ARCH model of Engle *et al.* (1990) illustrates another way in which consistency may break down. This model imposes the condition that there is a $p < q$, and, for each $h > 0$, $q \times 1$ vectors β_h^i, $i = 1, \ldots, p$, a $q \times q$ matrix Ω_h, and nonnegative scalar random processes $\{{}_h\lambda_t^i\}$, $i = 1, \ldots, p$, such that for all $t \geq 0$,

$$
{}_h\hat{\Omega}_{1:q,1:q,t} = \Omega_h + \sum_{i=1}^{p} \beta_h^i \beta_h^{i\prime} {}_h\lambda_t^i. \tag{4.1}
$$

It is easy to see that (4.1) implies that for any $s \geq 0$ and $t \geq 0$, ${}_h\hat{\Omega}_{1:q,1:q,t} - {}_h\hat{\Omega}_{1:q,1:q,s}$ is singular. This generally prevents consistent filtering, since the difference between arbitrary $\Omega_{1:q,1:q,t}$ and $\Omega_{1:q,1:q,s}$ need not be singular.

The model of Schwert and Seguin (1990) imposes similar restrictions on the estimated conditional covariance matrix:

$$
{}_h\hat{\Omega}_{1:q,1:q,t} = \Omega_h + \Sigma_h \cdot {}_h\sigma_t + \Lambda_h \cdot {}_h\sigma_t^2, \tag{4.2}
$$

where Ω_h, Σ_h, and Λ_h are $q \times q$ symmetric matrices and $\{{}_h\sigma_t^2\}$ is the conditional variance of returns on a market index. (4.2) constrains the $q(q+1)/2$ distinct elements of ${}_h\hat{\Omega}_{1:q,1:q,t}$ to lie in a one-dimensional subspace of $R^{q(q+1)/2}$ for all t. When $q > 1$, consistency breaks down if the model is misspecified.

Restrictions on conditional covariances matrices such as those imposed by Engle *et al.* (1990) and Schwert and Seguin (1990) are useful because they achieve parsimony. This considerable virtue comes with a price, namely the loss of consistency when the model is misspecified.

4.3. Failure of the Near-Diffusion Assumption

Consider again the $\hat{\Omega}_{1,1,T}(\Delta, M)$ discussed in the introduction, but with $\{X_{1,t}\}$ generated by a continuous time Poisson process instead of the diffusion (1, 1). The sample paths of $\{X_{1,t}\}$ are (almost surely) step functions with a finite number of steps over any finite time interval. For almost every t therefore, $\hat{\Omega}_{1,1,T}(\Delta, M) \to 0$ in probability as $\Delta \downarrow 0$ and $M \to \infty$. Yet the instantaneous conditional variance in the increments in $\{X_{1,t}\}$ is nonnegative, so $\hat{\Omega}_{1,1,T}(\Delta, M)$ does not consistently estimate $\Omega_{1,1,T}$. Similarly, if $\{X_t\}$ is a jump–diffusion, consistency for $\hat{\Omega}_{1:q,1:q,T}$ generated by misspecified ARCH models breaks down, since, except in the immediate aftermath of a jump, the ARCH models effectively ignore the contribution of the jump component to the instantaneous covariance matrix.[19]

The near-diffusion assumption is crucial for two reasons: First, it guarantees that the increments in $\{X_t\}$ are small over small time intervals (i.e., they are not too thick-tailed). Second, it guarantees that Ω_t changes by only small amounts over small time intervals. These two conditions effectively allowed us to apply laws of large numbers in estimating Ω_t. When these two assumptions break down, so, in general, does consistency. Unfortunately, so too will consistency for the other procedures commonly used for estimating time-varying conditional variances and covariances (e.g., Beckers, 1983; Cox and Rubinstein, 1985, sect. 6.1; French *et al.*, 1987; Fama and MacBeth, 1973; Schwert, 1989; Taylor, 1986).[20]

[19] The situation is not quite so bleak if the jumps occur in $\Omega_{1:q,1:q,t}$ but not in $X_{1:q,t}$. Recall that our consistency results did not assume ${}_h\hat{\Omega}_{1:q,1:q,0} = {}_h\Omega_{1:q,1:q,0}$, which is why we had ${}_h\hat{\Omega}_{1:q,1:q,t} - {}_h\Omega_{1:q,1:q,t} \to 0_{q \times q}$ in probability for *positive* t. As far as the filtering problem is concerned, a jump in $\Omega_{1:q,1:q,t}$ essentially resets the clock back in time 0, when ${}_h\hat{\Omega}_{1:q,1:q,0}$ and ${}_h\Omega_{1:q,1:q_0}$ were not equal. If the probability of a jump in ${}_h\Omega_{1:q,1:q,t}$ at any given t equals zero, then the consistency results of sections 2 and 3 should still hold—i.e., ${}_h\hat{\Omega}_{1:q,1:q,t} - {}_h\Omega_{1:q,1:q,t} \to 0_{q \times q}$ in probability (again, pointwise) at t. Of course, ${}_h\hat{\Omega}_{1:q,1:q,t}$ will not converge to ${}_h\Omega_{1:q,1:q,t}$ when t is a jump point, but the probability of this at any fixed t equals zero.

[20] The method in Bates (1991) may be an exception.

The possibility of large jumps in asset prices plays an important role in some asset pricing theories. For example, in a regime of fixed exchange rates, there is typically a large probability of no change in the exchange rate over a small interval and a small probability of a large move due to a currency revaluation (see, for example, Krasker's (1980) discussion of the "peso problem"). While the diffusion assumption is commonly made in asset pricing models, it is not without its dangers even apart from the peso problem. We know, for example, that asset prices cannot *literally* follow diffusion processes: markets are not always open (they are typically closed on evenings, weekends, and holidays), stocks do not trade at every instant even when the markets are open (leading to discrete price changes when they do trade), prices change in discrete units (one-eighths for most stocks), stocks in portfolios trade nonsynchronously, and prices are observed sometimes at the bid and sometimes at the asked. These and other facts of market microstructure make the diffusion assumption, if pushed too far, unrealistic.[21] Even if we could correct for such market microstructure considerations, it might be that "true" underlying asset prices do not change smoothly—as in the peso problem. Information may arrive in a lumpy manner better characterized by a mixed jump–diffusion process.

Our results are relevant to the extent that the estimation error becomes small for h large enough that market microstructure problems haven't destroyed the approximate validity of our diffusion assumption. Whether this is actually true is an open question. The usual large-sample asymptotics employed in time series analysis suffer from a similar defect: It is almost always unrealistic to assume that parameters (or hyperparameters in a random coefficients model) remain fixed *forever*—i.e., as the sample period and sample size T goes to infinity. If taken literally, the notion of consistency breaks down. Nevertheless, large-T consistency is a useful concept, partly because it imposes discipline in the choice of parameter estimators. Similarly, small-h consistency may help impose a useful discipline on the choice of ARCH models.

5. CONCLUSION

Traditionally, econometricians have regarded heteroskedasticity mainly as an obstacle to the efficient estimation of regression parameters. In recent years, however, financial economists have stressed that the behavior of

[21] Similar warnings are in order when diffusion models are applied in almost any field. For example, Brownian motion is often used by physicists to model particle motion, even though a particle cannot *literally* follow a Brownian motion: Brownian sample paths have unbounded variation almost surely, so a particle following a Brownian motion would, with probability one, travel an infinite distance in a finite time period, a physical impossibility.

conditional variances and covariances is important in and of itself. Unfortunately, traditional linear time series models are simply not capable of modelling dynamic conditional variances and covariances in a natural way. ARCH models have greatly expanded the ability of econometricians to model heteroskedastic time series, particularly financial time series.

If the near-diffusion assumption is approximately valid for high-frequency financial time series, our results may explain some features of estimated ARCH models: first, the tendency to find "persistent" volatility (e.g., IGARCH) when using high-frequency (e.g., daily) data, second, the small influence of misspecification in the conditional mean on the estimated conditional variance, and third, estimated models (especially in the multivariate case) that do not make sense as data-generating processes (i.e., in which the conditional covariance matrix loses nonnegative definiteness with positive probability.

Our results should be reassuring to users of ARCH models and suggest a reason for the forecasting success of ARCH: If the process generating prices is (approximately) a diffusion, then there is so much information about conditional second moments at high frequencies that even a misspecified model can be a consistent filter. This does not suggest, however, that the econometrician should be indifferent as to which ARCH model to use: A consistent filter may be a very poor long-term forecaster, and there may be considerable differences in the efficiency of different ARCH filters. These issues remain for future work.

APPENDIX

Proof of Theorem 2.1. See Ethier and Kurtz (1986, ch. 7, theorem 4.1) or Nelson (1990a, theorem 2.2).

Proof of Theorem 2.2. This theorem is a special case of Ethier and Nagylaki (1988, theorem 2.1). There are a few changes in notation: We have set ε_N, δ_N, $X^N(k)$, $Y^N(k)$, and $Z^N(k)$ in Ethier and Nagylaki equal to h, h^δ, ${}_hX_{kh}$, ${}_hY_{kh}$, $({}_hX_{kh}, {}_hY_{kh}, kh)$, respectively. By our (2.4), Ethier and Nagylaki's $h(x, y)$ function [see their (2.1)–(2.2)] is a vector of zeros. Our Theorem 2.2 now directly follows by Ethier and Nagylaki (1988, theorem 2.1). Q.E.D.

In proving the theorems in section 3, we need the following lemma.

LEMMA A.1. *Let* $\{\zeta_t\}_{[0,T]}$ *be generated by the stochastic integral equation*

$$\zeta_t = \zeta_0 + \int_0^t m(\zeta_s, s)\, ds + \int_0^t \Lambda^{1/2}(\zeta_s, s)\, dW_s, \qquad \text{(A.1)}$$

where ζ_0 is fixed, $\{W_s\}$ is a $q \times 1$ standard Brownian motion, $m(\cdot,\cdot)$ and $\Lambda(\cdot,\cdot)$ are continuous $(q \times 1)$- and $(q \times q)$-valued functions, respectively and where (A.1) *has a unique weak-sense solution. Next, let $g(t)$ be a $q \times 1$ function satisfying*

$$g(t) = o(t^{1/2}) \quad \text{as} \quad t \downarrow 0. \tag{A.2}$$

Then,

$$t^{-1/2}(\zeta_t - g(t) - \zeta_0) \xrightarrow{\mathrm{d}} \mathrm{N}(0_{q\times 1}, \Lambda(\zeta_0, 0)) \quad \text{as} \quad t \downarrow 0. \tag{A.3}$$

Further, let $f(\zeta, t)$ be a continuous function from R^{q+1} into R^1. Let there exist an $\varepsilon > 0$ such that

$$\limsup_{t \downarrow 0} \mathrm{E}\left|f(t^{-1/2}(\zeta_t - g(t) - \zeta_0), t)\right|^{1+\varepsilon} < \infty. \tag{A.4}$$

Then if $Z \sim \mathrm{N}(0_{q\times 1}, \Lambda(\zeta_0, 0))$,

$$\lim_{t \downarrow 0} \mathrm{E}[f(t^{-1/2}(\zeta_1 - g(t) - \zeta_0), t)] = \mathrm{E}[f(Z, 0)] < \infty. \tag{A.5}$$

Proof of Lemma A.1. Once we prove (A.3), (A.5) follows directly by Billingsley (1986, theorem 25.7, corollary a and theorem 25.12, corollary). To prove (A.3), note that by (A.2), $t^{-1/2}g(t) \to 0$ as $t \downarrow 0$. Next, rewrite (A.1) as

$$t^{-1/2}(\zeta_1 - \zeta_0) = t^{-1/2}\Lambda^{1/2}(\zeta_0, 0)W_t + t^{-1/2}\int_0^t m(\zeta_s, s)\, ds$$
$$+ t^{-1/2}\int_0^t [\Lambda^{1/2}(\zeta_s, s) - \Lambda^{1/2}(\zeta_0, 0)]\, dW_s. \tag{A.6}$$

Since $t^{-1/2}W_t \sim \mathrm{N}(0, I)$ and $t^{-1/2}g(t) \to 0_{q\times 1}$ as $t \downarrow 0$, (A.3) follows if the last two terms on the right-hand side of (A.6) converge in probability to zero as $t \downarrow 0$. The middle term is

$$\left|t^{-1/2}\int_0^t m(\zeta_s, s)\, ds\right| \le t^{1/2} \max_{0 \le s \le t} |m(\zeta_s, s)|, \tag{A.7}$$

which converges to a vector of zeros in probability as $t \downarrow 0$. Next, consider the last term. Define

$$Z_t \equiv \int_0^t [\Lambda^{1/2}(\zeta_s, s) - \Lambda^{1/2}(\zeta_0, 0)]\, dW_s. \tag{A.8}$$

The proof of (A.3) is complete if $t^{-1/2}Z_t$ converges in probability to a vector of zeros as $t \downarrow 0$. Letting $I(\cdot)$ be an indicator function, we have

$$t^{-1}Z_t \cdot Z_t' = I(\|Z_t\| < 1) \cdot t^{-1}Z_t \cdot Z_t' + I(\|Z_t\| \ge 1) \cdot t^{-1}Z_t \cdot Z_t'. \tag{A.9}$$

The second term on the right-hand side of (A.9) vanishes in probability as $t \downarrow 0$ since

$$\lim_{t \downarrow 0} \mathrm{E}\left[t^{-1} \cdot I(\|Z_t\| \geq 1) \mid \zeta_0\right] = 0_{q \times q} \tag{A.10}$$

[Arnold, 1973, p. 40, formula (a)]. Since Λ is continuous, the first term also vanishes [Arnold, 1973, p. 40, formula (b)]. Q.E.D.

Proof of Theorem 3.1. We need only verify Assumptions 1–7. Condition 3.1 includes Assumptions 1, 3, and 4. Assumption 3 and (1.1) directly imply Assumption 2 (see Arnold, 1974, p. 40). Define

$$h^{-1/2}{}_h\xi_{1:q,(k+1)h}$$
$$\equiv h^{-1/2}\left[{}_hX_{1:q,(k+1)h} - {}_hX_{1:q,kh} - h \cdot \hat{\mu}\left({}_hX_{kh}, {}_h\hat{\Omega}_{1:q,1:q,kh}\right)\right]. \tag{A.11}$$

Applying Condition 3.4 and Lemma A.1 we have for all $x \in R^{n+m}$,

$$\left[h^{-1/2}{}_h\xi_{1:q,(k+1)h} \mid {}_hX_{kh} = x\right] \Rightarrow \mathrm{N}\left(0, \Omega_{1:q,1:q}(x)\right), \tag{A.12}$$

as $h \downarrow 0$. By Condition 3.4 and Lemma A.1, the conditional absolute moments of $h^{-1/2}{}_h\xi_{1:q,kh}$ (given ${}_hX_{kh} = x$) of order less than $4 + \varepsilon$ converge to the corresponding moments of an $\mathrm{N}(0, \Omega_{1:q,1:q}(x))$. It is then easy to derive $c(y) = -A \cdot y$ and $d(y) = 0$, as required by Assumption 5. The differential equation of Assumption 6 is $dY/dt = -A \cdot y$ with unique solution $Y_t = \exp(-At)Y_0$ (see, e.g., Hochstadt, 1975). The solution is globally asymptotically stable if the eigenvalues of A have strictly positive real parts. To verify Assumption 7, define $\psi(y) \equiv y'y \cdot [1 - \exp(-y'y)]$ and

$$\rho(x, y) \equiv 1 + \omega(x) + \psi(y). \tag{A.13}$$

(2.18) is immediate and Condition 3.2 implies (2.20). Condition 3.3 implies

$$\limsup_{h \downarrow 0} \sup_{\|(x',y')\| \leq \eta} h^{-1}\mathrm{E}\left[\omega\left({}_hX_{(k+1)h}\right) - \omega(x) \mid {}_hX_{kh} = x, {}_hY_{kh} = y\right]$$
$$- \lambda\omega(x) \leq 0, \tag{A.14}$$

so to verify (2.21) it suffices that there is a $\lambda(\eta, h)$ satisfying (2.19) such that for every $\eta > 0$,

$$\limsup_{h \downarrow 0} \sup_{\|(x',y')\| \leq \eta} h^{-1}\mathrm{E}\left[\psi\left({}_hY_{i,(k+1)h}\right) - \psi(y) \mid {}_hX_{kh} = x, {}_hY_{kh} = y\right]$$
$$- \lambda(\eta, h)[1 + \psi(y)] \leq 0. \tag{A.15}$$

Using (1.1) and (3.1)–(3.2), we have

$${}_hY_{(k+1)h} = {}_hY_{kh} + {}_hD_{kh} + {}_hN_{kh}, \tag{A.16}$$

where ${}_hD_{kh}$ and ${}_hN_{kh}$ are, respectively, drift and noise terms defined by

$$ {}_hD_{kh} \equiv \mathrm{E}\big[{}_hY_{(k+1)h} - {}_hY_{kh} \mid {}_h\mathscr{L}_{kh}\big], \tag{A.17} $$

$$ {}_hN_{kh} \equiv {}_hY_{(k+1)h} - {}_hY_{kh} - {}_hD_{kh}. \tag{A.18} $$

Condition 3.4 guarantees that ${}_hD_{kh}$ is well-defined and that the conditional variance of ${}_hN_{kh}$ is bounded in compact regions of the state space. By Conditions 3.1, 3.4, and 3.5,

$$ \lim_{h \downarrow 0} \sup_{\|(x', y')\| \le \eta} h^{-1}\mathrm{E}\big[\psi({}_hY_{i,(k+1)h}) - \psi(y) \mid {}_hX_{kh} = x, {}_hY_{kh} = y\big] $$

$$ = -\frac{\partial\psi h^{\delta-1}}{\partial y}(Ay + o(1)) + \frac{\partial^2\psi}{2\partial y\,\partial y'} \cdot \big[O(1) + O(h^{2\delta-1})\big]. \tag{A.19} $$

Since the eigenvalues of A have strictly positive real parts, $-(\partial\psi/\partial y)Ay$ is negative for $y \neq 0_{m\times 1}$. Hence, the first term on the right-hand side of (A.19) dominates the second term (since it diverges to $-\infty$) except in a shrinking neighborhood of $y = 0_{m\times 1}$ (i.e., in neighborhoods of the form $\|y\| \le K \cdot h^{\Delta}$ for $K > 0$ and $0 < \Delta < \min\{1 - \delta, \delta\}$). In this shrinking neighborhood of $y = 0_{m\times 1}$, $\partial\psi/\partial y \to 0$ and $\partial^2\psi/\partial y\,\partial y' \to 0$, so the right-hand side of (A.19) is uniformly bounded on compacts, diverging to $-\infty$ except at $y = 0_{m\times 1}$. (A. 15) can therefore be satisfied with a $\lambda(\eta, h)$ satisfying (2.19). (2.21) now follows directly. Q.E.D.

Proof of Theorem 3.2. Assumptions 1–4 follow directly from Conditions 3.1–3.3. The proof that Assumptions 5 and 7 hold closely follows the proof of Theorem 3.1, with Condition 3.6 taking the place of Condition 3.4. The differential equation of Assumption 6 is

$$ dY/dt = \gamma \cdot m \cdot [\exp(-Y/2) - 1], $$

with solution

$$ Y_t = 2 \cdot \ln\big(1 - \exp[-\gamma m \cdot t/2] + \exp[\tfrac{1}{2}(Y_0 - \gamma m t)]\big), \tag{A.20} $$

which satisfies Assumption 6 as long as $\gamma > 0$. Q.E.D.

Proof of Theorems 3.3 and 3.4. Nearly identical to the proofs of Theorem 3.1 and Theorems 3.1–3.3, respectively. Q.E.D.

ACKNOWLEDGMENTS

I would like to thank Stuart Ethier, Dean Foster, Leo Kadanoff, G. Andrew Karolyi, Albert Kyle, Franz Palm, G. William Schwert, an editor (Robert Engle), two anonymous referees, participants in the 1989 NBER Summer Institute, the 1990 Conference on Statistical Models for Financial Volatility, the 1990 Paris ARCH conference, and at workshops at Northwestern,

Queen's University, the University of Chicago, and the University of Wisconsin at Madison for helpful comments. The usual disclaimer regarding errors applies. The Center for Research in Security Prices provided research support.

REFERENCES

Anderson, B. D. O., and Moore, J. B. (1979). "Optimal Filtering." Prentice-Hall, Englewood Cliffs, NJ.

Arnold, L. (1973). "Stochastic Differential Equations: Theory and Applications." Wiley, New York.

Baillie, R. T., and Bollerslev, T. (1989). The message in daily exchange rates: A conditional variance tale. *Journal of Business and Economic Statistics* **7**, 297–305.

Baillie, R. T., and Bollerslev, T. (1990). Intra-day and inter-market volatility in foreign exchange rates. *Review of Economic Studies* **58**, 565–585.

Bates, D. S. (1991). The crash of '87: Was it expected? The evidence from options markets. *Journal of Finance* **46**, 1009–1044.

Beckers, S. (1983). Variances of security price returns based on high, low, and closing prices. *Journal of Business* **56**, 97–112.

Billingsley, P. (1986). "Probability and Measure," 2nd ed. Wiley, New York.

Black, F., and Scholes, M. (1973). The pricing of options and corporate liabilities. *Journal of Political Economy* **81**, 637–654.

Bollerslev. T. (1986). Generalized autoregressive conditional heteroskedasticity. *Journal of Econometrics* **31**, 307–327.

Bollerslev. T., Engle, R. F., and Wooldridge, J. M. (1988). A capital asset pricing model with time varying covariances. *Journal of Political Economy* **96**, 116–131.

Bollerslev. T., Chou, R. Y., and Kroner, K. (1992). ARCH modeling in finance: A review of the theory and empirical evidence. *Journal of Econometrics* **52**, 5–60.

Chan, K., and Chung, Y. P. (1993). Intraday relationships among index arbitrage, spot and futures price volatility, and spot market volume: A transactions data test. *Journal of Banking and Finance* **17**, 663.

Cox, J., and Rubinstein, M. (1985). "Options Markets." Prentice-Hall, Englewood Cliffs, NJ.

Engle, R. F. (1982). Autoregressive conditional heteroskedasticity with estimates of the variance of United Kingdom inflation. *Econometrica* **50**, 987–1008.

Engle, R. F., and Bollerslev, T. (1986). Modelling the persistence of conditional variances, *Econometric Reviews* **5**, 1–50.

Engle, R. F., Ng, V., and Rothschild, M. (1990). Asset pricing with a factor ARCH covariance structure: Empirical estimates for Treasury bills. *Journal of Econometrics* **45**, 213–238.

Ethier, S. N., and Kurtz, T. G. (1986). "Markov Processes: Characterization and Convergence." Wiley, New York.

Ethier, S. N., and Nagylaki, T. (1988). Diffusion approximations of markov chains with two time scales and applications to population genetics. II. *Advances in Applied Probability* **20**, 525–545.

Fama, E. F., and MacBeth, J. D. (1973). Risk, return, and equilibrium: Empirical tests. *Journal of Political Economy* **81**, 607–636.

French, K. R., Schwert, G. W., and Stambaugh, R. F. (1987). Expected stock returns and volatility. *Journal of Financial Economics* **19**, 3–29

Gallant, A. R., Rossi, P. E., and Tauchen, G. (1990). "Stock Prices and Volume." University of Chicago, Chicago.

Glosten, L. R., Jagannathan, R., and Runkle, D. (1989). "Relationship Between the Expected Value and the Volatility of the Nominal Excess Return on Stocks." Northwestern University, Evanston, IL.

Granger, C. W. L., and Newbold, P. (1986). "Forecasting Economic Time Series." 2nd ed. Academic Press, New York.

Hardouvelis, G. A. (1990). Margin requirements, volatility and the transitory component of stock prices, *American Economic Review* **80**, 736–762.

Harvey, C. R. (1989). Time-varying conditional covariances in tests of asset pricing models. *Journal of Financial Economics* **24**, 289–318.

Higgins, M. L., and Bera, A. K. (1989). A class of nonlinear ARCH models. *International Economic Review* **33**, 137–158.

Hochstadt, H. (1975). "Differential Equations: A Modern Approach." Dover, New York.

Hsieh, D. A. and Miller, M. H. (1990). Margin regulation and stock market volatility. *Journal of Finance* **45**, 3–30.

Hull, J., and White, A. (1987). The pricing of options on assets with stochastic volatilities. *Journal of Finance* **42**, 281–300.

Krasker, W. S. (1980). The peso problem in testing for efficiency of foreign exchange markets. *Journal of Monetary Economics* **6**, 269–276.

Lintner, J. (1965). The valuation of risky assets and the selection of risky investments in stock portfolios and capital budgets. *Review of Economics and Statistics* **47**, 13–37.

Melino, A., and Turnbull, S. (1990). Pricing foreign currency options with stochastic volatility. *Journal of Econometrics* **45**, 239–266.

Merton, R. C. (1973). An intertemporal capital asset pricing model. *Econometrica* **41**, 867–887.

Merton, R. C. (1980). On estimating the expected return on the market. *Journal of Financial Economics* **8**, 323–361.

Nelson, D. B. (1990a). ARCH models as diffusion approximations. *Journal of Econometrics* **45**, 7–39.

Nelson, D. B. (1990b). Stationarity and persistence in the GARCH(1, 1) model. *Econometric Theory* **6**, 318–334

Nelson, D. B. (1991). Conditional heteroskedasticity in asset returns: A new approach. *Econometrica* **59**, 347–370.

Priestley, M. B. (1981). "Spectral Analysis and Time Series." Academic Press, London.

Ross, S. (1976). The arbitrage theory of capital asset pricing. *Journal of Economic Theory* **13**, 341–360.

Schwert, G. W. (1989). Why does stock market volatility change over time? *Journal of Finance* **44**, 1115–1154.

Schwert, G. W. (1990). Indexes of U.S. stock prices. *Journal of Business* **63**, 399–426.

Schwert, G. W. and Seguin, P. J. (1990). Heteroskedasticity in stock returns. *Journal of Finance* **45**, 1129–1156.

Shanken, J. (1990). Intertemporal asset pricing: An empirical investigation. *Journal of Econometrics* **45**, 99–120.

Sharpe, W. (1964). Capital asset prices: A theory of market equilibrium under conditions of risk. *Journal of Finance* **19**, 425–442.

Sims, C. (1984). "Martingale-like Behavior of Asset Prices and Interest Rates." University of Minnesota, Minneapolis, MN.

Stroock, D. W., and Varadhan, S. R. S. (1979). "Multidimensional Diffusion Processes." Springer-Verlag, Berlin.

Summers, L. H., and Summers, V. P. (1989). When financial markets work too well: A cautious case for a securities transactions tax. *Journal of Financial Services Research*, **3**, 261–286.

Taylor, S. (1986). "Modeling financial time series." Wiley, New York.

Wiggins, J. B. (1987). Option values under stochastic volatility: Theory and empirical estimates. *Journal of Financial Economics* **19**, 351–372.

7

FILTERING AND FORECASTING WITH MISSPECIFIED ARCH MODELS II: MAKING THE RIGHT FORECAST WITH THE WRONG MODEL*

DANIEL B. NELSON[†]
DEAN P. FOSTER[‡]

[†]*Graduate School of Business,*
University of Chicago,
Chicago, Illinois 60637
[‡]*Wharton School,*
University of Pennsylvania
Philadelphia, Pennsylvania 19104

I. INTRODUCTION

Since their introduction by Engle (1982), ARCH models have been widely (and quite successfully) applied in modeling financial time series; see, for example, the survey papers of Bollerslev *et al.* (1992, 1994). What accounts for the success of these models? A companion paper (Nelson, 1992) suggested one reason: When the data-generating process is well-approximated by a diffusion, high-frequency data contain a great deal of information about conditional variances, and as a continuous time limit is

* Reprinted with permission from Nelson, D. and Foster, D. P. "Filtering and Forecasting with misspecified ARCH models II" *Journal of Econometrics* **67** (1995), Elsevier Science SA, Lousanne, Switzerland.

approached, the sample information about conditional variances increases without bound. This allows simple volatility estimates formed, for example, by taking a distributed lag of squared residuals (as in Bollerslev's 1986, GARCH model), to consistently estimate conditional variances as the time interval between observations goes to zero. It is not surprising, therefore, that as continuous time is approached a sequence of GARCH models can consistently estimate the underlying conditional variance of a diffusion, even when the GARCH models are not the correct data-generating process.

As shown in Nelson (1992), this continuous time consistency holds not only for GARCH, but for many other ARCH models as well, and is unaffected by a wide variety of misspecifications. That is, when considered as a data-generating process, a given ARCH model might provide nonsensical forecasts—for example, by forecasting a negative conditional variance with positive probability, incorrectly forecasting explosions in the state variables, ignoring some unobservable state variables, or by misspecifying the conditional mean. Nevertheless, in the limit as continuous time is approached, such a model can provide a consistent estimate of the conditional variance in the underlying data-generating process.[1] [Our use of the term "estimation" corresponds to its use in the filtering literature rather in the statistics literature—i.e., the ARCH model "estimates" the conditional variance in the sense that a Kalman filter estimates unobserved state variables [See, e.g., Anderson and Moore (1979, ch. 2) or Arnold (1973, ch. 12]. In other words, while a misspecified ARCH model may perform disastrously in medium or long-term forecasting, it may perform well at filtering.

In this chapter, we show that under suitable conditions, a sequence of misspecified ARCH models may not only be successful at filtering, but at forecasting as well. That is, as a continuous time limit is approached, not only do the conditional covariance matrices generated by the sequence of ARCH models approach the true conditional covariance matrix, but the forecasts of the process and its volatility generated by these models converge in probability to the forecast generated by the true data-generating process.

The conditions for consistent estimation of the forecast distribution are considerably stricter than the conditions for consistent filtering; for example, *all* unobservable state variables must be consistently estimated and we must correctly specify the conditional mean and covariance matrix of *all* state variables as continuous time is approached. These conditions are developed in section 2.

[1] For a definition of "conditional variance" appropriate when the data are generated by a diffusion, see, e.g., Arnold (1973, pp. 39–40).

While the conditions in section 2 are stricter than the consistent filtering conditions developed in Nelson (1992), they are broad enough to accommodate a number of interesting cases. In section 3, we provide a detailed example, using a stochastic volatility model familiar in the options pricing literature (see, e.g., Wiggins, 1987; Hull and White, 1987; Scott, 1987; Melino and Turnbull, 1990). In this model, a stock price and its instantaneous returns volatility follow a diffusion process. While the stock price is observable at discrete intervals of length h, the instantaneous volatility is unobservable. We show that a suitably constructed sequence of ARCH models [in particular, AR(1) EGARCH models] can consistently estimate the instantaneous volatility and generate appropriate forecasts of the stock price and volatility processes as $h \downarrow 0$.

2. MAIN RESULTS

In this section we develop the conditions under which misspecified ARCH models can successfully extract correct forecast distributions as continuous time is approached. We will define different stochastic processes for each value of h. Our asymptotic results hold as h tends towards zero. For each value of h, consider a stochastic process $\{{}_hX_t\}$, which is a step function with jumps only at integer multiples of h—i.e., at times h, $2h$, $3h$, and so on. ${}_hX_t$ represents our $n \times 1$ *discretely observable* processes (for example a vector of stock prices). Unfortunately, ${}_hX_t$ is not Markovian. To remedy this, we will consider a companion process ${}_hU_t$ which is an $m \times 1$ *unobservable* process with jumps only at multiples of h. We will assume that the pair $\{{}_hX_t, {}_hU_t\}$ is Markovian with a probability measure P_h. In other words, $\{{}_hX_t\}$ is generated by a "stochastic volatility" model (see, e.g., Nelson, 1988; Jacquier *et al.*, 1994; Harvey and Shephard, 1993) with unobservable state variables $\{{}_hU_t\}$. These state variables will, in our examples, control the conditional covariance matrix of the increments in the $\{{}_hX_t\}$ process.

ARCH models provide a different remedy to make ${}_hX_t$ Markovian. An ARCH model recursively defines ${}_hU_t$ in such a fashion that it is a function only of ${}_hX_t$, ${}_hX_{t-h}$, and ${}_h\hat{U}_{t-h}$. The ARCH model then assigns a probability measure $\hat{\mathrm{P}}_h$ to $\{{}_hX_{t,}{}_h\hat{U}_t\}$ in such a fashion that the pair is Markovian. Formal definitions of P_h, $\hat{\mathrm{P}}_h$, and the ARCH updating rule for ${}_h\hat{U}_t$ are found in the appendix. In our notation, curly brackets indicate a stochastic process—e.g., $\{{}_hX_t, {}_hU_t\}_{[0,T]}$ is the sample path of ${}_hX_t$ and ${}_hU_t$ as (random) functions of time on the interval $0 \leq t \leq T$. We refer to values of a process at a *particular* time t by omitting the curly brackets—e.g., ${}_hX_\tau$ and $({}_hX_\tau, {}_hU_\tau)$ are, respectively, the (random) values taken by the $\{{}_hX_t\}$ and $\{{}_hX_t, {}_hU_t\}$ process at time τ.

We will assume that $\{{}_hX_t, {}_hU_t\}$ is the true Markov process, and $\{{}_hX_t, {}_h\hat{U}_t\}$ merely an approximation. Our goal is to determine how close forecasts made using the ARCH approximation $\hat{\mathrm{P}}_h(\cdot\,|\,{}_hX_t, {}_h\hat{U}_t\}$ are to forecasts based on the true model $\mathrm{P}_h(\cdot\,|\,{}_hX_t, {}_hU_t)$. To do this we must meld these two processes onto the same probability space, i.e., P_h and $\hat{\mathrm{P}}_h$ must assign probabilities to $\{{}_hX_t, {}_hU_t, {}_h\hat{U}_t\}$.

It is easy to extend $\hat{\mathrm{P}}_h$ to $\{{}_hX_t, {}_hU_t, {}_h\hat{U}_t\}$ by merely allowing ${}_hU_t = {}_h\hat{U}_t$ a.s. (almost surely) $[\hat{\mathrm{P}}_h]$. Thus, under $\hat{\mathrm{P}}_h$, $\{{}_hX_t, {}_hU_t, {}_h\hat{U}_t\}$ is a Markov process. Since ${}_h\hat{U}_t$ is a function of ${}_hX_t, {}_hX_{t-h}$, and ${}_h\hat{U}_{t-h}$ the triplet $\{{}_hX_t, {}_hU_t, {}_h\hat{U}_t\}$ is already Markovian under P_h. We can now ask questions about how these two measures compare. Note first that since P_h and $\hat{\mathrm{P}}_h$ use the same updating rule to generate $\{{}_h\hat{U}_t\}$, so given ${}_h\hat{U}_0$ and the sample path of $\{{}_hX_t\}$, the two measures assign (with probability one) the same value to ${}_h\hat{U}_t$. Nevertheless, the measures P_h and $\hat{\mathrm{P}}_h$ are mutually singular,[2] since $\hat{\mathrm{P}}_h[{}_hU_t = {}_h\hat{U}_t\ \forall t] = 1$ and in general $\mathrm{P}_h[{}_hU_t = {}_h\hat{U}_t\ \forall t] = 0$. As we shall see, despite this singularity, the two measures may be close in the forecasts they generate. The singularity of the two measures does not imply that distinguishing $\hat{\mathrm{P}}_h$ and P_h is trivial given a finite set of data on $\{{}_hX_t\}$, for the same reason that distinguishing ARCH and stochastic volatility models is nontrivial, since while $\{{}_hX_t\}$ and $\{{}_h\hat{U}_t\}$ are observable, $\{{}_hU_t\}$ is not. Of course, if ${}_hU_t$ were directly observable it would be a trivial matter to distinguish the models.

Our interest is in comparing forecasts made using information at a time T with the *incorrect* probability measure $\hat{\mathrm{P}}_h$ to those made using the *correct* measure P_h. Specifically, how can we characterize the forecasts generated by P_h and $\hat{\mathrm{P}}_h$ as $h \downarrow 0$? Under what circumstance do they become "close" as $h \downarrow 0$? Theorems 2.3 and 2.4 below compare the conditional forecast *distributions* generated by P_h and $\hat{\mathrm{P}}_h$, while Theorem 2.5 compares the conditional forecast *moments*. Finally, Theorem 2.6 compares the very long-term forecasts (i.e., the forecast stationary distributions) generated by P_h and $\hat{\mathrm{P}}_h$.

2.1. The Basic Setup[3]

Let $D([0,\infty), R^n \times R^{2m})$ be the space of functions from $[0,\infty)$ into $R^n \times R^{2m}$ that are continuous from the right with finite left limits. D is a metric space when endowed with the Skorohod metric (see Ethier and Kurtz, 1986, ch. 3, for formal definitions). Next, let $\mathscr{B}(\mathrm{E})$ be the Borel sets on a metric space E, and let ν_h and $\hat{\nu}_h$ be probability measures on $(R^{n+2m}, \mathscr{B}(R^n \times R^{2m}))$. Below, we will take ν_h and $\hat{\nu}_h$ to be, respectively, probability measures for the starting values $({}_hX_0, {}_hU_0, {}_h\hat{U}_0)$ under, respectively, the true data-generating process and the misspecified ARCH model.

[2] See Billingsley (1986, pp. 442–443).

[3] Our treatment here will be fairly informal: formal construction of the measures P_h and $\hat{\mathrm{P}}_h$ is carried out in the Appendix.

We will study processes $\{{}_hX_t, {}_hU_t, {}_h\hat{U}_t\}$. These processes are assumed to have jumps only at times $h, 2h, 3h, \ldots$.[4] The variables ${}_h\hat{U}_t$ can be computed from the previous values of ${}_hX_t$, and a startup value ${}_h\hat{U}_0$, so ${}_h\hat{U}_t$ is measurable with respect to the natural σ-field of the ${}_hX_t$ process and ${}_h\hat{U}_0$. Thus, the natural σ-field for all three variables is the union of the natural σ-field for $\{{}_hX_t, {}_hU_t\}$ and the σ-field generated by ${}_h\hat{U}_0$. Call this σ-field ${}_h\mathscr{F}_t$. We will write expectations taken under P_h as $\mathrm{E}_h[\cdot]$, and expectations taken under $\hat{\mathrm{P}}_h$ as $\hat{\mathrm{E}}[\cdot]$.

At time τ, we will want a forecast of the probability of a future event A. In other words, A depends on $\{{}_hX_t, {}_hU_t\}_{t \geq \tau}$, and we are interested in finding

$$\mathrm{P}_h(A \mid {}_h\mathscr{F}_\tau) = \mathrm{P}_h(A \mid {}_hX_\tau, {}_hU_\tau), \tag{2.1}$$

where the equality follows from the Markov property of $\{{}_hX_t, {}_hU_t\}$. This conditional probability is a random variable (for proof that it is well-defined, see Stroock and Varadhan, 1979, Theorem 1.1.6). To help our analysis it will be convenient to think of it as a function. Thus, we will informally[5] define the forecast function

$$F_h(A, x, u, \tau) \equiv \mathrm{P}_h(A \mid {}_hX_\tau = x, {}_hU_\tau = u). \tag{2.2}$$

Recall that under $\hat{\mathrm{P}}_h$, $\{{}_hX_t, {}_hU_t, {}_h\hat{U}_t\}$ is also a Markov chain. Thus,

$$\hat{\mathrm{P}}_h(A \mid {}_h\mathscr{F}_\tau) = \hat{\mathrm{P}}_h\big(A \mid {}_hX_\tau, {}_hU_\tau, {}_h\hat{U}_t\big) = \hat{\mathrm{P}}_h\big(A \mid {}_hX_\tau, {}_h\hat{U}_t\big), \tag{2.3}$$

where the first equality from Markov and the second from ${}_hU_t = {}_h\hat{U}_t$ a.s. under $\hat{\mathrm{P}}_h$. So, informally define the forecast function corresponding to $\hat{\mathrm{P}}_h$:

$$\hat{F}_h(A, x, u, \tau) \equiv \hat{\mathrm{P}}_h\big(A \mid {}_hX_\tau = x, {}_h\hat{U}_\tau = u\big). \tag{2.4}$$

The goal of this chapter is to show when $F_h(\cdot)$ is close to $\hat{F}_h(\cdot)$. How should "close" be defined of these two functions? A natural definition of "close" is that the functions when evaluated at the random variables $({}_hX_t, {}_hU_t)$ and $({}_hX_t, {}_h\hat{U}_t)$ respectively, should be close with high probability

[4] Our step function scheme follows Ethier and Kurtz (1986). Many other schemes would work just as well—for example, we could follow Stroock and Varadhan (1979, p. 267) and for $kh \leq t < (k+1)h$, set

$$\big({}_hX_t, {}_hU_t, {}_h\hat{U}_t\big) \equiv \big({}_hX_{kh}, {}_hU_{kh}, {}_h\hat{U}_{kh}\big) + h^{-1}(t - kh)\Big[\big({}_hX_{(k+1)h}, {}_hU_{(k+1)h}, {}_h\hat{U}_{(k+1)h}\big) - \big({}_hX_{kh}, {}_hU_{kh}, {}_h\hat{U}_{kh}\big)\Big].$$

This scheme makes $\{{}_hX_t, {}_hU_t, {}_h\hat{U}_t\}$ piecewise linear and continuous. Presumably splines should also work in the limit $h \downarrow 0$.

[5] If we worked with the transition probability functions and we placed some regularity conditions on them, this would be a perfectly valid definition. Since we are working with the processes instead of the semi-groups, we will have to worry about sets of measure zero. Formal definitions are found in the Appendix.

(i.e., convergent in probability under P_h). That is to say, for every $\zeta > 0$,

$$\lim_{h \downarrow 0} P_h\Big(|F_h(A, {}_hX_\tau, {}_hU_\tau, \tau) - \hat{F}_h(A, {}_hX_\tau, {}_h\hat{U}_\tau, \tau)| > \zeta\Big) = 0. \quad (2.5)$$

In other words, we first generate (using P_h of course) the underlying data, namely the sample path of $\{{}_hX_t, {}_hU_t\}$. Next, we use the ARCH recursive updating formula [which is identical in P_h and $\hat{P}_h$—see (A.1) in the Appendix for precise definitions] to generate the $\{{}_h\hat{U}_t\}$. Finally, at some time τ, we generate forecasts for the future path of $\{{}_hX_t, {}_hU_t\}$ first, using the true state variables $({}_hX_\tau, {}_hU_\tau)$ and the forecast function generated by the true probability P_h, and second, using the ARCH estimate ${}_h\hat{U}_\tau$ in place of ${}_hU_\tau$ and using the forecast function generated by the ARCH probability measure $\hat{P}_h$. We then compare the forecasts: If the difference between them converges to zero in probability under P_h as $h \downarrow 0$ for all well-behaved events, then we say that the forecasts generated by the ARCH model $\hat{P}_h$ are asymptotically correct. Unfortunately, we need to specify which events A are well-behaved for all $h > 0$. This technicality will be deferred.

2.2. The Basic Intuition

There are three natural steps in proving that $\hat{P}_h$ consistently estimates the forecast distribution for P_h over the interval $[\tau, \infty)$: First, we show that given ${}_h\hat{U}_\tau = {}_hU_\tau$, the forecasts generated at time τ by P_h and $\hat{P}_h$ become arbitrarily close as $h \downarrow 0$—i.e., for $\tau > 0$ and for every $(x, u) \in R^{n+m}$, $F_h(A, x, u, \tau) - \hat{F}_h(A, x, u, \tau) \to 0$ as $h \downarrow 0$. This first step, considered in detail in Nelson (1990), is concerned with the continuous limit properties of ARCH models as *data-generating processes*. The results are summarized in Lemma 2.1. The second step is to show that $\hat{P}_h$ is a consistent filter for P_h at time τ—i.e., that ${}_h\hat{U}_\tau - {}_hU_\tau \to 0$ in probability under P_h as $h \downarrow 0$. This second step is concerned with the *filtering* properties of misspecified ARCH models, considered in Nelson (1992). These results are summarized in Lemma 2.2. The third step is to show that the forecasts generated by the ARCH model are "smooth" in the underlying state variables, so that as ${}_h\hat{U}_\tau - {}_hU_\tau$ approaches zero, the forecasts generated by the ARCH model approach the forecasts generated by the correct model—i.e., $F_h(A, x, u, \tau) - \hat{F}_h(A, x, \hat{u}, \tau) \to 0$ as $u \to \hat{u}$ and $h \downarrow 0$. This step is carried out in the proof of Theorem 2.3 in the Appendix. The three steps, taken together, yield Theorems 2.3–2.4, the main results of this chapter. Theorems 2.5–2.6 extend these results.

2.3. Step 1. Convergence as Data-Generating Processes

The first set of assumptions assures that $\{{}_hX_t, {}_hU_t\}_{[0,T]}$ and $\{{}_hX_t, {}_h\hat{U}_t\}_{[0,T]}$ converge weakly to limit diffusions $\{X_t, U_t\}_{[0,T]}$ and $\{X_t, \hat{U}_t\}_{[0,T]}$ under P_h

and $\hat{\mathrm{P}}_h$ respectively as $h \downarrow 0$, where the limit processes $\{X_t, U_t\}_{[0,T]}$ and $\{X_t, \hat{U}_t\}_{[0,T]}$ are generated by the stochastic integral equations

$$(X_t', U_t')' = (X_0', U_0')' + \int_0^t \mu(X_s, U_s)\,\mathrm{d}s + \int_0^t \Omega^{1/2}(X_s, U_s)\,\mathrm{d}W_s, \quad (2.6)$$

$$\left(X_t', \hat{U}_t'\right)' = \left(X_0', \hat{U}_0'\right) + \int_0^t \hat{\mu}\left(X_s, \hat{U}_s\right)\mathrm{d}s + \int_0^t \hat{\Omega}^{1/2}\left(X_s, \hat{U}_s\right)\mathrm{d}W_s. \quad (2.6')$$

In (2.6), $\{W_t\}$ is an $(n+m) \times 1$ standard Brownian motion, $\mu(X_t, U_t)$ is the $(n+m) \times 1$ instantaneous drift per unit of time in $\{X_t, U_t\}$, and $\Omega(X_t, U_t\}$ is the $(n+m) \times (n+m)$ instantaneous conditional covariance matrix per unit of time of the increments in $\{X_t, U_t\}$. $[X_0, U_0]$ is taken to be random with a distribution π. In (6′), $\hat{\mu}$, $\hat{\Omega}$, $\hat{U}_t$, and $\hat{\pi}$ replace μ, Ω, U_t, and π. We call the probability measures on $D([0,\infty), R^n \times R^m)$ generated by (2.6) and (2.6′) P_0 and $\hat{\mathrm{P}}_0$.

Under certain regularity conditions, four characteristics completely define the distribution of $\{X_t, U_t\}$ in (2.6): the functions μ and Ω, the distribution π, and the almost sure continuity of the sample path of $\{X_t, X_t\}$. We achieve weak convergence of $\{{}_hX_t, {}_hU_t\}$ to $\{X_t, U_t\}$ by matching these four characteristics in the limit as $h \downarrow 0$ by making the drift, conditional covariance matrix, and time 0 distribution of $\{{}_hX_t, {}_hU_t\}$ converge to μ, Ω, and π, respectively, and by making the sizes of the jumps in $\{{}_hX_t, {}_hU_t\}$ converge in probability to zero at an appropriate rate. This holds for $(X_t, \hat{U}_t)$ in (2.6′) as well, making suitable substitutions for μ, Ω, t, and π. For further discussion and interpretation of these assumptions and relevant references to the probability literature, see Nelson (1990, 1992).

More notation: $X_{i,t}$ is the ith element of the vector X_t, and B_{ij} is the $(i-j)$th element of the matrix B. $\Rightarrow$ denotes weak convergence. $\|B\|$ is a norm of the $q \times r$ real matrix B defined by $\|B\| \equiv [\sum_{i=1,q} \sum_{j=1,r} B_{i,j}^2]^{1/2}$. $\mathrm{E}_{(h,x,u,kh)}[f]$ is the conditional expectation taken under P_h of f given $({}_hX_{kh}, {}_hU_{kh}) = (x, u)$. $\hat{\mathrm{E}}_{(h,x,u,kh)}[f]$ is the conditional expectation taken under $\hat{\mathrm{P}}_h$. For formal definitions that make these expectations functions of (x, u) rather than random variables, see the Appendix.

Assumption 1. *Under* P_h, $({}_hX_0, {}_hU_0, {}_h\hat{U}_0) \Rightarrow (X_0, U_0, \hat{U}_0)$ *as* $h \downarrow 0$, *where* $(X_0, U_0, \hat{U}_0)$ *has probability measure* ν_0. *Under* $\hat{\mathrm{P}}_h$, $({}_hX_0, {}_hU_0, {}_h\hat{U}_0) \Rightarrow (X_0, U_0, \hat{U}_0)$ *as* $h \downarrow 0$, *where* $(X_0, U_0, \hat{U}_0)$ *has probability measure* $\hat{\nu}_0$.

Assumption 2. *There exists an* $\varepsilon > 0$ *such that for every* $R > 0$ *and every* $k > 0$,

$$\lim_{h \downarrow 0} \sup_{\|x\| \leq R, \|u\| \leq R} h^{-1}\mathrm{E}_{(h,x,u,kh)}\left[\left|{}_hX_{i,(k+1)h} - {}_hX_{i,kh}\right|^{4+\varepsilon}\right] = 0 \quad (2.7)$$

for $i = 1, \ldots, n$, *and*

$$\lim_{h \downarrow 0} \sup_{\|x\| \le R, \|u\| \le R} h^{-1} \mathrm{E}_{(h,x,u,kh)}\left[|{}_hU_{i,(k+1)h} - {}_hU_{i,kh}|^{4+\varepsilon} \right] = 0 \quad (2.8)$$

a.s. $[\mathrm{P}_h]$ *for all* $i = 1, \ldots, m$. *Further* (7)–(8) *continue to hold when* $\mathrm{E}_h, \mathrm{P}_h$, *and* u *are replaced by* $\hat{\mathrm{E}}_h$, $\hat{\mathrm{P}}_h$, *and* $\hat{u}$ *respectively.*

Assumption 1 says that the distributions of the random starting points $({}_hX_0, {}_hU_0, {}_h\hat{U}_0)$ converge to (perhaps distinct) limit measures under P_h and $\hat{\mathrm{P}}_h$ as $h \downarrow 0$. Assumption 2 contains conditional moment restrictions guaranteeing that the jump sizes in $\{{}_hX_t, {}_hU_t, {}_h\hat{U}_t\}$ vanish to zero at an appropriate rate as $h \downarrow 0$. This is necessary because the sample paths of the limit diffusion are continuous with probability one.

Next, define the first and second conditional moments

$$\mu_h(x,u) \equiv h^{-1} \mathrm{E}_{(h,x,u,kh)}\left[\begin{bmatrix} {}_hX_{(k+1)h} - {}_hX_{kh} \\ {}_hU_{(k+1)h} - {}_hU_{kh} \end{bmatrix}\right],$$

$$\Omega_h(x,u) \equiv h^{-1}$$

$$\times \mathrm{E}_{(h,x,u,kh)}\left[\begin{bmatrix} {}_hX_{(k+1)h} - {}_hX_{kh} \\ {}_hU_{(k+1)h} - {}_hU_{kh} \end{bmatrix}\begin{bmatrix} {}_hX_{(k+1)h} - {}_hX_{kh} \\ {}_hU_{(k+1)h} - {}_hU_{kh} \end{bmatrix}'\right], \quad (2.9)$$

$$\hat{\mu}_h(x,u) \equiv h^{-1} \hat{\mathrm{E}}_{(h,x,\hat{u},kh)}\left[\begin{bmatrix} {}_hX_{(k+1)h} - {}_hX_{kh} \\ {}_h\hat{U}_{(k+1)h} - {}_h\hat{U}_{kh} \end{bmatrix}\right],$$

$$\hat{\Omega}_h(x,u) \equiv h^{-1}$$

$$\times \hat{\mathrm{E}}_{(h,x,\hat{u},kh)}\left[\begin{bmatrix} {}_hX_{(k+1)h} - {}_hX_{kh} \\ {}_h\hat{U}_{(k+1)h} - {}_h\hat{U}_{kh} \end{bmatrix}\begin{bmatrix} {}_hX_{(k+1)h} - {}_hX_{kh} \\ {}_h\hat{U}_{(k+1)h} - {}_h\hat{U}_{kh} \end{bmatrix}'\right]. \quad (2.9')$$

Note that we may ignore ${}_h\hat{U}_{kh}$ in the definitions of μ_h and Ω_h and may ignore ${}_hU_{kh}$ in the definitions of $\hat{\mu}_h$ and $\hat{\Omega}_h$. μ_h and Ω_h are the conditional drift and second moment matrix of the $\{{}_hX_t, {}_hU_t\}$ processes under P_h. Each is normalized by dividing by h. $\hat{\mu}_h$ and $\hat{\Omega}_h$ are the conditional drift and second moment matrix under $\hat{\mathrm{P}}_h$. Our next assumption requires them to converge, uniformly on compact studies of the state space, to appropriate limits.

ASSUMPTION 3. *There exist continuous* $(n+m) \times 1$ *functions* $\mu(x,u)$ *and* $\hat{u}(x,u)$ *and* $(n+m) \times (n+m)$ *continuous positive semidefinite func-*

tions $\Omega(x,u)$, *and* $\hat{\Omega}(x,u)$ *such that for every* $R > 0$,

$$\lim_{h \downarrow 0} \sup_{\|x\| \le R, \|u\| \le R} \|\mu_h(x,u) - \mu(x,u)\| = 0,$$

$$\lim_{h \downarrow 0} \sup_{\|x\| \le R, \|u\| \le R} \|\Omega_h(x,u) - \Omega(x,u)\| = 0, \tag{2.10}$$

$$\lim_{h \downarrow 0} \sup_{\|x\| \le R, \|u\| \le R} \|\hat{\mu}_h(x,u) - \hat{\mu}(x,u)\| = 0,$$

$$\lim_{h \downarrow 0} \sup_{\|x\| \le R, \|u\| \le R} \|\hat{\Omega}_h(x,u) - \hat{\Omega}(x,u)\| = 0. \tag{2.10$'$}$$

We also require that $\mu(x,u)$, $\hat{\mu}(x,u)$, $\Omega(x,u)$, $\hat{\Omega}(x,u)$, ν_0, and $\hat{\nu}_0$ completely characterize the distributions of the limit diffusions $\{X_t, U_t\}_{[0,T]}$ and $\{X_t, \hat{U}_t\}_{[0,T]}$.

ASSUMPTION 4. *For any choice of* π_0 *and* $\hat{\pi}_0$, *distributionally unique solutions exist to the stochastic integral equations* (2.6) *and* (2.6′).[6]

LEMMA 2.1. *Under Assumptions* 1–4, $\{_hX_t, {}_hU_t\}_{[0,\infty)} \Rightarrow \{X_t, U_t\}_{[0,\infty)}$ *under* P_h *as* $h \downarrow 0$, *where the initial distribution* π *is given by*

$$\pi(\Gamma) \equiv \nu_0(\Gamma \times R^m), \tag{2.11}$$

for every $\Gamma \in \mathscr{B}(R^n \times R^m)$. *If* π *and* ν_0 *in* (11) *are replaced with* $\hat{\pi}$ *and* $\hat{\nu}_0$, $\{_hX_t, {}_h\hat{U}_t\}_{[0,\infty)} \Rightarrow \{X_t, U_t\}_{[0,\infty)}$ *under* P_h *as* $h \downarrow 0$.

Proof. See Ethier and Kurtz (1986, ch. 7, Corollary 4.2) or Nelson (1990, Theorem 2.2).

As we defined forecast functions for the Markov processes $\{_hX_\tau, {}_hU_\tau\}$ and $\{_hX_t, {}_h\hat{U}_\tau\}$, we now (informally) define forecast functions for the Markov processes (2.6) and (2.6′):

$$F_0(A,x,u,\tau) \equiv \mathrm{P}_0[A \mid X_t = x, U_t = u], \tag{2.12}$$

$$\hat{F}_0(A,x,\hat{u},\tau) \equiv \hat{\mathrm{P}}_0\Big[A \mid X_t = x, \hat{U}_t = \hat{u}\Big], \tag{2.12$'$}$$

where again $A \in \mathscr{B}(D([\tau,\infty), R^n \times R^m))$, and $\mathrm{P}_0(\cdot)$ and $\hat{\mathrm{P}}_0(\cdot)$ are the probability measures corresponding to (6) and (6′) respectively.

Lemma 2.1 gives conditions under which P_h and $\hat{\mathrm{P}}_h$ are associated with well-behaved limit diffusions for $\{_hX_t, {}_hU_t\}$ and $\{_hX_t, {}_hU_t\}$ as $h \downarrow 0$. *Consistent estimation of the forecast distribution will also require that the drifts and conditional covariances of these diffusions are the same*—i.e., we require the following.

[6] See Ethier and Kurtz (1986, pp. 290–291) for formal definitions. Several sets of sufficient conditions for Assumption 4 are summarized in Appendix A of Nelson (1990).

ASSUMPTION 5. *For all* $(x, u) \in R^{n+m}$,

$$\hat{\mu}(x,u) = \mu(x,u) \quad and \quad \hat{\Omega}(x,u) = \Omega(x,u).$$

Assumption 5 says that the (misspecified) ARCH model generating $\hat{\mathrm{P}}_h$ correctly specifies the functional form of the first two conditional moments of ${}_hX_t$ and ${}_hU_t$. It is the most important additional assumption required to move from consistent filtering to consistent estimation of the forecast distribution. We have kept Assumption 5 separate from the first four assumptions, however, since we are also interested in characterizing the forecasts generated by $\hat{\mathrm{P}}_h$ when Assumption 5 is not satisfied (see Theorem 2.4 below).

The conditions of Lemma 2.1 (with the addition of Assumption 5) accomplish step 1—i.e., if ${}_hU_\tau = {}_h\hat{U}_\tau$, and if $\hat{\mu} = \mu$ and $\hat{\Omega} = \Omega$, the forecast distributions generated by $\hat{\mathrm{P}}_h$ and P_h at time τ become close (and both become close to the forecast distribution generated by the limit diffusion P_0) as $h \downarrow 0$.

2.4. Step 2. Consistent Filtering

DEFINITION. *We say that* $\{{}_h\hat{U}_t\}$ *(or, equivalently,* $\hat{\mathrm{P}}_h$*) is a consistent filter for* $\{{}_hU_t\}$ *at time* τ *under* $\{\mathrm{P}_h\}_{h \downarrow 0}$, *if for all* $\varepsilon > 0$,

$$\lim_{h \downarrow 0} \mathrm{P}_h\left[\left\|{}_h\hat{U}_\tau - {}_hU_\tau\right\| > \varepsilon\right] = 0.$$

In addition to Assumptions 1–4, three (quite technical) regularity conditions are required for consistent filtering. These are reproduced in the Appendix as Assumptions 6–8. Detailed discussion of these conditions can be found in Nelson (1992, sect. 2).

Lemma 2.2 accomplishes step 2.

LEMMA 2.2 (Ethier and Nagylaki, 1988). *Let Assumptions* 1–4 *and* 6–8 *hold. Then for every* τ, $0 < \tau < \infty$, $\{{}_h\hat{U}_t\}$ *is a consistent filter for* $\{{}_hU_t\}$ *at time* τ *under* $\{\mathrm{P}_h\}_{h \downarrow 0}$.

For proof, see Ethier and Nagylaki (1988) or Nelson (1992, Theorem 2.2).

2.5. Step 3. Consistent Estimation of Forecast Distributions

DEFINITION. *Let* ∂A *be the boundary of the set* A*—i.e., the set of all points in* $D([\tau, \infty), R^n \times R^m)$ *which are limit points both of* A *and of its complement. Let* $\mathscr{M}_\tau$ *be the correlation of sets* A *such that* $A \in \mathscr{B}(D([\tau, \infty),$

$R^n \times R^m$)) *and* $\mathrm{P}_0[\{X_t, U_t\}_{\tau \le t < \infty} \in \partial A \mid (X_t, U_t) = (x, u)] = 0$ *for all starting points* (x, u).[7]

DEFINITION. *We say that* $\{_h X_t, {}_h\hat{U}_t\}_{[\tau,\infty)}$ *consistently estimates the forecast distribution of* $\{_h X_t, {}_h U_t\}_{[\tau,\infty)}$ *at time* τ *if for every* $A \in \mathscr{M}_\tau$ *and every* $\varepsilon > 0$, $\mathrm{P}_h[|F_h(A, {}_h X_\tau, {}_h U_\tau, \tau) - \hat{F}_h(A, {}_k X_\tau, {}_h\hat{U}_\tau, \tau)| > \varepsilon] \to 0$ *as* $h \downarrow 0$.

Note that the definition of consistent estimation of the forecast distribution is only meaningful if $\{_h X_t, {}_h U_t\}_{[0,\infty)} \Rightarrow \{X_t, U_t\}_{[0,\infty)}$ under P_h as $h \downarrow 0$, where $\{X_t, U_t\}_{[0,\tau)}$ is generated by the diffusion (6). Assumptions 1–4, however, guarantee that it does.

THEOREM 2.3. *If Assumptions* 1–8 *are satisfied, then for every* τ *with* $0 < \tau < \infty$, $\{_h X_t, {}_h\hat{U}_t\}_{[\tau,\infty)}$ *consistently estimates the forecast distribution of* $\{_h X_t, {}_h U_t\}_{[\tau,\infty)}$ *as* $h \downarrow 0$.

For proof, see the Appendix.

Theorem 2.3 gives conditions for the difference between forecast distributions generated by the sequence of misspecified ARCH models and the forecasts generated by the "correct" model to vanish (in probability) at time τ as $h \downarrow 0$. In addition, both the true and ARCH forecast distributions approach the forecast distribution generated by the limit diffusion P_0. This holds for forecasts involving any of the state variables, whether observable or not. As indicated earlier, the proof of Theorem 2.3 consists of Lemmas 2.1 and 2.2, plus "smoothness" of the forecast distributions as functions of the unobserved state variables.

Together, the conditions of Theorem 2.3 ensure that the probabilities $\hat{\mathrm{P}}_h$ generated by the misspecified ARCH model provide both consistent filtering (for all $t > 0$) and consistent estimation of the forecast distribution for all time intervals $[\tau, \infty)$ for $0 < \tau < \infty$. In addition, they ensure that the time τ conditional probabilities generated by $\hat{\mathrm{P}}_h$ and P_h are well-approximated by the time τ conditional probability generated by the limit diffusion P_0.

What can we say about the forecasts generated by misspecified ARCH models when the conditions of Theorem 2.3 fail? Assumptions 1–4 and 6–8 allow a weaker characterization of the forecasts produced by the ARCH model.

THEOREM 2.4. *Let Assumptions* 1–4 *and* 6–8 *be satisfied. Then for every* $\varepsilon > 0$, *every* τ, $0 < \tau < \infty$, *and every* $A \in \mathscr{M}_\tau$,

$$\mathrm{P}_h\big[\big|F_h(A, {}_h X_\tau, {}_h U_\tau, \tau) - F_0(A, {}_h X_\tau, {}_h U_\tau, \tau)\big| > \varepsilon\big] \to 0, \quad (2.13)$$

$$\mathrm{P}_h\Big[\Big|\hat{F}_h\big(A, {}_h X_\tau, {}_h\hat{U}_\tau, \tau\big) - \hat{F}_0\big(A, {}_h X_\tau, {}_h U_\tau, \tau\big)\Big| > \varepsilon\Big] \to 0. \quad (2.14)$$

[7] As we did with the forecast functions, we are treating $\mathrm{P}_0[\cdot \mid (X_t, U_t) = (x, u)]$ as a function of (x, u) rather than as a random variable. See the Appendix for formal definitions.

For proof, see the Appendix.

Theorem 2.4 is easily summarized: when we drop Assumption 5, the ARCH model provides a *consistent filter* but uses the *wrong* limit diffusion [i.e., (2.6′) instead of (2.6)] to form the forecast.

2.6. Forecast Moments and Stationary Distributions

While Theorems 2.3 and 2.4 give convergence results for conditional *distributions*, they do not guarantee convergence of either forecast *moments* or of forecast *stationary distributions*. Theorem 2.5 considers convergence in probability of forecast expectations of well-behaved functions of $\{{}_hX_t, {}_hU_t\}_{[\tau,\infty)}$.

THEOREM 2.5. *Let* $0 < \tau < \infty$, *and let Assumptions* 1–4 *and* 6–8 *be satisfied. Let g be a continuous functional mapping* $D([\tau,\infty), R^n \times R^m)$ *into* R^1. *Let g satisfy the following uniform integrability condition: for every bounded* $\Lambda \subset R^{n+m}$,

$$\lim_{K\to\infty} \limsup_{h\downarrow 0} \sup_{(x,u)\in\Lambda} \hat{\mathrm{E}}_{(h,x,u,\tau)}\big[\big|g(\{{}_hX_t, {}_hU_t\}_{[\tau,\infty)})\big| \times I\big(\big|g(\{{}_hX_t, {}_hU_t\}_{[\tau,\infty)})\big| > K\big)\big] = 0. \quad (2.15)$$

$$\lim_{K\to\infty} \limsup_{h\downarrow 0} \sup_{(x,u)\in\Lambda} \mathrm{E}_{(h,x,u,t)}\big[\big|g(\{{}_hX_t, {}_hU_t\}_{[\tau,\infty)})\big| \times I\big(\big|g(\{{}_hX_t, {}_hU_t\}_{[\tau,\infty)})\big| > K\big)\big] = 0, \quad (2.16)$$

where $I(\cdot)$ *is an indicator function* [*i.e.*, $I(x \in H) = 1$ *if* $x \in H$ *and* $I(x \in H) = 0$ *otherwise*].

Define the forecast functions

$$G_h(g, x, u, \tau) \equiv \mathrm{E}_{(h,x,u,\tau)}\big[g\big(\{{}_hX_t, {}_hU_t\}_{[\tau,\infty)}\big)\big], \quad (2.17)$$

$$G_0(g, x, u, \tau) \equiv \mathrm{E}_{(0,x,u,\tau)}\big[g\big(\{X_t, U_t\}_{[\tau,\infty)}\big)\big], \quad (2.18)$$

$$\hat{G}_h(g, x, u, \tau) \equiv \hat{\mathrm{E}}_{(h,x,u,\tau)}\Big[g\Big(\big\{{}_hX_t, {}_h\hat{U}_t\big\}_{[\tau,\infty)}\Big)\Big], \quad (2.19)$$

$$\hat{G}_0(g, x, u, \tau) \equiv \hat{\mathrm{E}}_{(0,x,u,\tau)}\Big[g\Big(\big\{X_t, \hat{U}_t\big\}_{[\tau,\infty)}\Big)\Big]. \quad (2.20)$$

Then for every $\tau > 0$ *and every* $\varepsilon > 0$,

$$\mathrm{P}_h\Big[\Big|\hat{G}_h\big(g, {}_hX_\tau, {}_h\hat{U}_\tau, \tau\big) - \hat{G}_0\big(g, {}_hX_\tau, {}_h\hat{U}_\tau, \tau\big)\Big| > \varepsilon\Big] \to 0, \quad (2.21)$$

$$\mathrm{P}_h\big[\big|G_h(g, {}_hX_\tau, {}_hU_\tau, \tau) - G_0(g, {}_hX_\tau, {}_hU_\tau, \tau)\big| > \varepsilon\big] \to 0, \quad (2.22)$$

as $h \downarrow 0$. *If Assumption* 5 *is also satisfied, then for every* $\tau > 0$ *and for every* $\varepsilon > 0$,

$$\mathrm{P}_h\Big[\Big|\hat{G}_h\big(g, {}_hX_\tau, {}_h\hat{U}_\tau, \tau\big) - G_h\big(g, {}_hX_\tau, {}_hU_\tau, \tau\big)\Big| > \varepsilon\Big] \to 0, \quad (2.23)$$

as $h \downarrow 0$.

A sufficient condition for (15)–(16) *is that there exists an* $\varepsilon > 0$ *such that for every bounded* $A \subset R^{n+m}$,

$$\limsup_{h \downarrow 0} \sup_{(x,u) \in \Lambda} \hat{\mathrm{E}}_{(h,x,u,\tau)}\Big[\big|g\big(\{{}_hX_t, {}_hU_t\}_{[\tau,\infty)}\big)\big|^{1+\varepsilon}\Big] < \infty, \quad (2.15')$$

$$\limsup_{h \downarrow 0} \sup_{(x,u) \in \Lambda} \mathrm{E}_{(h,x,u,\tau)}\Big[\big|g\big(\{{}_hX_t, {}_hU_t\}_{[\tau,\infty)}\big)\big|^{1+\varepsilon}\Big] < \infty. \quad (2.16'')$$

For proof see the Appendix.

The uniform integrability conditions described in Theorem 2.5 are similar to standard conditions allowing integration to the limit (see, e.g., Billingsley, 1986, pp. 347–348).

The next result, due to Kushner (1984), gives conditions for convergence of forecast stationary distributions. It also gives conditions for moment boundedness, which sometimes can be used to verify (15′)–(16′).

THEOREM 2.6 (Kushner). *Let Assumptions* 1–4 *be satisfied, and let the diffusion* (2.6) *admit a unique invariant measure for* (X_t, U_t)*—i.e., there exists a random vector* (X_∞, U_∞) *on* R^{n+m} *such that for any probability measure* $\pi(\cdot)$ *for* (X_0, U_0), $(X_t, U_t) \Rightarrow (X_\infty, U_\infty)$ *as* $t \to \infty$ *under* P_0. *For each sufficiently small* $h > 0$, *let* P_h *also admit a unique invariant measure—i.e., for any probability measure* $\nu_h(\cdot)$ *for* $({}_hX_0, {}_hU_0)$ *in* (2.5), $({}_hX_t, {}_hU_t) \Rightarrow ({}_hX_\infty, {}_hU_\infty)$ *as* $t \to \infty$ *under* P_h. *Let there be a twice-continuously differentiable Liapunov function* $\zeta(x,u)$ *and positive numbers* h^*, Δ, *and* η *such that for all* h, $0 < h < h^*$,

$$\liminf_{\|[x'u']\| \to \infty} \zeta(x,u) = \infty, \quad (2.24)$$

$$\min_{[x'u'] \in R^{n+m}} \zeta(x,u) = 0, \quad (2.25)$$

$$\sup_{0 < h \le h^*} \mathrm{E}_h\big[\zeta({}_hX_0, {}_hU_0)\big] < \infty. \quad (2.26)$$

and for all $(x,u) \in R^{n+m}$ *and all* h, $0 < h < h^*$,

$$h^{-1}\mathrm{E}_{(h,x,u,\tau)}\big[\zeta\big({}_hX_{(k+1)h}, {}_hU_{(k+1)h}\big) - \zeta(x,u)\big] < \Delta - \eta \cdot \zeta(x,u). \quad (2.27)$$

Then $({}_hX_t, {}_hU_t) \Rightarrow (X_\infty, U_\infty)$ *under* P_h *as* $t \to \infty$ *and* $h \downarrow 0$, *and there exists a constant* K *and a sequence* $\eta_h \to \eta$ *as* $h \downarrow 0$ *such that for all* $t \geq 0$,

$$E_h[\zeta({}_hX_t, {}_hU_t)] \leq \eta^{-1}K(\Delta + \exp(-\eta_h \cdot t) \cdot \mathrm{E}_h[\zeta({}_hX_0, {}_hU_0)]). \quad (2.28)$$

If a second Liapunov function $\hat{\zeta}(x,u)$ *exists such that* (2.24)–(2.28) *hold when* E_h, P_h, P_0, $\zeta(x,u)$, $\nu_h(\cdot)$, ${}_hU_t$, *and* (2.6) *are replaced by* $\hat{\mathrm{E}}_h$, $\hat{\mathrm{P}}_h$, $\hat{\mathrm{P}}_0$, $\hat{\zeta}(x,u)$, $\hat{\nu}_h(\cdot)$, ${}_h\hat{U}_t$, *and* (2.6′), *respectively, then* $({}_hX_t, {}_h\hat{U}_t) \Rightarrow (X_\infty, U_\infty)$ *under* $\hat{\mathrm{P}}_h$ *as* $t \to \infty$ *and* $h \downarrow 0$, *and there exists a constant* K *and a sequence* $\hat{\eta}_h \to \eta$ *as* $h \downarrow 0$ *such that for all* $t \geq 0$,

$$\hat{\mathrm{E}}_h\left[\hat{\zeta}\left({}_hX_t, {}_h\hat{U}_t\right)\right] \leq \eta^{-1}\hat{K}\left(\Delta + \exp(-\hat{\eta}_h \cdot t) \cdot \hat{\mathrm{E}}_h\left[\hat{\zeta}\left({}_hX_0, {}_h\hat{U}_0\right)\right]\right). \quad (2.29)$$

For proof, see the Appendix.

In some applications, $\{{}_hX_t\}$ may be nonstationary even when $\{{}_hU_t\}$ is stationary. For example, $\{{}_hX_t\}$ may be the (nonstationary) cumulative return on a portfolio, while $\{{}_hU_t\}$ may be the (stationary) instantaneous conditional variance of this return. Theorem 2.6 can be adapted to this case: If there is no feedback from $\{{}_hX_t\}$ to $\{{}_hU_t\}$ so that $\{{}_hU_t\}$ is a stationary Markov process under P_h in its own right (i.e., without reference to ${}_hX_t$), replace "$({}_hX_t, {}_h\hat{U}_t)$" and "$({}_hX_t, {}_hU_t)$" with "$({}_h\hat{U}_t)$" and "$({}_kU_t)$" in the statement of Theorem 2.6, and it continues to hold.

The conditions of Theorem 2.6 are difficult to verify unless the innovations in $\{{}_hX_t, {}_h\hat{U}_t\}$ are bounded either above or below: Fortunately, in ARCH models they often are. For an ARCH example in which the conditions of Theorem 2.6 are verified [specifically, for GARCH(1, 1), in which the innovations in the conditional variance process are bounded below], see the proof of Theorem 2.3 in Nelson (1990). Sometimes, as in the example in the next section, it is possible to verify convergence of steady state distributions by other means.

3. A STOCHASTIC VOLATILITY MODEL

On first inspection, the assumptions underlying the results in section 2 are quite forbidding. In practice, however, they can often be verified. In this section we provide an example, based on a stochastic volatility model employed in the options pricing literature. Variations of this model have been investigated by Wiggins (1987), Scott (1987), Hull and White (1987), and Melino and Turnbull (1990). The ARCH approach to options pricing in a similar setup has been considered by, e.g., Kuwahara and Marsh (1992), Cao (1992), and Amin and Ng (1993). We propose an ARCH approximation to the model and show that it satisfies Lemmas 2.1 and 2.2 and Theorems 2.3, 2.5, and 2.6.

Let S_t be the time t price of a nondivided paying stock. σ_t is its instantaneous returns volatility. We assume that

$$d[\ln(S_t)] = (\mu - \sigma_t^2/2)\,dt + \sigma_t\,dW_{1,t}, \tag{3.1}$$

$$d[\ln(\sigma_t^2)] = -\beta[\ln(\sigma_t^2) - \alpha]\,dt + \Lambda \cdot dW_{2,t}. \tag{3.2}$$

$W_{1,t}$ and $W_{2,t}$ are standard Brownian motions with correlation ρ, i.e.,

$$\begin{bmatrix} dW_{1,t} \\ dW_{2,t} \end{bmatrix} [dW_{1,t}\ dW_{2,t}] = \begin{bmatrix} 1 & \rho \\ \rho & 1 \end{bmatrix} dt. \tag{3.33}$$

μ, Λ, β, and α are constants. If $\{\sigma_t\}$ were constant, $\{S_t\}$ would follow a geometric Brownian motion. (30)–(32) generalize this much-used model by allowing $\{\sigma_t\}$ to vary randomly, with $\{\ln(\sigma_t^2)\}$ following an Ornstein–Uhlenbeck process.

We assume that we observe $\{S_t\}$ at discrete intervals of length h, so for every t, ${}_hS_t \equiv S_{h[t/h]}$, where "$[x]$" is the integer part of x. In the notation of section 2, ${}_hX_t \equiv \ln({}_hS_t)$ and ${}_hU_t \equiv \ln({}_h\sigma_t^2)$. We take $S_0 > 0$ and $\sigma_0 > 0$ to be nonrandom. Because of the continuous time Markov structure, the discrete process $\{{}_hX_t, {}_hU_t\}$ is also Markovian.

Now consider using a conditionally normal AR(1) EGARCH model (see Nelson, 1991) to forecast when the data are generated by (3.1)–(3.3). The AR(1) EGARCH generates the fitted conditional variances ${}_h\hat{\sigma}_{kh}^2$ by the recursive formulae

$$\ln({}_h\hat{\sigma}_{(k+1)h}^2) = \ln({}_h\hat{\sigma}_{kh}^2) - \hat{\beta} \cdot h \cdot [\ln({}_h\hat{\sigma}_{kh}^2) - \hat{\alpha}] + h^{1/2} \cdot g({}_hZ_{(k+1)h}), \tag{3.4}$$

where ${}_h\hat{\sigma}_0^2 > 0$ is fixed for all h and

$$g(z) \equiv \theta \cdot z + \gamma[|z| - (2/\pi)^{1/2}], \tag{3.5}$$

$${}_hZ_{(k+1)h} \equiv h^{-1/2}[\ln({}_hS_{(k+1)h}) - \ln({}_hS_{kh}) - h \cdot \hat{\mu} + h \cdot {}_h\hat{\sigma}_{kh}^2/2] / {}_h\hat{\sigma}_{kh}. \tag{3.6}$$

Equations (30)–(35) completely specify the $\{{}_hS_{kh}, {}_h\hat{\sigma}_{kh}^2, {}_h\sigma_{kh}^2\}$ process under P_h. Next we construct the $\hat{\mathrm{P}}_h$ measure. Under $\hat{\mathrm{P}}_h$ the recursive updating formulae (3.4)–(3.6) continue to hold, but (3.1)–(3.3) are replaced with

$$\ln({}_hS_{(k+1)h}) = \ln({}_hS_{kh}) + (\hat{\mu} - {}_h\hat{\sigma}_{kh}^2/2) \cdot h + h^{1/2} \cdot {}_h\hat{\sigma}_{kh} \cdot {}_hZ_{(k+1)h}, \tag{3.7}$$

$${}_hZ_{kh} \sim \text{i.i.d. } \mathrm{N}(0,1), \tag{3.8}$$

$${}_h\hat{\sigma}_{kh}^2 = \sigma_{kh}^2 \text{ a.s. } [\hat{\mathrm{P}}_h]. \tag{3.9}$$

As in section 2, we create the continuous time processes $\{{}_hS_t, {}_h\sigma_t^2, {}_h\hat{\sigma}_t^2\}$ by interpolating the discrete processes $\{{}_hS_{kh}, {}_h\sigma_{kh}^2, {}_h\hat{\sigma}_{kh}^2\}$.

Note that (3.7) is simply a rearrangement of (3.6). Under P_h, (3.6) is a *definition* of ${}_hZ_{(k+1)h}$. Under $\hat{P}_h$, however, (3.4)–(3.9) define the transition probabilities for the $\{{}_hS_{kh}, {}_h\sigma_{kh}^2, {}_h\sigma_{kh}^2\}$ process. For our purposes, we don't need the random variables ${}_hZ_{kh}$ to be part of our sample space; they are merely used to generate the probability measure $\hat{P}_h$ over the stochastic process $\{{}_hS_{kh}, {}_h\sigma_{kh}^2, {}_h\hat{\sigma}_{kh}^2\}$. Of course, under the true probability measure P_h generated by (3.1)–(3.6), these recursively obtained ${}_hZ_{kh}$'s are not i.i.d. N(0, 1). Under the EGARCH measure $\hat{P}_h$, however, they are, and this affects the EGARCH forecast function. Note also that although the ARCH model $\hat{P}_h$ assumes that $\hat{\sigma}_0^2 = \sigma_0^2$, this needn't be true under P_h. This is why we required $\tau > 0$ in section 2—clearly consistent filtering is impossible at time 0 if $\hat{\sigma}_0^2 \neq \sigma_0^2$.

As shown in Nelson (1992), the main requirement for consistent filtering for this model is $\gamma > 0$. For consistent estimation of forecast distributions, however, we must match the first two conditional moments of the ARCH model considered as a data-generating process to the corresponding moments of the true data-generating process (3.1)–(3.3).

Under $\hat{P}_h$, $h^{1/2}g({}_hZ_{kh})$ is the innovation in $\ln({}_h\hat{\sigma}_{(k+1)h}^2)$, and has variance $h[\theta^2 + \gamma^2(1 - 2/\pi)]$. The instantaneous correlation of the increments in $\ln({}_hS_{kh})$ and $\ln({}_h\sigma_{kh}^2)$ under $\hat{P}_h$ is $\theta/[\theta^2 + \gamma^2(1 - 2/\pi)]^{1/2}$. Matching these conditional second moments (under $\hat{P}_h$ and P_h) requires that

$$\theta^2 + \gamma^2(1 - 2/\pi) = \Lambda^2, \tag{3.10}$$

$$\rho = \theta/\left[\theta^2 + \gamma^2(1 - 2/\pi)\right]^{1/2}, \tag{3.11}$$

which is easily accomplished by setting $\theta = \rho \cdot \Lambda$ and $\gamma = |\Lambda|(1 - \rho^2)^{1/2}/(1 - 2/\pi)^{1/2}$. The drifts of $\ln(S_t)$ and $\ln(\sigma_t^2)$ in (3.1)–(3.3) are $(\mu - \sigma_t^2/2)$ and $-\beta[\ln(\sigma_t^2) - \alpha]$ respectively. These are equal to the drifts in (3.4) and (3.7) if $\hat{\alpha} = \alpha$, $\hat{\beta} = \beta$, and $\hat{\mu} = \mu$.

THEOREM 3.1. *For each* $h > 0$, *let* $\{{}_hS_t, {}_h\sigma_t^2, {}_h\hat{\sigma}_t^2\}$ *be generated by* (3.1)–(3.6) *with* $\gamma > 0$. *Then,*

(*a*) *For every* τ, $0 < \tau < \infty$, $\{{}_h\hat{\sigma}_\tau^2\}$ *is a consistent filter for* $\{{}_h\sigma_\tau^2\}$.

If, in addition, (3.10)–(3.11) *are satisfied and* $\hat{\alpha} = \alpha$, $\hat{\beta} = \beta$, *and* $\hat{\mu} = \mu$, *then*:

(*b*) *For every* τ, $0 < \tau < \infty$, $\{{}_hS_t, {}_h\hat{\sigma}_t^2\}_{[\tau,\infty)}$ *consistently estimates the forecast distributions of* $\{{}_hS_t, {}_h\sigma_t^2\}_{[\tau,\infty)}$.

(*c*) *Let* $G(s_1, s_2, \sigma_1, \sigma_2)$ *be a continuous function from* R^4 *into* R^1 *satisfying*

$$|G(s_1, s_2, \sigma_1, \sigma_2)| < A + B \cdot |s_1|^a \cdot |s_2|^b \cdot |\sigma_1|^c \cdot |\sigma_2|^d, \tag{3.12}$$

for finite, nonnegative A, B, a, B, c, and d. Then the conditions of Theorem 2.5 are satisfied for $g(\{\ln({}_hS_t), {}_h\sigma_t\}_{[\tau,\infty)}) \equiv G(\ln({}_hS_t), \ln({}_hS_T), {}_h\sigma_t, {}_h\sigma_T)$ for every τ, t, and T, with $0 < \tau < t < T < \infty$.

(*d*) *If, in addition, $\beta > 0$, the stationary distributions of σ_t^2 and ${}_h\hat{\sigma}_t^2$ for each sufficiently small $h > 0$ exist, i.e., $\sigma_t^2 \Rightarrow \sigma_\infty^2$ as $t \to \infty$, and for each $h > 0$, ${}_h\hat{\sigma}_t^2 \Rightarrow {}_h\hat{\sigma}_\infty^2$ as $t \to \infty$. Further, ${}_h\hat{\sigma}_\infty^2 \Rightarrow \sigma_\infty^2$ as $h \downarrow 0$.*

For proof, see the Appendix.[8]

Parts (a) and (b) are applications of Lemma 2.2 and Theorem 2.3 respectively. Part (c) is an application of Theorem 2.5 for a class of $G(\cdot)$ functions encompassing moments and cross-moments of forecast stock returns (continuously compounded) and conditional variances. Part (d) yields the convergence in distribution of ${}_h\sigma_\infty^2$ to σ_∞^2. ${}_hS_\infty$ doesn't converge, since $\{S_t\}$ is nonstationary.

3.1. Nonuniqueness of Consistent Estimates of Forecast Distributions

It is important to note that there are many ARCH models which achieve consistent estimation of the forecast distribution of (3.1)–(3.3). For example, Nelson and Foster (1994) replace the $g(\cdot)$ of (3.5) with

$$g(z) \equiv \Lambda\Big[\rho z + \big[(1-\rho^2)/2\big]^{1/2} \cdot [z^2 - 1]\Big]. \tag{3.5$'$}$$

Though (3.5′) is not a model familiar in the ARCH literature, Nelson and Foster show that it has slightly better asymptotic filtering properties than (3.3) when (3.1)–(3.3) generate the date [i.e., $({}_h\hat{\sigma}_t^2 - {}_h\sigma_t^2)$ approaches zero at a slightly faster rate as $h \downarrow 0$]. Theorem 3.1 continues to hold for (3.5′). We leave the verification to the reader.

3.2. Consistent Filtering without Consistent Estimation of the Forecast Distribution

It is important to emphasize that the conditions in Theorem 3.1 are much stricter than would be required for consistent filtering alone. For example, the moment matching conditions are not needed for consistent filtering.

[8] At first glance, it is surprising that (3.4)–(3.8) can converge weakly to (3.1)–(3.3), since there are two noise terms—$dW_{1,t}$ and $dW_{2,t}$—driving (3.1)–(3.3) and two noise terms—${}_hZ_{1,kh}$ and ${}_hZ_{2,kh}$—driving the Euler approximation (3.1″)–(3.3″), but only *one* noise term $({}_hZ_{kh})$ driving (3.4)–(3.8). To grasp the intuition behind this, consider the following illustration: let η_k be i.i.d. N(0, 1), so $E|\eta_k| = (2/\pi)^{1/2}$. Although $[|\eta_k| - (2/\pi)^{1/2}]$ and η_k are certainly not *independent*, they are *uncorrelated*. By the central limit theorem therefore, $k^{-1/2}\Sigma_{j=1,k}\,\eta_j$ and $k^{-1/2}\Sigma_{j=1,k}[|\eta_j| - (2/\pi)^{1/2}]$ converge weakly to *independent* normal random variables. What happens in the continuous time limit of (3.4)–(3.8) is similar: ${}_hZ_{kh}$ and $g({}_hZ_{kh})$ are perfectly dependent but imperfectly correlated, so their partial sums converge weakly to the imperfectly correlated Brownian motions $W_{1,t}$ and $W_{2,t}$. For more detailed discussion see Nelson (1990, pp. 20–23).

For example, suppose we fail to set $\hat{\alpha} = \alpha$, $\hat{\beta} = \beta$, $\hat{\mu} = \mu$, or we select (θ, γ) with $\gamma > 0$ but not satisfying (3.10)–(3.11). In this case, the conditions of Lemmas 2.1–2.2 and Theorem 2.4 are satisfied, but the conditions of Theorem 2.3 are not. That is, the unobservable state U_τ is recovered as $h \downarrow 0$, but the time τ forecasts generated by $\hat{\mathrm{P}}_h$ approach the forecasts generated by the diffusion

$$d[\ln(S_t)] = (\hat{\mu} - \sigma_t^2/2)\,dt + \sigma_t\,dW_{1,t}, \tag{3.1''}$$

$$d[\ln(\sigma_t^2)] = -\hat{\beta}[\ln(\sigma_t^2) - \hat{\alpha}]\,dt + \Lambda^* \cdot dW_{2,t}, \tag{3.2''}$$

$$\begin{bmatrix} dW_{1,t} \\ dW_{2,t} \end{bmatrix} [dW_{1,t} \quad dW_{2,t}] = \begin{bmatrix} 1 & \rho^* \\ \rho^* & 1 \end{bmatrix} dt, \tag{3.3''}$$

where

$$\Lambda^* \equiv [\theta^2 + \gamma^2(1 - 2/\pi)]^{1/2}, \tag{3.10'}$$

$$\rho^* \equiv \theta/[\theta^2 + \gamma^2(1 - 2/\pi)]^{1/2}. \tag{3.12'}$$

Alternatively, suppose GARCH(1, 1) was used for filtering and forecasting (3.1)–(3.3). Under weak conditions on the GARCH coefficients, GARCH achieves consistent filtering (see Nelson, 1992), but the time τ forecasts generated by $\hat{\mathrm{P}}_h$ would approach the forecasts generated by the diffusion limit of GARCH(1, 1) [given in Nelson, 1990 (2.39)–(2.40)] rather than the forecasts generated by the true data generating process (3.1)–(3.3).

3.3. A Note on Multivariate Models

As we have seen, approximately correct forecasts may be generated by misspecified ARCH models provided at (1) the first two conditional moments of all state variables, observable and unobservable, are correctly specified and (2) the data-generating process is well approximated by a diffusion. (These are the most important conditions, though some other regularity conditions must also be satisfied.) Unfortunately, the task of correctly specifying the first two conditional moments is far from trivial, especially in the multivariate case.

Let ${}_hX_t$ be $n \times 1$. There are $(n + 1) \cdot n/2$ distinct elements of the conditional covariance matrix of ${}_hX_t$, so suppose ${}_hU_t$ is $m \times 1$, where $m \equiv (n + 1) \cdot n/2$. Even when n is large, consistent filtering of ${}_hU_t$ can be achieved by a wide variety of (very simple) ARCH models. For example, let ${}_hU_t \equiv \mathrm{vech}({}_h\Omega_t)$, where "vech" is the operator which stacks the upper triangle of a symmetric matrix into a vector and ${}_h\Omega_t$ is the time t

conditional covariance matrix of ${}_hX_t$ under P_h. For some $\alpha > 0$, define ${}_h\hat{U}_t$ by

$$
\begin{aligned}
{}_h\hat{U}_{(k+1)h} &= {}_h\hat{U}_{kh}(1 - h^{1/2}\alpha) \\
&\quad + h^{-1/2}\alpha \cdot \text{vech}\big[({}_hX_{(k+1)h} - {}_hX_{kh})({}_hX_{(k+1)h} - {}_hX_{kh})'\big]. \quad (3.13)
\end{aligned}
$$

This ARCH model (which is a special case of the model of Bollerslev *et al.*, 1988) has only one parameter (which can be arbitrarily fixed), yet it provides a consistent filter under quite mild regularity conditions (see Foster and Nelson, 1994; Nelson, 1992, pp. 71–73).

Consistent estimation of the forecast distribution, on the other hand, requires in addition correct specification of the conditional mean of $\{{}_hX_t\}$ (n moments), the conditional mean of $\{{}_hU_t\}$ (m moments), the conditional covariance matrix of $\{{}_hU_t\}$ ($[(m+1)\cdot m/2]$ moments), and the conditional covariances of $\{{}_hU_t\}$ and $\{{}_hX_t\}$ ($m \cdot n$ moments), a grand total of $n \cdot (n^3 + 6n^2 + 11n + 14)/8$ moments. When $n = 1$, as in the examples of this section, 4 moments must be matched. When $n = 2, 17$ moments must be matched, progressing to 215 moments for $n = 5$, 2155 moments for $n = 10$, and more than 13 million moments for $n = 100$. By way of comparison, the Center for Research in Security Prices (CRSP) at the University of Chicago recorded daily returns during 1991 for 16,854 stocks, requiring the matching of more than 10^{16} moments. Clearly, consistent estimation of the forecast distribution is problematic unless either n is very small or a drastically simplified structure exists for the first two conditional moments, making m much less than $(n+1)\cdot n/2$.

For example, let ${}_hX_t$ represent cumulative returns on n assets, and let $h \cdot {}_h\Omega_t$ be its conditional covariance matrix. In the Factor ARCH model of Engle *et al.* (1990), we have

$$
{}_h\Omega_t = \Omega + \sum_{i=1}^{m} \beta_i \beta_i' {}_h\lambda_t^i, \quad (3.14)
$$

where $m < n$, Ω is a (constant) $n \times n$ nonnegative definite symmetric matrix, and for $i = 1, \ldots, m$, β_i is an $n \times 1$ vector of constants and ${}_h\lambda_t^i$ is the conditional variance of a portfolio formed with the n assets in ${}_hX_t$. Here ${}_hU_t \equiv [{}_h\lambda_t^1, {}_h\lambda_t^2, \ldots, {}_h\lambda_t^m]'$ is $m \times 1$, with m less than n and much less than $n \cdot (n+1)/2$.

Although it is straightforward to find consistent filters for the portfolio variances ${}_h\lambda_t^i$, (3.14) implies that for $s \neq t$, ${}_h\Omega_s - {}_h\Omega_t$ has a rank of at most $m < n$. If the model is misspecified, there is no reason to expect this to hold, so ${}_h\hat{\Omega}_t \equiv \Omega + \Sigma_{i=1,m}\, \beta_i \beta_i' {}_h\hat{\lambda}_t^i$ need not approach ${}_h\Omega_t$ as $h \downarrow 0$. This illustrates the danger in assuming $m < n \cdot (n+1)/2$: Not only can consistent estimation of the forecast distribution break down if the model is misspecified, even consistent filtering can be lost. It would be useful to create classes of multivariate ARCH models which retain the parametric tightness and intuitive appeal of (3.14) while retaining the filtering robustness of (3.13).

4. CONCLUSION

Researchers using ARCH models have focused their energies on modelling the first two conditional moments of time series. If the true data-generating processes are near-diffusions, this emphasis is appropriate, since the first two conditional moments largely determine the behavior of the process. Near-diffusion stochastic volatility models and ARCH models therefore seem natural companions.[9] This chapter has shown that corresponding to many diffusion and near-diffusion models, there is a sequence of ARCH models that produce forecasts that are "close" to the forecasts generated by the true model. An ARCH model can produce reasonable forecasts in many cases, even when misspecified, *if the ARCH model correctly specifies the first two conditional moments of the* $\{_hX_t, {_hU_t}\}$ *process and if the data are generated by a near-diffusion.* This may be part of the reason for the broad empirical success of ARCH.

An important limitation of our results is our assumption that the sequence of ARCH models is selected a priori (and hence nonrandomly) rather than by an estimation procedure. Establishing consistency and asymptotic normality of maximum likelihood parameter estimates has proven notoriously difficult even for well-specified ARCH models, and is not likely to be any easier for misspecified ARCH models. Monte Carlo evidence reported in Schwartz *et al.* (1993) suggests that in large samples the filtering properties of misspecified ARCH models are affected little by parameter estimation. To our knowledge, no Monte Carlo studies of the forecasting properties of misspecified ARCH models with estimated parameters have been done.

Another important limitation of our results is that we have been concerned only with consistency, not with efficiency—i.e., we have had nothing to say about the *rates* at which the convergence in forecast distributions and moments takes place, or about the relative efficiency of different approximations in approximating forecast distributions. This we leave for future work.

[9] Of course, this is untrue if the near-diffusion assumption is invalid. See the discussion in Nelson (1992, sect. 4). Even if the near-diffusion assumption is valid, the entire instantaneous conditional distribution (i.e., not just the first two conditional moments) is important for the *efficiency* of an ARCH model in estimating conditional variances and covariances (see Nelson and Foster, 1994).

APPENDIX

Precise Definitions of the ARCH Updating Rule, the Measures, and Forecast Functions

To specify the ARCH updating rule for ${}_h\hat{U}_t$ we require that for each $h > 0$, there is a measurable function U_h from R^{2n+m} into R^m such that for all $(x, u, \hat{u})$,

$$\int_{\{\hat{u}^* = U_h(x^*, x, \hat{u})\}} \hat{\Pi}_h(x, u, \hat{u}, d(x^*, u^*, \hat{u}^*))$$

$$= \int_{\{\hat{u} = U_h(x^*, x, \hat{u})\}} \Pi_h(x, u, \hat{u}, d(x^*, u^*, \hat{u}^*)) = 1, \quad (A.1)$$

where for each $h > 0$, $\hat{\Pi}_h(\cdot)$ and $\Pi_h(\cdot)$ are transition functions on R^{n+2m}. The misspecified ARCH model generated by $\hat{\nu}_h$ and $\hat{\Pi}_h$ treats u and $\hat{u}$ as being equal almost surely. So for all $h > 0$ and all $(x, u, \hat{u}) \in R^{n+2m}$,

$$\int_{\{u^* = \hat{u}^*\}} \Pi_h(x, u, \hat{u}, d(x^*, u^*, \hat{u}^*)) = 1, \quad (A.2)$$

$$\int_{\{u^* = \hat{u}^*\}} \hat{\nu}_h(d(x^*, u^*, \hat{u}^*)) = 1. \quad (A.3)$$

No feedback from $\{{}_h\hat{U}_{kh}\}_{k=0,1,2,\ldots}$ into $\{{}_hX_{kh}, {}_hU_{kh}\}_{k=0,1,2,\ldots}$ is allowed—i.e., given ${}_hX_{kh}$ and ${}_hU_{kh}, {}_hX_{(k+1)h}$ and ${}_hU_{(k+1)h}$ are independent of ${}_h\hat{U}_{kh}$ under P_h, so for any $T \in \mathscr{B}(R^{n+m})$ and all h.

$$\Pi_h(x, u, \hat{u}, \Gamma \times R^m) = \Pi_h(x, u, 0_{m\times 1}, \Gamma \times R^m), \quad (A.4)$$

where $0_{m\times 1}$ is an $m \times 1$ vector of zeroes. Now let P_h and $\hat{P}_h$ be the probability measures on $D([0, \infty), R^n \times R^{2m})$ such that

$$P_h\left[\left({}_hX_0, {}_hU_0, {}_h\hat{U}_0\right) \in \Gamma\right] = \nu_h(\Gamma) \quad \text{for any} \quad \Gamma \in \mathscr{B}(R^n \times R^{2m}), \quad (A.5)$$

$$\hat{P}_h\left[\left({}_hX_0, {}_hU_0, {}_h\hat{U}_0\right) \in \Gamma\right] = \hat{\nu}_h(\Gamma) \quad \text{for any} \quad \Gamma \in \mathscr{B}(R^n \times R^{2m}), \quad (A.5')$$

$$P_h\left[\left({}_hX_t, {}_hU_t, {}_h\hat{U}_t\right) = \left({}_hX_{kh}, {}_hU_{kh}, {}_h\hat{U}_{kh}\right), \quad kh \le t < (k+1)h\right] = 1, \quad (A.6)$$

$$\hat{P}_h\left[\left({}_hX_t, {}_hU_t, {}_h\hat{U}_t\right) = \left({}_hX_{kh}, {}_hU_{kh}, {}_h\hat{U}_{kh}\right), \quad kh \le 1 < (k+1)h\right] = 1, \quad (A.6')$$

and for all $k \geq 0$ and $\Gamma \in \mathscr{B}(R^n \times R^{2m})$,

$$\mathrm{P}_h\Big[\big({}_hX_{(k+1)h}, {}_hU_{(k+1)h}, {}_h\hat{U}_{(k+1)h}\big) \in \Gamma \mid {}_h\mathscr{F}_{kh}\Big] = \Pi_h\big({}_hX_{kh}, {}_hU_{kh}, {}_h\hat{U}_{kh}, \Gamma\big), \tag{A.7}$$

almost surely under P_h, and almost surely under $\hat{\mathrm{P}}_h$,

$$\hat{\mathrm{P}}_h\Big[\big({}_hX_{(k+1)h}, {}_hU_{(k+1)h}, {}_h\hat{U}_{(k+1)h}\big) \in \Gamma \mid {}_h\mathscr{F}_{kh}\Big] = \hat{\Pi}_h\big({}_hX_{kh}, {}_hU_{kh}, {}_h\hat{U}_{kh}, \Gamma\big). \tag{A.7$'$}$$

For each $h > 0$, (A.5) specifies the distributions of the starting point $({}_hX_0, {}_hU_0, {}_h\hat{U}_0)$ under P_h. Equation (A.6) forms the continuous time process $\{{}_hX_t, {}_hU_t, {}_h\hat{U}_t\}_{0 \leq t}$ as a step function with jumps at times $0, h, 2h, \ldots$. (A.7) specifies the transition probabilities for the jumps in $\{{}_hX_t, {}_hU_t, {}_h\hat{U}_t\}_{0 \leq t}$. (A.5′)–(A.7′) play the same role for $\hat{\mathrm{P}}_h$.

Next we define the forecast functions. Unfortunately, the informal definitions of $F_h(A, x, u, \tau)$ and $\hat{F}_h(A, x, u, \tau)$ are inadequate, since the conditional expectations on the right-hand sides of (2.2) and (2.4) are random variables which are unique only up to probability 0 transforms. We will evaluate conditional probabilities under both P_h and $\hat{\mathrm{P}}_h$ *which generally have different sets of probability zero*, so we must describe the *version* of the conditional probabilities we are using; in particular, how we condition on events of the form $({}_hX_\tau, {}_hU_\tau) = (x, u)$. Since our interest is in forecasting $\{{}_hX_t, {}_hU_t\}$ rather than $\{{}_hX_t, {}_hU_t, {}_h\hat{U}_t\}$, we use the Markov structure of $\{{}_hX_t, {}_hU_t\}$ under both P_h and $\hat{\mathrm{P}}_h$ to "drop" $\{{}_h\hat{U}_t\}$. Let $A \in \mathscr{B}(D([\tau, \infty), R^n \times R^m))$. (Note: $R^n \times R^m$ *not* $R^n \times R^{2m}$!) The conditional probability under P_h that $\{{}_hX_t, {}_hU_t\}_{\tau \leq t \leq \infty} \in A$ is

$$\begin{aligned}
&\mathrm{P}_h\big[\{{}_hX_t, {}_hU_t\}_{\tau \leq t < \infty} \in A \mid {}_h\mathscr{F}_\tau\big] \\
&\quad = \mathrm{P}_h\Big[\{{}_hX_t, {}_hU_t\}_{\tau \leq t < \infty} \in A \mid {}_hX_\tau, {}_hU_\tau, {}_h\hat{U}_t\Big] \quad \text{a.s.} \quad [\mathrm{P}_h] \\
&\quad = \mathrm{P}_h\big[\{{}_hX_t, {}_hU_t\}_{\tau \leq t < \infty} \in A \mid {}_hX_\tau, {}_hU_\tau\big] \quad \text{a.s.} \quad [\mathrm{P}_h].
\end{aligned} \tag{A.8}$$

The first equality in (A.8) holds since $\{{}_hX_{kh}, {}_hU_{kh}, {}_h\hat{U}_{kh}\}$ is a Markov chain under P_h. The second equality follows from (A.4), so $\{{}_hX_{kh}, {}_hU_{kh}\}$ is also a Markov chain under P_h. Below, this allows us to define the conditional probability that $\{{}_hX_t, {}_hU_t\}_{\tau \leq t < \infty} \in A$ as a function of the time τ and the state variables ${}_hX_\tau$ and ${}_hU_\tau$.

Now for every $(x, u, \hat{u}, \tau) \in R^{n+2m+1}$, define measures $\mathrm{P}_{(h, x, u, \tau)}$ and $\hat{\mathrm{P}}_{(h, x, u, \tau)}$ on $(D([\tau, \infty), R^n \times R^m), \mathscr{B}(D([\tau, \infty), R^n \times R^m)))$ by replacing P_h,

$\hat{\mathrm{P}}_h$, and $k \geq 0$ in (A.6)–(A.7) and (A.6′)–(A.7′) with $\mathrm{P}_{(h,x,u,\tau)}$, $\hat{\mathrm{P}}_{(h,x,u,\tau)}$, and $k \geq \mathrm{Int}[\tau/h]$, and replacing (A.5)–(A.5′) with

$$\mathrm{P}_{(h,x,u,\tau)}\left[\left({}_hX_\tau, {}_hU_\tau, {}_h\hat{U}_\tau\right) = (x,u,u)\right] = 1, \tag{A.5''}$$

$$\hat{\mathrm{P}}_{(h,x,u,\tau)}\left[\left({}_hX_\tau, {}_hU_\tau, {}_h\hat{U}_\tau\right) = (x,u,u)\right] = 1. \tag{A.5'''}$$

It is without loss of generality in (A.5″) and (A.5‴) that ${}_hU_\tau = {}_h\hat{U}_\tau = u$, since the forecasts generated using these probabilities will regard only the future paths of $\{{}_hX_t, {}_hU_t\}$, and once ${}_hU_\tau$ is fixed, the value of ${}_h\hat{U}_\tau$, is irrelevant. Now define $F_h(A, x, u, \tau)$ and $\hat{F}_h(A, x, \hat{u}, \tau)$ by

$$F_h(A, x, u, \tau) \equiv \mathrm{P}_{(h,x,u,\tau)}\left[\{{}_hX_t, {}_hU_t\}_{\tau \leq t < \infty} \in A\right], \tag{A.9}$$

$$\hat{F}_h(A, x, u, \tau) \equiv \hat{\mathrm{P}}_{(h,x,u,\tau)}\left[\{{}_hX_t, {}_hU_t\}_{\tau \leq t < \infty} \in A\right]. \tag{A.10}$$

Note that $F_h(A, x, u, \tau)$ and $\hat{F}_h(A, x, u, \tau)$ are now functions of A, x, u, and τ, and are *not* random variables, since they are defined in terms of *unconditional* rather than *conditional* probabilities. [In essence we have constructed $F_h(A, x, u, \tau)$ and $\hat{F}_h(A, x, u, \tau)$ directly from the measures $\Pi_h(x, u, \hat{u}, \cdot)$ and $\hat{\Pi}_h(x, u, \hat{u}, \cdot)$ without resorting to a sample space.]

$\mathrm{P}_{(0,x,u,\tau)}$, $\hat{\mathrm{P}}_{(0,x,u,\tau)}$, and their associated forecast functions are defined analogously.

Additional Assumptions Required for Lemma 2.2

To keep the notation in line with Nelson (1992), it is useful to define the measurement error process $\{{}_hY_t\}$: For all $h > 0$ and all $t > 0$, define ${}_hY_t \equiv {}_h\hat{U}_t - {}_hU_t$. The following assumptions are adapted from Assumptions 5–7 in Nelson (1992).

ASSUMPTION 6. *For every $h > 0$ and $\delta > 0$ and every $(x, u, \hat{u})R^{n+2m}$, the following are well defined and finite*:

$$c_{h,\delta}(x,u,\hat{u}) \equiv h^{-\delta}\mathrm{E}_h\left[{}_hY_{(k+1)h} - {}_hY_{kh} \mid {}_hX_{kh} = x, {}_hU_{kh} = u, {}_h\hat{U}_{kh} = \hat{u}\right],$$

$$d_{h,\delta}(x,u,\hat{u}) \equiv h^{-\delta}\mathrm{E}_h\left[\left({}_hY_{(k+1)h} - {}_hY_{kk}\right)\left({}_hY_{(k+1)h} - {}_hY_{kh}\right)' \mid {}_hX_{kh} = x, {}_hU_{kh} = u, {}_h\hat{U}_{kh} = \hat{u}\right].$$

Further, there exists a function $c(x, u, \hat{u})$ with $c(x, u, \hat{u}) = 0$ whenever $u = \hat{u}$, such that for some δ, $0 < \delta < 1$, and for every $R > 0$,

$$\lim_{h \downarrow 0} \sup_{\|x\| \leq R, \|u\| \leq R, \|\hat{u}\| \leq R} \|c_{h,\delta}(x,u,\hat{u}) - c(x,u,\hat{u})\| = 0$$

and

$$\lim_{h \downarrow 0} \sup_{\|x\| \leq R, \|u\| \leq R, \|\hat{u}\| \leq R} \| d_{h,\delta}(x, u, \hat{u}) \| = 0.$$

ASSUMPTION 7. *For each* $(x, u, \hat{u}) \in R^{n+2m}$, *define the ordinary differential equation*

$$dY(t, x, u, \hat{u})/dt = c(x, u, [Y(t, x, u, \hat{u}) + u]),$$

with initial condition $Y(0, x, u, \hat{u}) = \hat{u} - u$. *Then* $0_{m \times 1}$ *(an* $m \times 1$ *vector of zeros) is a globally asymptotically stable solution for bounded values of* $(x, u, \hat{u})$*—i.e., for every* $R \geq 0$,

$$\lim_{t \to \infty} \sup_{\|x, u, \hat{u}\| \leq R} \| Y(t, x, u, \hat{u}) \| = 0_{m \times 1}.$$

ASSUMPTION 8. *There exists a nonnegative, twice differentiable function* $\rho(x, y, h)$ *and a positive function* $\lambda(R, h)$ *such that*

$$\lim_{R \to \infty} \liminf_{h \downarrow 0} \inf_{\|[x'y']\| \geq R} \rho(x, y, h) = \infty,$$

$$\limsup_{R \to \infty} \limsup_{h \downarrow 0} \lambda(R, h) < \infty,$$

$$\limsup_{h \downarrow 0} E_h[\rho({}_xX_0, {}_hY_0, h)] < \infty,$$

and for every $R > 0$ *and* $h > 0$,

$$\sup_{\|[x'u'\hat{u}']\| \leq R} h^{-1} \mathrm{E}_h\Big[\rho({}_hX_{(k+1)h}, {}_hY_{(k+1)h}, h) - \rho(x, \hat{\mu} - \mu, h) \mid {}_hX_{kh} = x, {}_hU_{kh} = u, {}_h\hat{U}_{kh} = \hat{u}\Big] - \lambda(R, h)\rho(x, \hat{u} - u, h) \leq 0.$$

Lemmas Needed in the Proof of Theorem 2.3

LEMMA A.1. *For any* $\tau \geq 0$, *define* $\mu_{h,\tau}$ *to be the probability measure for* $({}_hX_t, {}_hU_\tau)$ *generated by* P_h. *(So* $\mu_{h,\tau}$ *is a measure on* R^{n+m}.*) For any* $\tau < \infty$, *there exists an* $h^* > 0$ *such that* $\{\mu_{h,\tau}\}_{0 \leq h \leq h^*}$ *is uniformly tight—i.e., for any* $\delta > 0$, *there exists a compact* $\Lambda(\delta) \subset R^{n+m}$ *such that for all* h, $0 < h \leq h^*$, $\mathrm{P}_h[({}_hX_\rho, {}_hU_\tau) \in \Lambda(\delta)] > 1 - \delta$.

Proof. Follows directly from Lemma 2.1 and Dudley (1989, Proposition 9.3.4).

LEMMA A.2. *Let Assumptions* 1–4 *be satisfied. Let* $(x_h, u_h) \to (x, u)$ *as* $h \downarrow 0$. *Then* $\mathrm{P}_{(0, x_h, u_h, \tau)} \Rightarrow \mathrm{P}_{(0, x, u, \tau)}$, $\mathrm{P}_{(h, x_h, u_h, \tau)} \Rightarrow \mathrm{P}_{(0, x, u, \tau)}$, $\hat{\mathrm{P}}_{(0, x_h, u_h, \tau)} \Rightarrow$

$\hat{P}_{(0,x,u,\tau)}$, $P_{(h,x_h,u_h,\tau)} \Rightarrow \hat{P}_{(0,x,u,\tau)}$ *as* $h \downarrow 0$. *If Assumption* 5 *is also satisfied, then* $\hat{P}_{(h,x_h,u_h,\tau)} \Rightarrow P_{(0,x,u,\tau)}$ *as* $h \downarrow 0$. *In each case, this convergence is uniform on bounded subsets of* R^{n+m} [*i.e., for bounded* (x,u) *sets*].

Proof. See the proof of Stroock and Varadhan (1979, Theorem 11.2.3). Their requirement that "the martingale problem ... has exactly one solution" is implied by our Assumption 4 (see Ethier and Kurtz, 1986, ch. 5, Corollary 3.4).

Proof of Theorem 2.3. The theorem is equivalent to the following for every $A \in \mathscr{M}_\tau$, every $\delta > 0$, and every $\varepsilon > 0$, there exists an $h^* > 0$ such that for every h, $0 < h \leq h^*$,

$$P_h\left[\left|F_h(A, {}_hX_\tau, {}_hU_\tau, \tau) - \hat{F}_h\left(A, {}_hX_\tau, {}_h\hat{U}_\tau, \tau\right)\right| > \varepsilon\right] \leq \delta, \quad \text{(A.11)}$$

$$P_h\left[\left|F_h(A, {}_hX_\tau, {}_hU_\tau, \tau) - F_0(A, {}_hX_\tau, {}_hU_\tau, \tau)\right| > \varepsilon\right] \leq \delta, \quad \text{(A.12)}$$

$$P_h\left[\left|\hat{F}_h\left(A, {}_hX_\tau, {}_h\hat{U}_\tau, \tau\right) - \hat{F}_0\left(A, {}_hX_\tau, {}_hU_\rho, \tau\right)\right| > \varepsilon\right] \leq \delta. \quad \text{(A.13)}$$

We prove (A.11) first. By Lemma A.1 there exists a compact $\Lambda(\delta)$ and an $h^{**}(\delta) > 0$ such that $P_h[({}_hX_\tau, {}_hU_\tau) \in \Lambda(\delta)] > 1 - \delta/2$ when $h \leq h^{**}(\delta)$. By Lemma 2.2, ${}_hU_\tau - {}_h\hat{U}_\tau \to 0$ in probability under P_h as $h \downarrow 0$. Therefore for every $\eta > 0$ and $\delta > 0$, there exists an $h^{***}(\eta, \delta) > 0$ such that $P_h[\|{}_hU_\tau - {}_h\hat{U}_\tau\| > \eta] < \delta/2$ for $h < h^{***}(\eta, \delta)$. We then have for all $h \leq \min\{h^{**}(\delta), h^{***}(\eta, \delta)\}$,

$$P_h\left[\left|F_h(A, {}_hX_\tau, {}_hU_\tau, \tau) - \hat{F}_h\left(A, {}_hX_\tau, {}_h\hat{U}_\tau, \tau\right)\right| > \varepsilon\right] < \delta/2 + \delta/2$$
$$+ \sup_{(x,u)\in\Lambda(\delta),\, \|u-\hat{u}\|\leq\eta} I\left[\left|F_h(A, x, u, \tau) - \hat{F}_h(A, x, \hat{u}, \tau)\right| > \varepsilon\right], \quad \text{(A.14)}$$

where $I(\cdot)$ is the indicator function. By Lemma A.2, there exists an $\eta(\delta) > 0$ and an $h^{****}(\delta, \varepsilon) > 0$ such that $|F_h(A, x, u, \tau) - \hat{F}_h(A, x, \hat{u}, \tau)| \leq \varepsilon$ whenever $(x,u) \in \Lambda(\delta)$, $\|u - \hat{u}\| \leq \eta(\delta)$, and $h \leq h^{****}(\delta, \varepsilon)$. The bound in (A.14) is therefore achieved whenever $h \leq h^*(\delta, \varepsilon) \equiv \min\{h^{**}(\delta), h^{***}(\eta(\delta), \delta), h^{****}(\delta, \varepsilon)\}$.

Next we prove (A.12). Define $\Lambda(\delta)$ and $h^{**}(\delta)$ as above. For $h \leq h^{**}(\delta)$,

$$P_h\left[\left|F_h(A, {}_hX_\tau, {}_hU_\tau, \tau) - F_0(A, {}_hX_\tau, {}_hU_\tau, \tau)\right| > \varepsilon\right]$$
$$< \delta/2 + \sup_{(x,u)\in\Lambda(\delta)} I\left[|F_h(A, x, u, \tau) - F_0(A, x, u, \tau)| > \varepsilon\right]. \quad \text{(A.15)}$$

By Lemma A.2, the convergence of $P_{(h,x,u,\tau)} \Rightarrow P_{(0,x,u,\tau)}$ is uniform on compacts. Therefore the final term in (A.15), which restricts (x,u) to lie in $\Lambda(\delta)$ for all $h \leq h^*(\delta)$, vanishes uniformly in h, proving (A.12). Under

Assumption 5 and Lemma 2.2, $P_{(0,x,u,\tau)}$ and $\hat{P}_{(0,x,u,\tau)}$ are identical, so (A.13) follows from (A.11) and (A.12), proving the theorem.

Proof of Theorem 2.4. The proof is nearly identical to the proof of Theorem 2.3. The details are left to the reader.

Lemmas Needed in the Proof of Theorem 2.5

LEMMA A.3. *Let g be as in Theorem* 2.5, *but require in addition that g is bounded. Then the conclusion of Theorem* 2.5 *holds.*

Proof. Let $\{Z_n\}_{n\to\infty}$ be a sequence of random variables on some probability space, with $\{Z_n\}_{n\to\infty} \Rightarrow Z$, and let f be a bounded continuous functional mapping the space into R^1. Weak convergence implies (see, e.g., Billingsley, 1986, Theorem 25.8) that $E[f(Z_n)] \to E[f(z)]$ as $n \to \infty$. The proof of Lemma A.3 is essentially identical to the proof of Theorem 2.3, substituting $G_h(g, x, u, \tau)$ for $F_h(A, x, u, \tau)$, $G_0(g, x, u, \tau)$ for $F_0(A, x, u, \tau)$, $\hat{G}_h(g, x, u, \tau)$ for $\hat{F}_h(A, x, u, \tau)$, and $\hat{G}_0(g, x, u, \tau)$ for $\hat{F}_0(A, x, u, \tau)$.

LEMMA A.4. *Let the conditions of Lemma* 2.1 *hold. Define g as in Theorem* 2.5. *Then* (2.16) *implies that for every bounded* $\Lambda \subset R^{n+m}$,

$$\lim_{K\to\infty} \sup_{(x,u)\in\Lambda} E_{(0,x,u,\tau)}\Big[\big|g(\{X_t, U_t\}_{[\tau,\infty)})\big| \times I\big(\big|g(\{X_t, U_t\}_{[\tau,\infty)})\big| > K\big)\Big] = 0. \quad \text{(A.16)}$$

Proof. Let $\mathscr{A}$ be a metric space and let f be a mapping from $\mathscr{A} \times R^1$ into R^1. Then clearly

$$\limsup_{h\downarrow 0} \sup_{x\in\mathscr{A}} f(x,h) \geq \sup_{x\in\mathscr{A}} \limsup_{h\downarrow 0} f(x,h). \quad \text{(A.17)}$$

Now by (A.16) and (A.17), we have

$$\begin{aligned} 0 &= \lim_{K\to\infty} \limsup_{h\downarrow 0} \sup_{(x,u)\in\Lambda} E_{(h,x,u,\tau)}\Big[\big|g(\{{}_hX_t, {}_hU_t\}_{[\tau,\infty)})\big| \times I\big(\big|g(\{{}_hX_t, {}_hU_t\}_{[\tau,\infty)})\big| > K\big)\Big] \\ &\geq \lim_{K\to\infty} \sup_{(x,u)\in\Lambda} \limsup_{h\downarrow 0} E_{(h,x,u,\tau)}\Big[\big|g(\{{}_hX_t, {}_hU_t\}_{[\tau,\infty)})\big| \times I\big(\big|g(\{{}_hX_t, {}_hU_t\}_{[\tau,\infty)})\big| > K\big)\Big]. \quad \text{(A.18)} \end{aligned}$$

By Fatou's Lemma (Billingsley, 1986, Theorem 25.11),

$$\lim_{K\to\infty} \sup_{(x,u)\in\Lambda} \mathrm{E}_{(0,x,u,\tau)}\Big[\big|g(\{X_t, U_t\}_{[\tau,\infty)})\big| \times I\big(\big|g(\{X_t, U_t\}_{[\tau,\infty)})\big| > K\big)\Big] \geq 0. \quad \text{(A.19)}$$

Proof of Theorem 2.5. The theorem is equivalent to the following: Under (2.15)–(2.16) and Assumptions 1–4 and 6–8, there exists, for every $\delta > 0$ and $\varepsilon > 0$, an $h^* > 0$ such that for all h, $0 < h < h^*$,

$$\mathrm{P}_h\Big[\big|\hat{G}_h(g, {}_hX_\tau, {}_h, \tau) - \hat{G}_0(g, {}_hX_\tau, {}_hU_\tau, \tau)\big| > \varepsilon\Big] \leq \delta, \quad (2.21')$$

$$\mathrm{P}_h\big[\left|G_h(g, {}_hX_\tau, {}_hU_\tau, \tau) - G_0(g, {}_hX_\tau, {}_hU_\tau, \tau)\right| > \varepsilon\big] \leq \delta, \quad (2.22')$$

and if Assumption 5 is also satisfied,

$$\mathrm{P}_h\Big[\Big|\hat{G}_h\big(g, {}_hX_\tau, {}_h\hat{U}_\rho, \tau\big) - G_h(g, {}_hX_\tau, {}_hU_\tau, \tau)\Big| > \varepsilon\Big] \leq \delta. \quad (2.23')$$

First consider (2.23′). For any $K > 0$ we can rewrite g as

$$\begin{aligned} g = {} & (g + K)\cdot I(g < -K) + (g - K)\cdot I(g > K) \\ & + \max(-K, \min\{g, K\}). \end{aligned} \quad \text{(A.20)}$$

Define $\Lambda(\delta)$, η, h^{**}, and h^{***} as in the proof of Theorem 2.3. Then

$$\begin{aligned} & \mathrm{P}_h\Big[\Big|\hat{G}_h\big(g, {}_hX_\tau, {}_h\hat{U}_\tau, \tau\big) - G_h(g, {}_hX_\tau, {}_hU_\tau, \tau)\Big| > \varepsilon\Big] < \delta/2 + \delta/2 \\ & \quad + \sup_{(x,u)\in\Lambda(\delta),\, \|u-\hat{u}\|\leq\eta} I\Big[\big|G_h(g, x, u, \tau) - \hat{G}_h(g, x, \hat{u}, \tau)\big| > \varepsilon\Big] \end{aligned} \quad \text{(A.21)}$$

$$\begin{aligned} & \leq \delta/2 + \delta/2 \\ & \quad + \sup_{(x,u)\in\Lambda(\delta),\, \|u-\hat{u}\|\leq\eta} I\Big[|G_h((g+K)\cdot I(g<-K), x, u, \tau) \\ & \qquad\qquad - \hat{G}_h((g+K)\cdot I(g<-K), x, \hat{u}, \tau)| > \varepsilon/3\Big] \\ & \quad + \sup_{(x,u)\in\Lambda(\delta),\, \|u-\hat{u}\|\leq\eta} I\Big[|G_h((g-K)\cdot I(g>K), x, u, \tau) \\ & \qquad\qquad - \hat{G}_h((g-K)\cdot I(g>K), x, \hat{u}, \tau)| > \varepsilon/3\Big] \\ & \quad + \sup_{(x,u)\in\Lambda(\delta),\, \|u-\hat{u}\|\leq\eta} I\Big[|G_h(\max(-K, \min\{g, K\}), x, u, \tau) \\ & \qquad\qquad - \hat{G}_h(\max(-K, \min\{g, K\}), x, \hat{u}, \tau)| > \varepsilon/3\Big]. \end{aligned} \quad \text{(A.22)}$$

By (15) and (16), for any $\delta > 0$ and $\varepsilon > 0$, $K(\delta, \varepsilon)$ can be selected to make $|G_h((g + K(\delta, \varepsilon))\cdot I(g < -K(\delta, \varepsilon)), x, u, \tau) - \hat{G}_h((g + K(\delta, \varepsilon))\cdot I(g < -K(\delta, \varepsilon)), x, \hat{u}, \tau)| \leq \varepsilon/3$ for $(x, u) \in \Lambda(\delta)$, $\|u - \hat{u}\| \leq \eta$, and all sufficiently small h. The same is true of the $|G_h((g - K(\delta, \varepsilon)) \cdot I(g > K(\delta, \varepsilon)), x, u, \tau) - \hat{G}_h((g - K(\delta, \varepsilon)) \cdot I(g > K(\delta, \varepsilon)), x, \hat{u}, \tau)|$ term. For a given $K(\delta, \varepsilon)$, $\max(-K(\delta, \varepsilon), \min\{g, K(\delta, \varepsilon)\})$ is bounded and continuous for all $(x, u, \hat{u})$, so by Lemma A.3, the last term in (A.22) vanishes for sufficiently small h. This bounds (A.22) above by δ, proving (2.23′).

Next consider (22′). For $h < h^{**}$, we have

$$\begin{aligned}
&\mathrm{P}_h\big[\big|G_h(g, {}_hX_t, {}_hU_\tau, \tau) - G_0(g, {}_hX_\tau, {}_hU_\tau, \tau)\big| > \varepsilon\big] \\
&\quad < \delta/2 + \sup_{(x,u)\in\Lambda(\delta)} I\big[\big|G_h(g, x, u, \tau) - G_0(g, x, \hat{u}, \tau)\big| > \varepsilon\big] \\
&\quad \leq \sup_{(x,u)\in\Lambda(\delta)} I\big[\big|G_h((g + K)\cdot I(g < -K), x, u, \tau)\big| \\
&\qquad\qquad + \big|G_0((g + K)\cdot I(g < -K), x, u, \tau)\big| > \varepsilon/3\big] \\
&\qquad + \sup_{(x,u)\in\Lambda(\delta)} I\big[\big|G_h((g - K)\cdot I(g > K), x, u, \tau)\big| \\
&\qquad\qquad + \big|G_0((g - K)\cdot I(g > K), x, u, \tau)\big| > \varepsilon/3\big] \\
&\qquad + \delta/2 + \sup_{(x,u)\in\Lambda(\delta)} I\big[\big|G_h(\max(-K, \min\{g, K\}), x, u, \tau) \\
&\qquad\qquad - G_0(\max(-K, \min\{g, K\}), x, u, \tau)\big| > \varepsilon/3\big]. \qquad \text{(A.23)}
\end{aligned}$$

By (2.16) and Lemma A.4, for any $\delta > 0$ and any $\varepsilon > 0$, there is a $K(\delta, \varepsilon)$ such that $|G_h((g + K(\delta, \varepsilon)) \cdot I(g < -K(\delta, \varepsilon)), x, u, \tau)| + |G_0((g + K(\delta, \varepsilon)) \cdot I(g < -K(\delta, \varepsilon)), x, \hat{u}, \tau)| \leq \varepsilon/3$ whenever $(x, u) \in \Lambda(\delta)$ and h is sufficiently small. The same is true of the term $|G_h((g - K(\delta, \varepsilon)) \cdot I(g > K(\delta, \varepsilon)), x, u, \tau)| + |G_0((g - K(\delta, \varepsilon)) \cdot I(g > K(\delta, \varepsilon)), x, u, \tau)|$. By Lemma A.3 and the boundedness and continuity of $\max(-K(\delta, \varepsilon), \min\{g, K(\delta, \varepsilon)\})$, the last term on the right side of (A.23) also vanishes for sufficiently small h, proving (2.22′).

The proof of (2.21′) is similar to the proofs of (2.22′)–(2.23′). (2.15′) and (2.16′) follow by Billingsley (1986, 25.18).

Proof of Theorem 2.6. The theorem follows directly as a special case of Theorems 3 and 6 in Chapter 6 of Kushner (1984): both Kushner's Liapunov function $V(x, u)$ and his perturbed Liapunov function $V^\varepsilon(x, u)$ are set equal to our $\zeta(x, u)$ in (2.24)–(2.27). This makes Kushner's $\delta_\varepsilon \equiv 0$ in both cases. We set Kushner's ψ^c equal to the constant Δ. The moment conditions (2.26) and (2.27) imply the tightness of $\{{}_hX_t, {}_hU_t\}$ under P_h as $t \to \infty$ and $h \downarrow 0$ (see the discussion in Kushner, 1984, pp. 151, 157.) The proof then follows by Theorem 6 in Chapter 6 of Kushner. The proof for $\hat{\mathrm{P}}_h$ is identical.

Proof of Theorem 3.1. (a) To apply Lemma 2.2, we must verify Assumptions 1–4 and 6–8. It is convenient to do a change of variables in (30)–(31) by defining $X_t \equiv \ln(S_t)$ and $U_t \equiv \ln(\sigma_t^2)$. We then have

$$\begin{aligned} dX_t &= (\mu - \exp(U_t)/2)\,dt + \exp(U_t/2)\cdot dW_{1,t}, \\ dU_t &= -\beta[U_t - \alpha]\,dt + \Lambda\cdot dW_{2,t}. \end{aligned} \tag{A.24}$$

Assumption 1 is immediate, since ${}_hS_0, {}_h\sigma_0^2$, and ${}_h\hat{\sigma}_0^2$ are constant for all h. We next verify Assumption 2. Equation (2.8) is immediate for any $\varepsilon > 0$, since U_{t+h} is Gaussian with moments continuous in U_t (Arnold, 1973, Section 8.3). By (A.24),

$$\left(X_t - X_0 - \mu\cdot t - \tfrac{1}{2}\int_0^t \exp(U_s)\,ds\right) = \int_0^t \exp(U_s/2)\cdot dW_{1,s}.$$

By Karatzas and Shreve (1988, p. 163, Exercise 3.25),

$$h^{-1}E_{(h,x,u,\tau)}\left[\left|X_{t+h} - X_t - \mu\cdot h - \tfrac{1}{2}\int_t^{t+h}\exp(U_s)\cdot ds\right|^6\right]$$

$$\leq h\cdot 15^3\cdot E_{(h,x,u,\tau)}\int_t^{t+h}\exp(3\cdot U_s)\cdot ds. \tag{A.25}$$

Fubini's Theorem (e.g., Dudley, 1989, Theorem 4.4.5) allows us to interchange the expectations and integral in the right-hand side of (A.25). $\exp(3\cdot U_s)$ is lognormal and has a bounded expectation on $[t, t+h]$, bounding the expectation on the left. Since $\exp(U_s)$ has arbitrary finite moments (which are continous in U_t), Assumption 2 follows.

We next check Assumption 3: (2.10) is immediate from the definitions of μ and Ω in (3.1)–(3.2) and Assumption 2 (see, e.g., Arnold, 1973, p. 40). Equation (2.10′) follows as in the proof of Nelson (1990, Theorem 3.3), verifying Assumption 3. Assumption 4 also follows as in the proof of Nelson (1990, Theorem 3.3).

Assumptions 6, 7, and 8 are verified in Nelson (1992, Theorem 3.2) under Assumptions 1–4 and the following additional conditions:

First, there is an $\varepsilon > 0$ such that for every $\eta > 0$,

$$\lim_{h\downarrow 0}\ \sup_{\|x,u\|\leq\eta}\ h^{-1}E_{(h,x,u,kj)}\left[\|{}_hU_{(k+1)h} - u\|^{2+\varepsilon}\right] = 0. \tag{A.26}$$

Second, there is a twice continuously differentiable, nonnegative function $\omega(x,u)$ and a $\Theta > 0$ such that for every $\eta > 0$,

$$\liminf_{\|x,u\|\to\infty}\ \omega(x,u) = \infty. \tag{A.27}$$

$$\lim_{h\downarrow 0} \sup_{\|x,u\|\le\eta} h^{-1}E_{(h,x,u,kh)}\left(\left[|\omega({}_hX_{(k+1)h},{}_hU_{(k+1)h})\right] - \omega(x,u)|^{1+\Theta}\right) < \infty. \tag{A.28}$$

and there is a $\lambda > 0$, such that for all $(x,u) \in R^2$,

$$(\mu - e^{u/2})\frac{\partial\omega(x,u)}{\partial x} - \beta[u-\alpha]\frac{\partial\omega(x,u)}{\partial u} + \Lambda^2\frac{\partial^2\omega(x,u)}{2\,\partial u^2} + e^u\frac{\partial^2\omega(x,u)}{2\,\partial x^2} + \rho\Lambda e^{u/2}\frac{\partial^2\omega(x,u)}{\partial x\,\partial u} \le \lambda\omega(x,u). \tag{A.29}$$

Equation (A.26) is immediate, since U_t is Gaussian and therefore has arbitrary finite conditional moments. To verify (A.27)–(A.28), we set

$$\omega(x,u) \equiv 1 + (1 - \exp(-x^2))\cdot|x| + \exp(u) + \exp(-u).$$

Equation (A.27) is trivial. It is easy to verify that $\omega(x,u)$ is twice continuously differentiable on R^2 (including when $x = 0$) and that $-2 < \partial\omega/\partial x < 2$ and $-2 < \partial^2\omega/\partial\omega^2 < 2$, and $\partial^2\omega(x,u)/\partial x\partial u = 0$ for all (x,u). This bounds the left-hand side of (A.29) from above by

$$2\,|u| + 2\cdot\exp(u) - \beta\cdot[\exp(u) - \exp(-u)]\cdot(u-\alpha) + \Lambda^2[\exp(u) + \exp(-u)]/2, \tag{A.30}$$

which we require to be bounded above by the right-hand side of (A.29), which equals $\lambda(1 + |x|\cdot(1 - \exp(-x^2)) + \exp(u) + \exp(-u))$. Since $\beta \ge 0$, this holds for all $(x,u) \in R^2$ if we choose a sufficiently large λ. To verify (A.28), we choose $\Theta = 1$. The finiteness of the expected increments in X_t^2 follows from Assumption 2. Further, since $\pm U_t$ is normal (conditionally and unconditionally), $\exp(\pm U_t)$ is lognormal and its increments have finite expectation. (A.28) follows. Assumptions 6, 7, and 8 now follow from Nelson (1992, Theorem 3.2), concluding the proof of (a).

(b) Assumption 5 is verified in Nelson (1990, Sect. 3.3). Theorem 2.3 immediately follows, implying Theorem 3.1 (b).

(c) By (2.15′)–(2.16′) and the results of parts (a) and (b) of Theorem 2.5, it is sufficient that for every $\tau > 0$, every bounded $\Lambda \subset R^2$, and for all bounded nonnegative A, B, a, b, c, and d.

$$\limsup_{h\downarrow 0} \sup_{(\ln(s),\ln(\sigma))\in\Lambda} \hat{\mathrm{E}}_{(h,s,\sigma,\tau)}\left[\left|G(\ln({}_hS_t),\ln({}_hS_T),{}_h\sigma_t,{}_h\sigma_T)\right|^{1+\varepsilon}\right] < \infty, \tag{A.31}$$

$$\limsup_{h\downarrow 0} \sup_{(\ln(s),\ln(\sigma))\in\Lambda} \mathrm{E}_{(h,s,\sigma,\tau)}\left[\left|G(\ln({}_hS_t),\ln({}_hS_T),{}_h\sigma_t,{}_h\sigma_T)\right|^{1+\varepsilon}\right] < \infty. \tag{A.32}$$

We prove (A.32) first. This relation clearly holds when

$$\limsup_{h\downarrow 0} \sup_{(\ln(s),\ln(\sigma))\in\Lambda} \mathrm{E}_{(h,s,\sigma,\tau)} \times\left[|\ln({}_hS_t)|^a \cdot |\ln({}_hS_T)|^b \cdot \left(|{}_h\sigma_t|^c \cdot |{}_h\sigma_T|^d\right)\right] < \infty, \quad \text{(A.33)}$$

for arbitrary nonnegative τ, t, T, a, b, c, d (we can ignore the $1 + \varepsilon$ term since a, b, c, and d can be made arbitrarily large). Because of the Markov structure of $({}_hS_{kh}, {}_h\sigma_{kh}^2)$ and the arbitrariness of t and T, τ is irrelevant, so we set it equal to zero. Applying Hölder's inequality and the arbitrariness of a, b, c, and d, (A.32) holds if for arbitrary nonnegative a, b, and $t \geq 0$,

$$\limsup_{h\downarrow 0} \sup_{(\ln(s),\ln(\sigma))\in\Lambda} \mathrm{E}_{(h,s,\sigma,0)}\left[|\ln({}_hS_t)|^a\right] < \infty, \quad \text{(A.34)}$$

$$\limsup_{h\downarrow 0} \sup_{(\ln(s),\ln(\sigma))\in\Lambda} \mathrm{E}_{(h,s,\sigma,0)}\left[{}_h\sigma_t{}^b\right] < \infty. \quad \text{(A.35)}$$

The lognormality of $\sigma_t{}^2$ implies (A.35). Using Fubini's Theorem and Karatzas and Shreve (1988, p. 163, Exercise 3.25), (A.34) follows by the argument used to establish (A.25) above, proving (A.32).

We now turn to (A.31). By the same argument as in (A.34)–(A.35), (A.31) follows if for arbitrary nonnegative a, b, and $t \geq 0$,

$$\limsup_{h\downarrow 0} \sup_{(s,\ln(\hat{\sigma}))\in\Lambda} \hat{\mathrm{E}}_{(h,s,\hat{\sigma},0)}\left[|\ln({}_hS_t)|^a\right] < \infty, \quad \text{(A.36)}$$

$$\limsup_{h\downarrow 0} \sup_{(s,\ln(\hat{\sigma}))\in\Lambda} \hat{\mathrm{E}}_{(h,s,\hat{\sigma},0)}\left[{}_h\hat{\sigma}_t{}^b\right] < \infty. \quad \text{(A.37)}$$

We consider (A.37) first. Using (33) and (37) we write the expectation in (A.37) as

$${}_h\hat{\sigma}_0^{b(1-\beta h)^{k+1}/2} \cdot f(\alpha b, \beta h, k) \cdot \prod_{i=0,k} \hat{E}_h\left[\exp\left(bh^{1/2}(1-\beta h)^i \times g({}_hZ_{(k-i)h})/2\right)\right], \quad \text{(A.38)}$$

where $k \equiv [t/h]$, and

$$f(\alpha b, \beta h, k) \equiv \exp\left[\alpha b\left(1 - (1-\beta h)^{k+1}\right)/2\right]. \quad \text{(A.39)}$$

By Nelson (1991, Theorem A1.1), we have, for any real λ,

$$\mathrm{E}[e^{g(z)\lambda}] = \left\{\Phi(\gamma\lambda = \theta\lambda)\mathrm{e}^{\lambda^2(\theta+\lambda)^2/2} + \Phi(\gamma\lambda - \theta\lambda)\mathrm{e}^{\lambda^2(\gamma-\theta)^2/2}\right\} \cdot \mathrm{e}^{-\lambda\gamma(2/\pi)^{1/2}}, \quad \text{(A.40)}$$

where $\Phi(\cdot)$ is the cumulative distribution function of the standard normal, and γ and θ are as in (3.10)–(3.11). Considered as a function of $\lambda\gamma$ and $\lambda\theta$, it is easy to check that $\mathrm{E}[\mathrm{e}^{g(z)\lambda}] \ll \mathrm{E}[\mathrm{e}^{Y}]$, where $Y \sim \mathrm{N}(0, \lambda^2(\gamma + |\theta|)^2)$. Substituting into (A.38) yields

$$\hat{\mathrm{E}}_{(h,s,\hat{\sigma},0)}\left[{}_h\hat{\sigma}_t^{\,b}\right] \leq {}_h\hat{\sigma}_0^{\,b(1-\beta h)^{k+1}/2} \cdot f(\alpha b, \beta h, k)$$

$$\cdot \exp\left[b^2 h(\gamma + |\theta|)^2 \cdot \sum_{i=0,k} (1-\beta h)^{2i}/8\right].$$

Applying Taylor series expansions to $f(\cdot,\cdot,\cdot)$ and $(1-\beta h)^k$ and evaluating the sum $\Sigma_{i=1,k}(1-\beta h)^{2i}$, we obtain

$$\hat{\mathrm{E}}_{(h,s,\hat{\sigma},0)}\left[{}_h\hat{\sigma}_t^{\,b}\right] \leq (1 + O(h)) \cdot \hat{\sigma}^{\,b\cdot e^{-\beta t}/2} \cdot \exp(-b\beta\alpha(1 - \mathrm{e}^{-\beta t})/2)$$

$$\cdot \exp\left[\left(b^2(\gamma + |\theta|)^2/16\beta\right)\right]. \quad \text{(A.41)}$$

The bound (A.41) is uniform for bounded $(s, \hat{\sigma}, t)$ that holds for arbitrary nonnegative b. (A.37) follows.

Finally, we prove (A.36). Here, it helps to define the $\{{}_hZ_{kh}\}$ in (3.7) by ${}_hZ_{kh} \equiv h^{-1/2} \cdot [W_{kh} - W_{(k-1)h}]$, where W_t is a standard Brownian motion on $[-h, \infty)$. This allows us to write, for each positive h and t,

$$\ln({}_hS_t) - \ln({}_hS_0) - h[t/h]\mu + \int_0^{h[t/h]} {}_h\hat{\sigma}_s^{\,2}/2 \cdot ds = \int_0^{h[t/h]} {}_h\hat{\sigma}_s \cdot dW_s, \quad \text{(A.42)}$$

under $\hat{\mathrm{P}}_h$. Since ${}_h\hat{\sigma}_s^{\,2}$ has arbitrary finite moments, the moments of $\ln({}_hS_t)$ will be bounded if and only if the moments on the right-hand side of (A.42) are (A.36) will therefore be proven if we can show that for arbitrary positive a and t and arbitrary bounded $\Lambda \subset R^2$,

$$\limsup_{h\downarrow 0} \sup_{(\ln(s),\ln(\sigma))\in\Lambda} \hat{\mathrm{E}}_{(h,s,\sigma,0)}\left[\left|\int_0^{h[t/h]} {}_h\hat{\sigma}_t \cdot dW_s\right|^a\right] < \infty. \quad \text{(A.43)}$$

By Karatzas and Shreve (1988, p. 163, Exercise 3.25), (A.43) is bounded above by

$$\limsup_{h\downarrow 0} \sup_{(s,\ln(\sigma))\in\Lambda} [a(a-1)/2]^{a/2} \cdot t^{(a/2-1)} \hat{\mathrm{E}}_{(h,s,\sigma,0)} \int_0^{h[t/h]} {}_h\sigma_t^{\,a} \cdot ds. \quad \text{(A.44)}$$

Since ${}_h\hat{\sigma}_t$ is a step function, we can rewrite (A.44) as

$$\limsup_{h\downarrow 0} \sup_{(\ln(s),\ln(\sigma))\in\Lambda} [a(a-1)/2]^{a/2} \cdot t^{(a/2-1)}\hat{\mathrm{E}}_{(h,s,\sigma,0)} \times\left[\sum_{i=0,[t/h]} {}_h\hat{\sigma}_{ih}^{a} \cdot h\right]. \quad \text{(A.45)}$$

By in (A.41), the summation in (A.45) uniformly bounded for bounded (s, σ, t), proving (A.36) and completing the proof of (c).

(d) When $|\gamma| \geq |\theta|$, Theorem 2.6 can be employed to prove (d) (see the very similar application in the proof of Nelson, 1990, Theorem 2.3). To prove (d) in the general case, it is easier to employ the Lyapunov central limit theorem for triangular arrays (see, e.g., Billingsley, 1986, Theorem 27.3). When $\beta > 0$, the stationary distribution of $\ln(\sigma_t^2)$ is $\mathrm{N}[\alpha, \Lambda^2/2\beta]$ (see Arnold, 1973, Sect. 8.3). When the AR(1) EGARCH model is the data generating process we have

$$\ln\left({}_h\sigma_{kh}^2\right) = \alpha + (1-h\cdot\beta)^k\cdot\left[\ln\left({}_h\sigma_0^2\right)-\alpha\right] + \sum_{j=1,k}(1-h\cdot\beta)^{j-1}\cdot h^{1/2}\cdot g\left({}_hZ_{(k-j)h}\right).$$

If $|1-\beta h| < 1$, we let $k\to\infty$ to obtain the stationary distribution of $\ln({}_h\sigma_{kh}^2)$ as the distribution of

$$\ln\left({}_h\sigma_\infty^2\right) \equiv \alpha + \sum_{i=0,\infty}(1-h\cdot\beta)^i\cdot h^{1/2}\cdot g(Z_i),$$

where $\{Z_i\} \sim$ i.i.d. N(0, 1). From the normality of Z_i and the definition of $g(\cdot)$, it is clear that $g(Z_i)$ possesses finite moments of arbitrary finite order. We also have

$$\mathrm{Var}\left[\ln\left({}_h\sigma_\infty^2\right)\right] = \mathrm{Var}[g(Z_i)]/\left[2\beta-\beta^2\cdot h\right], \quad \text{(A.46)}$$

$$\mathrm{E}\left[(1-h\cdot\beta)^i\cdot h^{1/2}\cdot g(Z_i)\right]^4 = h^2E\left[g(Z_i)^4\right](1-h\cdot\beta)^{4i}. \quad \text{(A.47)}$$

The Lyapunov condition is satisfied if

$$0 = \lim_{h\downarrow 0}\sum_{i=0,\infty}\mathrm{E}\left[(1-h\cdot\beta)^i\cdot h^{1/2}\cdot g(Z_i)\right]^4/\left[\mathrm{Var}\left[\ln\left({}_h\sigma_\infty^2\right)\right]\right]^2.$$

But from (A.46)–(A.47),

$$\lim_{h\downarrow 0}\frac{h\cdot\left[4\beta^2+O(h)\right]\cdot\mathrm{E}\left[g(Z_i)^4\right]}{\left[\mathrm{Var}(g(Z_i))\right]^2\cdot[4\beta+O(h)]} = 0.$$

Applying the Lyapunov central limit theorem (Billingsley, 1986, Theorem 27.3) for triangular arrays, we have

$$\ln\left({}_h\sigma_\infty^2\right) \Rightarrow \mathrm{N}\left(\alpha, \mathrm{Var}\left[g(Z_i)\right]/2\beta\right).$$

We complete the proof by invoking (34).

ACKNOWLEDGMENTS

We thank Arnold Zellner and three anonymous referees for helpful comments. This material is based on work supported by the National Science Foundation under grants SES-9110131 and SES-9310683. We thank the Graduate School of Business of the University of Chicago, the Center for Research in Security Prices, and the William S. Fishman Research Scholarship for additional research support.

REFERENCES

Amin, K. I., and Ng, V. K. (1993). "ARCH Processes and Option Valuation." University of Michigan, Ann Arbor (unpublished).

Anderson, D. O. and Moore, J. B. (1979). "Optimal Filtering." Prentice-Hall, New York.

Arnold, L. (1973). "Stochastic Differential Equations: Theory and Applications." Wiley, New York.

Billingsley, P. (1986). "Probability and Measure," 2nd ed. Wiley, New York.

Bollerslev, T. (1986). Generalized autoregressive conditional heteroskedasticity. *Journal of Econometrics* **31**, 307–327.

Bollerslev, T. Engle, R. F., and Wooldridge, J. M. (1988). A capital asset pricing model with time varying covariances. *Journal of Political Economy* **96**, 116–131.

Bollerslev, T., Chou, R. Y., and Kroner, K. F. (1992). ARCH modeling in finance: A review of the theory and empirical evidence. *Journal of Econometrics* **52**, 5–60.

Bollerslev, T., Engle, R. F., and Nelson, D. B. (1994). ARCH models. *In* "Handbook of Econometrics" (R. F. Engle and D. McFadden, eds.), Vol. 4, Chapter 49, pp. 2959–3038. North-Holland Publ., New York.

Cao, C. Q. (1992). "Pricing Options with Stochastic Volatility: A General Equilibrium Approach," Working Paper. University of Chicago, Graduate School of Business, Chicago (unpublished).

Dudley, R. M. (1989). "Real Analysis and Probability." Wadsworth Brooks/Cole, Pacific Grove, CA.

Engle, R. F. (1982). Autoregressive conditional heteroskedasticity with estimates of the variance of United Kingdom inflation. *Econometrica* **50**, 987–1008.

Engle, R. F. Ng, V. K., and Rothschild, N. (1990). Asset pricing with a factor ARCH covariance structure: Empirical estimates for Treasury bills, *Journal of Econometrics* **45**, 213–238.

Ethier, S. N., and Kurtz, T. G. (1986). "Markov Processes: Characterization and Convergence." Wiley, New York.

Ethier, S. N., and Nagylaki, T. (1988). Diffusion approximations of Markov chains with two time scales and applications to population genetics. II. *Advances in Applied Probability* **20**, 525–545.

Foster, D. P., and Nelson, D. B. (1994). "Continuous Record Asymptotics for Rolling Sample Variance Estimators, "Working Paper. University of Chicago, Graduage School of Business, Chicago (unpublished).

Harvey, A. C., and Shephard, N. (1993). "The Econometrics of Stochastic Volatility," Working Paper. London School of Economics, London.

Hull, J., and White, A. (1987). The pricing of options on assets with stochastic volatilities. *Journal of Finance* **42**, 281–300.

Jacquier, E., Polson, N. G., and Rossi, P. E. (1994). Bayesian analysis of stochastic volatility models. *Journal of Business and Economic Statistics* **12**, 371–389.

Karatzas, I. and Shreve, S. E. (1988). "Brownian Motion and Stochastic Calculus." Springer-Verlag, New York.

Kushner, H. J. (1984). "Approximation and Weak Convergence Methods for Random Processes, with Applications to Stochastic Systems Theory." MIT Press, Cambridge, MA.

Kuwahara, H., and Marsh, T. A. (1992). The pricing of Japanese equity warrants. *Management Science* **38**, 1610–1641.

Melino, A., and Turnbull, S. M. (1990). Pricing foreign currency options with stochastic volatility. *Journal of Econometrics* **45**, 239–267.

Nelson, D. B. (1988). The time series behavior of stock market volatility and returns. Doctoral Dissertation, Massachussets Institute of Technology, Economics Department, Cambridge, MA (unpublished).

Nelson, D. B. (1990). ARCH models as diffusion approximations. *Journal of Econometrics* **45**, 7–38.

Nelson, D. B. (1991). Conditional heteroskedasticity in asset returns: A new approach. *Econometrica* **59**, 347–370.

Nelson, D. B. (1992). Filtering and forecasting with misspecified ARCH models I: Getting the right variance with the wrong model. *Journal of Econometrics* **52**, 61–90.

Nelson, D. B., and Foster, D. P. (1994). Asymptotic filtering theory for univariate ARCH models. *Econometrica* **62**, 1–41.

Schwartz, B. A., Nelson, D. B., and Foster, D. P. (1993). "Variance Filtering with ARCH Models: A Monte Carlo Investigation," Tel Aviv University, Tel Aviv (unpublished).

Scott, L. O. (1987). Option pricing when the variance changes randomly: Theory, estimation, and an application. *Journal of Financial and Quantitative Analysis* **22**, 419–437.

Stroock, D. W., and Varadhan, S. R. S. (1979). "Multidimensional Diffusion Processes." Springer-Verlag, Berlin.

Wiggins, J. B. (1987). Option values under stochastic volatility: Theory and empirical estimates. *Journal of Financial Economics* **19**, 351–372.

8

ASYMPTOTIC FILTERING THEORY FOR UNIVARIATE ARCH MODELS*

DANIEL B. NELSON[†]
DEAN P. FOSTER[‡]

[†] *Graduate School of Business*
University of Chicago
Chicago, Illinois 60637
[‡] *Wharton School*
University of Pennsylvania
Philadelphia, Pennsylvania 19104

1. INTRODUCTION

Most asset pricing theories relate expected returns on assets to their conditional variances and covariances. An enormous literature in empirical finance has documented that these conditional moments change over time. Practical experience (as in the 1929 and 1987 stock market crashes) reinforces this conclusion. Unfortunately, conditional variances and covariances are not directly observable, and researchers and market participants must use estimates of conditional second moments. To create these estimates, they rely on models which are, no doubt, misspecified. How accurate are these estimated variances and covariances? How can researchers estimate them more accurately?

Since their introduction by Engle (1982), ARCH models have become a widely used tool for estimating conditional variances and covariances

* Reprinted from *Econometrica* **62** (1994), 1–41.

(see the survey of Bollerslev *et al.*, 1992). Suppose that for each t, ξ_t is a (scalar) innovation in a time series model. Interpreted as a data generating mechanism, a univariate ARCH model assumes that

$$\mathrm{E}_{t-1}[\xi_t] = 0 \quad \text{and} \quad \mathrm{Var}_{t-1}[\xi_t] = \sigma_t^2, \qquad \text{with} \tag{1.1}$$

$$\sigma_t^2 \equiv \sigma^2(\xi_{t-1}, \xi_{t-2}, \ldots, t). \tag{1.2}$$

That is, σ_t^2 is the conditional variance of ξ_t given time $t-1$ information, and is a function of time and past ξ_t's.

Like all statistical and economic models, ARCH models are at best a rough approximation to reality: It is too much to hope that the models are "true." As we will see, however, we need not think of (1.1)–(1.2) as the true data generating mechanism in order for ARCH models to be useful in extracting conditional variances from data. Given an arbitrary sequence $\{\xi_t\}_{t=-\infty,\infty}$, we can use (1.2) to create a corresponding $\{\sigma_t^2\}_{t=-\infty,\infty}$ sequence: Under conditions developed below, this sequence may provide a good estimate of the true conditional variance of $\{\xi_t\}_{t=-\infty,\infty}$, even when the model (1.1)–(1.2) is misspecified. One can think of (1.2) as a *filter* through which we pass the data to produce an *estimate* of the conditional variance. We should note, however, that we use the term "estimation" as it is used in the filtering literature rather than as it is used in the statistics literature —i.e., the ARCH model "estimates" the true conditional variance in the same sense that a Kalman filter estimates unobserved state variables in a linear system.[1]

A previous paper, Nelson (1992), gave one likely reason for the empirical success of ARCH: When both observable variables and conditional variances change "slowly" relative to the sampling interval (in particular, when the data generating process is well approximated by a diffusion and the data are observed at high frequencies) then broad classes of ARCH models—even when misspecified—provide continuous-record consistent estimates of the conditional variances. That is, as the observable variables are recorded at finer and finer intervals, the conditional variance estimates produced by the (misspecified) ARCH model converge in probability to the true conditional variances.

This chapter builds on this earlier work by deriving the asymptotic *distribution* of the measurement error. This allows us to approximate the measurement accuracy of ARCH conditional variance estimates and compare the efficiency achieved by different ARCH models. We are also able to characterize the relative importance of different kinds of misspecification; for example, we show that misspecifying conditional means adds only trivially (at least asymptotically) to measurement error, while other factors

[1] See, e.g., the use of the term in Anderson and Moore (1979, Chapter 2) or Arnold (1973, Chapter 12).

(for example, capturing the "leverage effect," accommodating thick-tailed residuals, and correctly modelling the variability of the conditional variance process) are potentially much more important. Third, we are able to characterize a class of asymptotically optimal ARCH conditional variance estimates.

In section 2, we state the basic functional limit theorem we employ throughout the chapter. In section 3, we use this theorem to develop an asymptotic approximation for the measurement errors in an ARCH model's estimate of the conditional variances when the data are generated by a diffusion. The class of ARCH models considered is fairly broad, encompassing, for example, the GARCH(1, 1) model of Bollerslev (1986), the EGARCH model of Nelson (1991), and the model of Taylor (1986) and Schwert (1989). Section 4 derives asymptotically optimal ARCH conditional variance estimates in the diffusion case. Examples are provided. Section 5 expands the analysis of sections 3 and 4 to the case when the data are generated by a stochastic *difference* equation rather than a stochastic *differential* equation. Surprisingly, this change makes a considerable difference in the limit theorems and optimality theory. Section 6 compares the filtering properties of several commonly used ARCH models. Section 7 concludes. Proofs are gathered in the Appendix. Throughout the chapter, we consider only univariate ARCH models: our results can be extended to the multivariate case, but at some cost in complexity (see Nelson, 1993).

2. OVERVIEW OF CONTINUOUS RECORD ASYMPTOTICS

In examining the properties of the estimated conditional variances produced by ARCH models, it is very convenient to pass to continuous time: In discrete time, ARCH models are nonlinear stochastic *difference* equations, which are quite intractable. In continuous time, the models become stochastic *differential* equations, which are much easier to analyze. Often the nonlinearity vanishes as continuous time is approached, making the analysis even easier. Most importantly, we will see that when the movement in the state variables is well approximated by a diffusion, the difference between the true conditional variance and the variance estimate produced by a misspecified ARCH model vanishes in the continuous time limit. Continuous record asymptotics allow us to pin down the rate of convergence and derive the asymptotic distribution of the measurement error. This would not be possible with the usual large sample asymptotics, hence our resort to continuous record asymptotics.

The basic techniques we apply were developed by Stroock and Varadhan (1979) and have previously been applied to the study of ARCH models in Nelson (1990, 1992) and Nelson and Foster (1991). We refer the reader to these sources for detailed discussion. The basic intuition underlying the method is as follows: Consider the stochastic integral equation

$$X_t = X_0 + \int_0^t \mu(X_s)\, ds + \int_0^t \Omega^{1/2}(X_s)\, dW_s, \tag{2.1}$$

where $\{W_t\}$ is an $n \times 1$ standard Brownian motion, and $\mu(x) \in R^n$ and $\Omega^{1/2}(x) \in R^{n \times n}$ are continuous. For most of this chapter we take $n = 2$, but will state the results more generally. We take the initial value X_0 to be random with cumulative distribution function F. If the process in (2.1) is well-defined, then $\{X_t\}$ is a random function of t, mapping $[0, \infty)$ into R^n.[2] We can think of $\{X_t\}$ as a random variable taking a value in a function space, the space $D_{R^n}[0, \infty)$ of right continuous functions with finite left limits. $D_{R^n}[0, \infty)$ is a complete, separable metric space when equipped with the Skorohod topology (see Ethier and Kurtz, 1986, Chapter 3). Under certain regularity conditions, four features uniquely characterize the distribution of this random function $\{X_t\}$:

(a) the cumulative distribution F of the starting point X_0,
(b) the drift function $\mu(x)$,
(c) the diffusion function $\Omega(x) \equiv \Omega(x)^{1/2}(\Omega(x)^{1/2})'$ and
(d) the almost sure continuity of $\{X_t\}$ as a function of t.

We next consider for each $h > 0$ a *discrete time* $n \times 1$ Markov process $\{{}_hY_k\}_{k=0,\infty}$ and define for each $\Delta > 0$, each $h > 0$, and each integer $k \geq 0$,

$$\mu_{\Delta,h}(y) \equiv h^{-\Delta}\, \mathrm{E}[({}_hY_{k+1} - {}_hY_k) \mid {}_hY_k = y], \qquad \text{and} \tag{2.2}$$

$$\Omega_{\Delta,h}(y) \equiv h^{-\Delta}\, \mathrm{Cov}[({}_hY_{k+1} - {}_hY_k) \mid {}_hY_k = y]. \tag{2.3}$$

The initial value ${}_hY_0$ is random with cumulative distribution function F_h.

THEOREM 2.1 (Stroock–Varadhan). *Let* (a)–(d) *completely characterize the distribution of the diffusion* (2.1),[3] *and let*

(a′) $F_h(y) \to F(y)$ *as* $h \downarrow 0$ *at all continuity points of* F.

[2] In our notation, curly brackets indicate a stochastic process—e.g., $\{X_t\}_{[0,T]}$ is the sample path of X_t as a (random) function of time on the interval $0 \leq t \leq T$. We refer to values of a process at a *particular* time t by omitting the curly brackets—e.g., X_τ is the (random) value taken by the $\{X_t\}$ process at time τ.

[3] Formally, we require that the martingale problem associated with F, μ, and Ω be well-posed, or equivalently, that the stochastic integral equation (2.1) have a unique weak-sense solution. See Ethier and Kurtz (1986, pp. 290–291) or Stroock and Varadhan (1979). Several sets of sufficient conditions are summarized in Appendix A of Nelson (1990).

For some $\Delta > 0$ *and some* $\delta > 0$, *let*

(b′) $\mu_{\Delta,h}(y) \to \mu(y)$,
(c′) $\Omega_{\Delta,h}(y) \to \Omega(y)$, *and*
(d′) $h^{-\Delta}\,\mathrm{E}[\|{}_hY_{k+1} - {}_hY_k\|^{2+\delta} \mid {}_hY_k = y] \to 0$,

as $h \downarrow 0$, where $\|A\| \equiv [A'A]^{1/2}$, and where the convergence in (b′)–(d′) is uniform on every bounded y set.[4] For each $h > 0$, define the process $\{{}_hX_t\}$ by ${}_hX_t \equiv {}_hY_{[t\cdot h^{-\Delta}]}$ for each $t \geq 0$, where $[t \cdot h^{-\Delta}]$ is the integer part of $t \cdot h^{-\Delta}$. Then for any T, $0 < T < \infty$, $\{{}_hX_t\}_{[0,T]} \Rightarrow \{X_t\}_{[0,T]}$ as $h \downarrow 0$, where (2.1) defines $\{X_t\}$ and $\Rightarrow$ denotes weak convergence. Further, if F_h sets ${}_hY_0 = y_0$ with probability one for all h, the weak convergence of $\{{}_hX_t\}$ to $\{X_t\}$ is uniform on every bounded $\{y_0\}$ set.

For proof, see the Appendix.

(a′)–(d′) ensure that (a)–(d) are satisfied in the limit by the sequence of processes $\{{}_hX_t\}_{h\downarrow 0}$, thereby achieving weak convergence of $\{{}_hX_t\}$ to $\{X_t\}$ as $h \downarrow 0$.

3. CONTINUOUS RECORD ASYMPTOTICS FOR ARCH FILTERS: THE DIFFUSION CASE

We now turn to continuous record asymptotics for ARCH models. In this section we consider the case in which the data are generated in continuous time by the diffusion

$$d\begin{bmatrix} x_t \\ y_t \end{bmatrix} = \begin{bmatrix} \mu(x_t, y_t, t) \\ \kappa(x_t, y_t, t) \end{bmatrix} dt + \begin{bmatrix} \sigma(y_t)^2 & \rho(x_t, y_t, t)\Lambda(x_t, y_t, t)\sigma(y_t) \\ \rho(x_t, y_t, t)\Lambda(x_t, y_t, t)\sigma(y_t) & \Lambda(x_t, y_t, t)^2 \end{bmatrix}^{1/2} \times \begin{bmatrix} dW_{1,t} \\ dW_{2,t} \end{bmatrix}, \tag{3.1}$$

where $\{W_{1,t}, W_{2,t}\}$ is a bivariate standard Brownian motion, x_0 and y_0 are random variables independent of $\{W_{1,t}, W_{2,t}\}$, and $[\cdot]^{1/2}$ is a matrix square root (i.e., $\Omega^{1/2}$ is any matrix satisfying $\Lambda^{1/2}(\Lambda^{1/2})' = \Omega$). We assume that κ, μ, σ, ρ, and Λ are continuous and real valued, that $\Lambda(\cdot)$ and $\sigma(\cdot)$ are

[4] That is, on every set of the form $\{y\colon \|y\| < C\}$ for finite, positive C. We could, for example, write (b′) more formally as

$$\lim_{h\downarrow 0} \sup_{\|y\|<C} \|\mu_{\Delta,h}(y) - \mu(y)\| = 0 \text{ for every } C,\ 0 < C < \infty. \tag{b′}$$

positive, that $|\rho| \leq 1$, and that $\sigma(y)$ is differentiable and strictly increasing in y. We do not assume stationarity or ergodicity for (3.1), although we do assume that it has a unique weak-sense solution. None of our results will depend on the choice of matrix square root in (3.1).

We assume that x_t is observable at discrete time intervals of length h. y_t is the *unobservable* process controlling the instantaneous conditional variance of x_t, namely $\sigma(y_t)^2$. Our interest is in using an ARCH model to create an estimate $\tilde{y}_t$ of y_t given the discrete observations $(x_0, x_h, x_{2h}, \ldots, x_{[t/h]h})$. Our ARCH filtering theory is asymptotic in that we let h approach zero.

For reasons developed below, we also consider a variant of (3.1) in which the drifts in $\{x_t, y_t\}$ explode as $h \downarrow 0$:

$$d\begin{bmatrix} x_t \\ y_t \end{bmatrix} = \begin{bmatrix} \mu(x_t, y_t, t) \\ \kappa(x_t, y_t, t) \end{bmatrix} h^{-1/4}\, dt$$
$$+ \begin{bmatrix} \sigma(y_t)^2 & \rho(x_t, y_t, t)\Lambda(x_t, y_t, t)\sigma(y_t) \\ \rho(x_t, y_t, t)\Lambda(x_t, y_t, t)\sigma(y_t) & \Lambda(x_t, y_t, t)^2 \end{bmatrix}^{1/2}$$
$$\times \begin{bmatrix} dW_{1,t} \\ dW_{2,t} \end{bmatrix}. \tag{3.1'}$$

Define $\mathrm{E}_t[\cdot]$ to be the conditional expectation given time t information (i.e., given the σ-algebra generated by $\{x_\tau, y_\tau\}_{0 \leq \tau \leq t}$). We take expectations with respect to this information set, rather than with respect to the smaller information set observed by the econometrician, the σ-algebra generated by $x_0, x_h, x_{2h}, \ldots, x_{h\,t/h]}$. The $\{x_t, y_t\}$ process is Markovian, so if f in $\mathrm{E}_t[f]$ depends only on $\{x_\tau, y_\tau\}_{\tau \geq t}$, then $\mathrm{E}_t[f] = \mathrm{E}[f \mid x_t, y_t]$. We will similarly use $\mathrm{SD}_t[\cdot]$, $\mathrm{Corr}_t[\cdot]$, and $\mathrm{Var}_t[\cdot]$ for the conditional standard deviation, correlation, and variance.

We now define the normalized innovations in x_t and y_t:

$$\xi_{x,t+h} \equiv h^{-1/2}[x_{t+h} - x_t - \mathrm{E}_t[x_{t+h} - x_t]], \quad \text{and} \tag{3.2}$$

$$\xi_{y,t+h} \equiv h^{-1/2}[y_{t+h} - y_t - \mathrm{E}_t[y_{t+h} - y_t]]. \tag{3.3}$$

We consider the class of ARCH models which generate $\hat{y}_t$ (for t an integer multiple of h) by the recursion

$$\hat{y}_{t+h} = \hat{y}_t + h \cdot \hat{\kappa}(x_t, \hat{y}_t, t, h) + h^{1/2} \cdot g\big(\hat{\xi}_{x,t+h}, x_t, \hat{y}_t, t, h\big), \tag{3.4}$$

$$\hat{\xi}_{x,t+h} \equiv h^{-1/2}\big[x_{t+h} - x_t - h \cdot \hat{\mu}(x_t, \hat{y}_t, t, h)\big] \quad \text{under (3.1), and} \tag{3.5}$$

$$\hat{y}_{t+h} = \hat{y}_t + h^{3/4} \cdot \hat{\kappa}(x_t, \hat{y}_t, t, h) + h^{1/2} \cdot g\big(\hat{\xi}_{x,t+h}, x_t, \hat{y}_t, t, h\big), \tag{3.4'}$$

$$\hat{\xi}_{x,t+h} \equiv h^{-1/2}\big[x_{t+h} - x_t - h^{3/4} \cdot \hat{\mu}(x_t, \hat{y}_t, t, h)\big] \quad \text{under (3.1')}. \tag{3.5'}$$

$\hat{\kappa}(x_t, \hat{y}_t, t, h)$, $\hat{\mu}(x_t, \hat{y}_t, t, h)$, and $g(\hat{\xi}_{x,t+h}, x_t, \hat{y}_t, t, h)$ are functions selected by the econometrician. $\hat{\kappa}$ and $\hat{\mu}$ must be continuous in all arguments and g must be differentiable in $\hat{y}_t$, $\hat{\xi}_{x,t+h}$, and h almost everywhere and must possess one-sided derivatives everywhere. Just as μ and κ are the true drifts in x_t and y_t, $\hat{\mu}$ and $\hat{\kappa}$ are the econometrician's (presumably misspecified) specification of these drifts, and $\hat{\xi}_{x,t+h}$ is a residual obtained using $\hat{\mu}$ in place of μ. The ARCH model (3.4) treats this fitted residual $\hat{\xi}_x$ and fitted $\hat{y}$ as if they were the true residual ξ_x and the true y. In ARCH models, the driving noise term in the $\{y_t\}$ process is a function of the fitted residual in the observable process. $g(\hat{\xi}_x, x, \hat{y}, t, h)$ is the econometrician's specification of this function. Our chief interest is in the case in which y_t is unobservable, i.e., in which there are no functions $\hat{\mu}$, $\hat{\kappa}$, and g such that $g(\hat{\xi}_{x,t+h}, x_t, \hat{y}_t, t, h) = \xi_{y,t+h}$ almost surely for all t and h. Nevertheless, the ARCH model treats $h^{1/2} \cdot g(\hat{\xi}_x, x, \hat{y}, t, h)$ as if it were the true innovation in the y_t process. When $\hat{y}_t = y_t$, $h^{1/2} g(\cdot)$ is the noise term in the increment of $\hat{y}_t$. Under (3.1), the conditional mean and standard deviation of the increments in (x_t, y_t) over intervals of length h are $O(h)$ and $O(h^{1/2})$, respectively.

The class of ARCH models encompassed by (3.4)–(3.5) is fairly broad, encompassing, for example, GARCH(1, 1), which sets $y \equiv \sigma^2$, $\hat{\kappa} = \omega - \theta\sigma^2$, and $g \equiv \alpha \cdot (\hat{\xi}_x^2 - \hat{\sigma}^2)$, AR(1)EGARCH, which sets $y \equiv \ln(\sigma^2)$, $\hat{\kappa} = -\beta \cdot [\ln(\hat{\sigma}^2) - \alpha]$, and $g \equiv \theta\hat{\xi}_x/\hat{\sigma} + \gamma[|\hat{\xi}_x/\hat{\sigma}| - \mathrm{E}_t|\xi_x/\delta|]$, and the model of Taylor (1986, Chapter 4) and Schwert (1989), which sets $y \equiv \sigma$, $\hat{\kappa} \equiv \omega - \theta\hat{\sigma}$, and $g \equiv \alpha\hat{\sigma} \cdot (|\hat{\xi}_x/\hat{\sigma}| - \mathrm{E}_t|\xi_x/\sigma|)$.

We next define q_t, the normalized measurement error in y. For t an integer multiple of h, we set

$$q_t \equiv h^{-1/4}(\hat{y}_t - y_t). \tag{3.6}$$

For general t, we take $q_t \equiv q_{[t/h]h}$, making $\{q_t\}$ a random step function with jumps at time intervals of length h.

Why use such an ARCH model to extract estimates of the state rather than, say, a nonlinear Kalman filter (see, e.g., Kitagawa, 1987; Maybeck, 1982, Chapter 13)? First, since ARCH models are so widely used in practice, it seems reasonable to investigate their properties. Second, ARCH models are much more computationally tractable than standard nonlinear filters, which are typically infinite dimensional, and which involve extensive numerical integration (see, however, the recent work of Jacquier *et al.*, 1992). When the ARCH model is assumed to be the true data generating process, it is easily fit using maximum-likelihood methods (e.g., Engle, 1982). Finally, we will also see that the optimal ARCH filters are readily interpretable, and an explicit asymptotic distribution theory can be derived for these filters.

We next heuristically motivate the formal assumptions given below. We assume for the moment that $g(\hat{\xi}_x, x, \hat{y}, t, h)$ is differentiable, though in certain ARCH models of interest [notably EGARCH and the model of Taylor (1986) and Schwert (1989)] differentiability fails to hold everywhere, though it is still possible to verify Assumptions 1–2 in these models. Expanding $g(\hat{\xi}_x, x, \hat{y}, t, h)$ in a Taylor series around $\hat{\xi} = \xi$, $\hat{y} = y$, and $h = 0$, we obtain

$$\begin{aligned} g\left(\hat{\xi}_{x,t+h}, x_t, \hat{y}_t, t, h\right) &= g(\xi_{x,t+h}, x_t, y_t, t, 0) \\ &\quad + h \cdot [\partial g(\xi_{x,t+h}, x_t, y_t, t, 0)/\partial h] \\ &\quad + (\hat{y}_t - y_t) \cdot [\partial g(\xi_{x,t+h}, x_t, y_t, t, 0)/\partial y] \\ &\quad + \left(\hat{\xi}_{x,t+h} - \xi_{x,t+h}\right) \cdot [\partial g(\xi_{x,t+h}, x_t, y_t, t, 0)/\partial \xi] \\ &\quad + \text{higher order terms.} \end{aligned} \tag{3.7}$$

Utilizing the definitions of q_t and $\hat{\xi}_{x,t+h}$ allows us to rewrite (3.7) as a series expansion in terms of h, as

$$\begin{aligned} g\left(\hat{\xi}_{x,t+h}, x_t, \hat{y}_t, t, h\right) &= g(\xi_{x,t+h}, x_t, y_t, t, 0) \\ &\quad + h^{1/4} q_t \cdot [\partial g(\xi_{x,t+h}, x_t, y_t, t, 0)/\partial y] \\ &\quad + O(h^{1/2}), \qquad \text{and} \end{aligned} \tag{3.8}$$

$$\begin{aligned} g\left(\hat{\xi}_{x,t+h}, x_t, \hat{y}_t, t, h\right) &= g(\xi_{x,t+h}, x_t, y_t, t, 0) \\ &\quad + h^{1/2} q_t \cdot [\partial g(\xi_{x,t+h}, x_t, y_t, t, 0)/\partial y] \\ &\quad + h^{1/4} \cdot (\mu(x_t, y_t, t) - \hat{\mu}(x_t, y_t, t, h)) \\ &\quad \cdot [\partial g(\xi_{x,t+h}, x_t, y_t, t, 0)/\partial \xi] + O(h^{1/2}) \end{aligned} \tag{3.8$'$}$$

under (3.1) and (3.1′) respectively. Substituting into (3.4) and (3.4′) and using (3.6), we have under (3.1) and (3.1′)

$$\begin{aligned} q_{t+h} &= q_t + h^{1/2} q_t \cdot [\partial g(\xi_{x,t+h}, x_t, y_t, t, 0)/\partial y] \\ &\quad + h^{1/4}\left[g(\xi_{x,t+h}, x_t, y_t, t, 0) - \xi_{y,t+h}\right] + O(h^{3/4}) \qquad \text{and} \end{aligned} \tag{3.9}$$

$$\begin{aligned} q_{t+h} &= q_t + h^{1/2} q_t \cdot [\partial g(\xi_{x,t+h}, x_t, y_t, t, 0)/\partial y] \\ &\quad + h^{1/2}[\hat{\kappa}(x_t, y_t, t, h) - \kappa(x_t, y_t, t)] \\ &\quad + h^{1/2}[\mu(x_t, y_t, t) - \hat{\mu}(x_t, y_t, t, 0)] \\ &\quad \cdot [\partial g(\xi_{x,t+h}, x_t, y_t, t, 0)/\partial \xi] \\ &\quad + h^{1/4}\left[g(\xi_{x,t+h}, x_t, y_t, t, 0) - \xi_{y,t+h}\right] + O(h^{3/4}), \qquad \text{respectively.} \end{aligned} \tag{3.9$'$}$$

We also have

$$\begin{aligned} y_{t+h} &= y_t + \mathrm{E}_t \int_t^{t+h} \kappa(x_s, y_s, s)\, \mathrm{d}s + h^{1/2}\xi_{y,t+h} \\ &= y_t + h \cdot \kappa(x_t, y_t, t) + h^{1/2}\xi_{y,t+h} + o(h) \qquad \text{and} \qquad (3.10) \end{aligned}$$

$$\begin{aligned} x_{t+h} &= x_t + \mathrm{E}_t \int_t^{t+h} \mu(x_s, y_s, s)\, \mathrm{d}s + h^{1/2}\xi_{x,t+h} \\ &= x_t + h \cdot \mu(x_t, y_t, t) + h^{1/2}\xi_{x,t+h} + o(h) \qquad (3.11) \end{aligned}$$

under (3.1). (Under 3.1′), replace the $h \cdot \kappa$ and $h \cdot \mu$ by $h^{3/4}\kappa$ and $h^{3/4}\mu$.) Recall that Theorem 2.1 characterized the limit diffusion using the first two conditional moments of the state variables—in our case x_t, y_t, and q_t. Using the approximations (3.9)–(3.11) to characterize these conditional moments requires the following assumptions.

ASSUMPTION 1. *Uniformly on every bounded* (x, y, q, t) *set*

$$h^{-1/2}\, \mathrm{E}[q_{t+h} - q_t \mid x_t = x, y_t = y, q_t = q] \to A(x, y, t) - q \cdot B(x, y, t), \qquad (3.12)$$

and

$$h^{-1/2}\, \mathrm{Var}[q_{t+h} - q_t \mid x_t = x, y_t = y, q_t = q] \to C(x, y, t) \qquad (3.13)$$

as $h \downarrow 0$, *where*

$$\begin{aligned} A(x, y, t) &\equiv 0 \quad \textit{under } (3.1), \qquad \textit{and} \\ A(x, y, t) &\equiv [\hat{\kappa}(x, y, t, 0) - \kappa(x, y, t)] \\ &\quad + \lim_{h \downarrow 0} [\mu(x, y, t) - \hat{\mu}(x, y, t, h)] \\ &\quad \cdot \mathrm{E}[\partial g(\xi_{x,t+h}, x_t, y_t, t, h)/\partial \xi_x \mid x_t = x, y_t = y] \qquad (3.14) \end{aligned}$$

under (3.1′).

$$B(x, y, t) \equiv -\lim_{h \downarrow 0} \mathrm{E}[\partial g(\xi_{x,t+h}, x_t, y_t, t, h)/\partial y \mid x_t = x, y_t = y], \qquad (3.15)$$

$$C(x, y, t) \equiv \lim_{h \downarrow 0} \mathrm{E}\Big[\big(g(\xi_{x,t+h}, x_t, y_t, t, h) - \xi_{y,t+h}\big)^2 \mid x_t = x, y_t = y\Big]. \qquad (3.16)$$

Further $A(x, y, t)$, $B(x, y, t)$, *and* $C(x, y, t)$ *are twice continuously differentiable in* x *and* y.

To simplify notation we will often write A_t, B_t, and C_t for $A(x_t, y_t, t)$, $B(x_t, y_t, t)$, and $C(x_t, y_t, t)$.

ASSUMPTION 2. *For some* $\delta > 0$

$$E\left[\left|h^{-1/2}(y_{t+h} - y_t)\right|^{2+\delta} \mid x_t = x, y_t = y\right] \quad \textit{and} \tag{3.17}$$

$$E\left[\left|h^{-1/2}(x_{t+h} - x_t)\right|^{2+\delta} \mid x_t = x, y_t = y\right] \tag{3.18}$$

are bounded as $h \downarrow 0$, *uniformly on every bounded* (x, y, t) *set, and*

$$\limsup_{h \downarrow 0} E\left[\left|g\left(\hat{\xi}_{x,t+h}, x_t, \hat{y}_t, t, h\right)\right|^{2+\delta} \mid x_t = x, y_t = y, q_t = q\right] \tag{3.19}$$

is bounded uniformly on every bounded (x, y, q, t) *set.*

These assumptions are written in the most natural form for applying Theorem 2.1. As we will see in Section 4, they can be verified in many applications.

3.1. Changing the Time Scales

In (3.1), $\{x_t\}$, $\{y_t\}$, and $\{q_t\}$ are all scaled to be $O_p(1)$, while the first two conditional moments of $x_{t+h} - x_t$ and $y_{t+h} - y_t$ are $O_p(h)$ as $h \downarrow 0$. In (3.12)–(3.13), on the other hand, the first two conditional moments of $q_{t+h} - q_t$ are $O_p(h^{1/2})$. As $h \downarrow 0$, $\{q_t\}$ oscillates much more rapidly than $\{x_t, y_t\}$. If $\{q_t\}$ mean-reverts (which it will if $E_t[\partial g(\xi_{x,t+h}, x_t, y_t, t, 0)/\partial y] < 0$), it does so more and more rapidly as $h \downarrow 0$. As we pass from annual observations of x_t to monthly to daily to hourly (etc.), the rescaled measurement error q_t looks more and more like heteroskedastic white noise (i.e., *not* like a diffusion).[5]

To use the Stroock–Varadhan results to approximate the behavior of $\{q_t\}$ requires that we change the time scales, which Theorem 2.1 allows via our choice of Δ. Specifically, we choose a time T, a large positive number M, and a point in the state space (x, y, q, t), and condition on the event $(x_T, y_T, q_T) = (x, y, q)$. We then take the vanishingly small time interval $[T, T + M \cdot h^{1/2}]$ on our old time scale (i.e., calendar time) and stretch it into a time interval $[0, M]$ on a new, "fast" time scale. Formally, this involves using $\Delta = 1/2$ in place of $\Delta = 1$ in applying Theorem 2.1. On the usual calendar time scale, $\{x_t, y_t\}$ are a diffusion and $\{q_t\}$ is (asymptotically) white noise. On the new "fast" time scale, the $\{x_t, y_t, t\}$ process moves more and more slowly as $h \downarrow 0$, becoming constant at the values (x_T, y_T, T)

[5] The claim that q_t mean-reverts ever more quickly as $h \downarrow 0$ may seem counterintuitive, since if we use (3.15) to write q_t as an AR(1) process, the autoregressive coefficient is $(1 - h^{1/2}B(x, y, t))$ which approaches 1, not 0, as $h \downarrow 0$. Recall, however, that we are considering mean-reversion *per unit of calendar time*. A unit of calander time contains h^{-1} jumps in the state variables. So although the *first-order* autocorrelation in q_t approaches 1 as $h \downarrow 0$, the autocorrelations at *fixed calendar intervals* (say daily or monthly) approach 0.

in the limit as $h \downarrow 0$ while $\{q_t\}$ is (asymptotically) a diffusion. Thus we say that in the limit as $h \downarrow 0$, $\{q_t\}$ *operates on a faster natural time scale* than $\{x_t, y_t\}$. On the new time scale we require two time subscripts for the $\{q_t\}$ process, one giving the time T on the standard time scale and one giving the time elapsed since T on the "fast" time scale. We therefore write

$$q_{T,\tau} \equiv q_{T+\tau h^{1/2}}, \tag{3.20}$$

where the τ is the time index on the fast time scale. ($q_{T,\tau}$ also depends on the time T startup point (x, y, q, t), though we suppress this in the notation.) Our analysis of the measurement error process is therefore *local* in character: In a sense it treats the more slowly varying $\{x_t, y_t, t\}$ as constant at the values (x_T, y_T, T) and examines the behavior of the measurement error in the neighborhood of (x_T, y_T, T).

Figure 1 illustrates this changing of the time scales with artificially generated data. The upper-left panel plots a simulated $\{y_t\}_{t=0,h,2h,3h\ldots}$ series from a diffusion model and the corresponding $\{\hat{y}_t\}$ generated by an EGARCH model based on monthly observations of the observable $\{x_t\}_{t=0,h,2h,3h\ldots}$ series.[6] Time is measured in annual units so $h = 1/12$. In the lower-left panel the $\{y_t\}_{t=0,h,2h,3h\ldots}$ and $\{\hat{y}_t\}_{t=0,h,2h,3h\ldots}$ based on daily observations ($h = 1/264$) are plotted. The measurement error $\{\hat{y}_t - y_t\}$ is smaller for daily than for monthly data (the empirical variances in the simulation are .144 and 1.230, respectively) and is also less autocorrelated at fixed lags of calendar time (e.g., the autocorrelation at a 1 month lag is .219 with daily observations and .477 with monthly observations). We allow for the shrinking variance with the $h^{-1/4}$ term in the definition of q_t. The time deformation allows us to handle the changing serial correlation: For example, set $M \equiv 2^{1/2}$, $T = 20$, and $T + M \cdot h^{1/2} = 21$ for monthly data and call these $\tau = 0$ and $\tau = M$ on the "fast" time scale. As the left-hand panels indicate, the interval of *calendar* time between $\tau = 0$ and $\tau = M$ shrinks with h, from one year with monthly data to about $2\frac{1}{2}$ months with daily data. The right-hand panels plot the corresponding $q_{20,\tau}$ processes. The diffusion approximation we derive is for $q_{20,\tau}$ as $h \downarrow 0$.

Making this change of time scale, and conditioning on $(x_T, y_T, q_{T,0}, T)$ it is straightforward to derive the limit diffusion of $\{q_{T,\tau}\}_{h\downarrow 0}$:

$$dq_{T,\tau} = (A_T - B_T q_{T,\tau})\, d\tau + C_T^{1/2}\, dW_\tau, \tag{3.21}$$

where W_τ is a standard Brownian motion (on the "fast" time scale). Since the distributions of $\xi_{x,T+h}$ and $\xi_{y,T+h}$ are functions of y_T, x_T, T, and

[6] We used the Wiggins (1987) model, (4.20)–(4.22) below. The model was simulated using the Euler stochastic difference approximation with daily increments. We assumed 22 (trading) days per month. The parameter values used were $\mu = 0$, $\Lambda = 2.1$, $\alpha = -3.9$, $\beta = .825$, $\rho = -.69$, $y \equiv \ln(\sigma_t^2)$. The EGARCH models estimated were the asymptotically optimal EGARCH models for the diffusion—see Section 6 below.

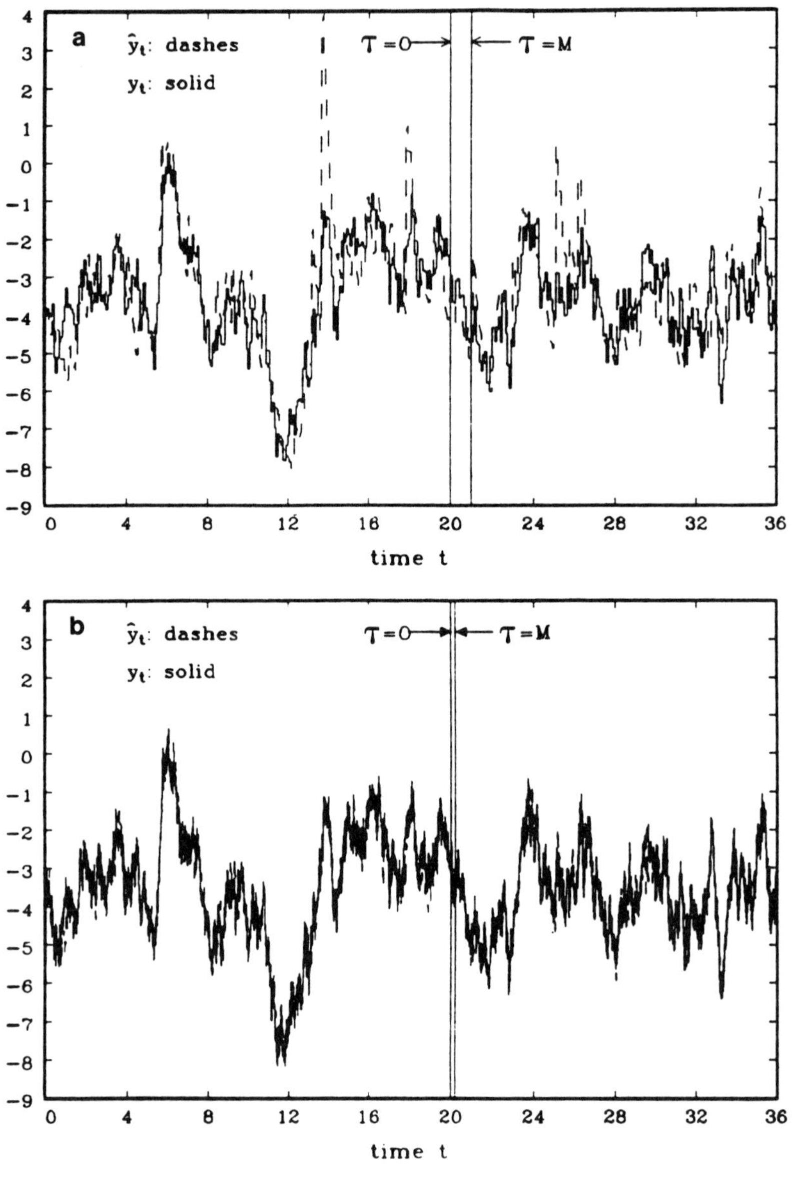
a
$\hat{y}_t$: dashes
y_t: solid
$\tau=0$
$\tau=M$
time t
b
$\hat{y}_t$: dashes
y_t: solid
$\tau=0$
$\tau=M$
time t

FIGURE 1

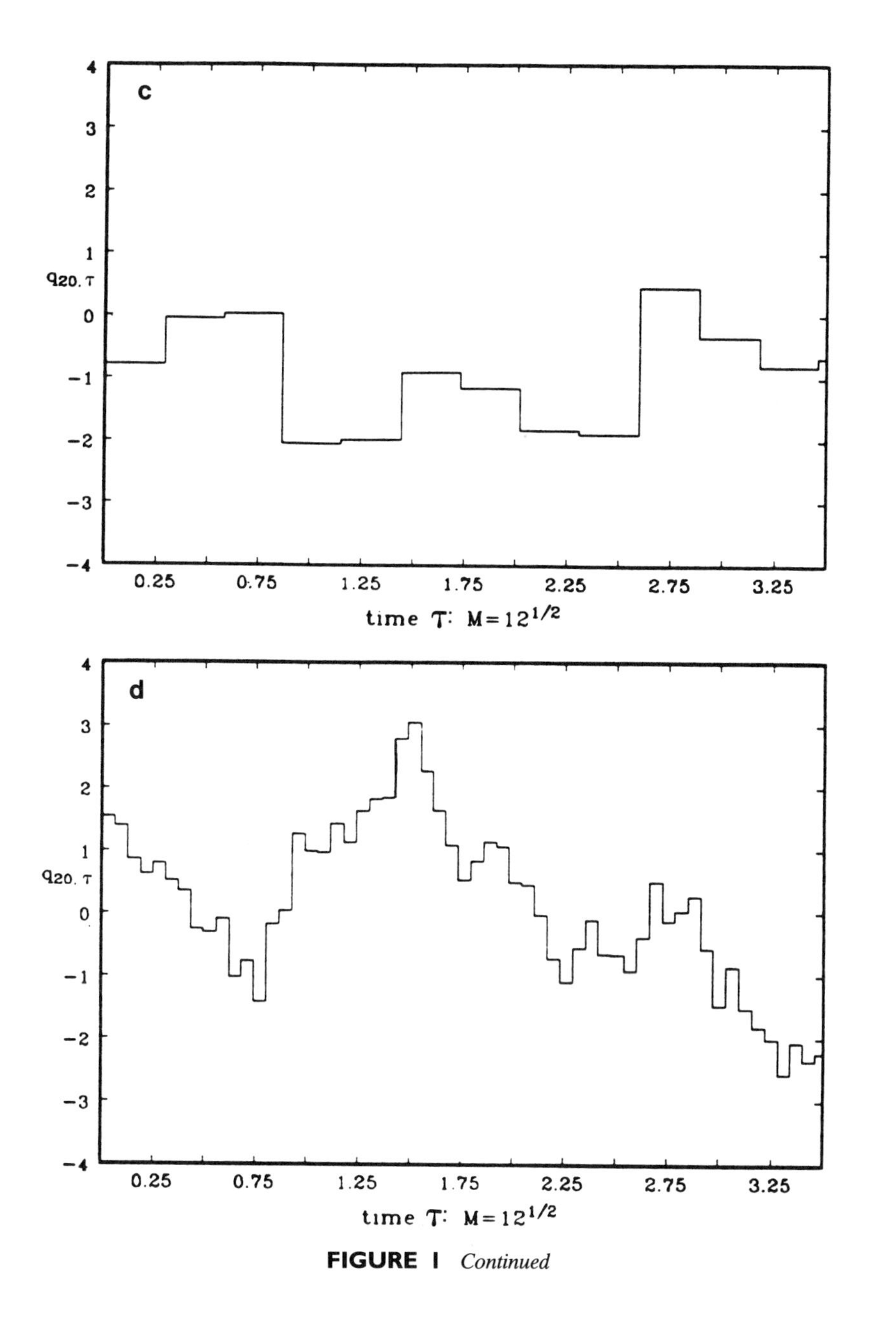

FIGURE 1 *Continued*

h, B_T, and C_T are functions only of y_T, x_T and T and are constant (conditional on y_T, x_T, and T) in the diffusion limit on the fast time scale. On the fast time scale, $\{q_{T,\tau}\}$ follows an Ornstein–Uhlenbeck process, the continuous time equivalent of a Gaussian AR(1) [see, e.g., Arnold (1973, Section 8.3) or Karatzas and Shreve (1988, p. 358)].

THEOREM 3.1. *Let Assumptions* 1–2 *be satisfied. Let* Θ *be any bounded, open subset of* R^4 *on which for some* $\varepsilon > 0$ *and all* $(x, y, q, T) \in \Theta$, $|A(x, y, T)| < 1/\varepsilon$, $\epsilon < B(x, y, T) < 1/\varepsilon$, *and* $C(x, y, T) < 1/\varepsilon$. *Then for every* $(x, y, q, T) \in \Theta$, $\{q_{T,\tau}\}_{[0, M]}$ *(conditional on* $(x_T, y_T, q_T) = (x, y, q)$*) converges weakly to the diffusion* (3.21) *as* $h \downarrow 0$. *This convergence is uniform on* Θ. *Further, for every* $(x_T, y_T, q_T, T) \in \Theta$,

$$[q_{T,M} \mid (x_T, y_T, q_{T,0}) = (x, y, q)]$$

$$\xrightarrow{d} N\left[A_T/B_T + e^{-M \cdot B_T}(q - A_T/B_T), C_T(1 - e^{-2M \cdot B_T})/2B_T\right], \quad (3.22)$$

where $\xrightarrow{d}$ *denotes convergence in distribution as* $h \downarrow 0$.

For proof, see the Appendix.

3.2. Interpretation

If $B_T > 0$, then for large M (recall that we can make M as large as we like as long as it is finite), $q_{T,M}$ given x_T and y_T, is approximately $N[A_T/B_T, C_T/2B_T]$. Though $\lim_{M \to \infty} \lim_{h \downarrow 0}[q_{T,M} \mid (x_T, y_T, q_{T,0}) = (x, y, q)] \stackrel{d}{=} N(A_T/B_T, C_T/2B_T)$, Theorem 2.1 does not allow us to interchange the limits. By the properties of limits, however [see Lemma 5.2 of Helland (1982)], we can say that if $M(h) \to \infty$ slowly enough as $h \downarrow 0$, then $q_{T,M(h)} \Rightarrow N(A_T/B_T, C_T/2B_T)$ as $h \downarrow 0$, formally justifying the $N(A_T/B_T, C_T/2B_T)$ as a large M asymptotic distribution, allowing us to abstract from the initial condition $q_{T,0}$.

At this point, four comments are in order.

First, under the conditions of Theorem 3.1, $[\hat{y}_t - y_t]$ is $O_p(h^{1/4})$. Although this *rate* of convergence ($O_p(h^{1/4})$) is the same throughout the state space (whenever the conditions of the theorem are satisfied), the asymptotic variance of the measurement error is a function of B_T and C_T and therefore of x_T, y_T, and T. The $O_p(h^{1/4})$ rate seems fairly slow, implying, for example, that in going from annual to daily returns data the standard deviation of the measurement error falls by about a factor of four. If the variance per unit of time were *constant*, we could achieve an $O_p(h^{1/2})$ rate of convergence (see, e.g., Merton, 1980), and the standard deviation of the measurement error would fall by a factor of 16. Our slower convergence rate results from the fact that the ARCH variance estimators are shooting at a rapidly oscillating target.

Second, Theorem 3.1 analyzes the local behavior of $\{q_{T,\tau}\}$ in the neighborhood of the time T state values (x_T, y_T, q_T). This analysis may not be particularly useful if q_T explodes as $h \downarrow 0$, i.e., if $[\hat{y}_T - y_T]$ converges to zero at a rate slower than $h^{1/4}$. A technical device gets us around this problem: In related work (Foster and Nelson, 1993), we show that under mild regularity conditions, rolling regression estimators achieve the $O_p(h^{1/4})$ convergence rate for $[\sigma(\hat{y}_t)^2 - \sigma(y_t)^2]$ and consequently for $\hat{y}_t - y_t$. A rolling regression at the beginning of a sample—say from dates $T - h^{1/2}$ to T—can be used to initialize the ARCH filter at time T.[7]

Third, Theorem 3.1 also allows us to characterize the asymptotic autocorrelation: using the autocorrelation function of the Ornstein–Uhlenbeck process (see Arnold (1973, Section 8.3) we have for small h and large positive τ and τ'

$$\operatorname{Corr}(q_{T+\tau h^{1/2}}, q_{T+\tau' h^{1/2}}) \approx \exp[-|\tau - \tau'| \cdot B_T]. \tag{3.23}$$

According to (3.23), $\{q_t\}$ is asymptotically white noise on the standard (calendar, slow) time scale, since the (asymptotic) serial correlation in the measurement errors vanishes except at lag lengths shrinking to zero at rate $O(h^{1/2})$. It is conditionally heteroskedastic white noise, however, since the asymptotic variance of g_t depends on x_t, y_t, and t.

Fourth, under (3.1), the asymptotic distribution of the measurement error process $\{q_t\}$ depends on $\rho(\cdot)$, $\sigma(\cdot)$, $\Lambda(\cdot)$, and $g(\cdot)$, but not on $\hat{\kappa}$, κ, μ, or $\hat{\mu}$. While errors in the drift terms $\hat{\kappa}$, κ, μ, or $\hat{\mu}$ affect $\{q_t\}$ for fixed $h > 0$, they are asymptotically negligible as $h \downarrow 0$. (3.1) is in a sense more natural than (3.1′) since it analyzes a *particular* diffusion while varying the sampling interval h. If the drift terms $\hat{\kappa}$, κ, μ, and $\hat{\mu}$ are large, however, asymptotics based on (3.1) may ignore an important source of measurement error (3.1′) blows up these drift terms at an appropriate rate to keep them from dropping out of the asymptotic distribution. This is analogous to the use of "near-integrated" processes in the analysis of unit root asymptotics (see, e.g., Phillips, 1988). In this "large drift" asymptotic, nonzero $[\hat{\kappa} - \kappa]$ and $[\mu - \hat{\mu}]$ create an asymptotic bias in $q_t \equiv h^{1/4}[\hat{y} - y]$, but do not affect its asymptotic variance. This confirms the intuition given in Nelson (1992) that seriously misspecified ARCH models may consistently extract conditional variances from data observed at higher and higher frequencies—i.e., consistency is not incompatible with (moderately) explosive (as $h \downarrow 0$) misspecification in the drifts.

To gauge the size of the bias, note that blowing up $[\hat{\kappa} - \kappa]$ and $[\mu - \hat{\mu}]$ at an $h^{-1/4}$ rate introduced an $O(h^{1/4})$ bias in the measurement error $[\hat{y} - y]$, suggesting that the effect of *non*exploding drifts introduces an $O(h^{1/2})$ bias.

[7] Formally, to guarantee that $q_t = O_p(h^{1/4})$ for all t, choose a large positive N and define $\hat{y}_T$ by $\hat{y}_T \equiv \hat{y}_T(\text{ARCH})$ whenever $|\hat{\sigma}_T^2(\text{ARCH}) - \hat{\sigma}_T^2(\text{Rolling Regression})| < h^{1/4}N$, and $\hat{y}_T \equiv \hat{\sigma}_T^2(\text{Rolling Regression})$ otherwise.

4. ASYMPTOTIC OPTIMALITY

In discussing optimal ARCH model selection, several warnings are in order.

First, we consider optimality only within the class of ARCH models given by (3.4)–(3.5) and subject to the regularity conditions in the Assumptions.

Second, we evaluate optimality in terms of the "large M" asymptotic bias A_T/B_T and asymptotic variance $C_T/2B_T$. As is well-known in standard large-sample asymptotics, minimizing asymptotic variance need not be the same as minimizing the limit of the variances.

Third, much the same difficulty arises in defining *globally* optimal ARCH filters as arises in defining globally optimal estimators in statistics: Just as the "optimal" estimate of a parameter Ψ is Ψ itself, the "optimal" ARCH model when $(y_T, x_T, T) = (y, x, t)$ is the model in which $\hat{y}_{T,\tau}$ is held *constant* at y. Even if y_t is randomly changing, the estimation error would be $O_p(h^{\Delta})$ (note the faster rate of convergence when $\Delta > 1/4$) in an $O(h^{\Delta})$ neighborhood of $y_t = y$. Obviously, in other regions of the state space such an estimate would perform disastrously. We call the ARCH model globally optimal if it eliminates the large M asymptotic bias A_T/B_T —even in the "fast drift" case—and minimizes the large M asymptotic variance $C_T/2B_T$ for every (x, y, t). Hence our optimally concept is patterned on the UMVAE (uniform minimum variance unbiased estimator) criterion.

4.1. Eliminating Asymptotic Bias

From (3.14) and Theorem 3.1, it is clear that asymptotic bias A_T/B_T can be eliminated by setting $\hat{\kappa}(x, y, t, h) = \kappa(x, y, t)$ and $\hat{\mu}(x, y, t, h) = \mu(x, y, t)$ for all (x, y, t). Though this choice of $\hat{k}(x, y, t, h)$ and $\hat{\mu}(x, y, t, h)$ is *sufficient* to eliminate asymptotic bias, it is not necessary, since bias from $\mu(x, y, t, h) \neq \mu(x, y, t)$ may exactly offset bias from $\hat{\kappa}(x, y, t, h) \neq \kappa(x, y, t)$.

4.2. Minimizing Asymptotic Variance

Suppose that the conditional density of $\xi_{x,T+h}$, say $f(\xi_{x,T+h} \mid x_T, y_T)$, is well-defined and is differentiable in y. Integrating by parts then allows us to write B_T as

$$\begin{aligned} B_T &= \lim_{h \downarrow 0} - \mathrm{E}_T[\partial g(\xi_{x,T+h}, x_T, y_T, T, h)/\partial y] \\ &= \lim_{h \downarrow 0} \mathrm{E}_T[g(\xi_{x,T+h}, x_T, y_T, T, h) \\ &\qquad \cdot \partial \ln[f(\xi_{x,T+h} \mid x_T, y_T, T, h)]/\partial y]. \end{aligned} \tag{4.1}$$

Under (3.1) or (3.1′) the increments in $\{x_{t+h} - x_t, y_{t+h} - y_t\}$ approach conditional normality as $h \downarrow 0$ (see, e.g., Stroock and Varadhan, 1979, pp. 2–4), and $(\xi_{x,t+h}, \xi_{y,t+h})$ is approximately (conditionally) bivariate normal with mean 0 and variances $\sigma(y_t)^2$ and $\Lambda(x_t, y_t, t)^2$ and correlation $\rho(x_t, y_t, t)$. If $f(\xi_{x,T+h} \mid x_T, y_T, T)$, $g(\xi_x, x, y, t)$, and their partial derivatives with respect to y are sufficiently well-behaved, the following will hold. Define.

$$\begin{bmatrix} \varepsilon_x \\ \varepsilon_y \end{bmatrix} \overset{\mathrm{d}}{=} \mathrm{N}\left[\begin{bmatrix} 0 \\ 0 \end{bmatrix}, \begin{bmatrix} \sigma(y)^2 & \rho(x,y,t)\Lambda(x,y,t)\sigma(y) \\ \rho(x,y,t)\Lambda(x,y,t)\sigma(y) & \Lambda(x,y,t)^2 \end{bmatrix}\right]. \tag{4.2}$$

Then

$$B(x,y,t) = \sigma(y)^{-3}\sigma'(y)E\left[\varepsilon_x^2 g(\varepsilon_x, x, y, t, 0)\right], \quad \text{and} \tag{4.3}$$

$$C(x,y,t) = E\left[\left(g(\varepsilon_x, x, y, t, 0) - \varepsilon_y\right)^2\right]. \tag{4.4}$$

Equations (4.2)–(4.4) turn out to be important technical conditions in deriving the asymptotic variance minimizer, as does a stronger version of (3.18).

ASSUMPTION 3. *For every* (x, y, t), (4.2)–(4.4) *hold, and for some* $\delta > 0$,

$$\limsup_{h \downarrow 0} E\left[\left|h^{-1/2}(x_{t+h} - x_t)\right|^{4+\delta} \mid x_t = x, y_t = y\right] \tag{4.5}$$

is uniformly bounded on every bounded (x, y, t) *set.*

The conditional density of $\xi_{x,t+h}$ can, in principle, be derived from the Kolmogorov backward or forward equations (see, e.g., Stroock and Varadhan, 1979, Chapters 2–3, 9–10). In practice, it is easier to check (4.2)–(4.5) directly then to establish convergence for the derivatives of the conditional density.

THEOREM 4.1. *Let Assumptions* 1–3 *hold. Under either* (3.1) *or* (3.1′), *the asymptotic variance* $C_T/2B_T$ *is minimized by setting*

$$g(\xi_x, x, y, t, h) = \left[\frac{\rho(x,y,t)\Lambda(x,y,t)\xi_x}{\sigma(y)}\right] + \left[\frac{\Lambda(x,y,t)\left[1 - \rho(x,y,t)^2\right]^{1/2}}{2^{1/2}}\right]\left[\frac{\xi_x^2}{\sigma(y)^2} - 1\right]. \tag{4.6}$$

This yields

$$C_T/2B_T = \frac{\left[1 - \rho(x_T, y_T, T)^2\right]^{1/2} \Lambda(x_T, y_T, T)\sigma(y_T)}{\left[2^{1/2}\sigma'(y_T)\right]},$$

and the corresponding asymptotic variance of $h^{-1/4}[\sigma(\hat{y})^2 - \sigma(y)^2]$ *equals* $[1 - \rho(x_T, y_T, T)^2]^{1/2}\Lambda(x_T, y_T, T)\sigma(y_T)^3\sigma'(y_T)8^{1/2}$.

For proof, see the Appendix.

Sometimes it is of interest to minimize $C_T/2B_T$ within a particular class of ARCH models (e.g., to find the optimal GARCH(1, 1) model). Accordingly, we consider models in which

$$g(\xi_x, x, y, t, h) = a(x, y, t, h) \cdot g^*(\xi_x, y, t, h), \tag{4.7}$$

where

$$\mathrm{E}_t[g^*(\xi_{x,t+h}, x_t, y_t, t, h)] = 0, \qquad \mathrm{Var}_t[g^*(\chi_{x,t+h}, x_t, y_t, t, h)] = 1, \tag{4.8}$$

and

$$\mathrm{E}_t[\partial g^*(\xi_{x,t+h}, x_t, y_t, t, 0)/\partial y] < 0.$$

We now treat $g^*(\cdot)$ as given and optimize over $a(\cdot)$.

THEOREM 4.2. *The asymptotic variance* $C_T/2B_T$ *is minimized subject to the constraints* (4.7)–(4.8) *by setting*

$$a(x, y, t, h) \equiv \Lambda(x, y, t). \tag{4.9}$$

The minimized $C_T/2B_T$ *equals*

$$\frac{\sigma(\gamma_T)^2[\Lambda(x_T, y_T, T) - \mathrm{Cov}_T[\xi_{t,T+h}, g^*(\xi_{x,T+h}, x_T, y_T, T, 0)]]}{\sigma'(y_T)\mathrm{E}_T[\xi^2_{x,T+h} g^*(\xi_{x,T+h}, x_T, y_T, T, 0)]} \tag{4.10}$$

and the corresponding asymptotic variance of $h^{-1/4}[\sigma(\hat{y}_T)^2 - \sigma(y_T)^2]$ *is*

$$4\sigma'(y_T)\sigma(y_T)^5 \times \frac{[\Lambda(x_T, y_T, T) - \mathrm{Cov}_T[\xi_{y,T+h}, g^*(\xi_{x,T+h}, x_T, y_T, T, 0)]]}{\mathrm{E}_T[\xi^2_{x,T+h} g^*(\xi_{x,T+h}, x_T, y_T, T, 0)]}$$

For proof, see the Appendix.

4.3. Interpretations

Since $\{x_t, y_t\}$ is generated by a diffusion, the increments $x_{t+h} - x_t$ and $y_{t+h} - y_t$ are approximately conditionally normal for small h. The first term in (4.6), $\rho(x, y, t)\Lambda(x, y, t)\xi_x/\sigma(y)$ is $\mathrm{E}[\xi_{y,t+h} \mid \xi_{x,t+h}, x_t, y_t]$ using the (limiting) conditionally normal distribution. Given the information in $\xi_{x,t+h}$ (4.6) optimally forecasts the innovation in y_{t+h}, $h^{1/2}\xi_{y,t+h}$.

To understand the second term in (4.6), consider y_t as an unknown parameter in the conditional distribution of $\xi_{x,t+h}$. Given y_t, $\xi_{x,t+h}$ is approximately $\mathrm{N}[0, \sigma(y_t)^2]$. We may write the (limiting) loglikelihood as

$$\ln(f(\xi_{x,t+h} \mid y_t)) = -.5 \cdot \ln(2\pi) - \ln[\sigma(y_t)] - .5 \cdot \xi_{x,t+h}^2/\sigma(y_t)^2, \tag{4.11}$$

so the score is

$$\frac{\partial \ln(f(\xi_{x,t+h} \mid x_t, y_t, y))}{\partial y} = \left[\frac{\xi_{x,t+h}^2}{\sigma(y_t)^2} - 1\right]\frac{\sigma'(y_t)}{\sigma(y_t)^2}. \tag{4.12}$$

For a given x, y, and t, the second term on the right-hand side of (4.6) is proportional to the score. As in maximum likelihood estimation, $\hat{y}_t$ is moved up when the score is positive and is moved down when the score is negative.[8]

Consider the problem of predicting y_{t+h} given $x_t, x_{t-h}, x_{t-2h}, \ldots$. There are two sources of uncertainty about y_{t+h}: first, uncertainty about $y_{t+h} - y_t$, i.e., uncertainty about *changes* in y_t. Second, there is uncertainty about the *level* of y_t. These two sources are asymptotically of the same order, $O_p(h^{1/4})$. The first term on the right side of (4.6) optimally extracts information about $y_{t+h} - y_t$ contained in $x_{t+h} - x_t$. The second term, in a manner analogous to maximum likelihood estimation, extracts information about y_t itself.

4.4. Conditional Moment Matching and the Connection with Consistent Forecasting

Nelson and Foster (1991) develop conditions under which the forecasts generated by a misspecified ARCH model approach the forecasts generated by the true model as a continuous time limit is approached. For example, suppose the diffusion (3.1)–(3.3) generates the data and the misspecified ARCH model (3.4)–(3.5) is used to estimate y_t and to make probabilistic forecasts about the future path of $\{x_t, y_t\}$. In particular,

[8] This is merely heuristic: Convergence in distribution of $[\xi_x, \xi_y]$ does not, of course, imply convergence of the conditional density or of its derivative, and we do not need to prove such convergence to verify Assumptions 1–3.

suppose the ARCH model is used in forecasting *as if* it were the true model—i.e., as if, instead of (3.1), we had

$$x_{t+h} = x_t + h \cdot \hat{\mu}(x_t, y_t, t, h) + h^{1/2} \cdot \xi_{x,t+h}, \tag{4.13}$$

$$y_{t+h} = y_t + h \cdot \hat{\kappa}(x_t, y_t, t, h) + h^{1/2} \cdot g(\xi_{x,t+h}, x_t, y_t, t, h), \tag{4.14}$$

where $\xi_{x,t+h} \mid x_t, y_t, t \sim \text{N}[0, \sigma(y_t)^2]$. Under what circumstances do forecasts generated by this misspecified model approach forecasts generated by (3.1)–(3.3) as $h \downarrow 0$? It turns out that these conditions are closely related to the conditions for asymptotically efficient filtering. In particular, the conditions for consistent forecasting include the first-moment-matching requirement that $\hat{\kappa}(x, y, t, h) = \kappa(x, y, t)$ and $\hat{\mu}(x, y, t, h) = \mu(x, y, t)$ for all (x, y, t). As we saw earlier, this is sufficient condition to eliminate asymptotic bias in the "large drift" case. We also require that the second moments are matched in the limit as $h \downarrow 0$–i.e., that

$$\text{Cov}_t\begin{bmatrix} \xi_x \\ \xi_y \end{bmatrix} = \text{Cov}_t\begin{bmatrix} \xi_x \\ g(\xi_x, x, y, t, 0) \end{bmatrix}$$

$$= \begin{bmatrix} \sigma(y)^2 & \rho(x,y,t)\Lambda(x,y,t)\sigma(y) \\ \rho(x,y,t)\Lambda(x,y,t)\sigma(y) & \Lambda(x,y,t)^2 \end{bmatrix}. \tag{4.15}$$

Using the approximate bivariate normality of ξ_x and ξ_y, it is easy to show that the asymptotically optical $g(\cdot)$ of Theorem 4.1 satisfies (4.15).

What accounts for this moment matching property of optimal ARCH filters? Recall that the first two conditional moments of $\{x_t, y_t\}$ (along with properties (a) and (d) of section 2) characterize the distribution of the process. Optimal ARCH filters make themselves *as much like the true data generating process as possible*. Since the first two conditional moments characterize the true data generating process in the continuous time limit, the optimal ARCH filters match these two moments. (This ignores the pathological case when biases arising from $\hat{\kappa} \neq \kappa$ and from $\hat{\mu} \neq \mu$ exactly cancel.)

Interestingly, misspecification in the drifts κ and μ has only a second-order ($O_p(h^{1/2})$ as opposed to $O_p(h^{1/4})$) effect on filtering (and hence on one-step ahead forecasting) but has an $O_p(1)$ effect on many-step-ahead forecasting performance. Over short time intervals, diffusions act like driftless Brownian motions, with the noise swamping the drift. In the medium and long term, however, the drift exerts a crucial impact on the process.

4.5. Invariance to the Definition of y_t

There is considerable arbitrariness in our definition of y_t. Suppose, for example, that we define $\tilde{y}_t \equiv \tilde{y}(y_t)$ and $\sigma_t^2 = \sigma(y_t)^2 = \sigma(\tilde{y}^{-1}(\tilde{y}_t))^2$ for some monotone increasing, twice continuously differentiable function $\tilde{y}(\cdot)$. We could then apply Ito's Lemma to (3.1), rewriting it as a stochastic integral equation in x_t and $\tilde{y}_t$. If the regularity conditions of Theorem 4.1 are satisfied, the theorem yields an asymptotically optimal filter producing an estimate, say $\hat{y}_t^{\sim}$, of $\tilde{y}_t$ and a corresponding estimate of σ_t^2, say $\hat{\sigma}_t^{\sim 2} \equiv \sigma(\tilde{y}^{-1}(\hat{y}_t^{\sim}))^2$ for the new system. Is it possible to make the asymptotic variance of $[\hat{\sigma}_t^{\sim 2} - \sigma_t^2]$ lower than the asymptotic variance of $[\hat{\sigma}_t^2 - \sigma_t^2]$ by a judicious choice of $\tilde{y}(\cdot)$? Using Ito's Lemma it is easy to verify that the answer is no, provided that the regularity conditions are satisfied for both the (x_t, y_t) and $(x_t, \tilde{y}_t)$ systems: the $\sigma'(y)$ in the asymptotic variance of $[\hat{\sigma}_t^2)\sigma_t^2]$ in Theorem 4.1 is replaced by $\sigma'(y)/[\partial\tilde{y}(y_t)/\partial y]$ in the $\tilde{y}$ system, but the $\Lambda(x, y, t)$ is replaced by $\Lambda(x, y, t) \cdot [\partial\tilde{y}(y_t)/\partial y]$, leaving the asymptotic variances of $[\hat{\sigma}_t^2 - \sigma_t^2]$ and $[\hat{\sigma}_t^{\sim 2} - \sigma_t^2]$ equal. Within the limits of the regularity conditions, the definition of y_t is arbitrary.

4.6. Examples

The asymptotically optimal ARCH filter of Theorem 4.1 looks unlike ARCH models commonly used in the literature. GARCH(1, 1) is an exception. Suppose the data are generated by the diffusion

$$dx_t = \mu \cdot dt + \sigma_t \, dW_{1,t}, \tag{4.16}$$

$$d\sigma_t^2 = \left(\omega - \theta\sigma_t^2\right) dt + 2^{1/2}\alpha\sigma_t^2 \, dW_{2,t}, \tag{4.17}$$

where $W_{1,t}$ and $W_{2,t}$ are independent standard Brownian motions. If we set $y_t \equiv \sigma_t^2$, this is in the form of (3.1). The asymptotically optimal filter of Theorem 4.1 sets $\hat{\mu} = \mu$ and

$$\hat{\sigma}_{t+h}^2 = \hat{\sigma}_t^2 + \left(\omega - \theta\hat{\sigma}_t^2\right) \cdot h + h^{1/2}\alpha\left(\hat{\xi}_{x,t+h}^2 - \hat{\alpha}_t^2\right), \tag{4.18}$$

which is recognizable as a GARCH(1, 1) when we rewrite (4.18) as

$$\hat{\sigma}_{t+h}^2 = \omega \cdot h + (1 - \theta \cdot h - \alpha \cdot h^{1/2})\hat{\sigma}_t^2 + h^{1/2}\alpha\hat{\xi}_{x,t+h}^2. \tag{4.19}$$

THEOREM 4.3. (4.16)–(4.18) *satisfy the conditions of Theorems* 3.1 *and* 4.1.

For proof, see the Appendix.

If $dW_{1,t}\, dW_{2,t} = \rho\, dt$, $\rho \neq 0$ (i.e., if $W_{1,t}$ and $W_{2,t}$ are correlated) GARCH(1, 1) is no longer optimal: The second moment matching condition (4.15) fails, since $\text{Corr}_t[\alpha(\xi_{x,t+h}^2 - \sigma_t^2), \xi_{x,t+h}]$ is zero, not ρ. A modification of GARCH(1, 1) ("Nonlinear Asymmetric GARCH") proposed by Engle and Ng (1993, Table 1) can be shown to be optimal in this case.

To further illustrate the construction of globally optimal ARCH models, we next consider two models from the option pricing literature. In each model, S_t is a stock price and σ_t is its instantaneous returns volatility. We observe $\{S_t\}$ at discrete intervals of length h. In each model, we have

$$dS_t = \mu S_t\, dt + S_t \sigma_t\, dW_{1,t}. \tag{4.20}$$

The first model (see Wiggins, 1987; Hull and White, 1987; Melino and Turnbull, 1990; Scott, 1987) sets

$$d\left[\ln(\sigma_t^2)\right] = -\beta\left[\ln(\sigma_t^2) - \alpha\right] dt + \psi \cdot dW_{2,t}, \tag{4.21}$$

where $W_{1,t}$ and $W_{2,t}$ are standard Brownian motions independent of (S_0, σ_0^2) with

$$\begin{bmatrix} dW_{1,t} \\ dW_{2,t} \end{bmatrix} [dW_{1,t} \quad dW_{2,t}] = \begin{bmatrix} 1 & \rho \\ \rho & 1 \end{bmatrix} dt. \tag{4.22}$$

μ, ψ, β, α, and ρ are constants.

Bates and Pennacchi (1990), Gennotte and Marsh (1993), and Heston (1993) propose a model which replaces (4.21) with

$$d\sigma_t^2 = -\beta\left[\sigma_t^2 - \alpha\right] dt + \psi \cdot \sigma_t \cdot dW_{2,t}. \tag{4.23}$$

Here σ_t^2 is generated by the "square root" diffusion popularized as a model of the short-term interest rate by Cox *et al.* (1985).

Now consider ARCH filtering of these models. Suppose we define $x_t \equiv \ln(S_t)$ and $y_t \equiv \ln(\sigma_t^2)$. We may then rewrite (4.20)–(4.21) as

$$dx_t = (\mu - \exp(y_t)/2)\, dt + \exp(y_t/2)\, dW_{1,t}, \tag{4.20$'$}$$

$$dy_t = -\beta[y_t - \alpha]\, dt + \psi \cdot dW_{2,t}. \tag{4.21$'$}$$

THEOREM 4.4. *The asymptotically optimal ARCH model for the model* (4.20′)–(4.21′) *and* (4.22) *is*

$$\hat{\mu}(x, \hat{y}) \equiv \mu - \exp(\hat{y})/2 \tag{4.24}$$

$$\hat{y}_{t+h} = \hat{y}_t - \beta[\hat{y}_t - \alpha] \cdot h + h^{1/2}\psi\Big[\rho\hat{\xi}_{x,t+h} \cdot \exp(-\hat{y}_t/2) + \left[(1-\rho^2)/2\right]^{1/2} \cdot \left(\hat{\xi}^2_{x,t+h} \cdot \exp(-\hat{y}_t) - 1\right)\Big]. \tag{4.25}$$

(4.20′)–(4.21′) *and* (4.24)–(4.25) *satisfy Assumptions* 1–3. *The resulting* (*minimized*) *asymptotic variance of*

$$h^{-1/4}\left[\sigma(\hat{y})^2 - \sigma(y)^2\right] \quad \textit{is} \quad \left[2(1-\rho^2)\right]^{1/2}\psi\sigma^4$$

For proof, see the Appendix.

Next consider the model given by (4.20) and (4.22)–(4.23). Using Ito's Lemma and $y \equiv \ln(\sigma^2)$, (4.23) becomes

$$dy_t = \left(-\beta + \exp(-y_t)\left[\beta\alpha - \psi^2/2\right]\right)dt + \psi \cdot \exp(-y_t/2) \cdot dW_{2,t}. \tag{4.23'}$$

The asymptotically optimal filter suggested by Theorem 4.1 is

$$\hat{y}_{t+h} = \hat{y}_t + \left(-\beta + \exp(-\hat{y}_t)\left[\beta\alpha - \psi^2/2\right]\right) \cdot h + h^{1/2}\psi\exp(-\hat{y}_t/2)$$
$$\cdot\left[\rho \cdot \hat{\xi}_{x,t+h} \cdot \exp(-\hat{y}_t/2) + \left[(1-\rho^2)/2\right]^{1/2}\right.$$
$$\left.\cdot\left(\hat{\xi}^2_{x,t+h} \cdot \exp(-\hat{y}_t) - 1\right)\right]. \tag{4.26}$$

Unfortunately, the regularity conditions break down at $y = -\infty$ (i.e., at $\sigma^2 = 0$). (In fact, the stochastic differential equation (4.23′) is not well-defined in this case, although (4.23) is.) This is not a problem in the theorem as long as the boundary $y_t = -\infty$ is unattainable in finite time. When $2\beta\alpha \ll \psi^2$, however, this boundary is attained in finite time with positive probability (see Cox *et al*., 1985). We therefore exclude this case.

THEOREM 4.5. *Let* $2\beta\alpha > \psi^2$. *The asymptotically optimal model for* (4.20′) *and* (4.22′)–(4.23′) *is given by* (4.24) *and* (4.26). *Equations* (4.20′), (4.22′), (4.23), *and* (4.26) *satisfy Assumptions* 1–3. *The resulting* (*minimized*) *asymptotic variance of* $h^{-1/4}[\sigma(\hat{y})^2 - \sigma(y)^2]$ *is* $[2(1-\rho^2)]^{1/2}\psi\sigma^3$.

For proof, see the Appendix.

The differences in the optimal filters for the two models are most easily understood in terms of the moment matching conditions. In (4.21) and its associated optimal ARCH model, the conditional variance of σ_t^2 rises linearly with σ_t^2, while in (4.23) the conditional variance of σ_t^2 rises linearly with σ_t^2. As we will see in section 6, most commonly used ARCH models effectively assume that the "variance of the variance" rises linearly with σ_t^4. If (4.23) generates the data, GARCH, EGARCH, and other such models will be very inefficient filters when σ^2 is very low or very high, since the $g(\cdot)$ functions in these ARCH models cannot match the ARCH and true "variance of the variance" everywhere in the state space.

5. NEAR-DIFFUSIONS

In this section we consider the case in which the data are generated in discrete time by the stochastic volatility model

$$\begin{bmatrix} x_{t+h} \\ y_{t+h} \end{bmatrix} = \begin{bmatrix} x_t \\ y_t \end{bmatrix} + \begin{bmatrix} \mu(x_t, y_t, t, h) \\ \kappa(x_t, y_t, t, h) \end{bmatrix} h + \begin{bmatrix} \xi_{x,t+h} \\ \xi_{y,t+h} \end{bmatrix}, \tag{5.1}$$

or, in the "fast drift" case analogous to (3.1′)

$$\begin{bmatrix} x_{t+h} \\ y_{t+h} \end{bmatrix} = \begin{bmatrix} x_t \\ y_t \end{bmatrix} + \begin{bmatrix} \mu(x_t, y_t, t, h) \\ \kappa(x_t, y_t, t, h) \end{bmatrix} h^{3/4} + \begin{bmatrix} \xi_{x,t+h} \\ \xi_{y,t+h} \end{bmatrix}, \tag{5.1′}$$

for some (small) $h > 0$, where

$$E_t \begin{bmatrix} \xi_{x,t+h} \\ \xi_{y,t+h} \end{bmatrix} = \begin{bmatrix} 0 \\ 0 \end{bmatrix},$$

$$\operatorname{Cov}_t \begin{bmatrix} \xi_{x,t+h} \\ \xi_{y,t+h} \end{bmatrix} = \begin{bmatrix} \sigma(y_t)^2 & \rho(x_t, y_t, t)\Lambda(x_t, y_t, t)\sigma(y_t) \\ \rho(x_t, y_t, t)\Lambda(x_t, y_t, t)\sigma(y_t) & \Lambda(x_t, y_t, t)^2 \end{bmatrix}. \tag{5.2}$$

We assume further that the process $\{x_t, y_t\}_{t=0,h,2h\ldots}$ is Markovian. Again, in (5.1)–(5.2), t is assumed to be a discrete multiple of h. To define the process for general t, set $(x_t, y_t) \equiv (x_{h[t/h]}, y_{h[t/h]})$. This makes each $\{x_t, y_t\}$ process a step function with jumps at discrete intervals of length h. E_t and Cov_t denote, respectively, expectation and covariances conditional on time t information—i.e., the σ-algebra generated by $\{x_\tau, y_\tau\}_{0 \le \tau \le t}$ or, equivalently for our purposes, by $(x_t, y_t, \tilde{y}_t, t)$. Note that the structure of the first two conditional moments of x_t and y_t in (5.1)–(5.2) are the same as in (3.1)–(3.2). In fact, under (5.1) and the regularity conditions assumed below, $\{x_t, y_t\}$ converges weakly to the diffusion (3.1)–(3.2) as $h \downarrow 0$. We therefore call such a $\{x_t, y_t\}$ process a *near-diffusion.*

5.1. Why the Near-Diffusion Case Is Important

At first glance, it would seem that there is little gain to generalizing the results of Section 3 to the near diffusion case. The intuition is this: The estimated conditional variance process $\{\sigma(\hat{y}_t)^2\}$ is a functional of the sample path of the $\{x_t\}$ of (5.1)–(5.2). The $\{x_t\}$ of (5.1)–(5.2) converges weakly to the $\{x_t\}$ of (3.1)–(3.2) as $h \downarrow 0$. If the mapping from $\{x_t\}$ to $\{\sigma(\hat{y}_t)^2\}$ is sufficiently well-behaved, the continuous mapping theorem should guarantee that $\{\sigma(\hat{y}_t)^2\}$ converges to the limit derived in Section 3 as $h \downarrow 0$, yielding the same results on efficiency, and etc., as in the diffusion case.

Unfortunately, this intuition is wrong, since the mapping from $\{x_t\}$ to $\{\sigma(\hat{y}_t)^2\}$ is not at all well-behaved in the sense required by the continuous mapping theorem. To see why, consider the case of i.i.d. residuals. Let

$$x_{t+h} = x_t + h^{1/2}\xi_{t+h} \tag{5.3}$$

where $x_0 = 0$ and for all t and all h, ξ_t is i.i.d. with mean 0 and variance σ^2. Consider the least squares estimator of σ^2, given at time $t+h$ by

$$\hat{\sigma}_{t+h}^2 \equiv (t/h)^{-1} \sum_{j=0,h,2h\ldots}^{t/h} h^{-1}[x_{(j+1)h} - x_{jh}]^2 = (t/h)^{-1} \sum_{j=0,h,2h\ldots}^{t/h} \xi_j^2. \tag{5.4}$$

Standard invariance arguments (e.g., using Donsker's theorem; see Jacod and Shiryaev, 1987) show that as $h \downarrow 0$, $\{x_t\}$ converges weakly to the limit process given by

$$x_t = \sigma W_{1,t}, \tag{5.5}$$

where $W_{1,t}$ is a standard Brownian motion. *This holds regardless of the distribution of* ξ_t, *provided* ξ_t *is i.i.d. with mean* 0 *and variance* σ^2. If $\mathrm{E}[\xi_t^4] < \infty$, $\{(t/h)^{1/2}(\hat{\sigma}_{t+h}^2 - \sigma^2)\}$ also converges weakly, to the process $\{\psi_t\}$ with

$$\psi_t \equiv \sigma^2\left(\mathrm{E}\left[\xi_t^4/\sigma^4\right] - 1\right)^{1/2} W_{2,t},$$

where $W_{2,t}$ is a second standard Brownian motion. *Note that the diffusion limit of* $\{x_t\}$ *does not depend on the distribution of* ξ_t/σ, *but the diffusion limit of* $\{\psi_t\}$ *does, through the kurtosis of* ξ_t.

Suppose, for example, that $\{x_t\}$ is a Brownian motion observed at time intervals of length h, so $\xi_t \sim \mathrm{N}(0, \sigma^2)$. Here $(\mathrm{E}[\xi_t^4/\sigma^4] - 1) = 2$. (This is the case analogous to (3.1)–(3.2).) Moving from the diffusion to the near-diffusion case by changing the distribution of ξ_t can have drastic consequences for $\hat{\sigma}_t$. For example, let ξ_t be i.i.d. Student's t with $v > 4$ degrees of freedom and variance σ^2. Then well-behaved diffusion limits exist for both $\{x_t\}$ and $\{\psi_t\}$. $\mathrm{E}[\xi_t^4/\sigma^4]$, however, is a decreasing function of v, so while the diffusion limit of $\{x_t\}$ does not depend on v, the limit of $\{\psi_t\}$ does. In particular, the efficiency of $\hat{\sigma}_t^2$ as an estimate of σ^2 is an increasing function of the degrees of freedom v. Even in this i.i.d. case, there is a crucial difference between the behavior of the variance estimate for the limit diffusion and the variance estimate for a sequence of process converging to the diffusion. As we will see below, this remains true in the more general ARCH case as well.[9]

Many empirical studies of asset market volatility have found that returns remain somewhat thick-tailed even after conditional heteroskedasticity is accounted for (e.g., Baillie and Bollerslev, 1989; Nelson, 1989, 1991). Near diffusions easily accommodate this. The diffusion case examined in section 3, on the other hand, effectively assumes conditional normality for sufficiently small h. The near-diffusion case is therefore likely to be practically important, and will allow us to consider optimality for different conditional distributions and the robustness of different ARCH models to the presence of conditionally thick-tailed residuals.

5.2. Main Results

THEOREM 5.1. *Let Assumptions* 1–2 *hold with* (5.1) *and* (5.1′) *replacing* (3.1) *and* (3.1′). *Then the statement of Theorem* 3.1 *holds.*

For proof, see the Appendix.

The interpretation of the optimal filter in terms of an estimation component and a forecasting component applies in the near-diffusion context also. We accordingly define the prediction component

$$P(\xi_x, x, y, t, h) \equiv E\left[\xi_{y,t+h} \mid (\xi_{x,t+h}, x_t, y_t, t, h) = (\xi_x, x, y, t, h)\right], \tag{5.6}$$

and the estimation (or score) component

$$S(\xi_x, x, y, t, h) \equiv \partial \ln[f(\xi_{x,t+h} \mid x, y, t, h)]/\partial y, \tag{5.7}$$

where $f(\xi_{x,t+h} \mid x, y, t, h)$ is the conditional density of $\xi_{t,t+h}$ given $(x_t, y_t) = (x, y)$. In the diffusion case of sections 3 and 4, $P(\cdot)$ is proportional to ξ_x and $S(\cdot)$ is proportional to $\xi_x^2 - \sigma^2$. This is *not* generally true in the near-diffusion case unless ξ_x and ξ_y are conditionally bivariate normal.

ASSUMPTION 4. *For every* (x, y, t, h), *the conditional densities* $f(\xi_x, \xi_y, y, t, h)$ *and* $f(\xi_x \mid x, y, t, h)$ *are well-defined and continuous in* $x, t,$

[9] Here is another heuristic: For Brownian motions, the instantaneous variance is the stochastic derivative of the quadratic variation of the process. Since Brownian motions are continuous with probability one, the Weierstrass theorem guarantees that they can be approximated to arbitrary accuracy with finite-order polynomials. Yet the quadratic variation of any polynomial is zero.

and h and $f(\xi_x \mid x, y, t, h)$ is continuously differentiable in y. Further, for some $\delta > 0$,

$$h^{-1}\,\mathrm{E}\Big[\left|\mathrm{P}(\xi_{x,t+h}, x_t, y_t, t, h)\right|^{2+\delta} \mid x_t = x, y_t = y\Big] \quad \text{and} \tag{5.8}$$

$$h^{-1}\,\mathrm{E}\Big[\left|S(\xi_{x,t+h}, x_t, y_t, t, h)\right|^{2+\delta} \mid x_t = x, y_t = y\Big] \tag{5.9}$$

are bounded as $h \downarrow 0$, uniformly on every bounded (x, y, t) set.

THEOREM 5.2. *$C_T/2B_T$ is minimized by setting*

$$g(\xi_x, x, y, t, h) = \mathrm{P}(\xi_x, x, y, t) + \omega(x, y, t) \cdot S(\xi_x, x, y, t), \tag{5.10}$$

where

$$\omega(x, y, t)$$

$$\equiv \frac{\left[\mathrm{Cov}_t(S, \mathrm{P})^2 + \left(\Lambda^2 - \mathrm{Var}_t(\mathrm{P})\right) \cdot \mathrm{Var}_t(S)\right]^{1/2} - \mathrm{Cov}_t(\mathrm{P}, S)}{\mathrm{Var}_t(S)} \tag{5.11}$$

(The arguments have been dropped in (5.11) to simplify notation.) The asymptotic variance achieved by any other $g(\cdot)$ function (say $\tilde{g}$) satisfying the constraints $\mathrm{E}_t[\tilde{g}] = 0$ and $B_T > 0$ is strictly higher unless $\tilde{g}(\xi_x, x, y, t, 0) = g(\xi_x, x, y, t, 0)$ with probability one.

The minimized $C_T/2B_T = \omega(x_T, y_T, T)$, and the corresponding asymptotic variance of $h^{-1/4}[\sigma(\hat{y})^2 - \sigma(y)^2]$ equals

$$4 \cdot \omega(x_T, y_T, T) \cdot \sigma(y_T)^2 [\sigma'(y_T)]^2.$$

For proof, see the Appendix.

THEOREM 5.3. *The asymptotic variance $C_T/2B_T$ is minimized subject to the constraints (4.7)–(4.8) by setting*

$$a(x, y, t, h) \equiv \Lambda(x, y, t). \tag{5.12}$$

The minimized $C_T/2B_T$ is $\Lambda[1 - \mathrm{Corr}_T(\xi_y, g^*)]/\mathrm{E}_T[g^* \cdot S]$, and the asymptotic variance of $h^{-1/4}[\sigma(\hat{y})^2 - \sigma(y)^2]$ is $4\sigma^2\sigma'^2\Lambda[1 - \mathrm{Corr}_T(\xi_y, g^*)]/\mathrm{E}_T[g^* \cdot S]$.

For proof, see the Appendix.

The interpretations of the optimal filter given in Section 4—i.e., moment matching, asymptotic irrelevance of transformations $\tilde{y}(y_t)$, and the prediction and estimation components of the optimal filter—continue to hold. To further understand the distinction between prediction component $\mathrm{P}(\xi_x, x, y, t)$ and the estimation component $S(\xi_x, x, y, t)$, consider first the case in which an ARCH model is the true data-generating process —i.e., in which the innovation in the y process, ξ_y, is a function of the innovation in x, ξ_x, and possibly the other state variables x, y, and t, say $\xi_y = g(\xi_x, x, y, t)$. Now $\xi_y = \mathrm{P}(\cdot)$, and $\omega(\cdot) = 0$. The innovations in y_t are observable, so it is not surprising that the asymptotic variance of the measurement error is zero.

Another polar case arises when $\mathrm{P}(\cdot) = 0$ with probability one, so ξ_x contains no information that helps predict ξ_y. This was true, for example, in (4.16)–(4.17), when the x_t and y_t were driven by independent Brownian motions. In this case, the asymptotic variance is $\Lambda/\mathrm{SD}_t(S)$. This is easily interpretable: Λ is the conditional standard deviation of y_t—the more locally variable y_t is, the less accurately it can be estimated. $\mathrm{Var}_t(S)$, on the other hand, is the filtering analog of the Fisher information—the smaller the Fisher information, the higher the asymptotic variance of the parameter estimates.

6. ANALYSIS OF SOME COMMONLY USED ARCH MODELS

6.1. GARCH(1, 1)

In the GARCH(1,1) model of Bollerslev (1986), we have $y \equiv \sigma^2$, $\hat{\kappa} \equiv \omega - \theta\sigma^2$, and $g \equiv a \cdot [\xi_r^2 - \sigma^2]$. As we saw in section 4, GARCH(1,1) is asymptotically optimal for the diffusion (4.16)–(4.17). More generally, suppose the data are generated by either (3.1), (3.1′), (5.1)–(5.2), or (5.1′)–(5.2) with $y \equiv \sigma^2$. By Theorem 5.2, GARCH(1,1) is optimal when for some α and all ξ_x, x, σ^2, and t, $\kappa(x, \sigma^2, t) = \omega - \theta\sigma^2$ and $\mathrm{P}(\xi_x, x, \sigma^2, t) + \omega(x, \sigma^2, t)S(\xi_x, \sigma^2, t) = \alpha[\xi_x^2 - \sigma^2]$. When GARCH(1,1) is the true data generating process, the GARCH(1,1) filter is (trivially) optimal and $\mathrm{P}(\xi_x, x, \sigma^2, t) = \alpha[\xi_x^2 - \sigma^2]$ while $\omega(x, \sigma^2, t) = 0$. Even when $\mathrm{P}(\xi_x, \sigma^2, t) = 0$, GARCH may be optimal provided ξ_x is conditionally normal.

Next consider minimizing $C_T/2B_T$ with a model of the form

$$g \equiv \alpha(x, \sigma^2, t) \cdot (\xi_x^2 - \sigma^2). \tag{6.1}$$

By Theorems 4.2 and 5.3, the optimal $\alpha(x, \sigma^2, t)$ equals $\Lambda(x, \sigma^2, t)/\mathrm{SD}_t(\xi_x^2)$. Suppose the conditional kurtosis of ξ_x is a constant K (this is always satisfied in the diffusion case, where $K = 3$, and is satisfied in many discrete stochastic volatility models as well). We may then write the (constrained) optimal g as

$$g \equiv \Lambda(x, \sigma^2, t)\sigma^{-2}(K - 1)^{-1/2} \cdot (\xi_x^2 - \sigma^2). \tag{6.2}$$

GARCH(1, 1) further constrains α to be constant, so clearly if GARCH(1, 1) is to be the constructed optimum in the class of models (6.1), Λ^2, the conditional variance of σ^2, must be proportional to σ^4. (Or, equivalently, the conditional variance of $\ln(\sigma^2)$ is constant.) As we will see, many other commonly used ARCH models effectively make the same assumption. The resulting (locally) minimized asymptotic variance of $h^{-1/4}[\hat{\sigma}^2 - \sigma^2]$ is

$$\mathrm{SD}_t\left(\xi_x^2/\sigma^2\right) \cdot \left[1 - \mathrm{Corr}_t\left(\xi_x^2, \xi_y\right)\right] \cdot \sigma^2 \cdot \Lambda. \tag{6.3}$$

That is, GARCH(1, 1) can more accurately measure σ_t^2 (1) the less locally variable σ_t^2 is (as reflected by Λ), (2) the lower the conditional kurtosis of ξ_x, (3) the lower σ_t^2, and (4) the more the true data-generating mechanism resembles GARCH(1, 1) (e.g., if GARCH(1, 1) is the data-generating process, $\mathrm{Corr}_t(\xi_x^2, \xi_y) = 1$).

6.2. The Taylor–Schwert Model

Davidian and Carroll (1987) argue (though not explicitly in a time series or ARCH context) that scale estimates based on absolute residuals are more robust to the presence of thick-tailed residuals than scale estimates based on squared residuals. Schwert (1989) applied the Davidian and Carroll intuition to ARCH models, conjecturing that estimating σ_t^2 with the square of a distributed lag of absolute residuals (as opposed to estimating it with a distributed lag of squared residuals, as in GARCH) would be more robust to ξ_x's with thick-tailed distributions. Taylor (1986, ch. 4) proposed a similar method. Hence, we consider the model $y \equiv \sigma$, $\hat{\kappa} \equiv \omega - \theta\hat{\sigma}$, and $g \equiv \alpha \cdot [|\hat{\xi}_x| - \mathrm{E}_t|\xi_x|]$, with the data generated by either (3.1), (3.1′), (5.1)–(5.2), or (5.1′)–(5.2). We also assume that the distribution of ξ_x/σ does not vary with x, σ, t, or h—i.e., that σ enters the distribution of ξ_x only as a scale parameter.

If we allow α to be a function of x, y, and t, the optimal α is $\Lambda/[\mathrm{SD}_t(|\xi_x|)]$. The minimized asymptotic variance of $h^{-1/4}[\hat{\sigma}_t^2 - \sigma_t^2]$ is

$$4 \cdot \mathrm{SD}_t(|\xi_x|) \cdot \sigma^3 \cdot \left[1 - \mathrm{Corr}_t\left(|\xi_x|, \xi_y\right)\right] \cdot \Lambda/\mathrm{E}_t|\xi_x|. \tag{6.4}$$

As in the GARCH case, *global* optimality of the Taylor–Schwert model in this class of models requires that α be constant. Again, when $\mathrm{E}_t[|\xi_x/\sigma|]$ is constant, this is equivalent to $\ln(\sigma^2)$ being conditionality homoskedastic in the diffusion limit.

Schwert's conjecture that this model is more robust than GARCH to conditionally thick-tailed ξ_x's can be rigorously justified. For example, compare the relative efficiencies of GARCH(1, 1), the Taylor–Schwert model, and the asymptotically optimal filter, supposing for the moment that $\mathrm{Corr}_t(\xi_x^2, \xi_y) = \mathrm{Corr}_t(|\xi_x|, \xi_y = 0$, and that ξ_x is conditionally Student's t with $K > 2$ degrees of freedom. It is important to note that (6.4) is not directly comparable to (6.3), since in (6.3) $y \equiv \sigma^2$ and Λ is the instantaneous standard deviation of σ^2, whereas in (6.4) $y \equiv \sigma$ and Λ is the instantaneous standard deviation of σ. In general the Λ in (6.3) equals 2σ times the Λ in (6.4). Making this adjustment, we can compare the

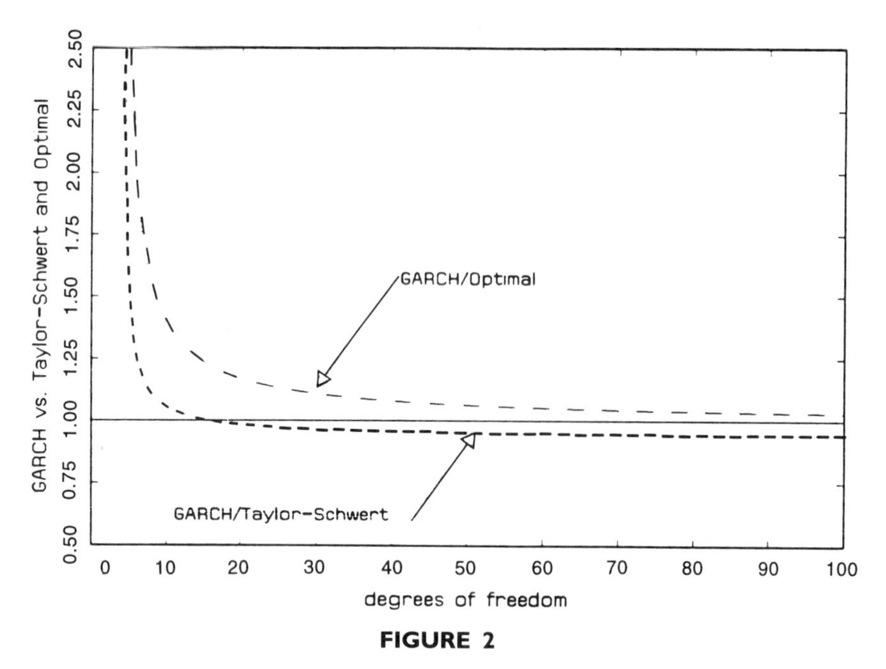

FIGURE 2

variances in (6.3), (6.4), and the variance achieved by the optimal filter. Figure 2 compares the minimized asymptotic variances of $h^{-1/4}[\hat{\sigma}_t^2 - \sigma_t^2]$ achieved by these filters, plotting both the ratio of the GARCH for the Taylor–Schwert variance and the ratio of GARCH to the optimal filter error variances over a range of K values. The GARCH error variance approaches infinity as $K \downarrow 4$, and is infinite for $K \leq 4$. The Taylor–Schwert model's error variance approaches infinity as $K \downarrow 2$ and is infinite for $K \leq 2$, but is finite for $K > 2$. The optimal filter's variance is bounded as $K \downarrow 2$. The horizontal line at 1 in Figure 1 divides the region in which the Taylor–Schwert error variance is lower (above the line) and the region in which the GARCH error variance is lower (below the line). For low degrees of freedom, K, the Taylor–Schwert model is much more efficient than GARCH, and remains more efficient until $K > 15.56$. Perhaps surprisingly, the variance of GARCH is only about 6.5% lower even as $K \to \infty$, so even under the most favorable circumstances, the efficiency gain from using GARCH is slight. When the ξ_x's are conditionally thick-tailed, the efficiency loss with GARCH can be dramatic.[10] As $K \to \infty$, the

[10] When ξ_x is conditionally distributed GED (Harvey, 1981), the analysis is similar: Taylor–Schwert performs much better than GARCH when ξ_x is conditionally thick-tailed. On the other hand, GARCH may perform substantially (up to 22.5%) better than Taylor–Schwert when ξ_x is much *thinner*-tailed than the normal. In the range of parameter estimates found for asset prices in the empirical literature (e.g., Nelson, 1989, 1991; Sentana, 1992) GARCH and Taylor–Schwert perform about equally.

GARCH and optimal error variances converge, but fairly slowly: Even for $K = 20$ the GARCH variance is about 13% higher than the optimal.

Our robustness results closely parallel those of Davidian and Carroll (1987). To see why, consider the optimal filter of Theorem 5.2, assuming for now that the prediction component $\mathrm{E}[\xi_{y,t+h} \mid \xi_{x,t+h}, x_t, y_t] = 0$ almost surely for all t. The optimal g is then proportional to the score $\partial \ln[f(\xi_x \mid x, y, t)]/\partial y$. A necessary condition in this case for GARCH to be optimal is that ξ_x is conditionally normal. Optimality of the Taylor–Schwert model in this case requires ξ_x to be conditionally double exponential, and hence thicker-tailed than the normal. Abandoning the assumption of optimality for either model, we see that GARCH uses a normal quasi-likelihood in estimating the level of y_t, while Taylor–Schwert uses a double exponential quasi-likelihood. More generally, the choice of an ARCH model embodies a choice of quasi-likelihood, a choice which can be analyzed in much the same way that Davidian and Carroll did in the case in which the conditional variance of the error term was a function (of known form) of observable variables.

Several papers in the ARCH literature have assumed conditionally Student's t errors and have treated the degrees of freedom as a parameter to be estimated. In modelling daily exchange rates, for example, Baillie and Bollerslev (1989) estimated degrees of freedom parameters ranging from 6.3 to 18.5, while Hsieh's (1989) estimates ranged from 3.1 to 6.5. In modelling daily stock price indices, Baillie and DeGennaro's (1990) estimated degrees of freedom ranged from 9.2 to 10.2.[11] More broadly, thick-tailed standardized residuals are the norm in empirical applications of ARCH (see Bollerslev *et al.*, 1992). This range of estimates, along with Figure 2, suggest the use of absolute residuals as opposed to squared residuals in estimating time varying volatilities in asset returns.

6.3. Exponential ARCH (EGARCH)

The exponential ARCH (EGARCH) model of Nelson (1991) was largely motivated by Black's (1976) empirical observation that stock volatility tends to rise following negative returns and to drop following positive returns. The EGARCH model exploits this empirical regularity by making the conditional variance estimate a function of both the size and the sign of lagged residuals. AR(1) EGARCH sets $y \equiv \ln(\sigma^2)$, $\hat{\kappa} \equiv -\beta \cdot [\hat{y} - \alpha]$, and $g \equiv \theta\hat{\xi}_x/\hat{\sigma} + \gamma[|\hat{\xi}_x/\hat{\sigma}| - \mathrm{E}_t|\xi_x/\sigma|]$.

[11] These models were estimated under the assumption that the estimated ARCH model was correctly specified. If this is literally true, there is clearly no efficiency gain in abandoning the true model. If the model is not correctly specified the robustness of the conditional variance estimates is important.

Again we assume the data are generated by either (3.1), (3.1′), (5.1)–(5.2), or (5.1′)–(5.2). If we make the simplifying assumptions that $E_t[\xi_x \cdot |\xi_x|] = 0$ and that the distribution of ξ_x/σ is independent of σ, it is easy to solve for the (locally) optimal γ and θ:

$$\gamma^* \equiv \Lambda\left[1 - \text{Corr}_t^2(\xi_x, \xi_y)\right]^{1/2} / \text{SD}_t(|\xi_x/\sigma|), \quad \text{and} \tag{6.5}$$

$$\theta^* \equiv \Lambda \cdot \text{Corr}_t(\xi_x, \xi_y). \tag{6.6}$$

For EGARCH to be optimal in this class requires that γ^* and θ^* be constant. It is straightforward to check that when the first two moments of ξ_x/σ and $|\xi_x/\sigma|$ are constant, constant γ^* and θ^* is equivalent to conditionally homoskedastic $\ln(\sigma_t^2)$ and constant conditional correlation between x_t and σ_t^2.

The minimized asymptotic variance of $h^{-1/4}[\hat{\sigma}_t^2 - \sigma_t^2]$ for this case is

$$2 \cdot \text{SD}_t(|\xi_x|) \cdot \Lambda \cdot \sigma^4 \cdot \left[\left(1 - \text{Corr}_t^2(\xi_x, \xi_y)\right)^{1/2} - \text{Corr}_t(\xi_y, |\xi_x|)\right] / \text{E}_t|\xi_x|. \tag{6.7}$$

If we adjust for the changed definition of y and therefore of $\Lambda(\cdot)$, we see that apart from the replacement of "1" with "$(1 - \text{Corr}^2(\xi_x, \xi_y))^{1/2}$," this is the minimized variance for the Taylor–Schwert model. The only difference between the two models in the $O_p(h^{1/4})$ error components is that the EGARCH takes advantage of conditional correlation in ξ_x and ξ_y (e.g., "leverage effects"). This suggests that we can significantly improve the variance estimators by exploiting correlations between changes in observable variables and changes in σ^2—the effects are "first order" appearing in the dominant component of the measurement error, the $O_p(h^{1/4})$ term—and so are likely to be important in practice. This, along with its relative robustness to conditionally thick-tailed ξ_x's, probably accounts for much of the empirical success of EGARCH in applications to stock market returns data (e.g., Pagan and Schwert, 1990).

6.4. Related Variants

Higgins and Bera (1992) nested GARCH and the Taylor–Schwert model in a class of "NARCH" (nonlinear ARCH) models, which set $\hat{\sigma}_t^{2\delta}$ equal to a distributed lag of past absolute residuals each raised to the 2δ power. Using a geometric lag, this corresponds to $y \equiv \sigma^{2\delta}$, $\hat{\kappa} \equiv \omega - \theta \cdot \hat{y}$, and $g \equiv \alpha \cdot [|\hat{\xi}_x|^{2\delta} - \text{E}_t[|\xi_x|^{2\delta}]]$.

The chief appeal of NARCH (as with Taylor–Schwert) is that when $\delta < 1$, it is more robust to conditionally thick-tailed ξ_x's than GARCH: NARCH limits the influence of large residuals essentially the same way as the l_p estimators employed in the robust statistics literature (see, e.g., Davidian and Carroll, 1987). While GARCH and Taylor–Schwert use normal and double exponential likelihoods, respectively, in their estimation components, NARCH uses a GED quasi-likelihood.[12]

Another variant is the threshold ARCH ("TARCH") model of Zakoian (1990). The locally minimized asymptotic measurement error variance (up to the $O_p(h^{1/4})$ terms) of this model are the same as EGARCH.

6.5. Similarities and Differences in $\hat{\kappa}(\cdot)$ and $a(\cdot)$ in These Models

Though the GARCH, EGARCH, Taylor–Schwert, Higgins–Bera, and Zakoian models have important differences, they have at least one similarity and potential limitation: For global asymptotic optimality of $a(\cdot)$ given $g^*(\cdot)$ (as in Theorem 5.3), each requires $\{\ln(\sigma_t^2)\}$ to be conditionally homoskedastic in the diffusion limit. This is clearest when we examine the diffusion limits of these models considered as data-generating processes. The diffusion limit for $\{\ln(\sigma_t^2)\}$ in the Higgins–Bera model as a data-generating process takes the form[13]

$$d\big[\ln(\sigma_t^2)\big] = \delta^{-1}\big(\omega\sigma_t^{-2\delta} - \theta^*\big)\,dt - \alpha^*\,dW_t \tag{6.8}$$

where W_t is a standard Brownian motion and θ^* and α^* are constants. The Taylor–Schwert and Zakoian models are a special case of (6.8) with $\delta = \frac{1}{2}$, and GARCH is a special case with $\delta = 1$. The AR(1) EGARCH diffusion limit for $\{\ln(\sigma_t^2)\}$ is given by (4.21). In (4.21), and for any $\delta > 0$ in (6.8), the conditional variance of the increments in $\{\ln(\sigma_t^2)\}$ is constant. A few ARCH models have been proposed that do not make this assumption [e.g., (4.26), the QARCH model of Sentana (1992), and the model of Friedman and Laibson (1989)]. Unfortunately, these models make similarly restrictive assumptions on the conditional second moments. Since the second moment matching condition (4.15) has a first-order effect on the measurement error variance, practitioners should probably parameterize $g(\cdot)$ in a way that allows (but does not force) the conditional variance of

[12] Davidian and Carroll (1987) considered the cases $\delta = 1, \frac{1}{2}, \frac{1}{3}, \frac{1}{4}, \frac{1}{6}$, and $\lim \delta \downarrow 0$ finding as we do that scale estimates using $\delta < 1$ are more robust in the presence of thick-tailed residuals than estimates using $\delta = 1$. Scale estimates using δ's close to 0 were sensitive to "inliers" rather than outliers.

[13] The limits for $\{\sigma_t^2\}$ in GARCH(1, 1) and AR(1) EGARCH as data-generating processes are given in Nelson (1990). The diffusion limit for $\{\sigma_t\}$ in the Taylor–Schwert model is given in Nelson (1992). We applied Ito's Lemma to convert the diffusion limits into stochastic differential equations for $\{\ln(\sigma_t^2)\}$.

$\{\ln(\sigma_t^2)\}$ and its instantaneous correlation with $\{x_t\}$ to vary with the level of σ_t^2 and ξ_t.

6.6. Jump – Diffusions and the Friedman – Laibson Model

Friedman and Laibson (1989) argue that stock movements have "ordinary" and "extraordinary" components. This motivated their modified ARCH (MARCH) model, which bounds $g(\cdot)$ to keep the "extraordinary" component from being too influential in determining $\hat{\sigma}_t^2$. The MARCH model is similar to GARCH, but with

$$\hat{\sigma}_{t+h}^2 = \omega \cdot h + \beta_h \cdot \hat{\sigma}_t^2 + h^{1/2} g(\hat{\xi}_x), \qquad \text{where} \tag{6.9}$$

$$\begin{aligned} g(\hat{\xi}_x) &\equiv \alpha > 0 \quad \text{if} \quad \gamma \cdot \hat{\xi}_x^2 \geq \pi/2, \\ &\equiv \alpha \cdot \sin\left[\gamma \cdot \hat{\xi}_x^2\right] \quad \text{if} \quad \gamma \cdot \hat{\xi}_x^2 < \pi/2. \end{aligned} \tag{6.10}$$

Though Friedman and Laibson's model is impossible to analyze using the methods of this chapter (it does not have a nontrivial diffusion limit), it can be understood as a robust filtering procedure: If ξ_x has occasional large outliers, least-squares based procedures such as GARCH will not estimate σ_t^2 efficiently. Much the same intuition comes from the near-diffusion results, in which the conditional distribution of ξ_x is allowed to be considerably thicker-tailed than the normal. In accord with Friedman and Laibson's (and Davidian and Carroll's) intuition, the thicker-tailed the conditional distribution of ξ_x, the less weight should be given to "large" observations, at least in the score component S of the optimal filter. For example, if (ξ_x, ξ_y) is conditionally bivariate t with $K > 2$ degrees of freedom, we have (in the notation of Theorem 5.2)

$$\mathrm{P} \equiv \rho(x, y, t) \Lambda(x, y, t) \xi_x / \sigma(y), \qquad \text{and} \tag{6.11}$$

$$S \equiv \frac{\sigma'(y)}{\sigma(y)} \left[\frac{K+1}{K-2} \cdot \frac{\xi_x^2 / \sigma(y)^2}{1 + \xi_x^2 / \left[(K-2)\sigma(y)^2\right]} \right]. \tag{6.12}$$

Given y, the score is bounded above by $(K+1)\sigma'(y)/\sigma(y)$. The lower the degrees of freedom, the tighter the bound. P, however, remains linear in ξ_x, as in the conditionally normal case.

Unfortunately, the near-diffusion assumption does not allow ξ_x and ξ_y to be *too* thick-tailed. In particular, ξ_x and ξ_y are assumed to have (conditionally) bounded $2 + \delta$ absolute moments, which is why we assumed $K > 2$ in the Student's t case. In the limit as $h \downarrow 0$, this effectively rules out the possibility that "lumpy" information arrival causes occasional large jumps in x_t or y_t. The assumption of a $2 + \delta$ absolute moment is also crucial for the robustness results in the chapter: For example, Engle and Ng (1993) show that even relatively "robust" ARCH models such as EGARCH may perform poorly in the presence of gross outliers. Such occasional large jumps may well be a feature of some financial time series. For example, using daily stock returns data, Nelson (1989, 1991) generated $\{\hat{\sigma}_t^2\}$ using an EGARCH model, but found occasional large outliers (i.e., more than five or six times $\hat{\sigma}_t$).[14]

It would be interesting to extend our results to allow the data to be generated by a jump–diffusion (or a near jump–diffusion). We suspect that this would lead to bounds on both P and S in the optimal $g(\cdot)$ function: If we fail to impose such a bound, $g(\hat{\xi}_x, x, y, t)$ will be enormous when a jump occurs, which may well make such jumps too influential in determining $\{\hat{y}_t\}$.

7. CONCLUSION

One widely voiced criticism of ARCH models (see, e.g., Campbell and Hentschel, 1993; Andersen, 1992) is that they are ad hoc—i.e., though they have been successful in empirical applications, they are *statistical* models, not *economic* models. This criticism, though correct, does not go far enough; Even as purely *statistical* models, ARCH models are ad hoc. In applied work, there has been considerable arbitrariness in the choice of ARCH models, despite (perhaps because of) the plethora of proposed ARCH specifications. Many models have been proposed, but few compared to the infinite potential number of ARCH models. How can we choose between these models? How do we design new models? We summarize the main implications of our results for the design of ARCH filters for near diffusions as follows.

Rule 1. Asymptotically optimal ARCH models are as "similar" to the true data-generating process as possible, in the sense that the first two conditional moments (as functions of the state variables and time) implied by the ARCH model considered as a data-generating process have the same functional form as in the true data-generating process.

The choice of an ARCH model therefore embodies an implicit assumption about the joint variability of the state variables x_t and y_t. GARCH(1, 1), AR(1) EGARCH, the Taylor–Schwert model. NARCH, and

[14] For example, the market drop on September 26, 1955 (in response to Eisenhower's heart attack) was 11 estimated conditional standard deviations. There were drops of about seven estimated conditional standard deviations on November 3, 1948, and June 26, 1950, in response to Truman's surprise re-election and the beginning of the Korean war, respectively. The Crash of October 19, 1987, was almost eight estimated conditional standard deviations, and the market rose seven estimated conditional standard deviations on July 6, 1955.

TARCH effectively assume that the conditional variance of σ_t^2 is linear in σ_t^4. Some ARCH models (e.g., GARCH, NARCH, Taylor–Schwert), effectively assume that increments in x_t and σ_t^2 are uncorrelated. EGARCH and TARCH assume a constant conditional correlation. It is probably wise to relax these constraints, since specification of the conditional second moments of $\{x_t, y_t\}$ affects the $O_p(h^{1/4})$ terms, and so is likely to be important in practice.

Rule 2. The optimality selected $g(\cdot)$ has two components. The first, $\mathrm{E}[\xi_{x,t+h} \mid \xi_{y,t+h}, x_t, y_t]$, forecasts *changes* in $\{y_t\}$, for example by taking advantage of "leverage effects." The second component of the optimal $g(\cdot)$ is proportional to the score of ξ_x, treating y as an unobserved parameter. This term estimates the *level* of $\{y_t\}$ in much the same way as a maximum likelihood estimator of a scale parameter in the i.i.d. case. The robustness results of Davidian and Carroll (1987) hold in the ARCH context: In particular, EGARCH and Taylor–Schwert are more robust than GARCH to conditionally thick-tailed ξ_x's. It is probably wise to design ARCH models to be robust to thick-tailed ξ_x's (perhaps by bounding $g(\cdot)$ as suggested by Friedman and Laibson), since conditional leptokurtosis seems to be the rule in financial applications of ARCH.

Rule 3. The asymptotic conditional mean of $[\hat{\sigma}_t^2 - \sigma_t^2]$ is zero when the drifts in $\{x_t, y_t\}$ are well-specified—i.e., when $\mu(x, y, t) = \hat{\mu}(x, y, t)$ and $\kappa(x, y, t) = \hat{\kappa}(x, y, t)$. Incorrect specification of the drifts creates an $O_p(h^{1/2}))$ effect on $[\hat{y}_t - y_t]$, but an $O_p(1)$ effect on the medium- and long-term forecasts generated by the ARCH model. If filtering rather than forecasting is of primary concern, specification of $\hat{\mu}(\cdot)$ and $\hat{\kappa}(\cdot)$ is probably less important than specification of $g(\cdot)$.

Our results could be extended in a number of interesting ways. In a sequel (see Nelson, 1993), we allow x_l and y_l to be vectors, and consider smoothing (i.e., allowing $\hat{y}_t$ to depend on leads and lags of x_t) as well as filtering. Other extensions are also possible. For example, versions of Theorem 2.1 are available which allow a jump diffusion limit for $\{x_t, y_t\}$, and which do not require $\{x_t, y_t, \hat{y}_t\}$ to be Markov (see, e.g., Jacod and Shiryaev, 1987, Ch. 9). Unfortunately, the regularity conditions are considerably more difficult to check than the coordinates of Theorem 2.1.[15]

The most important extension would be to allow the parameters of the ARCH model to be estimated by quasi-maximum likelihood methods. All this chapter's suggestions for ARCH model building rely on the conjecture that fitting misspecified parametric ARCH models by maximum likelihood selects (asymptotically as $h \downarrow 0$ and the span of calendar time goes to

[15] Foster and Nelson (1993) are able to drop the Markov assumption in analyzing rolling regressions and GARCH by using a central limit theorem for semi-martingales in place of Theorem 2.1. It isn't clear if this method can be successfully applied to broader classes of ARCH models.

infinity) the "best" available filter.[16] Monte Carlo experiments reported in Schwartz *et al.* (1993) suggest that this assumption is reasonable in practice. Unfortunately, however, formal asymptotic theory for ARCH parameter estimates has proven very difficult even for well-specified ARCH models, and is not likely to be any easier for misspecified models.

APPENDIX

Proof of Theorem 2.1. This is a modification of Stroock and Varadhan (1979, Theorem 11.2.3). The version of the result we cite is based on Ethier and Kurtz (1986, ch. 7, Corollary 4.2). We set Ethier and Kurtz's n equal to $h^{-\Delta}$. There are two changes from Ethier and Kurtz's version of the theorem: First, we define $\Omega_{\Delta,h}(y)$ as a covariance rather than as a conditional second moment. The conditional first and second moments are each $O(h^{-\Delta})$, so the difference between the conditional covariance and the conditional second moment vanishes at rate $O(h^{-2\Delta})$ as $h \downarrow 0$. Second, Ethier and Kurtz use truncated expectations in (2.2)–(2.3)—i.e.,

$$\mu_{\Delta,h}(y) \equiv h^{-\Delta}\, \mathrm{E}\big[({}_hY_{k+1} - {}_hY_k) \cdot \mathrm{I}(\|{}_hY_{k+1} - {}_hY_k\| < 1) \mid {}_hY_k = y\big], \quad \text{and} \tag{2.2'}$$

$$\Omega_{\Delta,h}(y) \equiv h^{-\Delta}\, \mathrm{Cov}\big[({}_hY_{k+1} - {}_hY_k) \cdot \mathrm{I}(\|{}_hY_{k+1} - {}_hY_k\| < 1) \mid {}_hY_k = y\big], \tag{2.3'}$$

where $\mathrm{I}(\cdot)$ is an indicator function. They then replace (d′) with the requirement that for every $\varepsilon < 0$,

$$h^{-\Delta}\mathrm{P}\big[\|{}_hY_{k+1} - {}_hY_k\| > \varepsilon \mid {}_hY_k = y\big] \to 0 \tag{d''}$$

as $h \downarrow 0$, uniformly on every bounded y set. To see that (2.2′)–(2.3′) and (d″) follow from (2.2)–(2.3) and (d′), see the proof of Nelson (1990,

[16] The basis for this conjecture is as follows. The probability measures of two diffusion processes are in general singular *unless the functional form of the conditional variances and covariances as functions of the state variables is identical in the two processes* (see, e.g., Liptser and Shiryaev, 1977, Chapter 7). The *drift* functions can differ, but the *variance* functions must be identical. Put another way, in continuous time, conditional variances and covariances completely dominate the likelihood. If this is also true for sequences of discrete time processes in the limit as continuous time is approached, then a misspecified ARCH model fit by MLE should, in the limit, select the parameters in the class of ARCH models fit which best match the conditional covariances—namely, the (constrained) optimal filter. It is also suggestive that the optimal filter matches the functional forms of the first two conditional moments to the true data-generating process.

Theorem 2.2). (Our version is a little simpler to state, but Ethier and Kurtz's is more general and is sometimes easier to verify. The moment conditions in this chapter could be weakened using Ethier and Kurtz's version.) For the uniformity of the weak convergence on bounded $\{y_0\}$ sets when ${}_hY_0$ is fixed, see Stroock and Varadhan (1979, Theorem 11.2.3).

Proof of Theorem 3.1. We employ Theorem 2.1 with $\Delta \equiv \frac{1}{2}$, treating x, y, and q as state variables and conditioning on (x_T, y_T, q_T). First, we consider the first two conditional moments of $(x_{t+h} - x_t)$, $(y_{t+h} - y_t)$, $(q_{t+h} - q_t)$, and verify that these increments vanish to zero in probability at an appropriate rate. Under (3.1)–(3.3), the first two conditional moments of $\{x_t, y_t\}$ and the $2 + \delta$ absolute moment are all $O(h)$ on bounded (x, y, t) sets. The "fast drift" assumption makes the first conditional moments $O(h^{3/4})$. This means that when we apply Theorem 2.1 using $\Delta \equiv \frac{1}{2}$, the first two conditional moments of $\{x_t, y_t\}$ converge, respectively, to a vector and matrix of zeros, since the conditional moments are $o(h^{1/2})$. By Assumption 1, the first two conditional moments of $q_{t+h} - q_t$ (normalized by $h^{-1/2}$) converge to $(A_t - B_t q_t)$ and C_t. The normalized covariances of the increments in q_t, x_t, and y_t converge to 0. This convergence is uniform on bounded (x, y, q, t) sets, as required. By Assumption 2, each element of the state vector has a bounded (uniformly on bounded (x, y, q, t) sets) $2 + \delta$ absolute moment. Since the drifts and variances of x_t and y_t are zeros in the limit, $x_{T,\tau}$ and $y_{T,\tau}$ are constant at x_T and y_T in the limit and A_t, B_t, and C_t are asymptotically constant in the diffusion limit on the fast time scale at their time T values.

We next verify that the diffusion limit has a unique weak-sense solution. Note first that the limit diffusion is clearly nonexplosive, since $x_{T,\tau}$ and $y_{T,\tau}$ are constants and $q_{T,\tau}$ follows an Ornstein–Uhlenbeck process. By Assumption 1, A_t, B_t, and C_t are twice continuously differentiable, so weak-sense uniqueness follows by Stroock and Varadhan (1979, Corollary 6.3.3 and Theorem 10.1.3). Weak convergence now follows by Theorem 2.1. (3.22) follows by the elementary properties of the Ornstein–Uhlenbeck.

Proof of Theorem 4.1. Assumption 3 allows us to solve the optimization problem as if $\xi_{x,t+h}$ and $\xi_{y,t+h}$ were conditionally bivariate normal with covariance matrix given by the last term on the right side of (4.15). The theorem now follows as a special case of Theorem 5.2 below.

Proof of Theorem 4.2. By Assumption 3 (dropping time subscripts and function arguments to ease notation), $C_T/2B_T = \mathrm{E}_T[g^2 - 2g\xi_y + \xi_y^2]/2\mathrm{E}_T[-\partial g/\partial y] = -\mathrm{E}_T[a^2 \cdot g^{*2} - 2a \cdot g^* \cdot \xi_y + \xi_y^2]/2a\,\mathrm{E}_T[\partial g^*/\partial y]$. Minimizing this with respect to a is equivalent to minimizing $(a + \Lambda^2/a)$. The first and second-order conditions then yield $a = \Lambda$. (4.10) follows.

Lemmas A.1 and A.2 are needed in the proofs of Theorem 4.3–4.5.

LEMMA A.1. *Let x_t be an $n \times 1$ diffusion, generated by*

$$x_t = x_0 + \int_0^t \mu(x_s, s)\, ds + \int_0^t \sigma(x_s, s)\, dW_s, \tag{A.1}$$

where $\mu(\cdot)$ is continuous and $n \times 1$, $\sigma(\cdot)$ is continuous and $n \times n$, and W_t is an n-dimensional standard Brownian motion. Then as $t \downarrow 0$, the limiting distribution of $t^{-1/2}(x_t - x_0)$ is $\mathrm{N}(0_{n\times 1}, \sigma(x_0,0)\sigma(x_0,0)')$. This still holds if the $\int_0^t \mu(x_s, s)\, ds$ in (A.1) is replaced by $t^{-1/4} \int_0^t \mu(x_s, s)\, ds$.

Proof of Lemma A.1. For every t and every τ, $0 \le \tau \le t$, define $W_{t,\tau} \equiv t^{-1/2} W_{\tau t}$ and $x_{t,\tau} \equiv t^{-1/2}(x_{t\tau} - x_0)$. For every t, $W_{t,\tau}$ is a standard Brownian motion on $\tau \in [0,1]$. We now rewrite (A.1) as

$$x_{t,\tau} = t^{1/2} \int_0^1 \mu\left(x_0 + t^{1/2} x_{t,s}, t \cdot s\right) ds + \int_0^1 \sigma\left(x_0 + t^{1/2} x_{t,s}, t \cdot s\right) dW_{t,s}. \tag{A.1$'$}$$

As $t \downarrow 0$, $t^{1/2}\mu(x_0 + t^{1/2}x_{t,s}, t \cdot s) \to 0$ and $\sigma(x_0 + t^{1/2}x_{t,s}, t \cdot s) \to \sigma(x_0, 0)$ uniformly on bounded $(x_0, x_{t,s})$ sets. Applying Stroock and Varadhan (1979, Theorem II.1.4), $x_{t,\tau}$ converges weakly to a Brownian motion with no drift and with diffusion matrix $\sigma(x_0,0)\sigma(x_0,0)'$. In the case in which $t^{-1/4}\int_0^t \mu(x_s, s)\, ds$ replaces $\int_0^t \mu(x_s, s)\, ds$ in (A.1), $t^{1/4}\int_0^1 \mu(x_0 + t^{1/2}x_{t,s}, t \cdot s)\, ds$ replaces $t^{1/2}\int_0^t \mu \times (x_0 + t^{1/2}x_{t,s}, t \cdot s)\, ds$ in (A.1$'$), and the convergence is unaffected. The lemma follows.

LEMMA A.2. *Let x_t be as in Lemma A.1. For every $p \ge 2$ there is a nonnegative, finite k such that for every x_0 and every $i = 1$ to n*

$$\begin{aligned}
&\left(\mathrm{E}_0 |t^{-1/2}(x_{i,t} - x_{i,0})|^p\right)^{1/p} \\
&\quad \le \left(t^{-1+p/2} \int_0^t \mathrm{E}_0 |\, \mu_i(x_s, s)|^p\, ds\right)^{1/p} \\
&\qquad + k \cdot \sum_{j=1,n} \left(t^{-1} \int_0^t \mathrm{E}_0 |\sigma_{i,j}(x_s, s)|^p\, ds\right)^{1/p},
\end{aligned} \tag{A.2}$$

where $x_{i,t}$ is the ith element of x_t, $\sigma_{i,j,t}$ is the i–jth element of $\sigma(x_t, t)$, and k is a constant depending only on p. When $t^{-1/4}\int_0^t \mu(x_s, s)\, ds$ replaces $\int_0^t \mu(x_s, s)\, ds$ in (A.1), $t^{-1+p/4}\int_0^t \mathrm{E}_0 |\, \mu_i(x_s, s)|^p\, ds$ replaces $t^{-1+p/2}\int_0^t \mathrm{E}_0 |\, \mu_i(x_s, s)|^p\, ds$ in (A.2).

Proof of Lemma A.2. From (A.1),

$$\left|t^{-1/2}(x_{i,t}-x_{i,0})\right|$$

$$=\left|t^{-1/2}\int_0^t \mu_i(x_s,s)\,ds+\sum_{j=1,n} t^{-1/2}\int_0^t \sigma_{i,j}(x_s,s)\,dW_{j,s}\right|$$

$$\leq t^{-1/2}\int_0^t |\mu_i(x_s,s)|\,ds+\sum_{j=1,n} t^{-1/2}\left|\int_0^t \sigma_{i,j}(x_s,s)\,dW_{j,s}\right|. \quad \text{(A.3)}$$

Applying Minkowski's inequality:

$$\left[\mathrm{E}_0\left|t^{-1/2}(x_{i,t}-x_{i,0})\right|^p\right]^{1/p}\leq\left[\mathrm{E}_0\left(t^{-1/2}\int_0^t|\mu_i(x_x,s)|\,ds\right)^p\right]^{1/p}$$

$$+\sum_{j=1,n}\left[\mathrm{E}_0\left(t^{-1/2}\left|\int_0^t\sigma_{i,j}(x_s,s)\,dW_{j,s}\right|\right)^p\right]^{1/p} \quad \text{(A.4)}$$

By the integral version of the means inequality (Hardy, *et al.*, (1951, Theorem 192) $\int_0^t|\mu_i(x_s,s)|\,ds\leq t^{(p-1)/p}(\int_0^t|\mu_i(x_s,s)|^p\,ds)^{1/p}$. By Karatzas and Shreve (1988, Exercise 3.25), $\mathrm{E}_0(|t^{-1/2}\int_0^t\sigma_{i,j}(x_s,s)\,dW_{j,s}|)^p\leq mt^{-1}\,\mathrm{E}_0\int_0^t|\sigma_{i,j}(x_s,s)|^p\,ds$, where m is a constant depending only on n and p. Substituting both these inequalities into (A.3) yields (A.2). Substituting $t^{-1/4}\mu_i(x_s,s)$ for $\mu_i(x_s,s)$ yields the remainder of the lemma.

Proof of Theorem 4.3. That (4.16)–(4.17) has a unique weak-sense solution was established in Nelson (1990). Next, note that $\xi_x=\hat{\xi}_x$, since $\hat{\mu}-\mu$. Employing the definitions of y, $\hat{y}$, ξ_x, ξ_y, and q, and using the "fast drift" convention (3.1′), we have

$$\xi_{x,t+h}=h^{-1/2}\int_t^{t+h}y_s^{1/2}\,dW_{1,t}, \quad \text{(A.5)}$$

$$\xi_{y,t+h}=\theta h^{-3/4}\int_t^{t+h}(\mathrm{E}_t y_s-y_s)\,ds+h^{-1/2}\int_t^{t+h}y_s\alpha 2^{1/2}\,dW_{2,t}, \quad \text{(A.6)}$$

$$q_{t+h}-q_t=h^{1/4}\left[\alpha\left(\xi^2_{x,t+h}-y_t-h^{1/4}q_t\right)-\xi_{y,t+h}\right]$$

$$-\theta h^{-1/2}\int_t^{t+h}\left(y_t-\mathrm{E}_t y_s+h^{1/4}q_t\right)ds, \quad \text{and} \quad \text{(A.7)}$$

$$\mathrm{E}_t[y_s]=\omega/\theta+(y_t-\theta/\omega)\left[\exp(-\theta h^{-1/4}(s-t))-1\right]. \quad \text{(A.8)}$$

By Lemma A.1, $(\xi_{x,t+h},\xi_{y,t+h})$ given time t information, converges in distribution to a bivariate normal with means of zero, no correlation, and variances y_t and $2\alpha y_t^2$. To verify Assumptions 1–3, however, we need convergence of moments (up to order $4 + \delta$ for ξ_x and $2 + \delta$ for ξ_y) as well. These moments (conditional on $(x_t = x, y_t = y, q_t = q)$) must be uniformly bounded on every bounded (x, y, q, t) set.

If there is a suitably bounded $2 + \delta$ moment for $|\xi_y|$, a bounded $4 + \delta$ moment for $|\xi_x|$ follows using Lemma A.2. For each $h > 0$, the $\{\sigma_t^2\}$ process (i.e., without x_t) satisfies standard Lipshitz and growth conditions and consequently (see Arnold, 1973, section 7.1) for every $\delta > 0$

$$\mathrm{E}_t\left[\sigma_{t+h}^{2+\delta}\right] \leq \left(1 + \sigma_t^{2+\delta}\right)\exp(C \cdot h^{1/2}) \tag{A.9}$$

where C is a constant depending on α, θ, and ω. Moment boundedness for $|\xi_{x,t+h}|^{4+\delta}$ now follows by Lemma A.2, satisfying Assumption 2 and (4.5).

Applying Lemma A.2 and (A.8) we now have

$$h^{-1/2}\,\mathrm{E}[q_{t+h} - q_t \mid x_t = x, y_t = y, q_t = q] \to -\alpha q, \tag{A.10}$$

$$h^{-1/2}\,\mathrm{Var}[q_{t+h} - q_t \mid x_t = x, y_t = y, q_t = q]$$

$$\to \mathrm{Var}\left[\alpha\left(\varepsilon_x^2 - y\right) - \varepsilon_y\right] = 4\alpha^2 y^2, \quad \text{and} \tag{A.11}$$

$$h^{-1/2}\,\mathrm{E}\left[|q_{t+h} - q_t|^3 \mid x_t = x, y_t = y, q_t = q\right] \to 0 \tag{A.12}$$

uniformly on bounded (x, y, q, t) sets as required, satisfying (3.13)–(3.16) and (4.2)–(4.4).

Proof of Theorem 4.4. That the system (4.20′), (4.21′), and (4.22) has a unique weak-sense solution is established in Nelson (1990). Under the "fast drift" convention (3.1′),

$$\xi_{x,t+h} = -h^{-3/4}(1/2)\int_t^{t+h}(\exp(y_s) - \mathrm{E}_t\exp(y_s))\,ds + h^{-1/2}\int_t^{t+h}\exp(y_s/2)\,dW_{1,s}, \tag{A.13}$$

$$\hat{\xi}_{x,t+h} = \xi_{x,t+h} + h^{-3/4}(1/2)\int_t^{t+h}\left(\mathrm{E}_t[\exp(y_s)] - \exp\left(y_t + h^{1/4}q_t\right)\right)ds, \tag{A.14}$$

$$\xi_{y,t+h} = -\beta \cdot h^{-3/4}\int_t^{t+h}(y_s - \mathrm{E}_t y_s)\,ds + h^{-1/2}\int_t^{t+h}\psi\,dW_{2,s}, \quad \text{and} \tag{A.15}$$

$$q_{t+h} - q_t = -h^{-1/2}\beta\int_t^{t+h}\left(y_t + h^{1/4}q_t - \mathrm{E}_t[y_s]\right)ds + h^{1/4}\left[g\left(\hat{\xi}_{x,t+h}, \hat{y}_t\right) - \xi_{y,t+h}\right] \tag{A.16}$$

where

$$g(\hat{\xi}, \hat{y}) \equiv \psi\left[\rho\hat{\xi}\cdot\exp(-\hat{y}/2) + \left[(1-\rho^2)/2\right]^{1/2}\cdot\left(\hat{\xi}^2\cdot\exp(-\hat{y}) - 1\right)\right]. \tag{A.17}$$

By Lemma A.1, given time t information $(\xi_{x,t+h}, \xi_{y,t+h})$ converges in distribution to a bivariate normal with means of zero, correlation ρ, and variances of $\exp(y_t)$ and ψ^2. The same is true of $(\hat{\xi}_{x,t+h}, \xi_{y,t+h})$. We also require convergence of moments up to order $4+\delta$ for $|\hat{\xi}_{x,t+h}|$ and $2+\delta$ for $|\xi_{y,t+h}|$. y_s is Gaussian (see, e.g., Arnold, 1973, Section 8.3), and for $s > t$, $y_s \mid y_t \sim$ normal with mean and variance

$$(\alpha + (y_t - \alpha))\exp(h^{-1/4}\beta(s-t))$$

and

$$\psi^2\left[(1-\exp(-2\beta h^{-1/4}(s-t)))/(2\beta h^{-1/4})\right].$$

This allows us to compute $\mathrm{E}_t[\exp(y_s)]$ and $\mathrm{E}_t[y_s]$ explicitly. For s with $t \le s \le t+h$, the conditional moments of arbitrary order of both y_s and $\exp(y_s)$ remain uniformly bounded on bounded y_t sets. (4.5) and Assumption 2 are therefore satisfied.

Substituting $(y_t + h^{1/4}q_t)$ for $\hat{y}_t$ in (A.16)–(A.17) and using the formulae for $\mathrm{E}_t[\exp(y_s)]$ and $\mathrm{E}_t[y_s]$ leads to

$$h^{-1/2}\,\mathrm{E}[q_{t+h} - q_t \mid x_t = x, y_t = y, q_t = q] \to -q\cdot\psi\left[(1-\rho^2)/2\right]^{1/2}, \tag{A.18}$$

and

$$h^{-1/2}\,\mathrm{Var}[q_{t+h} - q_t \mid x_t = x, y_t = y, q_t = q] \to 2\psi^2(1-\rho^2), \tag{A.19}$$

$$h^{-1/2}\,\mathrm{E}\left[|q_{t+h} - q_t|^{2+|}\,\S x_t = x, y_t = y, q_t = q\right] \to 0 \tag{A.20}$$

uniformly on bounded (x, y, q, t) sets, satisfying Assumption 1 and (4.2)–(4.4).

Proof of Theorem 4.5. To establish existence and uniqueness of a weak-sense solution to the system (4.22), (4.20′), and (4.23′) we first consider the system (4.20′) and (4.22)–(4.23). For the latter system, we apply Nelson (1990, Theorem A.1). Condition A of that theorem is clearly satisfied. For its nonexplosion condition, use $\varphi(x, y) \equiv 1 + x^2 + \sigma^4$. We conclude that (4.20′) and (4.22)–(4.23) have a unique weak-sense solution. When $\beta\alpha > \psi^2/2$ (or, in the fast drift case, whenever $\beta > 0$ and $\alpha > 0$) $\sigma_t^2 = 0$ is inaccessible (with probability one) in finite time, so the mapping from $\{x_t, \sigma_t^2\}$ to $\{x_t, \ln(\sigma_t^2)\}$ is almost surely uniformly continuous on $[0, T]$ for all $T < \infty$. The continuous mapping theorem then delivers a unique weak-sense solution for (4.22), (4.20′), and (4.23′).

Convergence in distribution of $(\hat{\xi}_{x,t+h}, \xi_{y,t+h})$ (and of $(\xi_{x,t+h}, \xi_{y,t+h})$) given time t information to a bivariate normal with mean $(0,0)'$, correlation ρ and variances $\exp(y_t)$ and $\psi^2 \exp(y_t)$ follows from Lemma A.1. We next check local boundedness of the moments. The conditional distribution of σ_s^2 given $\sigma_t^2 (s > t)$ is given by Cox *et al.* (1985, pp. 391–392). Using a formula for the noncentral chi-square distribution (see Johnson and Kotz, 1970, ch. 28 (1)) and the integral form of the Gamma function we obtain, for $v > -a$ and $s > t$,

$$\mathrm{E}_t\left[\sigma_s^{2a}\right] = c^{-a} \exp\left(-c \cdot \sigma_t^2 \cdot \mathrm{e}^{-\beta(s-t)}\right) \times \sum_{j=0}^{\infty} \frac{\left(c \cdot \sigma_t^2 \cdot \mathrm{e}^{-\beta(s-t)}\right)^j \Gamma(v+j+a)}{\Gamma(j+1) \cdot \Gamma(v+j)}, \tag{A.21}$$

where $v \equiv 2\beta h^{-1/4}\alpha/\psi^2$ and $c \equiv 2\beta h^{-1/4}/[\psi^2(1 - \exp(-\beta h^{-1/4}(s - t))]$ in the fast drift case and $v \equiv 2\beta\alpha/\psi^2$ and $c \equiv 2\beta/[\psi^2)t - \exp(-\beta(s - t))]$ otherwise. This can be rewritten as

$$\begin{aligned} \mathrm{E}_t\left[\sigma_s^{2a}\right] &= \frac{\Gamma(v+a)}{c^a \Gamma(v)} \exp\left(-c \cdot \sigma_t^2 \cdot \mathrm{e}^{-\beta(s-t)}\right) \\ &\quad \cdot M\left(a + v, v, c \cdot \sigma_t^2 \cdot \mathrm{e}^{-\beta(s-t)}\right), \\ &= \sigma_t^{2a}\left[1 + \sigma_t^{-2} O\left(|c|^{-1}\right)\right] \quad \text{as} \quad c \to 0, \end{aligned} \tag{A.22}$$

where $M(\cdot, \cdot, \cdot)$ is a confluent hypergeometric function. The last equality in (A.22) follows from Slater (1965, 13.1.4). Since $t \le s \le t + h$ for the relevant moments, $1 - \exp(-\beta(s - t)) = O(h)$ and $c^{-1} = O(s - t)$ as $h \downarrow 0$. $\mathrm{E}[\sigma_s^{2a} \mid \ln(\sigma_t^2) = y] - \sigma_t^{2a} \to 0$ uniformly on bounded y sets as $h \downarrow 0$, provided $v + a > 0$.

To bound the $4 + \delta$ conditional absolute moment of ξ_x, set $a = 2 + \delta/2$. To bound the $2 + \delta$ conditional absolute moment of ξ_y, set $a = -1 - \delta/2$. In the fast drift case, these moments are finite for sufficiently small h whenever $\alpha > 0$ and $\beta > 0$. Otherwise, we require $2\beta\alpha > \psi^2$. This satisfies Assumption 2 and (4.5). We now have, for the fast drift case (the standard case is similar),

$$\begin{aligned} \xi_{x,t+h} &= -h^{-3/4}(1/2)\int_t^{t+h} (\exp(y_s) - \mathrm{E}_t \exp(y_s))\, ds \\ &\quad + h^{-1/2} \int_t^{t+h} \exp(y_s/2)\, dW_{1,s}, \end{aligned} \tag{A.23}$$

$$\begin{aligned} \hat{\xi}_{x,t+h} &= \xi_{x,t+h} + h^{-3/4}(1/2) \\ &\quad \times \int_t^{t+h} \left(\mathrm{E}_t[\exp(y_s)] - \exp\left(y_t + h^{1/4} q_t\right)\right) ds, \end{aligned} \tag{A.24}$$

$$\xi_{y,t+h} = h^{-3/4}(\alpha\beta - \psi^2/2)\int_t^{t+h}(\exp(-y_s) - \mathrm{E}_t\exp(-y_s))\,ds$$

$$+ \psi h^{-1/2}\int_t^{t+h}\exp(-y_s/2)\,dW_{2,s}, \qquad \text{and} \qquad (A.25)$$

$$q_{t+h} - q_t = h^{-1/2}(\alpha\beta - \psi^2/2)\int_t^{t+h}(\exp(-\hat{y}_t) - \mathrm{E}_t[\exp(-y_s)])\,ds$$

$$+ h^{1/4}\left[g\left(\hat{\xi}_{x,t+h}, \hat{y}_t\right) - \xi_{y,t+h}\right], \qquad \text{where} \qquad (A.26)$$

$$g(\hat{\xi}, \hat{y}) \equiv \psi\cdot\exp(-\hat{y}/2)\Big[\rho\hat{\xi}\cdot\exp(-\hat{y}/2)$$

$$+\left[(1-\rho^2)/2\right]^{1/2}\cdot\left(\hat{\xi}^2\cdot\exp(-\hat{y}) - 1\right)\Big]. \qquad (A.27)$$

Applying Lemmas A.1–A.2, substituting $(y_t + h^{1/4}q_t)$ for $\hat{y}_t$ in (A.26)–(A.27), and using (A.22) leads to

$$h^{-1/2}\,\mathrm{E}[q_{t+h} - q_t \mid x_t = x, y_t = y, q_t = q]$$

$$\to -q\cdot\exp(-y/2)\psi\left[(1-\rho^2)/2\right]^{1/2}, \qquad (A.28)$$

$$h^{-1/2}\,\mathrm{Var}[q_{t+h} - q_t \mid x_t = x, y_t = y, q_t = q]$$

$$\to 2\psi^2(1-\rho^2)\exp(-y), \qquad \text{and} \qquad (A.29)$$

$$h^{-1/2}\,\mathrm{E}\left[|q_{t+h} - q_t|^{2+\delta} | x_t = x, y_t = y, q_t = q\right] \to 0 \qquad (A.30)$$

uniformly on bounded (x, y, q, t) sets, satisfying Assumption 1 and (4.2)–(4.4). Finally, apply the delta method to derive the asymptotic variance of $[\hat{\sigma}_t^2 - \sigma_t^2]$.

Proof of Theorem 5.1. Nearly identical to the proof of Theorem 3.1.

Before we prove Theorem 5.2, we present a heuristic derivation of the first-order condition. In the proof we verify global optimality.

Under Assumption 4 we may write $C_t/2B_t$ as

$$\frac{\int_{-\infty}^{\infty}\int_{-\infty}^{\infty}\left[g(\xi_x, x, y, t) - \xi_y\right]^2 f(\xi_x, \xi_y \mid x, y, t)\,d\xi_x\,d\xi_y}{2\int_{-\infty}^{\infty} g(\xi_x, x, y, t)\cdot S(\xi_x, x, y, t)\cdot f(\xi_x \mid x, y, t)\,d\xi_x}. \qquad (A.31)$$

We wish to minimize this with respect to $g(\cdot)\cdot$, subject to two constraints: first that $\mathrm{E}_t[g] = 0$, and second, that the denominator of (A.31) is nonnegative. For now we ignore these constraints since they are not binding at the solution (5.10)–(5.11). To derive the first-order conditions, we treat $g(\xi_x^*, x, y, t)$ for each (ξ_x^*, x, y, t) as a separate choice variable. Setting the partial derivative of (A.31) with respect to $g(\xi_x^*, x, y, t)$ equal to zero, dividing by $f(\xi_x^* \mid x, y, t)$ and multiplying by $\mathrm{Cov}_t[g \cdot S]$ yields

$$g(\xi_x^*, x, y, t) = \int_{-\infty}^{\infty} \xi_y f(\xi_y \mid \xi_x^*, x, y, t)\, d\xi_y$$

$$+ \frac{\int_{-\infty}^{\infty}\int_{-\infty}^{\infty}[g(\xi_x, x, y, t) - \xi_y]^2 f(\xi_x, \xi_y \mid x, y, t)\, d\xi_x, d\xi_y}{2\int_{-\infty}^{\infty}[g(\xi_x, x, y, t) \cdot S(\xi_x, x, y, t)] f(\xi_x \mid x, y, t)\, d\xi_x}$$

$$\cdot S(\xi_x^*, x, y, t) \tag{A.32}$$

$$= \mathrm{P}(\xi_x^*, x, y, t) + \omega(x, y, t) \cdot S(\xi_x^*, x, y, t) + \mathrm{P} + \omega S \tag{A.33}$$

for some function $\omega(x, y, t)$. Comparing (A.32)–(A.33) with (A.31), it is clear that $C_T/2B_T = \omega(x_T, y_T, T)$. Substituting $\mathrm{P} + \omega S$ for g in (A.33) and solving for ω leads to a quadratic in ω with two solutions:

$$\omega = \frac{\pm\left[\mathrm{Cov}_t(S, \mathrm{P})^2 + (\Lambda^2 - \mathrm{Var}_t(\mathrm{P})) \cdot \mathrm{Var}_t(S)\right]^{1/2} - \mathrm{Cov}_t(\mathrm{P}, S)}{\mathrm{Var}_t(S)} \tag{A.34}$$

The + solution is the only solution satisfying the constraint

$$0 \leq \int_{-\infty}^{\infty} g(\xi_x, x, y, t) \cdot S(\xi_x, x, y, t) \cdot f(\xi_x \mid x, y, t)\, d\xi_x,$$

leading to (5.10)–(5.11).

Proof of Theorem 5.2. Next, we verify that this $g(\cdot)$ is globally optimal. Dropping subscripts, we write this $g(\cdot)$ as $g \equiv \mathrm{P} + \omega S$. Now consider a perturbation of this function, $\tilde{g} \equiv \mathrm{P} + \omega S + H$, where H is a function of ξ_x, x, y, and t with $\mathrm{E}_t[H] = 0$ and $\mathrm{Cov}_t[\tilde{g}, S] > 0$ (these conditions force $\tilde{g}$ to obey the constraints $\mathrm{E}_t[\tilde{g}] = 0$ and $B_T > 0$). Our claim is that the asymptotic variance of $\tilde{g}$ is strictly higher than that of g unless $H = 0$ with probability one, or equivalently

$$\frac{\mathrm{E}_t\left[(\mathrm{P} + \omega S - \xi_y)^2\right]}{2 \cdot \mathrm{Cov}_t[\mathrm{P} + \omega S, S]} < \frac{\mathrm{E}_t\left[(\mathrm{P} + \omega S - \xi_y)^2\right] + \mathrm{E}_t[H^2] + 2 \cdot \mathrm{Cov}_t[H, \mathrm{P} + \omega S - \xi_y]}{2 \cdot \mathrm{Cov}_t[\mathrm{P} + \omega S, S] + 2 \cdot \mathrm{Cov}_t[H, S]} \tag{A.35}$$

for all such H. Recall that

$$\omega = C_t/2B_t = \mathrm{E}_t\left[(\mathrm{P} + \omega S - \xi_y)^2\right]_t \Big/ [2 \cdot \mathrm{Cov}_t(\mathrm{P} + \omega S, S)],$$

so the left side of (A.35) equals ω, and the $\mathrm{E}_t[(\mathrm{P} + \omega S - \xi_y)^2]$ term on the right of (A.35) equals $\omega \cdot [2 \cdot \mathrm{Cov}_t(\mathrm{P} + \omega S, S)]$. Making these substitutions, and using the positivity of both denominators in (A.35), (A.35) becomes

$$0 < \mathrm{E}_t[H^2] - 2 \cdot \left(\mathrm{Cov}_t[H, \xi_y] - \mathrm{Cov}_t[H, \mathrm{P}]\right). \qquad \text{(A.36)}$$

Recall that H is a function of ξ_x but not of ξ_y, and that $\mathrm{P} \equiv \mathrm{E}_t[\xi_y \mid \xi_x]$, so

$$\begin{aligned}
\mathrm{Cov}_t[H, \xi_y] &= \int_{-\infty}^{\infty} \int_{-\infty}^{\infty} H(\xi_x, x, y, t)\,\xi_y f(\xi_x, \xi_y \mid x, y, t)\, d\xi_x\, d\xi_y \\
&= \int_{-\infty}^{\infty} H(\xi_x, x, y, t) \left[\int_{-\infty}^{\infty} \xi_y f(\xi_y \mid \xi_x, x, y, t)\, d\xi_y \right] \\
&\quad \times f(\xi_x \mid x, y, t)\, d\xi_x \\
&= \int_{-\infty}^{\infty} H(\xi_x, x, y, t) \mathrm{E}[\xi_y \mid \xi_x, x, y, t] f(\xi_x \mid x, y, t)\, d\xi_x \\
&= \mathrm{Cov}_t[H, \mathrm{P}]. \qquad \text{(A.37)}
\end{aligned}$$

(A.35) is therefore equivalent to $0 < \mathrm{E}_t[H^2]$. The asymptotic variance of $h^{-1/4}[\sigma(\hat{y})^2 - \sigma(y)^2]$ follows by the delta method.

Proof of Theorem 5.2. Nearly identical to the proof of Theorem 4.2.

ACKNOWLEDGMENTS

This chapter is a revision of the working paper "Estimating Conditional Variances with Univariate ARCH Models: Asymptotic Theory." We thank two referees, a coeditor, Phillip Braun, John Cochrane, Enrique Sentana, reviewers for the NSF, and seminar participants at the University of Chicago, Harvard/M.I.T., Michigan State, Minnesota, the SAS Institute, Washington University, Wharton, Yale, the 1991 Midwest Econometrics Group meetings, the 1991 NBER Summer Institute, and the 1992 ASA Meetings for helpful comments. This material is based on work supported by the National Science Foundation under Grant #SES-9110131. We thank the Center for Research in Security Prices, the University of Chicago Graduate School of Business, and the William S. Fishman Research Scholarship for additional research support, and Boaz Schwartz for research assistance.

REFERENCES

Andersen, T. G. (1992). "A Model of Return Volatility and Trading Volume," Working Paper. Northwestern University, Evanston, IL.

Anderson, B. D. O., and Moore, J. B. (1979). "Optimal Filtering." Prentice-Hall, Englewood Cliffs, NJ.

Arnold, L. (1973). "Stochastic Differential Equations: Theory and Applications." Wiley, New York.

Baillie, R. T., and Bollerslev, T. (1989). The message in daily exchange rates: A conditional variance tale. *Journal of Business and Economic Statistics* **7**, 297–305.

Baillie, R. T., and DeGennaro, R. P. (1990). Stock returns and volatility. *Journal of Financial and Quantitative Analysis* **25**, 203–214.

Bates, D., and Pennacchi, G. (1990). "Estimating a Heteroskedastic State Space Model of Asset Prices," Working Paper. University of Illinois at Urbana/Champaign.

Black, F. (1976). Studies of stock market volatility changes. *Proceedings of the American Statistical Association, Business and Economic Statistics Section*, pp. 177–181.

Bollerslev, T. (1986). Generalized autoregressive conditional heteroskedasticity. *Journal of Econometrics* **31**, 307–327.

Bollerslev, T., Chou, R. Y., and Kroner, K. (1992). ARCH modeling in finance: A review of the theory and empirical evidence. *Journal of Econometrics* **52**, 5–60.

Campbell, J. Y., and Hentschel, L. (1993). No news is good news: An asymmetric model of changing volatility in stock returns. *Journal of Financial Economics* **31**, 281–318.

Cox, J. C., Ingersoll, J. E., Jr., and Ross, S. A., (1985). A theory of the term structure of interest rates. *Econometrica* **53**, 385–407.

Davidian, M., and Carroll, R. J. (1987). Variance function estimation. *Journal of the American Statistical Association* **82**, 1079–1091.

Engle, R. F. (1982). Autoregressive conditional heteroskedasticity with estimates of the variance of United Kingdom inflation. *Econometrica* **50**, 987–1008.

Engle, R. F., and Ng, V. K. (1993). Measuring and testing the impact of news on volatility. *Journal of Finance* **48**, 1749–1778.

Ethier, S. N., and Kurtz, T. G. (1986). "Markov Processes: Characterization and Convergence." Wiley, New York.

Foster, D. P., and Nelson, D. B. (1993). "Rolling Regressions," Working Paper. University of Chicago, Chicago.

Friedman, B. M., and Laibson, D. I. (1989). Economic implications of extraordinary movements in stock prices. *Brookings Papers on Economic Activity* **2**, 137–172.

Gennotte, G., and Marsh, T. A. (1993). Variations in economic uncertainty and risk premiums on capital assets. *European Economic Review* **37**, 1021–1041.

Hardy, G., Littlewood, J. E., and Pólya, G. (1951). "Inequalities," 2nd ed. Cambridge University Press, Cambridge, UK.

Harvey, A. C. (1981). "The Econometric Analysis of Time Series." Phillip Allan, Oxford.

Helland, I. S. (1982). Central limit theorems for martingales with discrete or continuous time, *Scandinavian Journal of Statistics* **9**, 79–94.

Heston, S. L. (1993). A closed form solution for optics with stochastic volatility with application to bond and currency options. *Review of Financial Studies* **6**, 327–344.

Higgins, M. L., and Bera, A. K. (1992). A class of nonlinear ARCH models. *International Economic Review* **33**, 137–158.

Hsieh, D. A. (1989). Modeling heteroskedasticity in daily foreign exchange rates. *Journal of Business and Economic Statistics* **7**, 307–317.

Hull, J., and White, A. (1987). The pricing of options on assets with stochastic volatilities. *Journal of Finance* **42**, 281–300.

Jacod, J., and Shiryaev, A. N. (1987). "Limit Theorems for Stochastic Processes." Springer-Verlag, Berlin.

Jacquier, E., Polson, N. G., and Rossi, P. E. (1992). "Bayesian Analysis of Stochastic Volatility Models," Working Paper. Cornell University, Ithaca, NY.

Johnson, N. L., and Kotz, S. (1970). "Continuous Univariate Distributions," Vol. 2. Wiley, New York.

Karatzas, I., and Shreve, S. E. (1988). "Brownian Motion and Stochastic Calculus." Springer-Verlag, New York.

Kitagawa, G. (1987). Non-Gaussian state space modeling of nonstationary time series [with discussion]. *Journal of the American Statistical Association* **82**, 1032–1063.

Liptser, R. S., and Shiryaev, A. N. (1977). "Statistics of Random Processes," Vol. 1. Springer-Verlag, New York.

Maybeck, P. S. (1982). "Stochastic Models, Estimation, and Control," Vol. 2. Academic Press, New York.

Melino, A., and Turnbull, S. (1990). Pricing foreign currency options with stochastic volatility. *Journal of Econometrics* **45**, 239–266.

Merton, R. C. (1980). On estimating the expected return on the market. *Journal of Financial Economics* **8**, 323–361.

Nelson, D. B. (1989). Modeling stock market volatility changes. *Proceedings of the American Statistical Association, Business and Economic Statistics Section*, pp. 93–98.

Nelson, D. B. (1990). ARCH models as diffusion approximations. *Journal of Econometrics* **45**, pp. 7–39.

Nelson, D. B. (1991). Conditional heteroskedasticity in asset returns: A new approach. *Econometrica* **59**, 347–370.

Nelson, D. B. (1992). Filtering and forecasting with misspecified ARCH models I: Getting the right variance with the wrong model. *Journal of Econometrics* **52**, 61–90.

Nelson, B. D. (1993). "Asymptotic Filtering and Smoothing Theory for Multivariate ARCH Models," Working Paper. University of Chicago, Chicago.

Nelson, D. B., and Foster, D. P. (1991). "Filtering and Forecasting with Misspecified ARCH Models II: Making the Right Forecast with the Wrong Model," Working Paper. University of Chicago, Chicago.

Pagan, A. R., and Schwert, G. W. (1990). Alternative models for conditional stock volatility. *Journal of Econometrics* **45**, 267–290.

Phillips, P. C. B. (1988). Regression theory for near-integrated time series. *Econometrica* **56**, 1021–1044.

Schwartz, B. A., Nelson, D. B., and Foster, D. P. (1993). "Variance Filtering with ARCH Models: A Monte Carlo Investigation," Working Paper. University of Chicago, Chicago.

Schwert, G. W. (1989). Why does stock market volatility change over time? *Journal of Finance* **44**, 1115–1154.

Scott, L. O. (1987). Option pricing when the variance changes randomly: Theory, estimation, and an application. *Journal of Financial and Quantitative Analysis* **22**, 419–438.

Sentana, E. (1992). "Quadratic ARCH Models: A Potential Reinterpretation of ARCH Models as Second Order Taylor Approximations," Working Paper. London School of Economics.

Slater, L. J. (1965). Confluent hypergeometric functions. *In* "Handbook of Mathematical Functions" (M. Abramowitz and I. Stegun, eds.), Chapter 13. Dover, New York.

Stroock, D. W., and Varadhan, S. R. S. (1979). "Multidimensional Diffusion Processes." Springer-Verlag, Berlin.

Taylor, S. (1986). "Modeling Financial Time Series." Wiley, New York.

Wiggins, J. B. (1987). Option values under stochastic volatility: Theory and empirical estimates. *Journal of Financial Economics* **19**, 351–372.

Zakoian, J. M. (1990). "Threshold Heteroskedastic Models," Working Paper. INSEE.

9

ASYMPTOTIC FILTERING THEORY FOR MULTIVARIATE ARCH MODELS*

DANIEL B. NELSON

Graduate School of Business
University of Chicago
Chicago, Illinois 60637

I. INTRODUCTION: ARCH AND STOCHASTIC VOLATILITY MODELS

It is widely understood that conditional moments of asset returns are time varying. Understanding this variation is crucial in all areas of asset pricing theory. Accordingly, an enormous literature has developed on estimating conditional variances and covariances in returns data. Many techniques have been employed for this purpose, but the most widely used are the class of ARCH (autoregressive conditionally heteroskedastic) models introduced by Engle (1982a) and expanded in many ways since (see the survey papers of Bollerslev *et al.*, 1992, 1994; Bera and Higgins, 1993). A popular alternative to ARCH models are stochastic volatility models (see Nelson, 1988; Harvey *et al.*, 1994; Watanabe, 1992; Taylor, 1994; Jacquier *et al.*, 1994).

* Reprinted with permission from Nelson, D. "Asymptotic Filtering Theory for Multivariate ARCH Models." *Journal of Econometrics*, **71** (1996), Elsevier Science SA, Lausanne, Switzerland.

In both ARCH and stochastic volatility models, we begin with an n-dimensional observable vector process we call X_t. In a typical application, X_t might consist of returns on a set of assets or a set of interest rates. X_t might also include state variables useful for predicting volatility such as implied volatilities calculated from options prices (e.g., Chiras and Manaster, 1978; Day and Lewis, 1992), trading volume (e.g., Karpoff, 1987; Gallant *et al.*, 1992), high–low spreads (e.g., Parkinson, 1980; Garman and Klass, 1980; Wiggins, 1991), or interest rates (e.g., Christie, 1982; Glosten *et al.*, 1993).

Associated with X_t is an m-dimensional process, Y_t, consisting of state variables that are *not directly observable*. Together, $(X_t, Y_t\}$ form a Markov process. For example, Y_t might include the unique elements of the conditional covariance matrix of X_{t+1}, or it may parameterize the conditional distribution of X_{t+1} more parsimoniously, as in a factor ARCH model. Y_t may also include higher order moments of X_{t+1}, such as conditional skewness or kurtosis.

In a *stochastic volatility model*, Y_t is a completely unobservable, latent state vector. To conduct prediction, filtering, or inference in such models requires that Y_t be integrated out using Bayes's Theorem. This task is generally impossible analytically (Shephard's, 1994, model seems a lone exception). Successful numerical integration has been carried out in a number of special cases, and numerical methods for these models are a very active area of research (e.g., Jacquier *et al.*, 1994; Danielsson, 1994; Kim and Shephard, 1993).

In an *ARCH model*, Y_t is *indirectly observable*, recoverable from the sample path of X_t, and a vector of initial states

$$[X_0, X_{-1}, X_{-2}, \ldots, X_{-k}, Y_0, Y_{-1}, \ldots, Y_{-j}]$$

via a recursive updating rule:

$$Y_t = Y[X_t, X_{t-1}, \ldots, X_{t-k}, Y_{t-1}, Y_{t-2}, \ldots, Y_{t-j}, t]. \tag{1.1}$$

A simple example of such a recursive updating scheme is GARCH (1, 1) (Bollerslev, 1986):

$$X_{t+1} = X_t + \xi_{t+1}, \tag{1.2}$$

$$Y_{t+1} = \omega + \beta Y_t + \alpha(X_t - X_{t-1})^2 = \omega + \beta Y_t + \alpha\xi_t^2. \tag{1.3}$$

Here a researcher assuming a GARCH (1, 1) specification would interpret Y_t as the conditional variance of ξ_{t+1}. Most ARCH models encountered in the literature can be put in the form (1.1), which we speak of variously as an ARCH model, a filter, or as a recursive updating rule. In practice, researchers using ARCH models typically assume that the model is "true" and that apart from errors in parameter estimates in finite samples, the conditional covariances produced by the model are "true." Under the (uncontroversial) view that all models are false, it is wise to ask how well

ARCH models perform when they are misspecified; for example, if the true model is stochastic volatility rather than ARCH. To do so, we interpret ARCH updating rules such as (1.1)–(1.3) as *filters* rather than as data generating processes (DGPs). It turns out that ARCH models have the remarkable robustness property that, under mild conditions, they are able to extract good estimates of the Y_t even when misspecified (see Nelson, 1992; Foster and Nelson, 1996; Nelson and Foster, 1994, 1995).

Nelson and Foster (1994a) (henceforward NF) specialize (1.1) to

$$\hat{Y}_t = Y\left[X_t, X_{t-1}, \hat{Y}_{t-1}, t\right], \tag{1.4}$$

with X_t and $\hat{Y}_t$ scalars, and $\hat{Y}_t$ regarded explicitly as an estimate of the unobservable state Y_t. They analyze, using continuous record asymptotics, the performance of simple recursive updating procedures of the form (1.4) when the true DGP is stochastic volatility with $\{X_t, Y_t\}$ a first-order Markov process. This chapter carries out this task in the case when X_t and Y_t are vectors. It is important to emphasize what the chapter does and does not deliver. First, *it delivers*

(a) A closed form for the asymptotic distribution of $(\hat{Y}_t - Y_t)$ given the DGP and the filter. Closed forms are a rarity in nonlinear filtering.
(b) A closed form for the asymptotically optimal filter given the DGP.
(c) Interpretation of each component of the optimal filter.
(d) Stochastic difference equations and discretely observed diffusions as DGPs.

It does not deliver

(a) A theory of parameter estimation: All parameters in both the DGP and the updating rule are taken as given.
(b) A theory for higher order Markov models or of filters of the general form (1.1): We restrict our attention to rules like (1.4), so, for example, GARCH (1, 1) is allowed, but not GARCH (1, 2). The reasons are discussed in section 2.

This exercise should cast light on both stochastic volatility and ARCH models. First, we derive simple, closed-form, approximately optimal filters for stochastic volatility models. For many of these models, implementing the exactly optimal filter via Bayes's Theorem is very difficult. Second, for researchers using ARCH models as DGPs, we show what features of an ARCH model make it robust (or not) to misspecification. Writing down new conditionally heteroskedastic models is very easy, as the plethora of ARCH models in the literature shows. What is missing is theoretical guidance on what is and is not important in building ARCH models that are robust to misspecification. Our results should provide some needed theoretical guidance on the design and specification of ARCH models.

Robustness to stochastic volatility alternatives is particularly important since these alternatives are so close to ARCH models. While we may be able to empirically distinguish a given ARCH model from a given stochastic volatility model, we are unlikely to be able to distinguish between ARCH and stochastic volatility models as groups; they are distributionally too close. To illustrate, consider the local scale model of Shephard (1994). Like other stochastic volatility models, the model has a latent state variable governing the conditional variance of an observable process. The difference is that in this model, the latent state can be integrated out in closed form. When it is, the result is a GARCH model. Of course, the stochastic volatility and GARCH versions of the model are observationally equivalent. Nelson (1990) provides many illustrations of how ARCH models can converge weakly (as DGPs) to stochastic volatility diffusion models as continuous time is approached. For example, AR(1)-EGARCH (Nelson, 1991) has the same diffusion limit as the stochastic volatility model used in Nelson (1988), Harvey *et al.* (1994), and Jacquier *et al.* (1994): Each converges to the diffusion model of Hull and White (1987) and Wiggins (1987), so the models can be very close distributionally.

In section 2, we derive the basic convergence result: Given the recursive updating rule (ARCH model) and the DGP, we derive the asymptotic distribution of the measurement error $\hat{Y}_t - Y_t$. We also derive the asymptotically optimal ARCH filter and explore its properties. Section 3 considers the special case in which the data are generated by a discretely observed diffusion. Section 4 develops examples of the main theorems, including analyses of some of the multivariate ARCH models in the literature. Section 5 is a brief conclusion.

2. FILTERING AND OPTIMALITY: MAIN RESULTS

The setup we consider below is a multivariate generalization of the case considered in NF. We keep similar notation. As in NF, there are two leading cases for the data generating processes, discrete-time stochastic volatility and discretely observed diffusion. We present the results for near-diffusions now and handle the diffusion case in Section 3.

In the stochastic volatility case, we assume that for $t = 0, h, 2h, \dots,$

$$\begin{bmatrix} X_{t+h} \\ Y_{t+h} \end{bmatrix} = \begin{bmatrix} X_t \\ Y_t \end{bmatrix} + \begin{bmatrix} \mu(X_t, Y_t, t, h) \\ \kappa(X_t, Y_t, t, h) \end{bmatrix} h^{\delta} + \begin{bmatrix} \xi_{X,t+h} \\ \xi_{Y,t+h} \end{bmatrix} h^{1/2} \tag{2.1}$$

$$E_t \begin{bmatrix} \xi_{X,t+h} \\ \xi_{Y,t+h} \end{bmatrix} = \begin{bmatrix} 0_{n\times 1} \\ 0_{m\times 1} \end{bmatrix}, \qquad \mathrm{Cov}_t \begin{bmatrix} \xi_{X,t+h} \\ \xi_{Y,t+h} \end{bmatrix} = \Omega(X_t, Y_t, t), \tag{2.2}$$

where δ equals either $\frac{3}{4}$ or 1.[1] By (1.2)–(2.2), $\{X_t, Y_t\}_{t=0,h,2h\ldots}$ is a stochastic difference equation, taking jumps at time intervals of length h. We assume $\{X_t, Y_t\}$ is a Markov process, although neither X_t or Y_t need be individually Markov. X_t is observable and $n \times 1$, while Y_t is unobservable and $m \times 1$. (X_0, Y_0) may be fixed or random. Our interest is in estimating $\{Y_t\}$ from the sample path of $\{X_t\}$.

Our analysis is asymptotic, in that h approaches 0. Conceptually, we are passing from annual data to monthly, daily, hourly, and on to a continuous time limit. In (2.1)–(2.2), t is assumed to be a discrete multiple of h. To define (X_t, Y_t) for general t, set $(X_t, Y_t) \equiv (X_{h[t/h]}, Y_{h[t/h]})$, where $[t/h]$ is the integer part of t/h. As in NF, E_t and Cov_t denote, respectively, the expectation vector and covariances matrix conditional on time t information—i.e., the σ-algebra generated by $\{X_\tau, Y_\tau\}_{0 \le \tau \le t}$ or, equivalently for our purposes, by $(X_t, Y_t, \hat{Y}_t)$. ($\hat{Y}_t$ is defined below.)

$\mu(\cdot)$ and $\kappa(\cdot)$ are continuous. $h^\delta\mu(\cdot)$ and $h^\delta\kappa(\cdot)$ are the time t conditional means of $(X_{t+h} - X_t)$ and $(Y_{t+h} - Y_t)$. If, for example, $(X_{t+h} - X_t)$ were a vector of asset returns, then $h^\delta\mu(\cdot)$ would represent a vector of expected returns. If Y_t parameterized the conditional covariance matrix, $h^\delta\kappa(\cdot)$ would govern mean-reversion in the covariances.

We call (2.1)–(2.2) a "near-diffusion," since when $\delta = 1$ and some mild regularity conditions are satisfied [e.g., Nelson (1990) or Ethier and Kurtz (1986, ch. 7, Theorem 4.1)], it converges weakly as $h \downarrow 0$ to the diffusion process

$$\mathrm{d}\begin{bmatrix} X_t \\ Y_t \end{bmatrix} = \begin{bmatrix} \mu(X_t, Y_t, t, 0) \\ \kappa(X_t, Y_t, t, 0) \end{bmatrix} dt + \Omega(X_t, Y_t, t)^{\underline{1/2}} dW_t, \tag{2.3}$$

where W_t is an $(n + m) \times 1$ standard Brownian motion.

As in NF, μ and κ drop out of the asymptotic distribution of the measurement error as $h \downarrow 0$ when $\delta = 1$: Expected returns and mean reversion in volatility are of negligible importance (asymptotically) to the filtering problem *unless they are very large*. If we believe that μ and κ might be important for $h > 0$, it is reasonable to create an asymptotic in which μ and κ do not wash out. Some empirical support for this view is provided by Lo and Wang (1995), who find, using daily returns data, that misspecification in conditional means may have an economically significant effect (i.e., changes of several percent in implied volatilities computed from options prices) on volatility estimates. To keep μ and κ in the asymptotic distribution, we make these terms asymptotically large by making the conditional mean term $h^\delta\mu$ and $h^\delta\kappa$ $O(h^{3/4})$ instead of $O(h)$.

[1] Notice that the scale factor $h^{1/2}$ on the ξ terms in (2.1) and (2.4) is missing in the univariate case presented in NF equations (5.1) and (5.1′). These scale terms are correct here but were inadvertently omitted in NF. In addition, the h^{-1} terms in NF (5.8)–(5.9) should be deleted.

We call this the "large drift" case. Unfortunately, $\delta = \frac{3}{4}$ is not, in general, compatible with a diffusion limit such as (2.3). Nevertheless, we call it a "near-diffusion."[2] Nothing is lost by using large drifts; the results for $\delta = 1$ are recovered by setting $\mu = \hat{\mu} = 0_{n\times 1}$ and $\kappa = \hat{\kappa} = 0_{m\times 1}$.

2.1. The Multivariate ARCH Model

We consider ARCH models (recursive updating rules, or filters) which generate estimates $\hat{Y}_t$ (for $t = 0, h, 2h, 3h, \ldots$) of the unobservable state variables Y_t by a recursion of the form (1.4)

$$\hat{Y}_{t+h} = \hat{Y}_t + h^{\delta} \cdot \hat{\kappa}\big(X_t, \hat{Y}_t, t, h\big) + h^{1/2} \cdot G\big(\hat{\xi}_{X,t+h}, X_t, \hat{Y}_t, t, h\big), \quad (2.4)$$

$$\hat{\xi}_{X,t+h} \equiv h^{-1/2}\big[X_{t+h} - X_t - h^{\delta} \cdot \hat{\mu}\big(X_t, \hat{Y}_t, t, h\big)\big]. \quad (2.5)$$

As in NF, $\hat{\kappa}(\cdot)$, $\hat{\mu}(\cdot)$, and $G(\cdot)$ are functions selected by the econometrician. G, $\hat{\kappa}$, and $\hat{\mu}$ are continuous in all arguments. Just as μ and κ are the true drifts in X_t and Y_t, $\hat{\mu}$, and $\hat{\kappa}$ are the econometrician's (possibly misspecified) specifications of these drifts. $\hat{\xi}_{X,t+h}$ is a residual obtained using $\hat{\mu}$ in place of μ. The ARCH model treats this fitted residual $\hat{\xi}_{X,t+h}$ as if it were the true residual $\xi_{X,t+h}$. When ARCH models are DGPs rather than filters, $\xi_{Y,t+h}$, the innovation in Y_{t+h}, is a function of $\xi_{X,t+h}$, X_t, Y_t, and t. $G(\cdot)$ is the econometrician's specification of this function. The ARCH model treats the fitted $\hat{Y}_t$ as if it were the true Y_t. Accordingly, we make the normalizing assumption that for all (X_t, Y_t, t, h), $E_t[G(\xi_{X,t+h}, X_t, Y_t, t, h)] = 0_{m\times 1}$.

2.2. The Asymptotic Distribution of the Measurement Error

Our next task is to derive the asymptotic distribution of $\hat{Y}_t - Y_t$, given a sequence (indexed by h as $h \downarrow 0$) of DGPs of the form (2.1)–(2.2) and a sequence of ARCH updating rules of the form (2.4)–(2.5). Below we emphasize heuristics over rigor. A rigorous proof of Theorem 2.1 below closely follows the proof of Theorem 3.1 in NF.

[2] Admittedly, setting $\delta = \frac{3}{4}$ is a bit unnatural, since it rules out a diffusion limit for $\{X_t, Y_t\}$ as $h \downarrow 0$. But asymptotics need not correspond to a natural data experiment to be useful, e.g., Phillips's (1988) "near integrated" processes, in which an AR root approaches 1 as the sample size grows, or the analysis of asymptotic local power, in which a sequence of alternative hypotheses collapse to the null as the sample size grows (Serfling, 1980, ch. 10).

To illustrate the intuition behind the passage to continuous time, for each $h > 0$ let $\{Z_t\}$ be a $k \times 1$ Markov random step function taking jumps at times h, $2h$, $3h$, and so on, with

$$Z_{t+h} - Z_t = h \cdot \mu(Z_t) + h^{1/2}\xi_{t+h}, \qquad \mathrm{Cov}_t\left(h^{1/2}\xi_{t+h}\right) = \Omega(Z_t) \cdot h,$$
$$\mathrm{E}_t[\,\xi_{t+h}] = 0_{k \times 1}. \tag{2.6}$$

Under weak conditions (see, e.g., Nelson, 1990), $\{Z_t\}$ converges weakly as $h \downarrow 0$ to the diffusion with the same first two conditional moments per unit of time—i.e., to

$$dZ_t = \mu(Z_t)\,dt + \Omega(Z_t)^{\underline{1/2}}\,dW_t, \tag{2.7}$$

where W_t is a $k \times 1$ standard Brownian motion.[3] The h in (2.6) is the discrete-time equivalent of dt in (2.7). In both (2.6) and (2.7) we interpret $\mu(Z_t, t)$ and $\Omega(Z_t, t)$, respectively, as the conditional mean (or "drift") and the conditional covariance of increments in Z_t *per unit of time*. A familiar example of this kind of convergence in financial economics is the binomial option pricing model of Cox *et al.* (1979), in which sequences of binomial models converge weakly to geometric Brownian motion as $h \downarrow 0$. Nelson (1990) reviews and applies these convergence theorems to show that many ARCH models converge weakly (as data generating processes not as filters) to stochastic volatility diffusion models as $h \downarrow 0$.

For our diffusion limits, we require rescaling of both the measurement error vector $(\hat{Y}_t - Y_t)$ and of time. To understand the role of such rescalings, consider the diffusion (2.7). First, to rescale Z define $M \equiv \alpha \cdot Z$ for $\alpha > 0$. By Ito's Lemma

$$dM_t = \alpha\mu(Z_t)\,dt + \alpha\Omega(Z_t)^{\underline{1/2}}\,dW_t. \tag{2.8}$$

So the effect of rescaling Z by α is to multiply the drift (per unit time) by α and the covariance per unit of time by α^2. Next consider the effect of rescaling the time index by defining $\tau \equiv \beta t$. Recall that $\mu(\cdot)$ and $\Omega(\cdot)$ are the drift and variance *per unit of time*, so with the time rescaling.

$$dZ_\tau = \beta^{-1}\mu(Z_\tau)\,d\tau + \beta^{-1/2}\Omega(Z_\tau)^{\underline{1/2}}\,dW_\tau \tag{2.9}$$

That is, rescaling the state vector Z moves the drift and standard deviation proportionately, whereas rescaling time moves the drift and variance

[3] Some notation: We use two different types of matrix square roots. When we write $F^{1/2}$ for a positive semidefinite F, we mean the unique positive semidefinite matrix with $F^{1/2}F^{1/2} = F$. When we write $F^{\underline{1/2}}$, we mean any matrix satisfying $F^{\underline{1/2}}F^{\underline{1/2}\prime} = F$. $F^{1/2}$ is unique (Horn and Johnson, 1985, Theorem 7.2.6). $F^{\underline{1/2}}$ need not be positive definite or symmetric, and is not in general unique: If U is orthonormal and real, $(F^{\underline{1/2}}U)(F^{\underline{1/2}}U)' = F$. When $F(z)$ is continuous, we assume $F^{\underline{1/2}}(z)$ is continuous. We can do so, since if $F(z)$ is continuous so is $F^{1/2}(z)$.

proportionately. The extra degree of freedom allowed by using both change of time and change of scale is crucial below.

Now consider the measurement error process $(\hat{Y}_t - Y_t)$. By (2.1) and (2.4)–(2.5),

$$\left(\hat{Y}_{t+h} - Y_{t+h}\right) = \left(\hat{Y}_t - Y_t\right) + h^{\delta} \cdot [\hat{\kappa} - \kappa] + h^{1/2}\left[G\left(\hat{\xi}_{X,t+h}, \hat{Y}_t, X_t, t\right) - \xi_{Y,t+h}\right]. \quad (2.10)$$

Assuming sufficient smoothness of $G(\cdot)$, we may expand it in a Taylor series around $(\hat{Y}_t, \hat{\xi}_{X,t+h}) = (Y_t, \xi_{X,t+h})$, obtaining

$$\begin{aligned}\left(\hat{Y}_{t+h} - Y_{t+h}\right) &= \left(\hat{Y}_t - Y_t\right) + h^{1/2}[G(\xi_{X,t+h}, Y_t) - \xi_{Y,t+h}] \\ &\quad + h^{1/2}\,\mathrm{E}_t\left[\frac{\partial G(\xi_{X,t+h}, Y_t)}{\partial Y}\right]\left(\hat{Y}_t - Y_t\right) \\ &\quad + h^{\delta}\left[\hat{\kappa} - \kappa + \mathrm{E}_t\left[\frac{\partial G(\xi_{X,t+h}, Y_t)}{\partial \xi_X}\right](\mu - \hat{\mu})\right] \\ &\quad + \text{higher order terms}. \end{aligned} \quad (2.11)$$

We suppress the X_t and t arguments in G and all arguments in μ, $\hat{\mu}$, κ, and $\hat{\kappa}$. $\partial G(\xi_{X,t+h}, X_t, Y_t, t, 0)/\partial Y$ is $m \times m$ with i, jth element $\partial G_i(\xi_{X,t+h}, X_t, Y_t, t, 0)/\partial Y_j$, and $\partial G(\xi_{X,t+h}, X_t, Y_t, t, 0)/\partial \xi_x$ is $m \times n$ with $i-j$th element $\partial G_i(\xi_{X,t+h}, X_t, Y_t, t, 0)/\partial \xi_{Xj}$.

Apart from $\hat{Y}_t - Y_t$ the terms on the second line of (2.11) can be treated as functions of (X_t, Y_t, t, h). So can $\mathrm{Cov}_t[G(\xi_{X,t+h}, Y_t) - \xi_{Y,t+h}]$. If these functions converge as $h \downarrow 0$, we may write

$$\begin{aligned}\left(\hat{Y}_{t+h} - Y_{t+h}\right) &= \left(\hat{Y}_t - Y_t\right) + h^{1/2}[G(\xi_{X,t+h}, Y_t) - \xi_{Y,t+h}] \\ &\quad - h^{1/2}B(X_t, Y_t, t)\left(\hat{Y}_t - Y_t\right) + h^{\delta}A(X_t, Y_t, t) \\ &\quad + \text{higher order terms}, \end{aligned} \quad (2.12)$$

where $A(X_t, Y_t, t)$, $B(X_t, Y_t, t)$ are the small h limits of $[\hat{\kappa} - \kappa + E_t[\partial G/\partial \xi_X](\hat{\mu} - \mu)]$ and $-\mathrm{E}_t[\partial G/\partial Y]$.

Now suppose we naively proceed to the diffusion limit, ignoring higher order terms and finding the diffusion whose first two conditional moments (per unit time) corresponded to the stochastic difference equation (2.12). This moment matching diffusion is

$$\begin{aligned} d\left(\hat{Y}_t - Y_t\right) &= \left[h^{\delta-1}A(X_t, Y_t, t) - h^{-1/2}B(X_t, Y_t, t)\left(\hat{Y}_t - Y_t\right)\right] dt \\ &\quad + C(X_t, Y_t, t)^{1/2}\, dW_t, \end{aligned} \quad (2.12')$$

with $C(X_t, Y_t, t)$ the small h limit of $\text{Cov}_t[G(\xi_{X,t+h}, Y_t) - \xi_{Y,t+h}]$. Clearly this will not do as a limit diffusion: The covariance matrix of the candidate limit diffusion is fine, but the drift explodes as $h \downarrow 0$. To remedy this, rescale $\hat{Y}_t - Y_t$ as $Q_t \equiv \alpha(\hat{Y}_t - Y_t)$ and rescale time as $\tau \equiv \beta t$. The candidate diffusion limit is now

$$\mathrm{d}Q_\tau = \beta^{-1}\left[\alpha h^{\delta-1}A(X_\tau, Y_\tau, \tau) - h^{-1/2}B(X_\tau, Y_\tau, \tau)Q_\tau\right] d\tau + \alpha\beta^{-1/2}C(X_\tau, Y_\tau, \tau)^{\underline{1/2}}\, dW_\tau, \tag{2.13}$$

where W_τ is an $(n+m) \times 1$ standard Brownian motion on the transformed time scale. Set $\alpha \equiv h^{-1/4}$ and $\beta \equiv h^{-1/2}$ and we have

$$dQ_\tau = \left[h^{\delta-3/4}A(X_\tau, Y_\tau, \tau) - B(X_\tau, Y_\tau, \tau)Q_\tau\right] d\tau + C(X_\tau, Y_\tau, \tau)^{\underline{1/2}}\, dW_\tau, \tag{2.14}$$

which is nicely behaved, allowing us to obtain a diffusion limit. The drift terms μ, $\hat{\mu}$, κ, and $\hat{\kappa}$ appear in the limit diffusion only through $A(X_t, Y_t, t)$. Hence, they drop out as $h \downarrow 0$ when $\delta = 1$ but appear when $\delta = 3/4$. As indicated above, we emphasize the $\delta = 3/4$ case, but results for the $\delta = 1$ case can be recovered by setting $A(\cdot) = 0$.

Notes: First, neither rescaling would have been enough alone—both were required. Second, an interval $[0, M]$ on our transformed time scale corresponds to the interval $[0, M \cdot h^{1/2}]$ on our original (calendar) time scale. Since $[0, M \cdot h^{1/2}]$ is vanishingly small as $h \downarrow 0$, we are effectively analyzing the measurement error process in a neighborhood of $t = 0$. X_t and Y_t are slow moving (like diffusions), so over the interval $[0, M \cdot h^{1/2}]$ are asymptotically constant at (X_0, Y_0). This has a fortuitous consequence: If $A(\cdot)$, $B(\cdot)$, and $C(\cdot)$ are continuous (below we assume they are) then, on the transformed time scale, the limit diffusion (2.14) is equivalent to

$$dQ_\tau = \left[A(X_0, Y_0, 0) - B(X_0, Y_0, 0)Q_\tau\right] d\tau + C(X_0, Y_0, 0)^{\underline{1/2}}\, dW_\tau, \tag{2.15}$$

which we recognize as a vector Ornstein–Uhlenbeck process, the continuous-time version of a Gaussian VAR(1), a well-studied process [see Karatzas and Shreve (1988, section 5.6) or Arnold (1973, section 8.2)], for which stationary distributions and the like are known.

The $B(\cdot)$ matrix governs mean-reversion in Q_τ: If the real parts of the eigenvalues of $B(\cdot)$ are all positive (which we require in Theorem 2.1 below), then Q_τ has a unique, Gaussian stationary distribution. If $A(\cdot)$ is a matrix of zeros, the mean of this distribution is a vector of zeros. $C(\cdot)$ is the covariance matrix of the driving noise term. $A(\cdot)$, $B(\cdot)$, and $C(\cdot)$ may be nonlinear functions of (X_t, Y_t, t), but we see from (2.15) that in a neighborhood of $(X_0, Y_0, 0)$, the normalized measurement error process

$Q_t \equiv h^{-1/4}(\hat{Y}_t - Y_t)$ is well approximated by a linear, Gaussian process. This makes possible the closed form solutions in NF and this chapter.

To analyze the measurement error process in the neighborhood of an arbitrary (X_T, Y_T, T), define $\tau \equiv (t - T)h^{-1/2}$. This maps the calendar interval $[T, T + M \cdot h^{1/2}]$ into $[0, M]$ on the transformed time scale. Sometimes it is important to keep track of both time scales, in which case we write, for example, $Q_{T,\tau} \equiv Q_{T+\tau h^{1/2}}$.

We now formally state our assumptions and results. The first assumption guarantees convergence of A_t, B_t, and C_t, and justifies ignoring the higher order terms in (2.12).

ASSUMPTION 1. *The following functions are well defined, continuous in t, and twice differentiable in X_t and Y_t:*

$$A(X,Y,t) \equiv \lim_{h \downarrow 0}\{[\hat{\kappa}(X,Y,t,h) - \kappa(X,Y,t,h)] + \mathrm{E}[\partial G(\xi_{X,t+h}, X_t, Y_t, t, h)/\partial \xi_X/(X_t, Y_t) = (X,Y)][\mu(X,Y,t,h) - \hat{\mu}(X,Y,t,h)]\}, \tag{2.16}$$

$$B(X,Y,t) \equiv -\lim_{h \downarrow 0} \mathrm{E}[\partial G(\xi_{X,t+h}, X_t, Y_t, t, h)/\partial Y/(X_t, Y_t) = (X,Y)], \tag{2.17}$$

$$C(X,Y,t) \equiv \lim_{h \downarrow 0} \mathrm{E}[(G(\xi_{X,t+h}, X_t, Y_t, t, h) - \xi_{Y,t+h}) \times (G(\xi_{X,t+h}, X_t, Y_t, t, h) - \xi_{Y,t+h})'/(X_t, Y_t) = (X,Y)]. \tag{2.18}$$

Further

$$h^{-1/2}\,\mathrm{E}[Q_{t+h} - Q_t/(X_t, Y_t, Q_t) = (X,Y,Q)] \to A(X,Y,t) - B(X,Y,t)\cdot Q, \tag{2.19}$$

$$h^{-1/2}\,\mathrm{Cov}[Q_{t+h} - Q_t/(X_t, Y_t, Q_t) = (X,Y,Q)] \to C(X,Y,t), \tag{2.20}$$

as $h \downarrow 0$ uniformly on every bounded (X, Y, Q, t) set.[4]

We often write A_t, B_t, and C_t for $A(X_t, Y_t, t)$, $B(X_t, Y_t, t)$, and $C(X_t, Y_t, t)$. Since diffusions have continuous paths almost surely, the next assumption asymptotically rules out jumps.

[4] That is, on every set of the form $\{(X, Y, Q\, t)\colon \|(X, Y, Q, t)\| < \Lambda\}$ for some finite, positive Λ. We could for example, write (2.20) more formally as "For every Λ, $0 < \Lambda < \infty$,

$$\lim_{h \downarrow 0} \sup_{\|(X,Y,Q,t)\| < \Lambda} \|h^{-1/2}\,\mathrm{Cov}[Q_{t+h} - Q_t \mid (X_t, Y_t, Q_t) = (X,Y,Q)] - C(X,Y,t)\| = 0." \tag{2.20'}$$

ASSUMPTION 2. *Define the matrix norm* $\|A\| \equiv [\text{Trace}(AA')]^{1/2}$. *For some* $\psi > 0$

$$\mathrm{E}\Big[\big\| h^{-1/2}(Y_{t+h} - Y_t)\big\|^{2+\psi} / (X_t, Y_t) = (X, Y)\Big], \tag{2.21}$$

$$\mathrm{E}\Big[\big\| h^{-1/2}(X_{t+h} - X_t)\big\|^{2+\psi} / (X_t, Y_t) = (X, Y)\Big], \tag{2.22}$$

$$\mathrm{E}\Big[\big\| G\big(\hat{\xi}_{X,t+h}, X_t, \hat{Y}_t, t, h\big)\big\|^{2+\psi} / (X_t, Y_t, Q_t) = (X, Y, Q)\Big] \tag{2.23}$$

are bounded as $h \downarrow 0$, *uniformly on every bounded* (X, Y, Q, t) *set.*

THEOREM 2.1. *Let Assumptions* 1–2 *be satisfied, let* $\delta = \frac{3}{4}$, *and let* $0 < T < \infty$, $0 < \tau < \infty$. *Let* Θ *be a bounded, open subset of* R^{n+2m+1} *on which for some* $\epsilon > 0$, *the real parts of all the eigenvalues of* $B(X, Y, T)$ *are bounded below by* ϵ. *Then for every* $(X, Y, Q, T) \in \Theta$, $\{Q_{T,\tau}\}_{[0,M]}$ (*conditional on* $(X_T, Y_T, Q_T) = (X, Y, Q)$) *converges weakly to the diffusion*

$$dQ_\tau = \big[A(X_T, Y_T, T) - B(X_T, Y_T, T) Q_\tau\big]\, d\tau + C(X_T, Y_T, T)^{1/2}\, dW_\tau \tag{2.15'}$$

as $h \downarrow 0$. *This convergence is uniform on* Θ. *For proof, see the Appendix.*

COROLLARY. *Under the conditions of Theorem* 2.1, $\forall\ (X_T, Y_T, Q_T, T) \in \Theta$ *and* $\forall \tau > 0$,

$$[Q_{T,\tau} / (X_T, Y_T, Q_T) = (X, Y, Q)] \Rightarrow N[b_{T,\tau}, V_{T,\tau}], \tag{2.24}$$

where $\Rightarrow$ *denotes weak convergence as* $h \downarrow 0$, *and*

$$\begin{aligned} b_{T,\tau} &= \exp[-B_T \cdot \tau] \cdot \left[Q + \int_0^\tau \exp[-B_T \cdot s] A_T\, ds\right] \\ &= \exp[-B_T \cdot \tau] \cdot \big[Q - B_T^{-1} A_T\big] + B_T^{-1} A_T, \qquad \text{and} \end{aligned} \tag{2.25}$$

$$V_{T,\tau} = \exp[-B_T \cdot \tau] \cdot \left[\int_0^\tau \exp[B_T \cdot s] C_T \exp[B_{T'} \cdot s]\, ds\right] \cdot \exp[-B_{T'} \cdot \tau], \quad \text{or equivalently} \tag{2.26}$$

$$\begin{aligned} \operatorname{vec}(V_{T,\tau}) = &-\exp[-(I_{m\times m} \otimes B_T + B_T \otimes I_{m\times m})\tau] \\ &\times \Big[[I_{m\times m} \otimes B_T + B_T \otimes I_{m\times m}]^{-1} \operatorname{vec}(C_T)\Big] \\ &+ [I_{m\times m} \otimes B_T + B_T \otimes I_{m\times m}]^{-1} \operatorname{vec}(C_T), \end{aligned} \tag{2.27}$$

where exp *is the matrix exponential,* $\otimes$ *is the Kronecker product, and* vec *is the operator that stacks the columns of a matrix into a column vector.* $V_{T,\tau}$ *is positive semidefinite and symmetric* $\forall \tau > 0$. *For proof, see the Appendix.*

As in NF, the theorem yields weak convergence for $\{Q_{T,\tau}\}_{[0,M]}$—i.e., for the time interval $[T, T + h^{1/2}M]$ on the standard time scale or $[0, M]$ on the "fast" time scale. Since this holds uniformly for every finite M, Lemma 5.2 of Helland (1982) guarantees that it also holds uniformly for M_h, where $M_h \to \infty$ sufficiently slowly as $h \downarrow 0$.[5] We then have $Q_{T,M_h} \Rightarrow N(b_T^*, V_T^*)$ as $h \downarrow 0$, where $\text{vec}(V_T^*) \equiv [I_{m\times m} \otimes B_T + B_T \otimes I_{m\times m}]^{-1}$ $\text{vec}(C_T)$ and $b_T^* \equiv B_T^{-1}A_T$. Note that, in general, deviation from zero of any element of either $\hat{\kappa}(X, Y, t, 0) - \kappa(X, Y, t, 0)$ or $\hat{\mu}(X, Y, t, 0) - \mu(X, Y, t, 0)$ creates asymptotic bias in other elements of $\hat{Y}$. V_T^* has two other representations which will prove useful (see Karatzas and Shreve, 1988, pp. 355–358):

$$B_T V_T^* + V_T^* B_T' = C_T \tag{2.28}$$

$$V_T^* = \int_0^\infty \exp[-B_T \cdot s] C_T \exp[-B_T' \cdot s] ds. \tag{2.29}$$

Because the matrix exponential in (2.29) is never singular (Bellman, 1970, p. 170) and the sum or integral of positive definite matrices is positive definite (Magnus and Neudecker, 1988, Chapter 11, Theorem 9), V_T can be singular only if C_T is.

To interpret the corollary, recall that $Q_{T,\tau} \equiv Q_{T+\tau h^{1/2}}$. Combining with (2.25)–(2.26),

$$Y_{T+\tau h^{1/2}} \approx \hat{Y}_{T+\tau h^{1/2}} - h^{1/4}\tilde{Q}_{T+\tau h^{1/2}},$$

$$\text{where } \left[\tilde{Q}_{T+\tau h^{1/2}} \mid \hat{Y}_T, Y_T, X_T\right] \sim N(b_{T+\tau h^{1/2}}, V_{T+\tau h^{1/2}}). \tag{2.30}$$

Of course, Y_T and Q_T are unobservable. If the filter has been running a "long" time (i.e., it was initialized at some time $t - M_h h^{1/2}$, where M_h goes to infinity sufficiently slowly as $h \downarrow 0$, then

$$Y_t \approx \hat{Y}_t - h^{1/4}Q_t, \qquad \text{where } Q_t \sim N(b_t, V_t), \tag{2.31}$$

where b_t and V_t can be evaluated at $(X_t, \hat{Y}_t, t)$, allowing us to characterize the uncertainty in Y_T given the information in X_t and $\hat{Y}_t$. This allows us to draw confidence bands. Note, however, that the information on Y_t in X_t and $\hat{Y}_t$ does not summarize all information in the sample path of $\{X_t\}_{0\le t\le T}$, since our filter is not an implementation of Bayes's Theorem. Implement-

[5] Helland's Lemma 5.2 does not specify the rate at which $M_h \to \infty$, only that there is such a rate. It may be very slow, e.g., $\ln[\ln(h^{-1})]$. In our example, M_h must increase at a slower rate than $h^{-1/2}$, or else (X_t, Y_t, t) would not be asymptotically constant on $[T, T + h^{1/2}M_h]$. We therefore impose $M_h = o(h^{-1/2})$ with the understanding that it may be much more slowly increasing.

ing Bayes's Theorem analytically here seems impossible—if it were possible, we would prefer to do so.[6]

2.3. Optimality

In NF, optimality was defined in terms of minimizing V_T^* while eliminating b_T^*. Since there is no bias–variance tradeoff, this is equivalent to minimizing the asymptotic mean squared error. A natural multivariate generalization is to minimize the matrix mean squared error, Trace$[b_T^* b_T^{*\prime} + V_T^*]$. We could just as easily minimize $u'[b_T^* b_T^{*\prime} + V_T^*]u$ for an arbitrarily selected $m \times 1$ vector u. None of the optimality results would be affected, because the proofs of the optimality results show that if b_0 and V_0 are the bias vector and error covariance matrix achieved by optimal filters proposed below, and b_1 and V_1 are the bias and covariance matrix achieved by any other filter then $[b_1 b_1' + V_1] - [b_0 b_0' + V_0]$ is positive semidefinite. Nor would it matter if we gave different weights to the bias and covariance: The optimal filter simply eliminates the bias.

As in NF, there are two sources of uncertainty at time t in estimating the value of Y_{t+h}: First, there is uncertainty about the first difference $Y_{t+h} - Y_t$. Second, there is uncertainty about the level of Y_t. The optimal filter turns out to have two terms. The first, which we call P, is the *prediction component* that extracts the information in $\xi_{X,t+h}$ regarding $Y_{t+h} - Y_t$. Formally,

$$P(\xi_X, X, Y, t) \equiv E\left[\xi_{Y,t+h} \mid (\xi_{X,t+h}, X_t, Y_t) = (\xi_X, X, Y)\right]. \quad (2.32)$$

The second term, which we call S, is the estimation or *score component*, the $m \times 1$ vector

$$\mathrm{S}(\xi_X, X, Y, t) \equiv \partial \ln\left[f(\xi_{X,t+h} \mid (X_t, Y_t) = (X, Y))\right] / \partial Y. \quad (2.33)$$

$f(\xi_{X,t+h} \mid (X_t, Y_t) = (X, Y))]$ is the time t conditional density of $\xi_{X,t+h}$. This score term extracts, in a manner analogous to maximum-likelihood estimation with Y_t treated as a parameter, information in $\xi_{X,t+h}$ on the *level* of Y_t.

More notation: When we take conditional expectations, we drop more time subscripts and write, e.g., $E_t[(\xi_Y - P)(\xi_Y - P)']$ for $\mathrm{E}_t[(\xi_{Y,t+h} - P_{t+h})(\xi_{Y,t+h} - P_{t+h})']$ and $\mathrm{E}_t[\mathrm{SS}']$ for $\mathrm{E}_t[\mathrm{S}_{t+h}\mathrm{S}_{t+h}']$.

To motivate the optimal filter, consider first the bias. If $\delta = 1$, there is no asymptotic bias in the measurement error. If $\delta = \frac{3}{4}$, setting $\hat{\kappa} = \kappa$ and

[6] Most numerical strategies for implementing Bayes's Theorem in stochastic volatility models involve draws of the Y_t process (e.g., Jacquier *et al.*, 1994). Nick Polson has suggested that one way to do this may be to condition on the sample path of Y_t and draw pseudo-sample paths of Y_t using the approximation implied by (2.15′). Finally, generate draws of Y_t sample paths using the Metropolis algorithm.

$\hat{\mu} = \mu$ is sufficient to eliminate it, since from (2.16) this ensures that $A(X, Y, t) = 0_{m \times 1}$. There is no variance–bias tradeoff, so we take $\hat{\kappa} = \kappa$, $\hat{\mu} = \mu$. We select $G(\cdot)$ to minimize the asymptotic covariance. It turns out the optimal $G(\cdot)$ equals the prediction component P plus a matrix weight times the score component S. The asymptotic covariance matrix of the Q_t process is governed by C_t, its instantaneous covariance matrix, and B_t, which governs mean reversion in Q_t. The smaller C_t and the larger B_t, the smaller is V_T^*, the asymptotic covariance of Q_t. $C_t = \mathrm{E}_t[\xi_Y - G)(\xi_Y - G)']$. Since P_{t+h} is the minimum mean squared error predictor of $\xi_{Y,t+h}$ given $\xi_{X,t+h}$, C_t would be minimized by setting $G_{t+h} = P_{t+h}$. There is a tradeoff with the B_t matrix, however. Recall $B_t = -\mathrm{E}_t[\partial G/\partial Y]$. If we integrate by parts, we see that $B_t = \mathrm{E}_t[GS']$, the covariance between G_{t+h} and the score term S_{t+h}. Mean reversion in Q_t is maximized by maximizing this covariance. The optimal filter balances the goals of minimizing C_t and maximizing B_t. We now formally state our assumptions.

Assumption 3. *For every h, $[\xi'_{X,t+h}, \xi'_{Y,t+h}]'$ possess conditional densities $f(\xi_{X,t+h}, \xi_{Y,t+h}/X_t, Y_t, t)$ and $f(\xi_{X,t+h}/X_t, Y_t, t)$. These conditional densities are continuous in X, Y, and t, and $f(\xi_X/X, Y, t)$ is continuously differentiable in Y almost everywhere, with one-sided partial derivatives everywhere, and for some $\psi > 0$*

$$\mathrm{E}\left[\|\mathrm{P}_{t+h}\|^{2+\psi} \mid X_t = X, Y_t = Y\right], \quad \text{and} \tag{2.34}$$

$$\mathrm{E}\left[\|\mathrm{S}_{t+h}\|^{2+\psi} \mid X_t = X, Y_t = Y\right] \tag{2.35}$$

are bounded uniformly on every bounded (X, Y, t) set as $h \downarrow 0$.

As usual, the $2 + \psi$ moment conditions asymptotically rule out jumps in the state variables as $h \downarrow 0$. To motivate the next assumption, recall that the limit diffusion (2.15′) is linear given (X, Y, T). Hence, matrix Riccati equations familiar from the literature on Kalman filtering and linear control (see Anderson and Moore, 1971, 1979) appear in the optimal filter.

Assumption 4. *There is a unique, positive semidefinite solution ω_T to the Riccati equation*:

$$\mathrm{E}_T[\mathrm{PS}']\omega_T + \omega_T \mathrm{E}_T[\mathrm{SP}'] + \omega_T \mathrm{E}_T[\mathrm{SS}']\omega_T = \mathrm{E}_T[(\xi_Y - \mathrm{P})(\xi_Y - \mathrm{P})']. \tag{2.36}$$

A sufficient condition for Assumption 4 is that $E_T[\mathrm{SS}']$ is positive definite (Kučera, 1972, Condition 3). This condition can be weakened (see Kučera, 1972; Lancaster and Rodman, 1980). Using (2.29) and Theorem 2.2, one can show that if $E_T[\mathrm{SS}']$ and $E_T[(\xi_Y - \mathrm{P})(\xi_Y - \mathrm{P})']$ are positive definite, so is ω_T. Anderson (1978) provides an algorithm for solving (2.36).

THEOREM 2.2. *Let Assumptions* 1–4 *be satisfied. If* $\delta = \frac{3}{4}$, *a set of sufficient conditions for* $\mathrm{Trace}[b_T^* b_T^{*\prime} + V_T^*]$ *to be minimized is that*

$$\lim_{h \downarrow 0} \hat{\kappa}(X,Y,T,h) - \kappa(X,Y,T,h) = 0, \tag{2.37}$$

$$\lim_{h \downarrow 0} \hat{\mu}(X,Y,T,h) - \mu(X,Y,T,h) = 0, \quad \text{and} \tag{2.38}$$

$$G(\xi_X, X, Y, T, h) = \mathrm{P}(\xi_X, X, Y, T) + \omega_T \mathrm{S}(\xi_X, X, Y, T), \tag{2.39}$$

with ω_T *the positive semidefinite solution to* (2.36). *If* $\delta = 1$, (2.39) *alone is sufficient. The minimized* $V_T^* = \omega_T$ *and the minimized* $b_T^* b_T^{*\prime} = 0_{m \times m}$. *Let* $\tilde{G}$ *satisfy the conditions of Theorem* 2.1 *and let* $\tilde{V}_T$ *be the asymptotic error covariance matrix delivered by Theorem* 2.1 *using* $\tilde{G}$ *in place of the* G *in* (2.39). *Then* $\tilde{V}_T = \omega_T$ *if and only if* $\tilde{G}(\xi_X, X, Y, T, h) = G(\xi_X, X, Y, T, h)$ *almost surely. For proof, see the Appendix.*

Notes on Theorem 2.2:

(i) $\mathrm{E}_t[\mathrm{SP}']$, $\mathrm{E}_t[\mathrm{SS}']$, $\mathrm{E}_t[(\xi_Y - \mathrm{P})(\xi_Y - \mathrm{P})']$, and ω_t are continuous matrix functions of X_t, Y_t, and t. For $t \in [T, T + h^{1/2}M_h]$, $(X_t, Y_t, t) \rightarrow (X_T, Y_T, T)$ as $h \downarrow 0$. In the diffusion limit on the transformed time scale, (X_t, Y_t, t) are constants evaluated at (X_T, Y_T, T). For our purposes, therefore, $\mathrm{E}_t[\mathrm{SP}']$, $\mathrm{E}_t[\mathrm{SS}']$, $\mathrm{E}_t[(\xi_Y - \mathrm{P})(\xi_Y - \mathrm{P})']$, and ω_t are asymptotically equivalent to $\mathrm{E}_T[\mathrm{SP}']$, $\mathrm{E}_T[\mathrm{SS}']$, $\mathrm{E}_T[(\xi_Y - \mathrm{P})(\xi_Y - \mathrm{P})']$, and ω_T when $T \le t \le T + h^{1/2}M_h$.

(ii) The functions $G(\cdot)$, $\mathrm{P}(\cdot)$, $\mathrm{S}(\cdot)$, and the like were defined using expectations taken conditional on (X_t, Y_t, t). Y_t, however, is not in the econometrician's information set, so the functions are evaluated at $(X_t, \hat{Y}_t, t)$. This was explicitly accounted for in the asymptotics.

(iii) Recall that the drift terms $\mu(\cdot)$, $\hat{\mu}(\cdot)$, $\kappa(\cdot)$, and $\hat{\kappa}(\cdot)$ drop out of the asymptotic unless we make the (somewhat unnatural) assumption of asymptotically large drift ($\delta = \frac{3}{4}$). This suggests that these drift terms are of second-order importance in the filtering problem. This does not mean that these terms are unimportant components of ARCH models. Nelson and Foster (1995) show that, although these terms are of second-order importance for filtering, they are of first-order importance in multi-step forecasting.

(iv) In the $\delta = 1$ case the asymptotic bias $b_T = 0_{m \times 1}$; and in the $\delta = \frac{3}{4}$ case, $b_T = B_T^{-1} A_T = B_T^{-1}[(\hat{\kappa} - \kappa) + \mathrm{E}_T[\partial G / \partial \xi_X](\mu - \hat{\mu})]$. $\mu = \hat{\mu}$ and $\hat{\kappa} = \kappa$ are *sufficient* but not *necessary* conditions for the elimination of this bias. In the forecasting results of Nelson and Foster (1995); $\mu = \hat{\mu}$ and $\hat{\kappa} = \kappa$ are necessary conditions for the ARCH model to generate appropriate forecasts as $h \downarrow 0$, further motivating our choice of $\mu = \hat{\mu}$ and $\hat{\kappa} = \kappa$.

(v) The assumptions of Theorems 2.1 and 2.2 are not as general as we would like. The first-order Markov structure immediately rules out, for example, fractionally integrated models (Ding *et al.*, 1993; Baillie *et al.*, 1993). As is well-known, finite-order Markov models can be written in first-order Markov form, but this does not usually help in our setup. Suppose, for example, that X_{t+h} is a scalar with conditional variance σ_t^2, and that $\ln(\sigma_t^2)$ is a linear ARMA(2, 1). To write it in first-order Markov form, suppose we are able to decompose $\ln(\sigma_t^2)$ into the sum of two linear AR(1) components, $y_{1,t}$ and $y_{2,t}$. Here, $\ln(\sigma_t^2) = y_{1,t} + y_{2,t}$ appears in the score term S_{t+h} but $y_{1,t}$ and $y_{2,t}$ do not appear individually, so $E_t[SS']$ is singular. In most cases, this prevents a solution to the Riccati equation. The B_T matrix of Theorem 2.1 is also singular, so both Theorems 2.1 and 2.2 break down. Similar problems arise for many higher order Markov processes: To avoid singular $E_t[SS']$, we generally need *all* elements of Y_t to enter the score S_{t+h} directly in a nondegenerate way. Whether this chapter's results can be extended to higher order models is an open question.

2.4. An Important Special Case

In NF, a closed form was available for ω_T. In the multivariate case, unfortunately, there often is not. Theorem 2.3 considers an important special case.

THEOREM 2.3.

(*a*) *Let* $E_T[SS']$ *be positive definite and let* $E_T[PS'] = 0_{m\times m}$. *Then the unique positive semidefinite solution to the Riccati equation of Assumption* 4 *is*[7]

$$\omega_T = (E_T[SS'])^{-1/2}\Big[(E_T[SS'])^{1/2}(E_T[(\xi_Y - P)(\xi_Y - P)'])$$

$$\times (E_T[SS'])^{1/2}\Big]^{1/2}(E_T[SS'])^{-1/2}, \quad (2.40)$$

(*b*) *In addition to the assumptions of Theorem* 2.2, *let the joint conditional density of* $[\xi_{X,t+h}, \xi_{Y,t+h}]$ *be elliptically symmetric. Then* $P_{t+h} = E_t[\xi_Y \xi_X'](E_t[\xi_X \xi_X'])^{-1}\xi_{X,t+h}$ (*i.e., given* X_t *and* Y_t, *the prediction component* P_{t+h} *is linear in* $\xi_{X,t+h}$) *and* $E_t[PS'] = 0_{m\times m}$. *For proof, see the Appendix.*

[7] $E_T[SS']$ and $E_T[(\xi_Y - P)(\xi_Y - P)']$ can typically be computed analytically as functions of the state variables. Given numerical values of $E_T[SS']$ and $E_T[(\xi_Y - P)(\xi_Y - P)']$, fast algorithms are available to compute ω_T in (2.40)—for example, using the MatrixPower command in version 2.2 of Mathematica (Wolfram Research, 1992). Given routines for computing eigenvalues and eigenvectors [for example, the eigrg2 and eigrs2 commands in Gauss (Aptech Systems, 1992)], it is also easy to write code to compute matrix square roots of real symmetric matrices.

The intuition behind (b) is straightforward. Let $Z \sim N(0, \Omega)$. $\mathrm{E}[Z_i \mid Z_k \ldots Z_m]$ is computed as a linear regression, and the differential with respect to Ω of the log of the density is $\frac{1}{2}$ Trace$(\mathrm{d}\Omega)\Omega^{-1}(ZZ' - \Omega)\Omega^{-1}$ (Magnus and Neudecker, 1988, Section 15.3) which is orthogonal to any linear combination of the Z_i's. While the functional form of the score is different for other elliptically symmetric distributions, orthogonality between the Z's and the score for parameters of the covariance matrix still holds, as does the linear regression structure of the conditional expectations (Cambanis *et al.*, 1981; Mitchell, 1989).

The elliptically symmetric case includes, for example, the multivariate normal, multivariate t, as well as the case in which $\{X_t, Y_t\}$ is generated by a discretely observed diffusion (see Theorem 3.1).

2.5. Fisher Information and Unpredictable Components

To interpret the matrix Riccati equation of Assumption 4, it helps to relate it to the familiar concepts of maximum likelihood and optimal prediction. Consider the simplest case, in which $\mathrm{E}_T[\mathrm{SS}']$ and $\mathrm{E}_T[(\xi_Y - \mathrm{P})(\xi_Y - \mathrm{P})']$ are diagonal and $\mathrm{E}_T[\mathrm{PS}'] = 0_{m \times m}$. Then (2.40) becomes

$$\omega_T = \left(\mathrm{E}_T[\mathrm{SS}']\right)^{-1/2}\left(\mathrm{E}_T\left[(\xi_Y - \mathrm{P})(\xi_Y - \mathrm{P})'\right]\right)^{1/2}. \qquad (2.41)$$

$\mathrm{E}_T[\mathrm{SS}']$ is the filtering analog of the Fisher information: The larger the Fisher information, the smaller is ω_T. $\mathrm{E}_T[(\xi_Y - \mathrm{P})(\xi_Y - \mathrm{P})']$, on the other hand, is a measure of the variance in the innovations variance in Y_{t+h} *after the predictable component* $\mathrm{P}(\xi_{X,t+h}, X_t, Y_t, t)$ *has been removed*. The larger this residual variance, the more unpredictably variable is $\{Y_t\}$ and the larger is ω_T. This argument holds in the general case in which $\mathrm{E}_T[\mathrm{SS}']$ and $\mathrm{E}_T[(\xi_Y - \mathrm{P})(\xi_Y - \mathrm{P})']$ may or may not be diagonal and $\mathrm{E}_T[\mathrm{PS}']$ may not be a matrix of zeros.

THEOREM 2.4. *Let* $\mathrm{E}_T[\mathrm{SS}']$ *and* $\mathrm{E}_T[(\xi_Y - \mathrm{P})(\xi_Y - \mathrm{P})']$ *be positive definite. Let* Δ *be* $m \times m$ *with* Δ *positive semidefinite and* $\Delta \neq 0_{m\times m}$, *let* ζ *be a scalar, and define* $\omega_T(\zeta)$ *implicitly by*

$$\mathrm{E}_T[\mathrm{PS}']\omega_T(\zeta) + \omega_T(\zeta)\mathrm{E}_T[\mathrm{SP}'] + \omega_T(\zeta)\mathrm{E}_T[\mathrm{SS}']\omega_T(\zeta)$$
$$= \mathrm{E}_T\left[(\xi_Y - \mathrm{P})(\xi_Y - \mathrm{P})'\right] + \zeta \cdot \Delta. \qquad (2.36')$$

Then $\mathrm{d}\omega_T(\zeta)/\mathrm{d}\zeta$ *evaluated at* $\zeta = 0$ *equals*

$$\int_0^{*} \exp\left[-s(\mathrm{E}_T[\mathrm{PS}'] + \omega_T(0)\mathrm{E}_T[\mathrm{SS}'])\right]\Delta$$
$$\times \exp\left[-s(\mathrm{E}_T[\mathrm{SP}'] + \mathrm{E}_T[\mathrm{SS}']\omega_T(0))\right]ds \qquad (2.42)$$

which is positive semidefinite and nonnull. Similarly, let β be $m \times m$ with β positive semidefinite and $\beta \neq 0_{m\times m}$, and let ψ be a scalar. Define $\omega_T(\psi)$ implicitly by

$$\mathrm{E}_T[\mathrm{PS}']\omega_T(\psi) + \omega_T(\psi)\mathrm{E}_T[\mathrm{SP}'] + \omega_T(\psi)(\mathrm{E}_T[\mathrm{SS}'] + \psi\cdot\beta)\omega_T(\psi) = \mathrm{E}_T[(\xi_Y - \mathrm{P})(\xi_Y - \mathrm{P})'] \quad (2.36'')$$

Then $\mathrm{d}\omega_T(\psi)/\mathrm{d}\psi$ evaluated at $\psi = 0$ equals

$$-\int_0^\infty \exp(-s(\mathrm{E}_T[\mathrm{PS}'] + \omega_T(0)\mathrm{E}_T[\mathrm{SS}']))\beta \times \exp(-s(\mathrm{E}_T[\mathrm{SP}'] + \mathrm{E}_T[\mathrm{SS}']\omega_T(0)))ds \quad (2.43)$$

which is negative semidefinite and nonnull. For proof, see the Appendix.

2.6. Conditional Moment Matching

One important aspect of the NF filtering results which is preserved in the multivariate setting is moment matching—i.e., for both the true data generating process and the ARCH model *interpreted as a data generating process*, the first two conditional moments are functions of the time t and the state variables X_t and Y_t. In NF's optimal filter these functions are identical in the ARCH model and the true DGP. When $\delta = \frac{3}{4}$, Theorem 2.2 required matching the first conditional moments—i.e., $\hat{\kappa}(X,Y,T,0) = \kappa(X,Y,T,0)$ and $\hat{\mu}(X,Y,T,0) = \mu(X,Y,T,0)$ (although this was a sufficient, not a necessary condition to eliminate asymptotic bias). Under a very mild additional assumption, the second moments are matched, whether $\delta = \frac{3}{4}$ or 1.

THEOREM 2.5. *Let the conditions of Theorem 2.2 hold, with $G(\cdot)$ given by* (2.39). *Then*

$$\mathrm{E}_t[G(\xi_{X,t+h}, X_t, Y_t, t, 0)G(\xi_{X,t+h}, X_t, Y_t, t, 0)'] = \mathrm{E}_t[\xi_{Y,t+h}\xi'_{Y,t+h}]. \quad (2.44)$$

Also, given X and t, let $\xi_X[f(\xi_X/X, y, t)]$ be uniformly continuous in Y on $(\xi_X, Y) \in R^{n+m}$. Then

$$\mathrm{E}_t[G(\xi_{X,t+h}, X_t, Y_t, t, 0)\xi'_{X,t+h}] = \mathrm{E}_t[\xi_{Y,t+h}\xi'_{X,t+h}]. \quad (2.45)$$

For proof, see the Appendix.

Notes:

(i) It is very important to note that second moment matching, while *necessary* for optimal filtering, is *not sufficient*. In general, there will be many ARCH models matching the first two moments, only one of which is the asymptotically optimal filter.

(ii) In a sense, the asymptotically optimal ARCH model makes itself as much like the true data generating process as possible. In filtering, as we have noted, the first moment terms μ, $\hat{\mu}$, κ, and $\hat{\kappa}$ are only second-order important. Because it matches both of the first two conditional moments, the optimal filter performs well at both filtering and forecasting (see Nelson and Foster, 1995). Other moment-matching ARCH models may also perform well at forecasting and adequately at filtering. In the continuous-time limit framework, only the first two conditional moments matter for multi-step forecasting.

(iii) Moment matching clearly has important implications for the design of ARCH models: It is hardly surprising, for example, that mean-reversion in volatility should be important for forecasting (if not so much for filtering). Many ARCH models have accommodated conditional correlation between ξ_X and ξ_Y to account, for example, for the negative conditional correlation between stock volatility and stock returns (i.e., the "leverage effect"). Modelling the conditional covariance matrix of Y_t has received less attention. For example, GARCH (Bollerslev, 1986), EGARCH (Nelson, 1991), the Taylor–Schwert model (Taylor, 1986; Schwert, 1989), TARCH (Zakoian, 1990) and many other univariate ARCH models assume that the log of the conditional variance of X_t is homoskedastic in the diffusion limit (see Nelson, 1990; Nelson and Foster 1994). This may or may not be appropriate—see the empirical section of Bollerslev *et al.* (1994) for some evidence that it may not be appropriate for U.S. stock returns data. Other models [for example, a Cox *et al.* (1985) model for the conditional variance] would be homoskedastic in the conditional standard deviation (i.e., the square root of the conditional variance). To be homoskedastic in these levels, one would have to keep the variance bounded away from zero, for example by adding a $1/\sigma_t$ drift term.

In models of short-term interest rates, it has been widely documented that interest yields are positively related to interest rate volatility (e.g., Chan *et al.*, 1992). This makes GARCH, EGARCH, and many related models inappropriate for modelling interest rate volatility, since they do not account for the effect of yields on volatility. A better choice is the ARCH model of Brenner *et al.* (1994) (altered slightly to admit a diffusion limit), in which

$$r_{t+h} = r_t + a \cdot h + b \cdot hr_t + h^{1/2}\xi_{r,t+h} \tag{2.46}$$

$$\text{Var}_t[\xi_{r,t+h}] = Y_t r_t^{2\gamma}, \qquad Y_{t+h} = \omega \cdot h + \beta \cdot Y_t + \alpha \cdot h^{1/2}[\xi_{r,t+h}^2 - Y_t r_t^{2\gamma}] \tag{2.47}$$

Even this model assumes that the variance innovations term $[\xi_{r,t+h}^2 r_t^{-2\gamma} - Y_t]$ is conditionally uncorrelated with the interest rate innovations term $\xi_{r,t+h}$, which may be inappropriate.

(iv) Moment matching has a practical application as a shortcut in computing the optimal filter. Once the functional forms of S and P are known, ω can be computed via the moment matching condition, which may be easier than solving the matrix Riccati equation. The model of Bollerslev *et al.* (1988) analyzed in section 4 is an example.

2.7. Change of Variables

So far, we have considered only optimality *given* the definitions of the state variables X_t and Y_t. Is there an optimal way to define the state variables? For example, in GARCH, Y_t is the conditional variance of X_{t+h}, while in EGARCH, Y_t is the log of the conditional variance. Is this difference asymptotically important? Formally, suppose we define the new state variables

$$\Upsilon_t \equiv \Upsilon(X_t, Y_t, t), \chi_t \equiv \chi(X_t, t), \tag{2.48}$$

where the functions $\Upsilon(X, Y, t)$ and $\chi(X, t)$ are twice continuously differentiable with $\partial\Upsilon(X, Y, t)/\partial Y$ and $\partial_\chi(X, t)/\partial X$ nonsingular. Given t, there is a one-to-one mapping from χ_t to X_t, and given t and X_t, there is a one-to-one mapping from Υ_t to Y_t, so the σ-algebras generated by $\{X_t, Y_t\}$ and $\{X_t\}$, respectively, are the same as those generated by $\{\chi_t, \Upsilon_t\}$ and $\{\chi_t\}$. Like X_t, χ_t is $n \times 1$ and observable. Like Y_t, Υ_t is $m \times 1$ and unobservable.

If we first construct asymptotically optimal estimates of Y_t using Theorem 2.2 and then apply the delta method (Serfling, 1980, Section 3.3, Theorem A), we derive the asymptotic distribution of $\Xi_t \equiv h^{-1/4}[\Upsilon(X_t, \hat{Y}_t, t) - \Upsilon(X_t, Y_t, t)]$ as

$$\Xi_t \Rightarrow N\left[0_{m\times 1}, \left[\frac{\partial\Upsilon}{\partial Y}\,\omega\,\frac{\partial\Upsilon}{\partial Y'}\right]\right], \tag{2.49}$$

where the i, jth element of $\partial\Upsilon/\partial Y$ is $\partial\Upsilon_i/\partial Y_j$.

Suppose, however, that we change variables *first* and then complete the optimal filter for the $\{\chi_t, \Upsilon_t\}$ process using Theorem 2.2. Is it possible, by a judicious choice of $\Upsilon(X, Y, t)$ and $\chi(X, t)$ to achieve an asymptotic covariance matrix of $h^{-1/4}[\hat{\Upsilon}_t - \Upsilon_t]$ *smaller* than the matrix in (2.49)? In yet another parallel to maximum-likelihood estimation, the answer is no.

THEOREM 2.6. *Define the state variables* Υ_t *and* χ_t *as in* (2.48). *In addition, let* $\xi_X f(\xi_X/X, Y, t)$ *be uniformly continuous in* Y *on* $(\xi_X, Y) \in$

R^{n+m}, and let there be a $\psi > 0$ such that

$$\mathrm{E}\Big[\big\|h^{-1/2}(\Upsilon_{t+h} - \Upsilon_t)\big\|^{2+\psi} \mid (\chi_t, \Upsilon_t) = (\chi, \Upsilon)\Big] \quad \text{and} \tag{2.50}$$

$$\mathrm{E}\Big[\big\|h^{-1/2}(\chi_{t+h} - \chi_t)\big\|^{2+\psi} \mid (\chi_t, \Upsilon_t) = (\chi, \Upsilon)\Big] \tag{2.51}$$

are bounded as $h \downarrow 0$, uniformly on every bounded (χ, Υ, t) set. Then the asymptotic distribution of $h^{-1/4}[\hat{\Upsilon}_t - \Upsilon_t]$ achieved by the asymptotically optimal filter for this system is given by (2.49). *For proof, see the Appendix.*

So we are not able to improve the asymptotic performance of the filter via a change of variables. For nonzero h such transformations may be important. Nelson and Schwartz (1992) and Schwartz (1994) show that, in Monte Carlo experiments, transformations of the state variables which reduce or eliminate the dependence of ω_T on Y_t substantially improve the asymptotic approximation for $h > 0$. This, of course, has many parallels in the literature on maximum likelihood going back at least to Fisher (1921).

If the assumptions of Theorem 2.1 are not satisfied (if, for example, the limit data-generating process is a jump–diffusion rather than a diffusion), transformations are important for another reason: The "near-diffusion" assumption guarantees that increments in the state variables $\{X_t, Y_t\}$ are small over small time intervals. This allowed us to approximate the increments in the transformed process $\{\Upsilon_t, \chi_t\}$ using a two-term Taylor series expansion of the functions $\Upsilon(\cdot)$ and $\chi(\cdot)$. If jumps are present as $h \downarrow 0$ (i.e., if the conditional distribution of ξ_X and ξ_Y is too thick-tailed) such an expansion is invalid, and the global, rather than just the local, properties of $\Upsilon(\cdot)$ and $\chi(\cdot)$ become relevant—for example, NF show that, under our regularity conditions, the filtering performance of the EGARCH model of Nelson (1991) is relatively robust to the presence of thick-tailed errors. However, as Engle and Ng (1993) point out, EGARCH arrives at $\hat{\sigma}_t$ by *exponentiating* a function of rescaled lagged residuals, so when the residual is huge (e.g., October 19, 1987) the two-term Taylor series approximation may break down and, because of the tail behavior of exp(·), the "robust" EGARCH model may become nonrobust.

3. DIFFUSIONS

We now consider the case in which the data are generated by a diffusion with $\{X_t\}$ observable at discrete intervals of length h:

$$d\begin{bmatrix} X_t \\ Y_t \end{bmatrix} = \begin{bmatrix} \mu(X_t, Y_t, t) \\ \kappa(X_t, Y_t, t) \end{bmatrix} h^{\delta-1}\, dt + \Omega(X_t, Y_t, t)^{\underline{1/2}}\, dW_t \tag{3.1}$$

where X_t and $\mu(X_t, Y_t, t)$ are $n \times 1$ and Y_t and $\kappa(X_t, Y_t, t)$ are $m \times 1$, W_t is an $(n+m) \times 1$ standard Brownian motion, and $\Omega(X_t, Y_t, t)$ is $(n+m) \times (n+m)$. (X_0, Y_0) may be random but are independent of $\{W_t\}_{0 \le t < \infty}$. We assume that (3.1) has a unique weak-sense solution for $0 < h < 1$. The $h^{\delta - 1}$ term in (3.1) plays the same role as h^{δ} in (2.2); namely, that $\delta = \frac{3}{4}$ allows the "large" drifts discussed in section 2.

As in the near-diffusion case, our interest is in estimating $\{Y_t\}$ using an ARCH model of the form (2.4)–(2.5). As we will see, the results for the diffusion case are identical to those for the near-diffusion case when $[\xi'_{X,t+h}, \xi'_{Y,t+h}]'$ is conditionally multivariate normal; i.e., for the model

$$\begin{bmatrix} X_{t+h} \\ Y_{t+h} \end{bmatrix} = \begin{bmatrix} X_t \\ Y_t \end{bmatrix} + \begin{bmatrix} \mu(X_t, Y_t, t) \\ \kappa(X_t, Y_t, t) \end{bmatrix} h^{\delta} + \begin{bmatrix} \xi_{X,t+h} \\ \xi_{Y,t+h} \end{bmatrix} h^{1/2}, \qquad \text{where} \quad (3.2)$$

$$\begin{bmatrix} \xi_{X,t+h} \\ & | X_t, Y_t \\ \xi_{Y,t+h} \end{bmatrix} \sim N\left[0_{(m+n)\times 1}, \Omega(X_t, Y_t, t)\right]. \quad (3.3)$$

THEOREM 3.1. *For each h, $0 < h < 1$, let the diffusion* (3.1) *possess a unique weak-sense solution, and let $\mu(\cdot)$, $\kappa(\cdot)$, and $\Omega(\cdot)$ be continuous. Define $G(\cdot)$, $\hat{\mu}$, and $\hat{\kappa}$ as in Theorem* 2.1, *where G has two partial derivatives with respect to ξ_X and Y almost everywhere, and let the absolute values of G and these two partial derivatives be bounded above by some polynomial in ξ_X with the coefficients of the polynomial continuous in X_t, Y_t, and t. Then the results of section* 2 *that are valid for the conditionally normal stochastic volatility model* (3.2)–(3.3) *hold for the diffusion model* (3.1). *For proof, see the Appendix.*

Notes:

(i) Essentially, Theorem 3.1 lets us treat a discretely observed diffusion (3.1) as if it were generated by the conditionally normal stochastic difference equation (3.2)–(3.3), the Euler approximation to (3.1). Since the multivariate normal is elliptically symmetric, the results of Theorem 2.3 hold.

(ii) The major complication that arises in the proof of Theorem 3.1 is that the moments required in Assumptions 1–3, in particular the $2 + \psi$ absolute moments required to asymptotically rule out jumps, may not exist. Even when these moments do exist, proving so can be tedious (e.g., the proofs of Theorems 4.3–4.5 in NF), since most of the stochastic volatility diffusion models fail the usual "growth condition" (Arnold, 1973, Chapter 6, 6.2.5) that guarantees the existence of arbitrary finite conditional moments. Fortunately, weaker moment conditions are available, involving truncated conditional means and covariances in place of the usual means and covariances. In the place of bounds on the $2 + \psi$ absolute moments used to asymptotically rule out jumps, direct limits on jump probabilities are imposed. For example, if $\{Z_t\}_{0, h, 2h \ldots}$ is a Markov chain, the moment conditions employed in the results of section 2 were of the form $h^{-\Delta} \mathrm{E}_t[Z_{t+h} - Z_t] \to \mu(Z_t)$, $h^{-\Delta} \mathrm{Cov}_t[Z_{t+h} - Z_t] \to \Omega(Z_t)$, and $h^{-\Delta} \mathrm{E}_t \|Z_{t+h} - Z_t\|^{2+\psi} < \infty$ for some $\psi > 0$ and some $\Delta > 0$. Such moment conditions are convenient if $\{Z_t\}$ is generated by a stochastic difference equation. The weaker conditions take the form $h^{-\Delta} \mathrm{E}_t[(Z_{t+h} - Z_t) I(\|Z_{t+h} - Z_t\| < 1)] \to \mu(Z_t)$, $h^{-\Delta} \mathrm{Cov}_t[(Z_{t+h} - Z_t) I(\|Z_{t+h} - Z_t\| < 1)] \to \Omega(Z_t)$, and $\forall \epsilon > 0$, $h^{-\Delta} \mathrm{Probability}_t[\|Z_{t+h} - Z_t\| > \epsilon] \to 0$, where $I(\cdot)$ is an indicator function. This latter set of conditions avoids the need for the conditional moments to exist and is ideally suited to cases (such as discretely observed diffusion) in which the existence of such moments is difficult to prove. This approach is taken in the proof of Theorem 3.1 in the Appendix.

(iii) Theorem 3.1 illustrates the important way in which near-diffusions and diffusions differ: Consider a sequence of stochastic volatility near-diffusions converging weakly to a given diffusion. If we think of the sample paths as random variables on function spaces (in our case, $D_{n+m}[0, \infty)$, the space of $n + m$-valued functions of t which are continuous from the right with finite left limits) the distributions of these random processes become arbitrarily close as $h \downarrow 0$. Nevertheless, in spite of arbitrary closeness *in distribution*, the filtering properties of the diffusion and the near-diffusion may be *very* different, since the time t conditional density of (X_{t+h}, Y_{t+h}) may be very far from conditionally normal. Convergence in distribution implies, for example, that for any fixed times $0 < t(1) < t(2) < \cdots < t(k) < \infty$, ($k$ finite) the joint distribution of $[(X_{t(1)}, Y_{t(1)}, \ldots, (X_{t(k)}, Y_{t(k)})]$ implied by the near-diffusion converges as $h \downarrow 0$ to the distribution implied by the limit diffusion. Convergence in distribution *does not imply*, however, that the h-step ahead transition probabilities implied by the near diffusion (i.e., from (X_t, Y_t) to (X_{t+h}, Y_{t+h})) converge in any sense to the h-step ahead transition probabilities implied by the diffusion, which become conditionally normal as $h \downarrow 0$. Clearly this is related to the standard result that convergence in distribution need not imply convergence of densities. Inference and filtering depend crucially on the h-step ahead transition densities. Weak convergence does not. This has important consequences in practice. As illustrated in NF, section 6, if the true conditional density is Student's t rather than normal, assuming conditional normality leads to a very inefficient filter even when the degrees of freedom of the t are quite high.

4. EXAMPLES

We next turn to selected examples of the results of the first three sections.

4.1. Some Multivariate ARCH Models

Many multivariate ARCH models have appeared in the literature. In all of these models we begin with an $n \times 1$ observable process $\{X_t\}_{t=0,h,2h\ldots}$ and its associated innovations process $\{\xi_{X,t}\}_{t=0,h,2h\ldots}$. We assume that $E_t[\xi_{X,t+h}] = 0_{m\times 1}$ and $\mathrm{Cov}_t[\xi_{X,t+h}] = \Omega_t$, where Ω_t is determined by a vector of unobservable state variables Y_t. The fitted residuals $\hat{\xi}_{X,t}$ are formed by

$$\hat{\xi}_{X,t+h} \equiv \left[X_{t+h} - X_t - h^{\delta}\hat{\mu}\left(X_t, \hat{Y}_t, t\right)\right]h^{-1/2} \tag{4.1}$$

The multivariate ARCH model with the simplest filtering properties is the factor ARCH model of Engle *et al.* (1990). The model posits that certain linear combinations of the $X_t s$ drive the conditional covariance matrix. For some $K < n$ and $k = 1$ to K, let $\hat{Y}_{k,t}$ be the ARCH estimate of the conditional variance of $\alpha_k X_{t+h}$, where α_k is nonstochastic and $1 \times n$. The model estimates $\mathrm{Cov}_t[\xi_{X,t+h}]$ by

$$\hat{\Omega}_t = \overline{\Omega} + \sum_{k=1}^{K} \beta_k \beta_k' \hat{Y}_{k,t}, \tag{4.2}$$

where $\overline{\Omega}$ is constant and positive semidefinite. In the simplest version of factor ARCH, the $\alpha_k X_{t+h}$ are modelled using univariate ARCH

$$\hat{Y}_{k,t+h} = \hat{Y}_{k,t} + h^{\delta}\hat{\kappa}_k\left(\alpha_k X_t, \hat{Y}_{k,t}, t\right) + h^{1/2}G_k\left(\alpha_k \hat{\xi}_{X,t+h}, \alpha_k X_t, \hat{Y}_{k,t}, t\right) \tag{4.3}$$

What are the filtering properties of the model under misspecification? Clearly for the model to perform well as a filter, the DGP had better exhibit the same factor structure; i.e.,

$$X_{t+h} = X_t + h^{\delta}\mu(X_t, Y_t, t) + h^{1/2}\xi_{X,t+h}, \qquad \text{where} \tag{4.4}$$

$$\Omega_t = \overline{\Omega} + \sum_{k=1}^{K} \beta_k \beta_k' Y_{k,t}, \qquad \text{with } \mathrm{Var}_t[\alpha_k \xi_{X,t+h}] = Y_{k,t}, \tag{4.5}$$

otherwise even consistent estimation of $\mathrm{Cov}_t[\xi_{X,t+h}]$ is impossible, although consistent estimation of $\mathrm{Cov}_t[\alpha_k \xi_{X,t+h}]$ may be. It should be clear that (4.1)–(4.3) correspond to an asymptotically optimal filter if the $\alpha_k X_{t+h}$

follow independent univariate stochastic volatility processes with optimal filters corresponding to (4.3). Even if the $(\alpha_k X_{t+h}, Y_{k,t})$ are dependent across k, Theorem 2.1 may still apply to the system of filters (4.2)–(4.3) provided (4.4)–(4.5) hold.

GARCH Generalizations

We now consider multivariate GARCH models. Like GARCH, these models estimate conditional covariances using averages of terms like $\xi_{X,t+h}\xi'_{X,t+h}$. The first multivariate GARCH model we consider is due to Bollerslev *et al.* (1988). This model sets $\hat{\xi}_{X,t}$ as in (4.1) and estimates the conditional covariance matrix of X_{t+h} via the recursion

$$\hat{\Omega}_{t+h} = \gamma_h + \beta_h \odot \hat{\Omega}_t + \alpha_h \odot \left(\hat{\xi}_{X,t} \hat{\xi}'_{X,t} \right) \tag{4.6}$$

where $\odot$ is the Hadamard (i.e., element-by-element) product. If $\hat{\Omega}_t$, γ_h, β_h, and α_h are all nonnegative definite, then as a consequence of the Schur Product Theorem (Horn and Johnson, 1985, Theorem 7.5.3), $\hat{\Omega}_{t+h}$ is nonnegative definite. The conditions of Theorem 2.1 require that all the elements of α_h be positive, so we make this assumption. In this model, the ijth element of $\hat{\Omega}_t$ is a function of only the ijth elements of lagged $\hat{\xi}_{X,t}\hat{\xi}'_{X,t}$. To put this filter in the form (2.4)–(2.5), define $\hat{Y}_t \equiv \text{vech}(\hat{\Omega}_t)$, where vech is the operator that stacks the lower triangle of a matrix into a vector. Now set $\gamma_h \equiv h^{\delta} \cdot \gamma$, $\alpha_h \equiv h^{1/2}\alpha$, and $\beta_h \equiv I - h^{1/2}\alpha - h^{\delta} \cdot \theta$, where $\delta = \frac{3}{4}$ or 1, θ is symmetric (but not necessarily positive semidefinite), and α and γ are symmetric and nonnegative definite with all elements of α positive. Now apply the vech operator to (4.6):[8]

$$\begin{aligned} \hat{Y}_{t+h} = \hat{Y}_t &+ h^{\delta} \cdot \text{vech}(\gamma) - h^{\delta} \cdot \text{vech}(\theta) \odot \hat{Y}_t \\ &+ h^{1/2}\, \text{vech}(\alpha) \odot \left[\text{vech}\left(\hat{\xi}_{X,t+h} \hat{\xi}'_{X,t+h} \right) - \hat{Y}_t \right] \end{aligned} \tag{4.6'}$$

This is in the form of (2.4)–(2.5), with

$$\begin{aligned} \hat{\kappa}\left(\hat{Y}_t\right) &= \text{vech}(\gamma) - \text{vech}(\theta) \odot \hat{Y}_t \\ G\left(\hat{\xi}_{X,t+h}, \hat{Y}_t \right) &= \text{vech}(\alpha) \odot \left[\text{vech}\left(\hat{\xi}_{X,t+h} \hat{\xi}'_{X,t+h} \right) - \hat{Y}_t \right]. \end{aligned} \tag{4.7}$$

[8] The constant correlations model of Bollerslev (1990) is a variant on this model, obtained by replacing the vech operator with the operator that stacks the diagonal elements of a square matrix into a vector. The conditional correlations are assumed constant.

Now suppose the DGP is a stochastic volatility model with $\{X_t, Y_t\}$ generated by (4.4) and

$$Y_{t+h} = Y_t + h^{\delta}\kappa(X_t, Y_t, t) + h^{1/2}\xi_{Y,t+h},$$

$$\text{where } \operatorname{vech}(\operatorname{Cov}_t[\xi_{X,t+h}]) = \operatorname{vech}(\Omega_t) = Y_t \quad (4.8)$$

When are the conditions of Theorem 2.1 satisfied so the filter, optimal or not, achieves the $h^{1/4}$ rate of convergence? When are the conditions of Theorem 2.2 satisfied, so the filter is asymptotically optimal? We first must evaluate the A_t, B_t, and C_t of Assumption 1:

$$A_t = \hat{\kappa}(Y_t) - \kappa(X_t, Y_t, t), B_t = \operatorname{diagrv}(\operatorname{vech}(\alpha)),$$

$$C_t = \operatorname{Cov}_t\left[\xi_{Y,t+h} - \operatorname{vech}(\alpha)\odot[\operatorname{vech}(\xi_{X,t+h}\xi'_{X,t+h})] - Y_t\right], \quad (4.9)$$

where diagrv(z) is the diagonal matrix with the elements of the vector z on the main diagonal. Note that even when $\mu \neq \hat{\mu}$ and $\delta = \frac{3}{4}$, μ and $\hat{\mu}$ do not introduce asymptotic bias, since $E_t[\partial G/\partial \xi_X] = 0_{m \times n}$. If we set $\mathrm{E}_t[X_{t+h} - X_t] = h^{5/8}\mu(\cdot)$ instead of $h^{3/4}\mu(\cdot)$, μ and $\hat{\mu}$ would remain in the asymptotic, since one of the higher order terms in (2.11) would no longer drop out. So in this model we might describe the effect of μ and $\hat{\mu}$ as third order for filtering, although it is still first order for forecasting. All the requirements of Theorem 2.1 are clearly satisfied as long as $\mathrm{E}_t[\|\xi_{X,t+h}\|^{4+\psi}]$ is bounded on bounded (X_t, Y_t, t) sets for some $\psi > 0$.

When are the conditions of Theorem 2.2 satisfied, so the filter is asymptotically optimal? Clearly we require $G = \mathrm{P} + \omega \mathrm{S}$, and if $\delta = \frac{3}{4}$, $\hat{\kappa}(Y_t) = \kappa(X_t, Y_t, t)$. One way for the filter to be optimal is for the GARCH model to be the true DGP. If $\xi_{Y,t+h} = G_{t+h} = \operatorname{vech}(\alpha)\odot[\operatorname{vech}(\xi_{X,t+h}\xi'_{X,t+h}) - Y_t]$, then $\mathrm{P}_{t+h} = \xi_{Y,t+h} = G_{t+h}$ and the covariance matrix V_T of Theorem 2.1 is a matrix of zeros. (We use Theorem 2.1 rather than Theorem 2.2 because we have violated Assumption 3's requirement of joint conditional densities for $(\xi_{X,t+h}, \xi_{Y,t+h})$.)

There are more interesting cases in which the filter can be optimal; note that the multivariate GARCH model estimates conditional covariances by averaging lagged $\xi_{X,t+h}\xi_{X,t+h}$'s, somewhat like an MLE estimator of Ω in an i.i.d. normal setting. Pursuing this analogy, suppose $Z \sim N(0_{n\times 1}, \Omega)$ with $Y = \operatorname{vech}(\Omega)$. Then using the results in Magnus and Neudecker (1988, section 8.4) we may write the score vector with respect to Y as $\frac{1}{2}D'(\Omega^{-1} \otimes \Omega^{-1})D[\operatorname{vech}(ZZ') - Y]$, where D is the $n^2 \times m$ duplication matrix (see Magnus and Neudecker, 1988, Section 3.8). This score looks strikingly like the G function of (4.7). Now define $\Lambda(\Omega)$ to be the $m \times m$ covariance matrix of $\operatorname{vech}(\alpha)\odot\operatorname{vech}(ZZ' - \Omega)$. Using subscripts to denote matrix or vector elements, we can evaluate the terms in $\Lambda(\Omega)$ using the relation $\mathrm{E}[Z_i \cdot Z_j \cdot Z_k \cdot Z_m] = \Omega_{ij}\Omega_{km} + \Omega_{ik}\Omega_{jm} + \Omega_{im}\Omega_{jk}$

[Anderson, 1984, Section 2.6 (26)]. Once again, let $\kappa(X_t, Y_t, t) = \hat{\kappa}(Y_t)$. Invoking the moment matching Theorem 2.5, we see that the filter (4.6′) is asymptotically optimal when the data are generated by

$$X_{t+h} = X_t + h^{\delta}\mu(X_t, Y_t, t) + h^{1/2}\xi_{X,t+h}$$

$$Y_{t+h} = Y_t + \left[\text{vech}(\gamma) - \text{vech}(\theta) \odot Y_t\right] h^{\delta} + h^{1/2}\xi_{Y,t+h},$$

where

$$\begin{bmatrix} \xi_{X,t+h} \\ \\ \xi_{Y,t+h} \end{bmatrix} \mid X_t, Y_t \sim N\left[\begin{bmatrix} 0_{n\times 1} \\ 0_{m\times 1} \end{bmatrix}, \begin{bmatrix} \Omega(Y_t) & 0_{n\times m} \\ 0_{m\times n} & \Lambda(Y_t) \end{bmatrix}\right] \tag{4.10}$$

$\xi_{X,t+h}$ and $\xi_{Y,t+h}$ are conditionally independent, with $\mathrm{P}_{t+h} = 0_{m\times 1}$ and $G_{t+h} = \omega_t \mathrm{S}_{t+h}$. Using (4.7), we see that

$$\omega_t = 2\left[D'\left(\Omega_t^{-1} \otimes \Omega_t^{-1}\right)D\right]^{-1} \text{diagrv}(\text{vech}(\alpha)).$$

Using Theorem 3.1, we see that the filter is also asymptotically optimal when the filter is generated by the diffusion

$$X_t = X_0 + \int_0^t h^{\delta-1}\mu(X_s, Y_s, s)\,ds + \int_0^t \Omega(Y_s)^{1/2}\,dW_{X,s}$$

$$Y_t = Y_0 + \int_0^t h^{\delta-1}[\text{vech}(\gamma) - \text{vech}(\theta) \odot Y_s]\,ds + \int_0^t \Lambda(Y_s)^{1/2}\,dW_{Y,s}, \tag{4.11}$$

where $W_{X,t}$ and $W_{Y,t}$ are independent $n \times 1$ and $m \times 1$ standard Brownian motions. (4.11) is the diffusion limit[9] of the Bollerslev, Engle, and Wooldridge model taken as a data generating process [i.e., the sort of diffusion limits derived in Nelson (1990)]. Although the ARCH model in this particular case is the optimal filter for its diffusion limit, this is not true generally. By Theorem 3.1, the diffusion limits can be treated as conditionally normal stochastic difference equations. The given ARCH models may or may not be optimal under conditional normality. For example, AR(1)-EGARCH has the Wiggins (1987) and Hull and White (1987) model as its diffusion limit (see Nelson, 1990). The optimal filter for this model, given in NF, pp. 20–21, is not EGARCH.

[9] For proof that (4.11) is a valid diffusion limit, see the proof of Theorem 4.2 in the NBER technical working paper version of this chapter (#162, August, 1994).

Next, consider the BEKK model of Engle and Kroner (1994), in which

$$\hat{\Omega}_{t+h} = \gamma_h + \beta_h \odot \hat{\Omega}_t + \sum_{k=1}^{K} A'_{h,k} \hat{\xi}_{X,t+h} \hat{\xi}'_{X,t+h} A_{h,k} + \sum_{k=1}^{K} B'_{h,k} \hat{\Omega}_t B_{h,k} \tag{4.12}$$

where γ_h and B_h are nonnegative definite, and the $A_{h,k}$ and $B_{h,k}$ are arbitrary $n \times n$ matrices.[10] The $\hat{\Omega}_{t+h}$ is positive semidefinite whenever $\hat{\Omega}_t$ is.

To put the filter in the form (2.4)–(2.5), we first let $\gamma_h = h^{\delta}\gamma$, $\beta_h = I_{n \times n} - \alpha h^{1/2} - \theta h^{\delta}$, $A_{h,k} = h^{1/4} A_k$, and $B_{h,k} = h^{\delta/2} B_k$, where the A_k and B_k are arbitrary $n \times n$ matrices, γ and α are $n \times n$ and positive semidefinite, and θ is $n \times n$ and symmetric. Applying vech($\cdot$) to (4.12) and rearranging;

$$\begin{aligned} \hat{Y}_{t+h} = {} & \hat{Y}_t - \text{diagrv}(\text{vec}(\alpha h^{1/2} + h^{\delta}\theta)) D\hat{Y}_t \\ & + \sum_{k=1}^{K} \left[h^{\delta} B'_k \otimes B'_k + h^{1/2} A'_k \otimes A'_k \right] D\hat{Y}_t \\ & + h^{\delta}\, \text{vech}(\gamma) + h^{1/2} \left[\sum_{k=1}^{K} A'_k \otimes A'_k \right] D\left[\text{vech}\left(\hat{\xi}_{X,t+h} \hat{\xi}_{X',t+h} \right) - \hat{Y}_t \right] \end{aligned} \tag{4.13}$$

So $G(\hat{\xi}_{X,t+h}, \hat{Y}_t) = [\Sigma_{k=1,K} A'_k \otimes A'_k] D[\text{vech}(\hat{\xi}_{X,t+h} \hat{\xi}'_{X,t+h}) - \hat{Y}_t]$ and

$$\begin{aligned} \hat{\kappa}_h(\hat{Y}) = {} & \text{vech}(\gamma) - \text{diagrv}(\text{vec}(\alpha h^{1/2-\delta} + \theta)) D\hat{Y} \\ & + \Sigma_{k=1,K} \left[B'_k \otimes B'_k + h^{1/2-\delta} A'_k \otimes A'_k \right] DY. \end{aligned} \tag{4.14}$$

A problem arises from the terms in $\hat{\kappa}$ multiplied by $h^{1/2-\delta,}$ which explode as $h \downarrow 0$, violating the conditions of Theorems 2.1 and 2.2. The only way to avoid this problem is for the two terms to cancel; i.e., for diagrv(vec(α)) $= \Sigma_{k=1,K}[A'_k \otimes A'_k]$, which in turn requires each of the A_k matrices to be diagonal. If we make this assumption, the model simplifies

$$\begin{aligned} \hat{Y}_{t+h} = {} & \hat{Y}_t + h^{\delta}\, \text{vech}(\gamma) - h^{\delta}\, \text{diagrv}(\text{vech}(\theta)) \hat{Y}_t \\ & + h^{\delta} \sum_{k=1}^{K} [B'_k \otimes B'_k] D\hat{Y}_t + h^{1/2}\, \text{diagrv}(\text{vech}(\alpha)) \\ & \times \left[\text{vech}\left(\hat{\xi}_{X,t+h} \hat{\xi}'_{X,t+h} \right) - \hat{Y}_t \right] \end{aligned} \tag{4.15}$$

[10] For notational convenience, we have added the $\beta_h \odot \Omega_t$ term to the BEKK model as presented in Engle and Kroner (1994). As Engle and Kroner show, this term can be effectively added to their equation (2.2) by a suitable choice of the $B_{h,k}$'s.

which is the same as the Bollerslev, Engle, and Wooldridge filter apart from the $\hat{\kappa}$ term. Not surprisingly, it satisfies the conditions of Theorem 2.1 when the elements of α are positive and for some $\psi > 0$, $E_t[\|\xi_{X,t+h}\|^{4+\psi}$ is bounded on bounded (X_t, Y_t, t) sets. Like the earlier model, the filter is asymptotically optimal; e.g., when the BEKK model is the true DGP, or when

$$X_{t+h} = X_t + h^{\delta}\mu(X_t, Y_t, t) + h^{1/2}\xi_{X,t+h}$$

$$Y_{t+h} = Y_t + h^{\delta}[\text{vech}(\gamma) - \text{vech}(\theta) \odot Y_t]$$

$$+ h^{\delta} \sum_{k=1}^{K} B_k' \otimes B_k' DY_t + h^{1/2}\xi_{Y,t+h}, \qquad \text{where}$$

$$\begin{bmatrix} \xi_{X,t+h} \\ \\ \xi_{Y,t+h} \end{bmatrix} | X_t, Y_t \sim N\left[\begin{bmatrix} 0_{n\times 1} \\ 0_{m\times 1} \end{bmatrix} \begin{bmatrix} \Omega(Y_t) & 0_{n\times m} \\ 0_{m\times n} & \Lambda(Y_t) \end{bmatrix}\right] \tag{4.16}$$

or when the DGP is the discretely observed diffusion:

$$X_t = X_0 + \int_0^t h^{\delta-1}\mu(X_s, Y_s, s)\,ds$$

$$+ \int_0^t \Omega(Y_s)^{1/2}\, dW_{X,s}$$

$$Y_t = Y_0 + \int_0^t h^{\delta-1}\left[\text{vech}(\gamma) - \text{vech}(\theta) \odot Y_s + \sum_{k=1}^{K} B_k \otimes B_k DY_s\right] ds$$

$$+ \int_0^t \Lambda(Y_s)^{1/2}\, dW_{Y,s}, \tag{4.17}$$

The analysis of Kroner and Ng's (1993) General Dynamic Covariance Model is similar: The $G(\cdot)$ function is the same as in the previous two models, but the $\hat{\kappa}$ function differs.

4.2. Conditional Heterokurticity

Most ARCH models assume a constant shape to the conditional distribution—e.g., a conditional Student's t distribution with fixed degrees of freedom. There is evidence, however, that the shapes of the conditional distributions of asset returns are time varying. In Figure 1, we plot

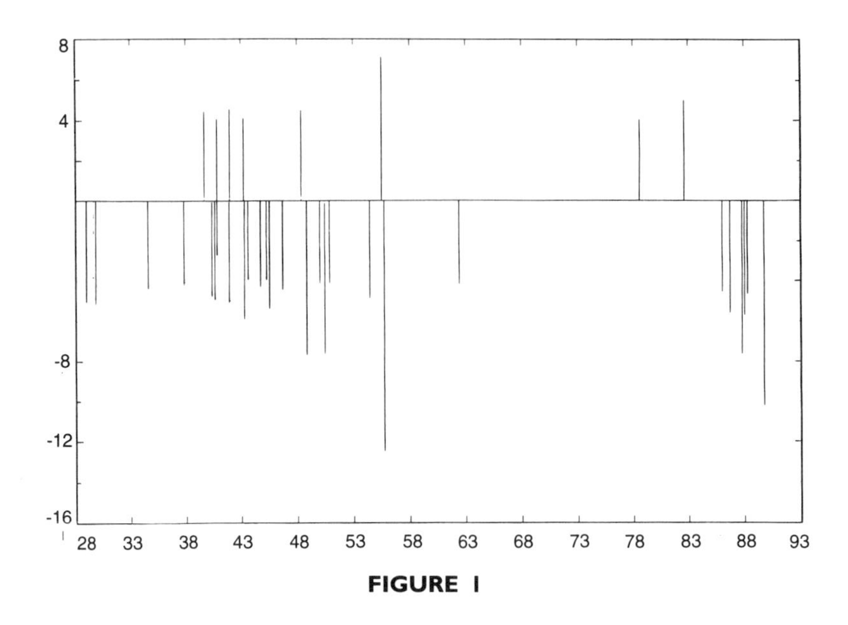

FIGURE 1

standardized residuals $\hat{\xi}_{X,t}/\hat{\sigma}_t$ exceeding four in absolute value, where the $\hat{\sigma}_t$'s are generated by a univariate EGARCH model fit to S & P 500 daily returns from January 1928 through December 1990. If the EGARCH model is correctly specified (or if it is a relatively efficient filter), the standardized residuals should be approximately i.i.d. It is clear that the large residuals clump together over time; there are many more outliers in the 1940s, 1950s, and late 1980s than during other periods. The conditional skewness may also be changing over time.[11]

To keep things relatively simple, we next consider a model with changing conditional kurtosis but constant (zero) conditional skewness. The (scalar) observable state variable x_t exhibits both time-varying conditional variance (governed by the unobservable state variable y_t) and time-varying conditional tail thickness (governed by the unobservable state variable ν_t). We assume that $[X'_{t+h}, Y'_{t+h}]'$ is conditionally multivariate t[12]

[11] For evidence of time-varying higher order conditional moments found using other methodologies, see Bates (1991, 1993), Turner and Weigel (1992), and Hansen (1994).

[12] There are a number of "multivariate t" distributions—here we mean the multivariate t with a common denominator and common degrees of freedom (see Johnson and Kotz, 1972, sections 37.3–37.4).

with $2 + \exp(\nu_t)$ degrees of freedom:

$$\begin{bmatrix} x_{t+h} \\ y_{t+h} \\ \nu_{t+h} \end{bmatrix} = \begin{bmatrix} x_t \\ y_t \\ \nu_t \end{bmatrix} + \begin{bmatrix} \mu(x_t, y_t, \nu_t) \\ \kappa_y(x_t, y_t, \nu_t) \\ \kappa_\nu(x_t, y_t, \nu_t) \end{bmatrix} h^\delta + \begin{bmatrix} \xi_{x,t+h} \\ \xi_{y,t+h} \\ \xi_{\nu,t+h} \end{bmatrix} h^{1/2} \tag{4.18}$$

$$\begin{bmatrix} \xi_{x,t+h} & \\ \xi_{y,t+h} & | \, x_t, y_t, \nu_t \\ \xi_{\nu,t+h} & \end{bmatrix}$$

$$\sim \text{MVT}_{2+e^{\nu_t}} \left[\begin{bmatrix} 0 \\ 0 \\ 0 \end{bmatrix}, \begin{bmatrix} e^{y_t} & e^{y_t/2}\rho_1\Lambda_1 & e^{y_t/2}\rho_2\Lambda_2 \\ e^{y_t/2}\rho_1\Lambda_1 & \Lambda_1^2 & \rho_3\Lambda_1\Lambda_2 \\ e^{y_t/2}\rho_2\Lambda_2 & \rho_3\Lambda_1\Lambda_2 & \Lambda_2^2 \end{bmatrix} \right] \tag{4.19}$$

To satisfy the regularity conditions and to ensure that the conditional variances are well defined, the degrees of freedom are bounded below by 2. To enforce positive semidefiniteness on the conditional covariance matrix, let $|\rho_j| \leq 1$ for $j = 1, 2, 3$ and $1 - \rho_1^2 - \rho_2^2 - \rho_3^2 + 2\rho_1\rho_2\rho_3 \geq 0$.[13]

Our interest is in estimating both y_t and υ_t. Without the aid of Theorem 2.2, it is by no means clear how to proceed: A GARCH estimate of σ_t^2, for example, has an immediate and intuitive interpretation as a smoothed empirical variance, but how does one empirically smooth conditional kurtosis? Application of Theorem 2.2, however, yields Theorem 4.1.

THEOREM 4.1. *The asymptotically optimal filter for* (4.18)–(4.19) *is*

$$G_{t+h} \equiv \text{P}_{t+h} + \omega_t \text{S}_{t+h}, \qquad \text{where} \tag{4.20}$$

$$\omega_t = (\text{E}_t[\text{SS}'])^{-1/2}\Big[(\text{E}_t[\text{SS}'])^{1/2}(\text{E}_t[(\xi_Y - \text{P})(\xi_Y - \text{P})]$$

$$\times (\text{E}_t[\text{SS}'])^{1/2}\Big]^{1/2}(\text{E}_t[\text{SS}'])^{-1/2}, \tag{4.21}$$

$$\text{P}_{t+h} \equiv \text{E}_t\left[\begin{bmatrix} \xi_{y,t+h} \\ \xi_{\nu,t+h} \end{bmatrix} \,|\, \xi_{x,t+h} \right] = \begin{bmatrix} \rho_1\Lambda_1 \\ \rho_2\Lambda_2 \end{bmatrix} \xi_{x,t+h} \, e^{-y_t/2} \tag{4.22}$$

[13] This is not an ARCH model according to the definition of Andersen (1992), but it is by the definition of Bollerslev *et al.* (1994), although they did not focus on higher order moments. Certainly, time-varying higher order moments are "heteroskedastic" in the sense of McCulloch (1985). Hansen (1994) would call this an "autoregressive conditional density" model.

$$
S_{t+h} = \begin{bmatrix} \dfrac{(3+e^{\nu_t})\,\xi_{x,t+h}^2}{2(\xi_{x,t+h}^2 + e^{y_t+\nu_t})} - \dfrac{1}{2} \\[2ex] \dfrac{e^{\nu_t}}{2}\left[\psi\left(\dfrac{3+e^{\nu_t}}{2}\right) - \psi\left(1+\dfrac{e^{\nu_t}}{2}\right)\right] \\[2ex] -\dfrac{e^{\nu_t}}{2}\ln\left(1+\dfrac{\xi_{x,t+h}^2}{e^{y_t+\nu_t}}\right) + \dfrac{(3+e^{\nu_t})\,\xi_{x,t+h}^2}{2(\xi_{x,t+h}^2 + e^{y_t+\nu_t})} - \dfrac{1}{2} \end{bmatrix} \tag{4.23}
$$

$$
E_t[(\xi_Y - P)(\xi_Y - P)'] = \begin{bmatrix} (1-\rho_1^2)\Lambda_1^2 & (\rho_3 - \rho_1\rho_2)\Lambda_1\Lambda_2 \\ (\rho_3 - \rho_1\rho_2)\Lambda_1\Lambda_2 & (1-\rho_2^2)\Lambda_2^2 \end{bmatrix} \tag{4.24}
$$

$$
E_t[SS']
$$

$$
= \begin{bmatrix} \dfrac{2+e^{\nu_t}}{2(5+e^{\nu_t})} & \dfrac{3}{e^{2\nu_t}+8e^{\nu_t}+15} \\[2ex] \dfrac{3}{e^{2\nu_t}+8e^{\nu_t}+15} & \dfrac{e^{2\nu_t}}{4}\left[\psi'\left(1+\dfrac{e^{\nu_t}}{2}\right) - \psi'\left(\dfrac{3+e^{\nu_t}}{2}\right)\right] \\ -\dfrac{(e^{\nu_t}-1)(e^{\nu_t}+6)}{2(e^{\nu_t}+3)(e^{\nu_t}+5)} & \end{bmatrix} \tag{4.25}
$$

where $\mu = \hat{\mu}$, $\kappa_y = \hat{\kappa}_y$, $\kappa_\nu = \hat{\kappa}_\nu$. *Here* $\psi(\cdot)$ *is the Digamma*[14] *function* $\psi(x) \equiv d[\ln \Gamma(x)]/dx$ *for* $x > 0$. *For proof, see the Appendix.*

Clearly one could not arrive at (4.20)–(4.25) by ad hoc modelling strategies. Ignoring conditional heterokurticity can have important consequences. Suppose, for example, that we ignore ν_t, and estimate y_t using the filter corresponding to a conditionally normal distribution. When the conditional degrees of freedom $2 + \exp(\nu_t)$ exceed 4, the filter still achieves an $h^{1/4}$ rate of convergence for $(\hat{y}_t - y_t)$, but its efficiency can be very bad (see NF, Figure 2 and pp. 27–28), a conclusion reinforced by the Monte Carlo experiments of Schwartz (1994). When the conditional degrees of freedom drop below four, the filter breaks down altogether.

It may seem surprising that higher order conditional moments can be consistently estimated in a diffusion limit framework, since distributionally,

[14] For given values of ν, $\psi(\cdot)$ and its derivative $\psi'(\cdot)$ can be easily computed using, for example, the PolyGamma function in Mathematica (Wolfram Research, 1992). It is also easy to write code for computing $\psi(\cdot)$ and $\psi'(\cdot)$ using the asymptotic approximations and recurrence relations given in Davis (1964, formulas 6.3.6, 6.3.18, 6.4.6, and 6.4.12).

diffusions are characterized by their first two conditional moments.[15] To see why higher order conditional moments can be extracted, note that if the conditional degrees of freedom process has a diffusion limit, it has (in the limit) continuous sample paths almost surely and so is asymptotically constant (as is the conditional variance process) over a vanishingly small time interval $[T, T + M_h h^{1/2}]$. Yet, over that vanishingly small interval, we see a growing number $[M_h h^{1/2}]$ of realizations of $h^{-1/2}(x_{t+h} - x_t)$, each of which is (asymptotically as $h \downarrow 0$) conditionally Student's t with a constant variance and constant degrees of freedom. We can estimate the variance and degrees of freedom in this case just as we can in the i.i.d. case. The asymptotically optimal filter carries this estimation out.

5. CONCLUSION

Most empirical research on conditional variances and covariances uses parametric ARCH models.[16] Is the filtering theory relevant to the design of these models? I believe the answer is definitely yes. As is frequently observed, all models are misspecified; but many ARCH models possess a remarkable robustness property, so that even when misspecified they may perform well at both filtering (this chapter and NF) and forecasting (Nelson and Foster, 1995). In my opinion, this robustness is largely responsible for the empirical success of ARCH. Robustness can be much improved, however, by applying the lessons of filtering theory. Thirteen years after the publication of Engle (1982a), nothing could be easier than writing down a conditionally heteroskedastic time series model. Writing down a model that is robust to an interesting array of misspecifications is harder. Since the implications of economic theory for the behavior of volatility are sparse, striving for filtering and forecasting robustness places a useful check (one of the few available) on ad hockery in model design.

The results of this chapter (and of Nelson and Foster, 1994b) show that certain features of the ARCH model are particularly important in making filtering and forecasting performance robust to misspecification:

(a) Correct specification of the second conditional moment (for accurate filtering). Specifying the conditional second moments of the states correctly requires, for example, correctly modelling explanatory variables entering the conditional moments of the Y variables, such as the effect of interest rates on interest rate volatility.

[15] More precisely, diffusions with unique weak-sense solutions are characterized by their first two conditional moments, sample path continuity, and the distribution of the initial starting point.

[16] These comments apply equally to semi-nonparametric models (Gallant, *et al.*, 1992, 1993), which rely heavily on a good choice of the leading (parametric) term.

(b) Correct specification of the first two conditional moments (for accurate forecasting).
(c) The prediction component P, which captures correlation between the observable and unobservable processes, for example, the "leverage effect" in stock volatility.
(d) The score component S, which, in a manner similar to maximum likelihood, captures information on Y_t contained in the residual $\xi_{X,t+h}$. Failure to account for heavy-tailed or skewed errors via appropriate specification of the score may lead to poor filter performance.

APPENDIX

Proof of Theorem 2.1. Substantially identical to the proof of Theorem 3.1 in NF.

Proof of the Corollary. The results are taken from Karatzas and Shreve (1988, section 5.6). Their (6.12)–(6.13) give the differential equations for $b_{T,\tau}$ and $V_{T,\tau}$:

$$db_{T,\tau}/d\tau = B_T b_{T,\tau} + A_T, \tag{A.1}$$

$$dV_{T,\tau}/d\tau = -B_T V_{T,\tau} - V_{T,\tau} B'_T + C_T \tag{A.2}$$

with initial conditions $b_{T,0} = Q$ and $V_{T,0} = 0_{m\times m}$. The unique solutions are given in (2.25)–(2.26): (2.25) is their (6.10) and (2.26) is their (6.14′). To go from (A.2) to (2.27), take the vec of both sides of (A.2) using the rule for evaluating the vec of a product of two matrices [see Magnus and Neudecker, 1988, Chapter 2, section 4, equation (7)]. This yields

$$(d/dt)\text{vec}(V_{T,\tau}) \\ = -[I_{m\times m} \otimes B_T + B_T \otimes I_{m\times m}]^{-1}\text{vec}(V_{T,\tau}) + \text{vec}(C_T), \tag{A.3}$$

a linear vector o.d.e. with solution (2.27).

Proof of Theorem 2.2. Since there is no variance–bias tradeoff, we may consider $b_T^* b_T^{*\prime}$ and V^* separately. Clearly, (2.37)–(2.38) are sufficient to eliminate $b_T^* b_T^{*\prime}$, so we turn our attention to V^*. We first guess a solution and then verify it. To arrive at the guess (2.39), we differentiate $\text{Trace}[b_T b'_T + V_T^*]$ with respect to $G(\xi_X, \cdot)$, drop the $d\xi_X$, $d\xi_y$ terms, (naively) treating $G(\xi_X, \cdot)$ as a separate choice variable for each ξ_X. This yields (2.39). To verify global optimality, we need the following lemma, a consequence of the law of iterated expectations.

LEMMA A.1. *Let $\theta(\cdot)$ be a $k \times 1$ integrable vector function of ξ_X, X, Y, and t. Then $E_t[\theta'(\xi_X, X, Y, t)\xi'_Y] = \text{E}_t[\theta'(\xi_X, X, Y, T, \tau)\text{P}']$.*

Proof of Lemma A.1.

$$\mathrm{E}_t[\theta\xi_Y] = \int_{-\infty}^{\infty} \cdots \int_{-\infty}^{\infty} \theta' \xi_Y f(\xi_X, \xi_Y \mid X, Y, T, \tau) d\xi_Y \, d\xi_X \tag{A.4}$$

$$= \int_{-\infty}^{\infty} \cdots \int_{-\infty}^{\infty} \theta' \left[\int_{-\infty}^{\infty} \cdots \int_{-\infty}^{\infty} \xi_Y f(\xi_Y \mid \xi_X, X, Y, t) d\xi_Y \right]$$

$$\times (\xi_X \mid X, Y, t) d\xi_X \tag{A.5}$$

$$= \int_{-\infty}^{\infty} \cdots \int_{-\infty}^{\infty} \theta' \mathrm{P} f(\xi_X \mid X, Y, t) d\xi_X = \mathrm{E}_t[\theta' \mathrm{P}]. \tag{A.6}$$

Now we continue with the proof of Theorem 2.2. We drop time subscripts when it should cause no confusion. Integrating by parts yields $\mathrm{E}_t[\partial G_i / \partial Y_j] = -\mathrm{E}_t[G_i \mathrm{S}_j]$, so

$$B = E_t[G\mathrm{S}']. \tag{A.7}$$

Now consider another choice of G, say $\tilde{G} \equiv \mathrm{P} + \omega\mathrm{S} + H$, where H is a function of ξ_X, X, Y, T, and t with $\tilde{G}$ satisfying the conditions of Theorem 2.1. If the real parts of the eigenvalues of $\tilde{B} \equiv \mathrm{E}_t[\tilde{G}\mathrm{S}'] = \mathrm{E}_t[\mathrm{PS}'] + \omega \mathrm{E}_t[\mathrm{SS}'] + \mathrm{E}_t[H\mathrm{S}']$ are all positive (as required by Theorem 2.1) there is a bounded $\tilde{V}$ that is the asymptotic covariance matrix of the measurement error process (see Lancaster and Tismenetsky, 1985, Chapter 12.3, Theorem 3). If $\tilde{V}$ is unbounded its trace is clearly larger than that of ω, so we assume that the real parts of the eigenvalues of $\tilde{B}$ are strictly positive. By (2.28),

$$\tilde{B}\tilde{V} + \tilde{V}\tilde{B}' - \tilde{C} = 0, \tag{A.8}$$

where by (2.18), $\tilde{C} = \mathrm{E}_t[(\mathrm{P} + \omega_T \mathrm{S} + H - \xi_Y)(P + \omega_T \mathrm{S} + H - \xi_Y)']$. Similarly, ω satisfies

$$B\omega + \omega B' - C = 0, \tag{A.9}$$

where $B \equiv \mathrm{E}_t[\mathrm{PS}'] + \omega_T \mathrm{E}_t[\mathrm{SS}']$ and $C \equiv \mathrm{E}_t[(\mathrm{P} + \omega\mathrm{S} - \xi_Y)(\mathrm{P} + \omega\mathrm{S} - \xi_Y)']$. Subtracting (A.9) from (A.8), substituting for B, C, $\tilde{B}$, and $\tilde{C}$, and simplifying (employing Lemma A.1) leads to

$$\tilde{B}(\tilde{V} - \omega) + (\tilde{V} - \omega)\tilde{B}' = \mathrm{E}_t[HH'], \tag{A.10}$$

which is an equation of the same form as (2.28), and has (see Lancaster and Tismenetsky, 1985, Chapter 12.3, Theorem 3) a solution of the same form as (2.29):

$$(\tilde{V} - \omega) = \int_0^{\infty} \exp[-\tilde{B} \cdot s] \mathrm{E}_t[HH'] \exp[-\tilde{B}' \cdot s] ds. \tag{A.11}$$

The right-hand side of (A.11) is clearly positive semidefinite. $\tilde{V}$ therefore exceeds ω by a positive semi-definite matrix. Because of the nonsingularity of $\exp[-\tilde{B} \cdot s]$, $\tilde{V} = \omega$ if and only if $\mathrm{E}_t[HH']$ is a matrix of zeros. The Riccati equation follows by substituting for B and C in (A.9).

Proof of Theorem 2.3.

(a) Equation (2.36) becomes

$$\mathrm{E}_T[(\xi_Y - \mathrm{P})(\xi_Y - \mathrm{P})'] = \omega_T \mathrm{E}_T[\mathrm{SS}']\omega_T. \qquad (2.36')$$

Pre- and postmultiplying by $\mathrm{E}_T[\mathrm{SS}']^{1/2}$, we obtain

$$\mathrm{E}_T[\mathrm{SS}']^{1/2}\mathrm{E}_T[(\xi_Y - \mathrm{P})(\xi_Y - \mathrm{P})']\mathrm{E}_T[\mathrm{SS}']^{1/2} = \mathrm{E}_T[\mathrm{SS}']^{1/2}\omega_T \mathrm{E}_T[\mathrm{SS}']^{1/2}\mathrm{E}_T[\mathrm{SS}']^{1/2}\omega_T \mathrm{E}_T[\mathrm{SS}']^{1/2}. \qquad (\mathrm{A}.12)$$

Taking symmetric matrix square roots,

$$\left[\mathrm{E}_T[\mathrm{SS}']^{1/2}\mathrm{E}_T[(\xi_Y - \mathrm{P})(\xi_Y - \mathrm{P})']\mathrm{E}_T[\mathrm{SS}']^{1/2}\right]^{1/2} = \mathrm{E}_T[\mathrm{SS}']^{1/2}\omega_T \mathrm{E}_{sT}[\mathrm{SS}']^{1/2}. \qquad (\mathrm{A}.13)$$

Pre- and postmultiplying by $\mathrm{E}_T[\mathrm{SS}']^{-1/2}$ yields (2.40).

(b) That $\mathrm{P}_{t+h} = \mathrm{E}_T[\xi_Y \xi_X'](\mathrm{E}_T[\xi_X \xi_X'])^{-1}\xi_{X,t+h}$ is Cambanis *et al.* (1981, Corollary 5). $\mathrm{E}_T[\mathrm{PS}'] = 0_{m \times m}$ follows from Mitchell [1989, (2.4) and (2.7)].

To prove Theorem 2.4 we need a lemma.

LEMMA A.2. *Let* $\mathrm{E}_T[\mathrm{SS}']$ *and* $\mathrm{E}_T[(\xi_Y - \mathrm{P})(\xi_Y - P)']$ *be positive definite. Then the matrix* $-(\mathrm{E}_T[\mathrm{PS}'] + \omega_T \mathrm{E}_t[\mathrm{SS}'])$ *is stable—i.e., the real parts of its eigenvalues are strictly negative.*

Proof of Lemma A.2. By Kučera (1972, Theorem 3), a sufficient condition for $-(\mathrm{E}_T[\mathrm{PS}']\omega_T \mathrm{E}_T[\mathrm{SS}'])$ to be stable is that there exist real matrices M_1 and M_2 such that

$$-\mathrm{E}_T[\mathrm{SP}'] + \mathrm{E}_T[\mathrm{SS}'] \cdot M_1 \quad \text{and} \qquad (\mathrm{A}.14)$$

$$-\mathrm{E}_T[\mathrm{PS}'] + \mathrm{E}_T[(\xi_Y - \mathrm{P})(\xi_Y - \mathrm{P})'] \cdot M_2 \qquad (\mathrm{A}.15)$$

are stable. For $i = 1, 2$, we choose $M_i = -I_{m \times m}/k_i$, where k_i is a small positive number. Equations (A.14)–(A.15) are equivalent to the stability of $-\mathrm{E}_T[\mathrm{SP}'] \cdot k_1 - \mathrm{E}_T[\mathrm{SS}'\}$ and $-\mathrm{E}_T[\mathrm{PS}'] \cdot k_2 - \mathrm{E}_t[(\xi_Y - \mathrm{P})(\xi_Y - \mathrm{P})']$. Consider these as functions of k_i. (Recall $-\mathrm{E}_T[(\xi_Y - \mathrm{P})(\xi_Y - \mathrm{P})']$ and $-\mathrm{E}_T[\mathrm{SS}']$ are stable by assumption.) These functions are analytic, so their eigenvalues are continuous functions of k_i (see Lancaster and Tismenetsky (1985, Chapter 11.7, Theorem 1 and Exercise 6). This implies their stability for sufficiently small k_i.

Proof of Theorem 2.4. Define $\omega_T(\zeta)$ as in (2.36′). Differentiating at $\zeta = 0$,

$$(\mathrm{E}_T[\mathrm{PS}'] + \omega_T(0)\mathrm{E}_T[\mathrm{SS}'])d\omega_T(\zeta)/d\zeta + (d\omega_T(\zeta)/d\zeta)(\mathrm{E}_T[\mathrm{SP}'] + \mathrm{E}_T[\mathrm{SS}']\omega_T(0)) = \Delta. \quad (\mathrm{A.18})$$

Equation (2.42) now follows from Lemma A.2 and from Lancaster and Tismenetsky (1985, Chapter 12.3, Theorem 3). The $d\omega_T(\zeta)/d\zeta$ is nonnegative definite and nonnull, since Δ is and since matrix exponentials are nonsingular. The proof of (2.43) is essentially identical.

Proof of Theorem 2.5. We have

$$\begin{aligned}\mathrm{E}_t[GG'] &= \mathrm{E}_t[(\mathrm{P} + \omega\mathrm{S})(\mathrm{P} + \omega\mathrm{S})'] \\ &= \mathrm{E}_t[\mathrm{PP}' + \omega\mathrm{SS}'\omega + \omega\mathrm{SP}' + \mathrm{PS}'\omega]. \end{aligned} \quad (\mathrm{A.19})$$

Subtracting (2.36) from (A.19) and simplifying leads to

$$\mathrm{E}_t[GG'] = \mathrm{E}_t[\mathrm{PP}'] + \mathrm{E}_t[(\xi_Y - \mathrm{P})(\xi_Y - \mathrm{P})'] = \mathrm{E}_t[\xi_Y\xi_Y'], \quad (\mathrm{A.20})$$

proving (2.45). To prove (2.46), note that

$$\mathrm{E}_t[G\xi_X'] = \mathrm{E}_t[\mathrm{P}\xi_X'] + \omega_T\mathrm{E}_t[\mathrm{S}\xi_X'], \qquad \text{but} \quad (\mathrm{A.21})$$

$$\begin{aligned}\mathrm{E}_t[\mathrm{S}\xi_X'] &= \int_{-\infty}^{\infty}\cdots\int_{-\infty}^{\infty} \frac{\partial \ln(f(\xi_X \mid X, Y))}{\partial Y}\, \xi_X' f(\xi_X \mid X, Y)\, d\xi_X \\ &= \int_{-\infty}^{\infty}\cdots\int_{-\infty}^{\infty} \frac{\partial f(\xi_X \mid X, Y)}{\partial Y}\, \xi_X'\, d\xi_X \\ &= \frac{\partial}{\partial Y}\int_{-\infty}^{\infty}\cdots\int_{-\infty}^{\infty} f(\xi_X \mid X, Y)\, \xi_X'\, d\xi_X = 0. \end{aligned} \quad (\mathrm{A.20})$$

The interchange of limits in (A.20) is allowed by uniform convergence. So $\mathrm{E}_t[G\xi_X'] = \mathrm{E}_t[\mathrm{P}\xi_X'] = \mathrm{E}_t[\xi_Y\xi_X']$ by Lemma A.1, proving (2.46).

Proof of Theorem 2.6. To simplify the presentation of the proof, we take $\mu = \hat{\mu} = 0_{n\times 1}$ and $\kappa = \hat{\kappa} = 0_{m\times 1}$. Since the asymptotically optimal filter eliminates asymptotic bias, these terms are not of direct interest. We have

$$\Upsilon_{t+h} - \Upsilon_t = \Upsilon(X_{t+h}, Y_{t+h}, t + h) - \Upsilon(X_t, Y_t, t), \qquad \text{and} \quad (\mathrm{A.21})$$

$$\chi_{t+h} - \chi_t = \chi(X_{t+h}, t + h) - \chi(X_t, t) \quad (\mathrm{A.22})$$

Expanding (A.21)–(A.22) in Taylor series around X_t, Y_t and t, we have

$$\Upsilon_{t+h} = \Upsilon_t + \frac{\partial \Upsilon}{\partial t} \cdot h + \frac{\partial \Upsilon}{\partial X}(X_{t+h} - X_t) + \frac{\partial \Upsilon}{\partial Y}(Y_{t+h} - Y_t)$$

$$+ O\left(\|(X_{t+h} - X_t, Y_{t+h} - Y_t, h)\|^2\right)$$

$$= \Upsilon_t + \frac{\partial \Upsilon}{\partial t} \cdot h + \left[\frac{\partial \Upsilon}{\partial X} \xi_{X,t+h} + \frac{\partial \Upsilon}{\partial Y} \xi_{Y,t+h}\right] h^{1/2}$$

$$+ O\left(\|(X_{t+h} - X_t, Y_{t+h} - Y_t, h)\|^2\right) \quad (A.23)$$

$$\chi_{t+h} = \chi_t + \frac{\partial \chi}{\partial t} \cdot h + \frac{\partial \chi}{\partial X}(X_{t+h} - X_t) + O\left(\|(X_{t+h} - X_t, h)\|^2\right)$$

$$= \chi_t + \frac{\partial \chi}{\partial t} \cdot h + \left[\frac{\partial \chi}{\partial X} \xi_{X,t+h}\right] h^{1/2} + O\left(\|(X_{t+h} - X_t, h)\|^2\right) \quad (A.24)$$

where the partial derivatives are evaluated at (X_t, Y_t, t). By Assumption 2 and (2.51)–(2.52) and the corollary to Billingsley (1986, Theorem 25.12), the moments of higher order terms (normalized by $h^{1/2}$) vanish to zero as $h \downarrow 0$, as do the $\partial \Upsilon / \partial t$ and $\partial \chi / \partial t$ terms.

We can now write $\xi_{\Upsilon,t+h}$ and $\xi_{\chi,t+h}$ as

$$\xi_{\Upsilon,t+h} = \left[\frac{\partial \Upsilon}{\partial X} \xi_{X,t+h} + \frac{\partial \Upsilon}{\partial Y} \xi_{Y,t+h}\right] + \text{higher order term},$$

$$\xi_{\chi,t+h} = \frac{\partial \chi}{\partial X} \xi_{X,t+h} + \text{higher order term}, \quad (A.25)$$

where the higher order terms, along with their relevant moments, disappear as $h \downarrow 0$. So

$$P_{\Upsilon,t+h} \equiv \mathrm{E}\left[\xi_{\Upsilon,t+h} \mid \xi_{\chi,t+h}, \chi_t, \Upsilon_t\right] = \mathrm{E}\left[\xi_{\Upsilon,t+h} \mid \xi_{X,t+h}, X_t, Y_t\right]$$

$$\text{as } h \downarrow 0, \quad \mathrm{P}_{\Upsilon,t+h} \to \mathrm{E}\left[\frac{\partial \Upsilon}{\partial X} \xi_{X,t+h} + \frac{\partial \Upsilon}{\partial Y} \xi_{Y,t+h} \mid \xi_{X,t+h}, X_t, Y_t\right]$$

$$= \frac{\partial Y}{\partial X} \xi_{X,t+h} + \frac{\partial \Upsilon}{\partial Y} \mathrm{P}_{t+h} \quad (A.26)$$

where as in Theorem 2.2, $\mathrm{P}_{t+h} \equiv \mathrm{E}[\xi_{Y,t+h} \mid \xi_{X,t+h}, X_t, Y_t]$. We have as $h \downarrow 0$

$$\mathrm{E}_t[(\xi_\Upsilon - \mathrm{P}_\Upsilon)(\xi_\Upsilon - \mathrm{P}_\Upsilon)'] \to \left[\frac{\partial \Upsilon}{\partial Y}\right], \quad \mathrm{E}_t[(\xi_Y - \mathrm{P})(\xi_Y - \mathrm{P})']\left[\frac{\partial \Upsilon}{\partial Y'}\right]. \quad (A.27)$$

Bounded $2 + \psi$ absolute conditional moments guarantee convergence of the relevant conditional moments. The conditions of the implicit function theorem are satisfied, so given t,

$$\begin{bmatrix} d\Upsilon \\ d\chi \end{bmatrix} = \begin{bmatrix} \dfrac{\partial \Upsilon}{\partial Y} & \dfrac{\partial \Upsilon}{\partial X} \\ 0_{n\times m} & \dfrac{\partial \chi}{\partial X} \end{bmatrix} \begin{bmatrix} dY \\ dX \end{bmatrix} \quad \text{and}$$

$$\begin{bmatrix} dY \\ dX \end{bmatrix} = \begin{bmatrix} \left(\dfrac{\partial \Upsilon}{\partial Y}\right)^{-1} & -\left(\dfrac{\partial \Upsilon}{\partial Y}\right)^{-1} \dfrac{\partial \Upsilon}{\partial X} \left(\dfrac{\partial \chi}{\partial X}\right)^{-1} \\ 0_{n\times m} & \left(\dfrac{\partial \chi}{\partial X}\right)^{-1} \end{bmatrix} \begin{bmatrix} d\Upsilon \\ d\chi \end{bmatrix}. \tag{A.28}$$

So, given t, we may treat (X_t, Y_t) as implicit functions of (χ_t, Υ_t). Similarly, given (X_t, Y_t, t) or (χ_t, Υ_t, t), we may treat $\xi_{X,t+h}$ as an implicit function of $\xi_{\chi,t+h}$. Call $\varphi(\xi_{\chi,t+h} \mid \chi_t, \Upsilon_t)$ the conditional density of $\xi_{\chi,t+h}$ given χ_t and Υ_t. Applying the change of variables formula and the chain rule,

$$\varphi(\xi_{\chi,t+h} \mid \chi_t, \Upsilon_t) = f(\xi_{x,t+h} \mid X_t, Y_t) \left\|\frac{\partial \chi}{\partial X}\right\|^{-1}, \qquad \text{so the score is} \tag{A.29}$$

$$\frac{\partial \ln[\varphi(\xi_{\chi,t+h} \mid \chi_t, \Upsilon_t)]}{\partial \Upsilon} = \frac{\partial Y}{\partial \Upsilon'} \frac{\partial \ln[f(\xi_{X,t+h} \mid X_t, Y_t)]}{\partial Y} = \frac{\partial Y}{\partial \Upsilon'} \mathrm{S}_{t+h}, \tag{A.30}$$

where S_{t+h} is as in Theorem 2.2. So the Fisher information matrix for $\{\chi_t, \Upsilon_t\}$ is

$$\frac{\partial Y}{\partial \Upsilon'} \mathrm{E}_t[\mathrm{SS}'] \frac{\partial Y}{\partial \Upsilon}, \tag{A.31}$$

where $\mathrm{E}_t[\mathrm{SS}']$ is the Fisher information matrix for the $\{X_t, Y_t\}$ system. We also have

$$\begin{aligned} \mathrm{E}_t[\mathrm{P}_\Upsilon \mathrm{S}'_\Upsilon] &\to \frac{\partial \Upsilon}{\partial X} \mathrm{E}_t[\xi_X \mathrm{S}'] \frac{\partial Y}{\partial \Upsilon} + \frac{\partial \Upsilon}{\partial Y} \mathrm{E}_t[\mathrm{PS}'] \frac{\partial Y}{\partial \Upsilon} \\ &= \frac{\partial \Upsilon}{\partial Y} \mathrm{E}_t[\mathrm{PS}'] \frac{\partial Y}{\partial \Upsilon}, \end{aligned} \tag{A.32}$$

where the last equality holds since $E_t[\xi_X \mathrm{S}'] = 0_{m\times m}$—see the proof of Theorem 2.5. Let $\overline{\omega}$ be the solution to the Riccati equation for the (χ, Υ)

system. From (A.27), and (A.31)–(A.32),

$$\frac{\partial \Upsilon}{\partial Y} \mathrm{E}_t[\mathrm{PS}'] \frac{\partial Y}{\partial \Upsilon} \overline{\omega} + \overline{\omega} \frac{\partial Y}{\partial \Upsilon'} \mathrm{E}_t[\mathrm{SP}'] \frac{\partial \Upsilon}{\partial Y'} + \overline{\omega} \frac{\partial Y}{\partial \Upsilon'} \mathrm{E}_t[\mathrm{SS}'] \frac{\partial Y}{\partial \Upsilon} \overline{\omega}$$

$$= \frac{\partial \Upsilon}{\partial Y} \mathrm{E}_t[(\xi_Y - \mathrm{P})(\xi_Y - \mathrm{P})'] \frac{\partial \Upsilon}{\partial Y'}. \quad \text{(A.33)}$$

The theorem is proven if we can show that $\overline{\omega} = [\partial \Upsilon / \partial Y] \omega [\partial \Upsilon / \partial Y']$, where ω is as in Theorem 2.2. By (A.28) $[\partial Y / \partial \Upsilon]^{-1} = \partial \Upsilon / \partial Y$. Substituting for $\overline{\omega}$ in (A.33) and simplifying leads to

$$\frac{\partial \Upsilon}{\partial Y} \mathrm{E}_t[\mathrm{PS}'] \omega \frac{\partial \Upsilon}{\partial Y} + \frac{\partial \Upsilon}{\partial Y} \omega \mathrm{E}_t[\mathrm{SP}'] \frac{\partial \Upsilon}{\partial Y'} + \frac{\partial \Upsilon}{\partial Y} \omega \mathrm{E}_t[\mathrm{SS}'] \omega \frac{\partial \Upsilon}{\partial Y'}$$

$$= \frac{\partial \Upsilon}{\partial Y} \mathrm{E}_t[(\xi_Y - \mathrm{P})(\xi_Y - \mathrm{P})'] \frac{\partial \Upsilon}{\partial Y'}. \quad \text{(A.34)}$$

Left multiplying by $[\partial \Upsilon / \partial Y]^{-1} = \partial Y / \partial \Upsilon$ and right multiplying by $[\partial \Upsilon / \partial Y']^{-1} = \partial Y / \partial \Upsilon'$ recovers the Riccati equation of Theorem 2.2. By assumption 4, ω is unique, so $\overline{\omega}$ is too.

To prove Theorem 3.1, we first need two lemmas.

LEMMA A.3 (Ethier and Kurtz, 1986, ch. 7, Corollary to Theorem 4.1). *Let there be a unique weak-sense solution to the $m \times 1$ stochastic integral equation*

$$Z_t = Z_0 + \int_0^t \mu(Z_s)\, ds + \int_0^t \Omega(Z_s)^{1/2} dW_s, \quad \text{(A.35)}$$

where $\{W_t\}$ is an $n \times 1$ standard Brownian motion independent of Z_0, $\mu(x) \in R^n$ and $\varphi(x) \in R^{n \times n}$ are continuous, and Z_0 is random with cdf F. Now consider $\forall h > 0$ a discrete time $n \times 1$ Markov process $\{_hY_k\}_{k=0,\infty}$ and define $\forall \Delta > 0$, $\forall h > 0$, and $\forall$ integer $k \geq 0$.

$$\mu_{\Delta,h}(y) \equiv h^{-\Delta} \mathrm{E}\big[(_hY_{k+1} - {_hY_k}) \cdot I(\|_hY_{k+1} - {_hY_k}\| < 1) \mid {_hY_k} = y\big], \quad \textit{and} \quad \text{(A.36)}$$

$$\Omega_{\Delta,h}(y) \equiv h^{-\Delta} \mathrm{Cov}\big[(_hY_{k+1} - {_hY_k}) \cdot I(\|_hY_{k+1} - {_hY_k}\| < 1) \mid {_hY_k} = y\big], \quad \text{(A.37)}$$

where $I(\cdot)$ is the indicator function. The initial value $_hY_0$ is random with cdf F_h. Let

(a′) $F_h(y) \Rightarrow F(y)$ as $h \downarrow 0$, and for some $\Delta > 0$, let
(b′) $\mu_{\Delta,h}(y) \to \mu(y)$,
(c′) $\Omega_{\Delta,h}(y) \to \Omega(y)$, and for every ϵ, $0 < \epsilon < 1$,
(d′) $h^{-\Delta} \mathrm{P}[\|_hY_{k+1} - {_hY_k}\| > \epsilon \mid {_hY_k} = y] \to 0$

as $h \downarrow 0$, uniformly on every bounded y set. For each $h > 0$, define the process $\{_hZ_t\}$ by $_hX_t \equiv {}_hY_{[t \cdot h^{-\Delta}]}$ for each $t \geq 0$, where $[t \cdot h^{-\Delta}]$ is the integer part of $t \cdot h^{-\Delta}$. Then, for any T, $0 < T < \infty$, $\{_hZ_t\}_{[0,T]} \Rightarrow \{Z_t\}_{[0,T]}$ as $h \downarrow 0$, where (A.35) defines $\{Z_t\}$.

We stated Lemma A.3 for the case in which μ and Ω did not depend on t. The lemma remains true in the time inhomogeneous case: Simply make t an element of Z_t.

We also need the following result, which adapts arguments in Friedman (1975, Chapter 5) and Arnold (1973, section 7.1) to the large drift case. For notational simplicity, we state and prove it for the time-homogeneous case, but it is true for the inhomogeneous case (again, make t an element of λ_t—but do not multiply its drift by $h^{-1/4}$ in the $\delta = \frac{3}{4}$ case!).

LEMMA A.4. *For every h, $0 < h < 1$, let there be a unique weak-sense solution to*

$$\lambda_t = \lambda_0 + h^{-1/4} \int_0^t \nu(\lambda_s)\, ds + \int_0^t \varphi(\lambda_s)^{1/2}\, dW_s, \qquad \text{(A.38)}$$

with $\nu(\cdot)$ and $\varphi(\cdot)$ continuous and $n \times 1$, and $\{W_t\}$ an $n \times 1$ standard Brownian motion independent of λ_0. Following Stroock and Varadhan (1979, Theorem 10.1.1), for each $N > 0$ and each h, $0 < h < 1$, let $\lambda_{N,t}$ satisfy the weak-sense unique stochastic integral equation

$$\lambda_{N,t} = \lambda_{N,0} + h^{-1/4} \int_0^t \nu_N(\lambda_{N,s})\, ds + \int_0^t \varphi_N(\lambda_{N,s})^{1/2}\, dW_s, \quad \text{(A.39)}$$

where $\nu_N(\cdot)$ and $\varphi_N(\cdot)$ are bounded and continuous with $\nu_N(\lambda) = \nu(\lambda)$ and $\varphi_N(\lambda) = \varphi(\lambda)$ on the set $\{\lambda: \|\lambda\| \leq N\}$. Then for every integer $j \geq 1$, there is a continuous function $M_j(\lambda, N)$ and an $h^* > 0$ such that $\forall h$, $0 < h < h^*$, $\mathrm{E}(\|\lambda_{N,t+h} - \lambda_{N,t}\|^{2j} \mid \lambda_{N,t}) \leq M_j(\lambda_{N,t}, N) h^j$ almost surely.

Proof of Lemma A.4. Note first that for every $N > 0$ and every $h > 0$, the drift and diffusion coefficients of (A.39) are bounded and continuous, so there is a K_N such that

$$\|\nu_N(\lambda)\|^2 + \|\varphi_N(\lambda)\|^2 \leq K_N^2(1 + \|\lambda\|^2). \qquad \text{(A.40)}$$

Assuming $h < 1$, (A.40) implies

$$\|h^{-1/4}\nu_N(\lambda)\|^2 + \|\varphi_N(\lambda)\|^2 \leq h^{-1/2}K_N^2(1 + \|\lambda\|^2). \qquad \text{(A.41)}$$

This is the familiar "growth condition" (Arnold, 1973, Theorem 6.2.2), which is satisfied for the $\{\lambda_{N,t}\}$ process even though it may not be satisfied for the $\{\lambda_t\}$ process. We do not need to assume the usual Lipschitz condition, since weak (as opposed to strong) existence and uniqueness of solutions to (A.38)–(A.39) is enough for our purposes. Keeping $h^{-1/4}$ outside the integral and carrying out the steps in the proof of Arnold (1973, 7.1.3) leads to

$$\mathrm{E}_s\big[\|\lambda_{N,t}\|^{2j}\big] \le \big(1 + \|\lambda_{N,s}\|^{2j}\big)\exp\big[2j(2j+1)K_n^2 h^{-1/4}(t-s)\big], \quad 0 \le s < t. \tag{A.42}$$

Now

$$\mathrm{E}_t\|\lambda_{N,t+h} - \lambda_{N,t}\|^{2j} \le \mathrm{E}_t\left[h^{-1/4}\int_t^{t+h}\|\nu_N(\lambda_{N,s})\|\,ds + \left\|\int_t^{t+h}\varphi(\lambda_{N,s})\,dW_s\right\|\right]^{2j}. \tag{A.43}$$

By Jensen's inequality and the convexity of x^{2j} for positive integer j,

$$\le 2^{2j-1}h^{-j/2}\mathrm{E}_t\left[\int_t^{t+h}\|\nu_N(\lambda_{N,s})\|\,ds\right]^{2j} + 2^{2j-1}\,\mathrm{E}_t\left\|\int_t^{t+h}\varphi_N(\lambda_{N,s})\,dW_s\right\|^{2j}. \tag{A.44}$$

By the integral means inequality (Hardy *et al.*, 1952, Theorem 192), Jensen's inequality and (A.40)

$$2^{2j-1}h^{-j/2}\,\mathrm{E}_t\left(\int_t^{t+h}\|\nu_N(\lambda_{N,s})\|\,ds\right)^{2j} \le 2^{3j-2}h^{3j/2-1}K_N^{2j}\int_t^{t+h}\mathrm{E}_t\big(1 + \|\lambda_{N,s}\|^{2j}\big)\,ds \tag{A.45}$$

Substituting from (A.42) into the integrand and integrating yields, for small h,

$$2^{2j-1}h^{-j/2}\,\mathrm{E}_t\left(\int_t^{t+h}\|\nu_N(\lambda_{N,s})\|\,ds\right)^{2j} \le h^{3j/2}\big(1 + \|\lambda_{N,t}\|^{2j}\big)2^{3j}K_N^{2j}, \tag{A.46}$$

bounding the first term on the right-hand side of (A.44). To bound the second term, we first apply Friedman (1975, Corollary 4.6.4). We now bound $E_t\,|\varphi_{N,s}|^{2j}$ using (A.40), substitute into the integrand and integrate, obtaining

$$2^{2j-1}\mathrm{E}_t \left\| \int_t^{t+h} \varphi_N(\lambda_{N,s})\, dW_s \right\|^{2j}$$

$$\leq 2^{2j-1} h^{j-1} \left[\frac{4j^3}{2j-1} \right]^j \int_t^{t+h} \mathrm{E}_t \left\| \varphi_N(\lambda_{N,s}) \right\|^{2j} ds \quad \text{(A.47)}$$

$$2^{2j-1}\,\mathrm{E}_t \left\| \int_t^{t+h} \varphi_{N,s}\, dW_s \right\|^{2j} \leq 2^{2j-1} h^j K_N^{2j} \left[\frac{4j^3}{2j-1} \right]^j \left(3 + 2\left\| \lambda_{N,t} \right\|^{2j} \right) \quad \text{(A.48)}$$

completing the proof of the lemma.

Proof of Theorem 3.1. We prove the Theorem for $\delta = \frac{3}{4}$ (the $\delta = 1$ case is similar but simpler.) As indicated in the text, we use Lemma A.3 in place of Theorem 2.1 of NF. Note that the conditional moments are truncated by the $I(\|{}_hY_{k+1} - {}_hY_k\| < 1)$ term. This implies that only the local properties of the $\{{}_hY_k\}$ process enter. Suppose, for example, that ${}_hY_k \equiv \lambda_{kh}$, where λ_t is the diffusion in Lemma A.4. Then provided that $N > 1 + \|\lambda_{kh}\|$, all the conditional moments in (b′)–(d′) are exactly the same as if we had defined ${}_hY_k \equiv \lambda_{N,kh}$, where $\{\lambda_{N,kh}\}$ is the diffusion defined by (A.39). In fact, if $\|\lambda_0\| < N$, the transition probabilities for $\{\lambda_{N,t}\}$ and $\{\lambda_t\}$ are the same on the interval $[0, \tau_N]$, where the stopping time $\tau_N \equiv \inf\{t > 0: \|\lambda_t\| > N\}$ (see Stroock and Varadhan, 1979, Theorem 10.1.1). Since the required convergence in (b′)–(d′) is uniform on compact subsets of R^{n+2m} (not on all of R^{n+2m}) it is enough to prove that the theorem holds for the diffusion $\{X_{N,t}, Y_{N,t}\}$ for all $N > 0$. Let $\lambda_{N,t}$ be as in Lemma A.4, and write $\mu_{N,t}$ for $\mu_N(X_t, Y_t, t)$ and similarly with $\kappa_{N,t}$ and $\Omega_{N,t}$. We have

$$d\begin{bmatrix} X_{N,t} \\ Y_{N,t} \end{bmatrix} = h^{-1/4} \begin{bmatrix} \mu_{N,t} \\ \kappa_{N,t} \end{bmatrix} dt + \begin{bmatrix} c_{N11,t} & c_{N12,t} \\ c_{N21,t} & c_{N22,t} \end{bmatrix} \begin{bmatrix} dW_{1,t} \\ dW_{2,t} \end{bmatrix} \quad \text{(A.49)}$$

$$\text{where} \begin{bmatrix} c_{N11,t} & c_{N12,t} \\ c_{N21,t} & c_{N22,t} \end{bmatrix} \equiv \Omega_{N,t}^{1/2}. \qquad \text{So} \quad \text{(A.50)}$$

$$X_{N,t+h} = X_{N,t} + h^{-1/4} \int_t^{t+h} \mu_{N,s}\, ds + \int_t^{t+h} (c_{N11,s}\, dW_{1,s} + c_{N12,s}\, dW_{2,s}) \quad \text{(A.51)}$$

$$Y_{N,t+h} = Y_{N,t} + h^{-1/4} \int_t^{t+h} \kappa_{N,s}\, ds + \int_t^{t+h} (c_{N21,s}\, dW_{1,s} + c_{N22,s}\, dW_{2,s}) \quad \text{(A.52)}$$

That $\mathrm{E}_t \|h^{-1/2}(X_{N,t+h} - X_t)\|^{2j}$ and $\mathrm{E}_t \|h^{-1/2}(Y_{N,t+h} - Y_t)\|^{2j}$ are bounded for $j \geq 1$ is immediate from Lemma A.4. That

$$h^{-1/2}\mathrm{P}\big[\|(X_{N,t+h} - X_t)', (Y_{N,t+h} - Y_t)'\| > \epsilon\big]$$

converges to zero uniformly on compact subsets of R^{n+m} for every $\epsilon > 0$ now follows from Markov's inequality [Billingsley 1986, (5.27)]. Next, take $\|(X_t', Y_t')\| < N - 1$, and define $\hat{\xi}_{N,X,t+h}$ and $\xi_{N,Y,t+h}$ as

$$\hat{\xi}_{N,X,t+h} \equiv h^{-3/4} \int_t^{t+h} (\mu_{N,s} - \hat{\mu}_t)\, ds$$

$$+ h^{-1/2} \int_t^{t+h} (c_{N,11,s}\, dW_{1,s} + c_{N,12,s}\, dW_{2,s}) \quad \text{(A.53)}$$

$$\xi_{N,Y,t+h} \equiv h^{-1/2} \int_t^{t+h} (c_{N21,s}\, dW_{1,s} + c_{N22,s}\, dW_{2,s}) \quad \text{(A.54)}$$

Now define $\bar{\xi}_{N,X,t=h} \equiv h^{-1/2} \int_t^{t+h} (C_{N,11,t}\, dW_{1,t} + C_{N,12,t}\, dW_{2,s})$. Note that given time t information, $\bar{\xi}_{N,X,t+h}$ is Gaussian, since $C_{N,11}$ and $C_{N,12}$ are held constant in the integrand at their time t values. Next, expand $Q_{t+h} - Q_t$ around $\bar{\xi}_{N,X,t+h} = \bar{\xi}_{N,X,t+h}$ and $\hat{Y}_t = Y_t$.

$$Q_{t+h} - Q_t = h^{1/4} G\big(\bar{\xi}_{X,t+h}, X_t, Y_t, t\big)$$

$$- h^{-3/4} \int_t^{t+h} (C_{N,21,s}\, dW_{1,s} + C_{N,22,s}\, dW_{2,s})$$

$$+ h^{1/2} \frac{\partial G}{\partial Y} Q_t + h^{1/4} \frac{\partial G}{\partial \xi_X}\big(\hat{\xi}_{N,X,t+h} - \bar{\xi}_{X,t}\big)$$

$$+ h^{-1/2} \int_t^{t+h} \big(\hat{\kappa}_t - \kappa_{N,s}\big)\, ds + \text{higher order terms}, \quad \text{(A.55)}$$

where the partial derivatives are evaluated at $(\bar{\xi}_{X,t+h}, X_t, Y_t, t)$. As $h \downarrow 0$, $h^{-1/2} \int_t^{t+h} (C_{N,21,s}\, dW_{1,s} + C_{N,22,s}\,(dW_{2,s})$ converges weakly to the conditionally Gaussian random variable $\bar{\xi}_{Y,t+h} \equiv h^{-1/2} \int_t^{t+h} (C_{21,t}\, dW_{1,s} + C_{22,t}\, dW_{2,s})$ (NF Lemma (A.1)). By Lemma A.4, for any N, $h^{-1/2} \int_t^{t+h} (C_{N,21,s}\, dW_{1,s} + C_{N,22,s}\, dW_{2,s})$ has arbitrary finite moments, so its conditional moments converge to the corresponding conditional moments of $\bar{\xi}_{Y,t+h}$ (Billingsley, 1986, Corollary to Theorem 25.12). Similarly, $\xi_{N,X,t+h}$ (and its conditional moments) converges to $\bar{\xi}_{X,t+h}$ (and its conditional moments). This moment boundedness (along with the bounds on G and its derivatives) ensures that the expectation and variances of the higher order terms in the Taylor series expansion vanish as $h \downarrow 0$. Finally, the conditional moment bounds also ensures that

$$h^{-1}\mathrm{E}_t\left[\int_t^{t+h}(\hat{\kappa}_t - \kappa_{N,s})\,ds\right]$$

and

$$h^{-1}\mathrm{E}_t\left[\int_t^{t+h}(\hat{\mu}_t - \mu_{N,s})\,ds\right]$$

converge, respectively, to $(\hat{\kappa}_t - \kappa_t)$ and $(\hat{\mu}_t - \mu_t)$. Define$I_{\epsilon,t+h} \equiv 1$ if

$$\|(Q_{t+h} - Q_t)', (X_{t+h} - X_t)', (Y_{t+h} - Y_t)'\| < \epsilon$$

and $\equiv 0$ otherwise.Then

$$\mathrm{E}_t[(Q_{t+h} - Q_t)\cdot I_{1,t+h}]$$
$$\rightarrow (\hat{\kappa}_t - \kappa_t) + \mathrm{E}_t\left[\frac{\partial G}{\partial \xi_X}\right](\mu_t - \hat{\mu}_t) + \mathrm{E}_t\left[\frac{\partial G}{\partial Y}\right]Q_t \quad \text{(A.56)}$$

$$\mathrm{E}_t[(Q_{t+h} - Q_t)(Q_{t+h} - Q_t)'\cdot I_{1,t+h}]$$
$$\rightarrow \mathrm{E}_t\Big[(G(\xi_{X,t+h}, X_t, Y_t, t) - \xi_{Y,t+h})\big(G\big(\hat{\xi}_{X,t+h}, X_t, Y_t, t\big) - \bar{\xi}_{Y,t+h}\big)'\Big]$$

and $\forall\epsilon > 0$, $\mathrm{E}_t[I_{\epsilon,t+h}] \rightarrow 1$ by Markov's inequality. Because the moment bounds delivered by Lemma A.4 are uniform on compact (X, Y) sets, the convergence of these conditional moments is uniform on compacts, as required by Lemma A.3, completing the proof.

Proof of Theorem 4.1. First we need a minor modification of NF Theorem 2.1. Condition (d′) required the existence of a $\Delta > 0$ and $\psi > 0$ such that $\forall\Lambda > 0$. We can replace this with the condition that $\exists\Delta > 0$ such that $\forall\Lambda > 0\exists\psi_\Lambda > 0$ such that

$$\lim_{h\downarrow 0}\sup_{\|y\|<\Lambda} h^{-\Delta}\,\mathrm{E}\left[\|{}_hY_{k+1} - {}_hY_k\|^{2+\psi} \mid {}_hY_k = y\right] \rightarrow 0. \quad \text{(A.57)}$$

$$\lim_{h\downarrow 0}\sup_{\|y\|<\Lambda} h^{-\Delta}\,\mathrm{E}\left[\|{}_hY_{k+1} - {}_hY_k\|^{2+\psi_\Lambda} \mid {}_hY_k = y\right] \rightarrow 0. \quad \text{(A.58)}$$

The proof of NF Theorem 2.1 is unaffected by this change. (This allows us to define the conditional degrees of freedom as $2 + e^{\nu}$ rather than as, say, $2.01 + e^{\nu}$.)

Equations (4.22)–(4.24) are routinely derived from the multivariate t density. Once the assumptions of Theorem 2.2 are verified, (4.20)–(4.21) and Assumption 4 follow from Theorem 2.3. To derive (4.25) we made use of Prudnikov *et al.* (1986, p. 505, formulas 53 and 56), Davis (1964, formulas 6.2.1 and 6.3.5), and the integrator in Mathematica (Wolfram Research, 1992). The existence of the conditional densities in Assumption 3 is obvious. For bounded ν_t, the conditional degrees of freedom are bounded away from both 2 (assuring that (2.34) is satisfied) and ∞ (bounding S_{t+h} and so assuring that (2.35) is satisfied). The $2 + \psi$ moment boundedness of S_{t+h} and P_{t+h} in turn assures that (2.23) is satisfied. A_t is trivially $0_{m \times 1}$. B_t and C_t are derived for the optimal filter in the proof of Theorem 2.2. (2.19)–(2.20) follow from carrying out the Taylor series expansion in (2.11) and making use of the $2 + \psi$ moment boundedness of P_{t+h} and S_{t+h}.

ACKNOWLEDGMENTS

This is a revision of parts of an earlier working paper, "Asymptotic Filtering and Smoothing Theory for Multivariate ARCH Models." I would like to thank the editor, three referees, Yacine Aït-Sahalia, Andrea Bertozzi, Tim Bollerslev, George Easton, Lars Hansen, Mahesh Maheswaran, Peter Rossi, Boaz Schwartz, and seminar participants at Brigham Young, Chicago, the Conference on Stochastic Volatility, the Econometric Society Summer Meetings, Harvard/M.I.T., Minnesota, the Multivariate Financial Time Series Conference, the NBER Asset Pricing Group, Northwestern, Princeton, Utah, Yale, Washington University, and Wisconsin (Madison) for helpful discussions. I am particularly grateful to Dean Foster. This material is based on work supported by the National Science Foundation under grants SES-9110131 and SES-9310683. The Center for Research in Security Prices provided additional research support.

REFERENCES

Andersen, T. G. (1992). "Volatility," Working Paper. Northwestern University, Evanston, IL.

Anderson, B. D. O. (1978). Second-order convergent algorithms for the steady state Riccati equation. *International Journal of Control* **28**, 295–306.

Anderson, B. D. O., and Moore, J. B. (1971). "Linear Optimal Control." Prentice-Hall, Englewood Cliffs, NJ.

Anderson, B. D. O., and Moore, J. B. (1979). "Optimal Filtering." Prentice-Hall, Englewood Cliffs, NJ.

Anderson, T. W. (1984). "An Introduction to Multivariate Statistical Analysis," 2nd ed. Wiley, New York.

Aptech Systems (1992). "GAUSS, Version 3.0." Aptech Systems, Maple Valley, WA.

Arnold, L. (1973). "Stochastic Differential Equations: Theory and Applications." Wiley, New York.

Baillie, R. T., Bollerslev, T., and Mikkelsen, H. O. A. (1993). "Fractionally Integrated Generalized Autoregressive Conditional Heteroskedasticity, "Working Paper. Michigan State University, East Lansing.

Bates, D. S. (1991). The crash of '87: Was it expected? The evidence from options markets. *Journal of Finance* **46**, 1009–1044.

Bates, D. S. (1993). "Jumps and Stochastic Volatility: Exchange Rate Processes Implicit in PHLX Deutschemark Options," Working Paper. Wharton School, Philadelphia.

Bellman, R. (1970). "Introduction to Matrix Analysis," 2nd ed. McGraw-Hill, New York.

Bera, A. K., and Higgins, M. L. (1993). A survey of ARCH models: Properties, estimation, and testing. *Journal of Economic Surveys* **7**, 305–366.

Billingsley, P. (1986). "Probability and Measure," 2nd ed. Wiley, New York.

Bollerslev, T. (1986). Generalized autoregressive conditional heteroskedasticity. *Journal of Econometrics* **31**, 307–327.

Bollerslev, T. (1990). Modelling the coherence in short-run nominal exchange rates: A multivariate generalized ARCH approach. *Review of Economics and Statistics* **72**, 498–505.

Bollerslev, T., Engle, R. F., and Wooldridge, J. M. (1988). A capital asset pricing model with time-varying covariances. *Journal of Political Economy* **96**, 116–131.

Bollerslev, T., Chou, R. Y., and Kroner, K. (1992). ARCH modeling in finance: A review of the theory and empirical evidence. *Journal of Econometrics* **52**, 5–60.

Bollerslev, T., Engle, R. F., and Nelson, D. B. (1994). ARCH models. *In* "The Handbook of Econometrics" (R. F. Engle and D. L. McFadden, eds.,), Vol. 4, Chapter 49, pp. 2959–3038. North-Holland Publ., Amsterdam.

Brenner, R. J., Harjes, R. H., and Kroner, K. F. (1994). "Another Look at Alternative Models of the Short-term Interest Rate," Working Paper. University of Arizona, Tucson.

Cambanis, S., Huang, S., and Simons, G. (1981). On the theory of elliptically contoured distributions. *Journal of Multivariate Analysis* **11**, 368–385.

Chan, K. C., Karolyi, G. A., Longstaff, F. A., and Sanders, A. B. (1992). An empirical comparison of alternative models of the short-term interest rate. *Journal of Finance* **47**, 1209–1227.

Chiras, D. P., and Manaster, S. (1978). The information content of option prices and a test of market efficiency. *Journal of Financial Economics* **6**, 213–234.

Christie, A. A. (1982). The stochastic behavior of common stock variances: Value, leverage and interest rate effects. *Journal of Financial Economics* **10**, 407–432.

Cox, J. C., Ross, S. A., and Rubinstein, M. (1979). Options pricing: A simplified approach. *Journal of Financial Economics* **7**, 229–263.

Cox, J. C., Ingersoll, J. E., Jr., and Ross, S. A. (1985). A theory of the term structure of interest rates. *Econometrica* **53**, 385–407.

Danielsson, J. (1994). Stochastic volatility in asset prices: Estimation with simulated maximum likelihood. *Journal of Econometrics* **64**, 375–400.

Davis, P. J. (1964). Gamma function and related functions. *In* "Handbook of Mathematical Functions" (M. Abramowitz and I. A. Stegun, eds.), pp. 253–293. Dover, New York.

Day, T. E., and Lewis C. M. (1992). Stock market volatility and the information content of stock index options. *Journal of Econometrics* **52**, 267–288.

Ding, Z., Granger, C. W. J., and Engle, R. F. (1993). A long memory property of stock returns and a new model. *Journal of Empirical Finance* **1**, 83–106.

Engle, R. F. (1982a). Autoregressive conditional heteroskedasticity with estimates of the variance of United Kingdom inflation. *Econometrica* **50**, 987–1008.

Engle, R. F. (1982b). Comment on "Bayesian analysis of stochastic volatility models." *Journal of Business and Economic Statistics* **12**, 395–396.

Engle, R. F., and Kroner, K. F. (1994). Multivariate simultaneous generalized ARCH. *Econometric Theory*. **11**, 122.

Engle, R. F., and Ng, V. (1993). Measuring and testing the impact of news on volatility. *Journal of Finance* **48**, 1749–1778.

Engle, R. F., Ng, V., and Rothschild, M. (1990). Asset pricing with a factor-ARCH covariance structure: Empirical estimates for treasury bills. *Journal of Econometrics* **45**, 213–238.

Ethier, S. N., and Kurtz, T. G. (1986). "Markov Processes: Characterization and Convergence." Wiley, New York.

Fisher, R. A. (1921). On the probable error of a coefficient of correlation deduced from a small sample. *Metron* **1**, 1.

Foster, D. P., and Nelson, D. B. (1996). Continuous record asymptotics for rolling sample variance estimators. *Econometrica* **64**, 139–174.

Friedman, A. (1975). "Stochastic Differential Equations with Applications," Vol. 1. Academic Press, New York.

Gallant, A. R., Rossi, P. E., and Tauchen, G. (1992). Stock prices and volume. *Review of Financial Studies* **5**, 199–242.

Gallant, A. R., Rossi, P. E., and Tauchen, G. (1993). Nonlinear dynamic structures. *Econometrica* **61**, 871–907.

Garman, M. B., and Klass, M. J. (1980). On the estimation of security price volatilities from historical data. *Journal of Business* **53**, 67–78.

Glosten, L. R., Jagannathan, R., and Runkle, D. (1993). On the relation between the expected value and the volatility of the nominal excess return on stocks. *Journal of Finance* **48**, 1779-1801.

Hansen, B. E. (1994). Autoregressive conditional density estimation. *International Economic Review* **35**, 705–730.

Hardy, G., Littlewood, J. E., and Pólya, G. (1952). "Inequalities," 2nd ed. Cambridge University Press, Cambridge, U.K.

Harvey, A., Ruiz, E., and Shephard, N. (1994). Multivariate stochastic variance models. *Review of Economic Studies* **61**, 247–264.

Helland, I. S. (1982). Central limit theorems for martingales with discrete or continuous time. *Scandinavian Journal of Statistics* **9**, 79–94.

Horn, R. A., and Johnson, C. R. (1985). "Matrix Analysis." Cambridge University Press, Cambridge, U.K.

Hull, J., and White, A. (1987). The pricing of options on assets with stochastic volatilities. *Journal of Finance* **42**, 281–300.

Jacquier, E., Polson, N. G., and Rossi, P. E. (1994). Bayesian analysis of stochastic volatility models. *Journal of Business and Economic Statistics* **12**, 371–389.

Johnson, N. L., and Kotz, S. (1972). "Continuous Multivariate Distributions." Wiley, New York.

Karatzas, I., and Shreve, S. E. (1988). "Brownian Motion and Stochastic Calculus." Springer-Verlag, New York.

Karpoff, J. (1987). The relation between price changes and trading volume: A survey. *Journal of Financial and Quantitative Analysis* **22**, 109–126.

Kim, S., and Shephard, N. (1993). "Stochastic Volatility: New Models and Optimal Likelihood Inference," Working Paper. Nuffield College, Oxford.

Kroner, K. F., and Ng, V. K. (1993). "Modelling the Time Varying Comovement of Asset Returns," Working Paper. University of Arizona, Tucson.

Kučera, V. (1972). A contribution to matrix quadratic equations. *IEEE Transactions on Automatic Control* **AC-17**, 344–356.

Lancaster P., and Rodman, L. (1980). Existence and uniqueness theorems for the algebraic riccati equation. *International Journal of Control* **32**, 285–309.

Lancaster P., and Tismenetsky, M. (1985). "The Theory of Matrices," 2nd ed. Academic Press, San Diego, CA.

Lo, A. W., and Wang, J. (1995). Implementing option pricing formulas when asset returns are predictable. *Journal of Finance* **50**, 87–129.

Magnus, J. R., and Neudecker, H. (1988). "Matrix Differential Calculus." Wiley, New York.

McCulloch, J. H. (1985). On Heteros* edasticity. *Econometrica* **53** (2), 483.

Mitchell, A. F. S. (1989). The information matrix, skewness tensor and σ-connections for the general multivariate elliptic distribution. *Annals of the Institute of Statistical Mathematics* **41**, 289–304.

Nelson, D. B. (1988). The time series behavior of stock market volatility and returns. Doctoral Dissertation, Massachusetts Institute of Technology, Economics Department, Cambridge, MA (unpublished).

Nelson, D. B. (1990). ARCH models as diffusion approximations. *Journal of Econometrics* **45**, 7–38.

Nelson, D. B. (1991). Conditional heteroskedasticity in asset returns: A new approach. *Econometrica* **59**, 347–370.

Nelson, D. B. (1992). Filtering and forecasting with misspecified ARCH models I: Getting the right variance with the wrong model. *Journal of Econometrics* **52**, 61–90.

Nelson, D. B., and Foster, D. P. (1994). Asymptotic filtering theory for univariate ARCH models. *Econometrica* **62**, 1–41.

Nelson, D. B., and Foster, D. P. (1995). Filtering and forecasting with misspecified ARCH models II: Making the right forecast with the wrong model. *Journal of Econometrics* **67**, 303.

Nelson, D. B., and Schwartz, B. A. (1992). Filtering with ARCH: A monte carlo experiment, *Proceedings of the Americal Statistical Association, Business and Economic Statistics Section*, pp. 1–6.

Parkinson, M. (1980). The extreme value method for estimating the variance of the rate of return. *Journal of Business* **53**, 61–65.

Phillips, P. C. B. (1988). Regression theory for near-integrated time series. *Econometrica* **56**, 1021–1044.

Prudnikov, A. P., Brychkov, Y. A., and Marichev, O. I. (1986). "Integrals and Series," Vol. 1. Gordon & Breach, New York.

Schwartz, B. A. (1994). Essays on ARCH filtering and estimation. Doctoral Dissertation, University of Chicago, Graduate School of Business, Chicago (unpublished).

Schwert, G. W. (1989). Why does stock market volatility change over time? *Journal of Finance* **44**, 1115–1154.

Serfling, R. J. (1980). "Approximation Theorems of Mathematical Statistics." Wiley, New York.

Shephard, N. (1994). Local scale models: State space alternatives to integrated GARCH processes. *Journal of Econometrics* **60**, 181–202.

Stroock, D. W., and Varadhan, S. R. S. (1979). "Multidimensional Diffusion Processes." Springer-Verlag, Berlin.

Taylor, S. J. (1986). "Modeling Financial Time Series." Wiley, New York.

Taylor, S. J. (1994). Stochastic volatility: A review and comparative study. *Mathematical Finance* **4**, 183–204.

Turner, A. L., and Weigel, E. J. (1992). Daily stock market volatility: 1928–1989. *Management Science* **38**, 1586–1609.

Watanabe, T. (1992). "Alternative Approach to Conditional Heteroskedasticity in Stock Returns: Approximate Non-Gaussian Filtering," Working Paper. Yale University, New Haven, CT.

Wiggins, J. B. (1987). Option values under stochastic volatility: Theory and empirical estimates. *Journal of Financial Economics* **19**, 351–372.

Wiggins, J. B. (1991). Empirical tests of the bias and efficiency of the extreme-value variance estimator for common stocks. *Journal of Business* **64**, 417–432.

Wolfram Research (1992). "Mathematica, Version 2.2." Wolfram Research, Champaign, IL.

Zakoian, J. M. (1990). "Threshold Heteroskedastic Models," Working Paper. INSEE.

10

CONTINUOUS RECORD ASYMPTOTICS FOR ROLLING SAMPLE VARIANCE ESTIMATORS*

DEAN P. FOSTER[†]
DANIEL B. NELSON[‡]

[†]*University of Pennsylvania*
Philadelphia, Pennsylvania 19104
[‡]*University of Chicago*
Graduate School of Business
Chicago, Illinois 60637

1. INTRODUCTION

Most asset pricing theories relate expected returns on assets to their conditional variances and covariances. See, for example, the review of the ARCH literature in Bollerslev *et al.* (1992). It is widely recognized that these conditional moments change over time. Unfortunately, conditional covariances are not directly observable, so in tests of asset-pricing theories researchers must use estimates of conditional second moments. Similarly, market participants use estimates of conditional variances and covariances in hedging, option pricing, and in many other aspects of portfolio selection. How accurate are these estimated variances and covariances? How can they be estimated more accurately?

* Reprinted from Foster, D. P., and Nelson, D. "Continuous Record Asymptotics for Rolling Sample Variance Estimators." *Econometrica*.

If conditional variances and covariances were *constant* over time, then standard statistical techniques would yield the answer to these questions. When conditional heteroskedasticity is present, these techniques will not suffice. In fact, as we see in section 2, statistical methods that assume constant variances and covariances even over short time intervals present a misleadingly optimistic picture of how accurate the measurement is.

Although there are many strategies for estimating time-varying variances and covariances, among the most popular have been (a) chopping the returns data into blocks of time and treating conditional variances and covariances as constant within each block (e.g., Merton, 1980; Poterba and Summers, 1986; French *et al.*, 1987) and (b) the rolling regression approach of Officer (1973) and Fama and Macbeth (1973).

The appeal of such strategies is clear: On the one hand, they allow for the possibility (almost a certainty in economic applications!) that the parameters of the process evolve randomly over time. On the other hand, they impose little structure on the precise *way* in which the parameters evolve. All of these strategies accommodate random evolution in parameters by estimating the value of the parameters at time t using only data "near" t. For example, Fama and Macbeth (1973) estimated conditional betas at data t using only the returns data for a period of five to eight years prior to data t—a "rolling regression."[1]

As Fama and Macbeth explain it, this estimation strategy "reflects a desire to balance the statistical power obtained with a large sample from a stationary process against potential problems caused by any non-constancy of the β_i." The most important "the statistical power obtained with a large sample" is, the more inclined a researcher should be to use a *long* string of data in the rolling regression. On the other hand, minimizing the "potential problems caused by any non-constancy of the β_i" points toward using a *short* period for the rolling regression.

Fama and Macbeth's choice of a five- to seven-year window was motivated by the work of Fisher (1970) and Gonedes (1973), who found that this window length gave the best out-of-sample forecasting performance for individual stocks. In related work, Fisher (1970) and Fisher and Kamin (1985) develop approximate distributions for measurement errors in betas and optimal weighting schemes under the assumption that conditional betas are random walks independent of market returns.[2]

[1] These estimation strategies are also popular on Wall Street: See, for example, the Merrill Lynch (1986) beta book, which uses a five-year rolling regression with monthly data to estimate betas. Rolling regressions are also used in estimating conditional means (see, for example, Banerjee *et al.*, 1992), although our results do not apply directly to this case.

[2] There is a large literature on random coefficient regression, of which the work of Fisher (1970) and Fisher and Kamin (1985) is an application (see, for example, Chow, 1984, and the references therein).

In this chapter, we extend these theoretical results to a much broader class of data-generating processes. In section 2, we show how, under weak assumptions, to approximate the distribution of measurement errors in estimated conditional variances and covariances. These results are broad enough to accommodate not only one- and two-sided rolling regressions, but also more general weighting schemes such as the ARCH(p) model of Engle (1982) and one of the multivariate extensions proposed by Bollerslev *et al.* (1988).[3] In section 3, we characterize optimal window lengths and optical weights to use in rolling regressions. Section 4 considers estimation of conditional betas. In section 5, we provide an empirical example. Section 6 is a brief conclusion. The proofs are collected in the Appendix.

2. ASYMPTOTIC DISTRIBUTIONS

To illustrate the intuition behind our approximation method, consider the following simple case; suppose the data are generated by the diffusion

$$dX_t = \mu(X_t, \sigma_t)\, dt + \sigma_t \cdot dW_{1,t} \tag{2.1}$$

$$d\sigma_t^2 = \lambda(X_t, \sigma_t)\, dt + \Lambda(X_t, \sigma_t) \cdot dW_{2,t}, \tag{2.2}$$

where $W_{1,t}$ and $W_{2,t}$ are (possibly correlated) standard Brownian motions, X_t and σ_t^2 are scalars, and $\Lambda(\cdot, \cdot)$, $\Lambda(\cdot, \cdot, \cdot)$, and $\mu(\cdot, \cdot, \cdot)$, are continuous, with $\Lambda(\cdot, \cdot)$ strictly positive.

Our assumption that $\Lambda(\cdot, \cdot)$ is strictly positive separates our approach from that of the nonparametric literature.

Suppose that the $\{X_t\}$ process is observable but $\{\sigma_t^2\}$ is not. How can we use the information in the sample path of $\{X_t\}$ to estimate the path of $\{\sigma_t^2\}$? It is well known that, as a diffusion is observed at finer and finer time intervals (say of length h), its conditional variance at any instant can be approximated with ever greater accuracy, until in th limit, as $h \to 0$, it is known exactly. To understand why, note first that because σ_t^2 in (2.1)–(2.2) is generated by a diffusion, it is continuous (with probability one) as a function of time. This implies that for every $\epsilon > 0$ and every $t > 0$ there exists, with probability one, a random $\delta(t) > 0$ such that

$$\sup_{t-\delta(t) \le s \le t} |\sigma_s^2 - \sigma_t^2| < \epsilon. \tag{2.3}$$

That is, over suitably small time intervals, the change in σ_t^2 can be made as small as we like. Now choose a small constant $\delta > 0$ and chop the

[3] Asymptotic measurement error distributions for conditional variances generated by other ARCH models (which cannot be accommodated by the methods in this chapter), are given in Nelson and Foster (1994) and Nelson (1994).

interval $[t - \delta, t]$ into M equal pieces. We then estimate σ_t^2 by

$$\hat{\sigma}_t^2(\delta, M) \equiv \delta^{-1} \sum_{j=1}^{M} \left(X_{t-(j-1)\delta/M} - X_{t-j\delta/M}\right)^2. \tag{2.4}$$

Equation (2.4) is a standard one-sided rolling regression in which we act as if μ_t were identically zero. When δ is small, μ_t and σ_t^2 are effectively constant, so when we condition on $\mu_{t-\delta}$ and $\sigma_{t-\delta}^2$, the normalized increments $(M/\delta)^{1/2}[X_{t-(j-1)\delta/M} - X_{t-j\delta M}]$ are approximately i.i.d. $N(0, \sigma_{t-\delta}^2)$. Under suitable moment conditions, the tails of these normalized increments are well behaved (i.e., not too thick), allowing us to apply a Law of Large Numbers yielding $[\hat{\sigma}_t^2(\delta, M) - \sigma_t^2] \to 0$ in probability as $\delta \to 0$ and $M \to \infty$. Failing to correct for the nonzero drifts in X_t and σ_t^2 does not interfere with consistency—the effect of drift terms on $\hat{\sigma}_t^2(\delta, M)$ vanishes as $M \to \infty$ and $\delta \to 0$.

Though quite a special case, (2.1)–(2.4) illustrate the basic intuition underlying our results: As $M \to \infty$ and $\delta \to 0$, the normalized increments in X_t become approximately i.i.d. with zero conditional mean, finite conditional variance, and sufficiently thin tails, allowing us to apply a Law of Large Numbers to estimate σ_t^2. As we see below, it is possible—in a far more general setting—to apply a Central Limit Theorem to develop an asymptotic normal distribution for the measurement error $[\hat{\sigma}_t^2(\delta, M) - \hat{\sigma}_t^2]$.

We will now introduce the notation need for our theorems. For each $h > 0$, consider a random vector step function ${}_hX_t \in \mathbf{R}^k$, which makes jumps only at times 0, h, $2h$, and so on. Assume that ${}_hX_t$ is a random process with an (almost surely) finite conditional covariance matrix. Formally, ${}_hX_t$ is a locally square integrable semi-martingale (see, e.g. Jacod and Shiryaev, 1987, Chapters 1–2). We take ${}_hX_t$ to be adapted to the filtration $\{{}_h\mathscr{F}_t\}$, where $\{{}_h\mathscr{F}_t\}$ is increasing and right continuous. ${}_hX_t \in \mathbf{R}^k$ can be decomposed into a "predictable" part and a martingale part; i.e., the Doob–Meyer decomposition:

$${}_h\Delta X_\tau \equiv {}_hX_\tau - {}_hX_{\tau-h} = {}_h\mu_\tau h + ({}_hM_{\tau+h} - {}_hM_\tau) = {}_h\mu_\tau \,\Delta\tau + \Delta_h M_\tau,$$

where ${}_h\mu_t \in \mathbf{R}^k$ is ${}_h\mathscr{F}_{t-h}$ measurable, and $\Delta_h M_t \in \mathbf{R}^k$ is a local martingale difference array with an (almost surely) finite conditional covariance matrix. Further, to make our sums look like integrals, we set $\Delta\tau = h$, and $\Delta_h M_\tau \equiv {}_hM_\tau - {}_hM_{\tau-h}$.

The conditional covariance matrix of ${}_h\Delta X_\tau$ per unit of time is the $k \times k$ matrix ${}_h\Omega_\tau = ({}_h\Omega_{(ij)\tau})$. In other words,

$$\mathrm{E}\left({}_h\Delta M_\tau \cdot {}_h\Delta M_\tau^T \mid {}_h\mathscr{F}_{\tau-h}\right) = {}_h\Omega_\tau \,\Delta\tau,$$

where ${}_h\Omega_\tau$ is ${}_h\mathscr{F}_{\tau-h}$ is measurable.

Our interest is in estimating ${}_h\Omega_t$ when it randomly evolves over time. Just as the change in ${}_hX_\tau$ can be decomposed into a drift component (i.e., a component that is predictable one step ahead) and a martingale component, so, we assume, can the change in ${}_h\Omega_\tau$:

$$\Delta_h\Omega_\tau = {}_h\lambda_\tau\, \Delta\tau + {}_h\Delta M^*_\tau,$$

where ${}_h\lambda_\tau$, the instantaneous drift in ${}_h\Omega_\tau$, is ${}_h\mathscr{F}_{\tau-2h}$ measurable, and ${}_hM^*_\tau$ is a $k \times k$ matrix-valued local martingale with respect to the filtration ${}_h\mathscr{F}_{\tau-h}$. Further,

$$\mathrm{E}\big({}_h\Delta M^*_{(ij)\tau} \cdot {}_h\Delta M^*_{(kl)\tau} \mid {}_h\mathscr{F}_{\tau-2h}\big) = {}_h\Lambda_{(ijkl)\tau}\, \Delta\tau.$$

So ${}_h\Lambda_\tau$ is ${}_h\mathscr{F}_{\tau-2h}$ measurable. ${}_h\lambda_t$ and ${}_h\Lambda_t$ are, respectively, the drift and variance per unit of time in the conditional variance process ${}_h\Omega_t$. Since ${}_h\Omega_t$ is a $k \times k$ matrix, its drift ${}_h\lambda_t$ is as well. The "variance of the variance" process Λ_τ is a $k \times k \times k \times k$ tensor. As we see below, the more variable the ${}_h\Omega_t$ process (as measured by ${}_h\Lambda_t$), the less accurately it can be measured.

The class of data-generating processes encompassed in this setup is very large, including, for example, discrete-time stochastic volatility models (e.g., Melino and Turnbull, 1990), diffusions observed at discrete intervals of length h (e.g., Wiggins, 1987; Hull and White, 1987), ARCH models (e.g., Bollerslev *et al.*, 1992), and many random coefficient models (Chow, 1984).

As is well known for standard regressions, the efficiency of least squares covariance matrix estimates depends to a considerable extent on tail thickness of the noise terms (see, e.g., Davidian and Carroll, 1987). This is true for rolling regressions as well. To motivate our next bit of notation, suppose for the moment that the Δ_hX_t's were i.i.d., scalar draws from a distribution with mean zero and variance Ω. If we estimate Ω using T observations by

$$\hat{\Omega} = T^{-1} \sum_{t=1}^{T} (\Delta_hX_t)^2,$$

the variance of $\hat{\Omega}$ is $T^{-1}\,\mathrm{Var}[(\Delta_hX_t)^2]$. That is, the sample variance of $\hat{\Omega}$ depends on the *fourth* moments of the Δ_hX_t's. When ${}_h\Omega_t$ randomly evolves over time, we require an analogous measure of the *conditional* tail thickness of Δ_hX_t. Accordingly, we define ${}_hB_\tau$, a $k \times k$ matrix-valued martingale by the following martingale difference array:[4]

$${}_h\Delta B_\tau = h^{-1/2}\big({}_h\Delta M_\tau \cdot {}_h\Delta M_\tau^T - {}_h\Omega_\tau\, \Delta\tau\big).$$

[4] The reason for the $h^{-1/2}$ in the definition of B is to keep ${}_hB = O_p(1)$. Therefore, the notation will remind us of the size of various integrals. In other words, for M and M^*, we have the usual "size" condition that ${}_h\Delta M^2 = O(\Delta\tau)$, and ${}_h\Delta M^{*2} = O(\Delta\tau)$, and now this also holds for the B process: ${}_h\Delta B^2 = {}_h\theta\,\Delta\tau = O(\Delta\tau)$.

The ${}_hB_\tau$ is essentially an empirical second-moment process with its conditional mean removed each period to make it a martingale. We next define the conditional variance process for ${}_hB_\tau$, the $k \times k \times k \times k$ tensor process ${}_h\theta_\tau$ with

$$ {}_h\theta_{(ijkl)\tau}\, \Delta\tau = \mathrm{E}\big({}_h\Delta B_{(ij)\tau} \cdot {}_h\Delta B_{(kl)\tau} \mid {}_h\mathscr{F}_{\tau-h}\big). $$

The $\theta_{(ijkl)t}$ is closely related to the multivariate conditional fourth moment of $\Delta_h M_t$:

$$ \begin{aligned} \theta_{(iiii)t} &= \mathrm{E}\Big[\big({}_h\Delta X_{i,t} - {}_h\mu_{i,t} \cdot h\big)^4 - {}_h\Omega^2_{(ii)t} \mid {}_h\mathscr{F}_{t-h}\Big] \\ &= \mathrm{E}\Big[\Delta_h M^4_{i,t} - {}_h\Omega^2_{(ii)t} \mid {}_h\mathscr{F}_{t-h}\Big]. \end{aligned} $$

The $\theta_{(iiii)\tau}/\Omega^2_{(ii)t}$ is the conditional coefficient of kurtosis less one of the ith variable at time τ. If $\Delta X_{i,t}$ is conditionally normal, then $\theta_{(iiii)\tau} = 2\Omega^2_{(ii)\tau}$.

We next define

$$ {}_h\rho_{(ijkl)\tau} \equiv \operatorname{corr}\big(\Delta_h B_{(ij)\tau}, \Delta_h M^*_{(kl)\tau-\Delta\tau} \mid {}_h\mathscr{F}_{\tau-h}\big). $$

The ${}_h\rho_t$ is the conditional correlation between the innovations in the empirical second moment process ${}_hB_\tau$ and the innovations in the conditional variance process Ω_τ. The behavior of ${}_h\rho_{(ijkl)\tau}$ is an important determinant of our ability to measure ${}_h\Omega_t$ accurately. To see why, suppose that ${}_h\Omega_t$ is generated by a diagonal multivariate GARCH model as in Bollerslev *et al.* (1988). In this case ${}_h\Omega_t$ equals a distributed lag of the outer product of residual vectors and therefore ${}_h\rho_{(iiii)t} = 1$. In this case, rolling regressions can estimate ${}_h\Omega_t$ arbitrarily well, since $\Delta_h\Omega_t$ is *perfectly* correlated with elements of $\Delta_h X_t\, \Delta_h X_t^T$. That is, when we see $\Delta_h X_t$ this tells us all we need to know about the change in ${}_h\Omega_t$. On the other hand, suppose that ${}_h\Omega_t$ is generated by a diffusion observable at intervals of length h. In this case ${}_h\rho_{(ijkl)t} = 0$, and though $\Delta_h X_t\, \Delta_h X_t^T$ contains information about the *level* ${}_h\Omega_t$, it in general contains no information about *changes* in ${}_h\Omega_t$. The case where ${}_h\rho < 0$ is a sort of "reverse GARCH" case, in which larger than expected residuals cause variance to drop. Our results are able to accommodate this case, although it seems unlikely to be practically relevant. In general, however, the higher $|{}_h\rho_{(ijkl)t}|$, the more accurately measurable is ${}_h\Omega_{(ij)t}$. Unfortunately, we will have to assume a value for ρ because we will see that it is not identifiable.

The estimator we will study is

$$ {}_h\hat{\Omega}_{(ij)T} \equiv \sum_\tau {}_hw_{(ij)(\tau-T)}\big[{}_h\Delta X_{(i)\tau} - h \cdot {}_h\hat{\mu}_{(i)\tau}\big]\big[{}_h\Delta X_{(j)\tau} - h \cdot {}_h\hat{\mu}_{(j)\tau}\big], \quad (2.5) $$

where ${}_h\hat{\mu}$ is an estimate of ${}_h\mu$, and ${}_h\Omega_{(ij)}$ is the ijth component of ${}_h\Omega$, ${}_h\hat{\mu}_{(i)}$ is the ith component of ${}_h\hat{\mu}$, and ${}_hw_{\tau-T}$ is a $k \times k$ weighting matrix for which $\sum {}_hw_{(ij)(\tau-T)}\, \Delta\tau = 1$. For now both the conditional mean estimate ${}_h\hat{\mu}_t$ and the weights ${}_hw_{(ij)t}$ as exogenously given, though below we consider data-dependent selection of ${}_hw_{(ij)t}$.

A special case of the above is the standard flat-weight rolling regression motivated by the following argument. $E(\Delta M)^2/\Delta\tau = \Omega$, so it seems reasonable that if we average terms like $\Delta\hat{M}^2/\Delta\tau$, we should get a good approximation to Ω. So, the rolling regression estimator of Ω is defined as

$$ {}_h\hat{\Omega}_T \equiv [(n+m)h]^{-1} \sum_{\tau=T-nh}^{\tau=T+(m-1)h} [{}_h\Delta X_\tau - h_h\hat{\mu}_\tau][{}_h\Delta X_\tau - h_h\hat{\mu}_\tau]^T. $$

Thus the weights are equal over some region. So

$$ {}_hw_{(ij)\tau} = \begin{cases} \dfrac{1}{(n+m)h} & -nh \le \tau < mh \\ 0 & \text{otherwise}. \end{cases} \tag{2.6} $$

For example, when $n = m = kh^{-1/2}$ for some constant k (which when $\rho = 0$ will turn out to be the asymptotically optimal way of choosing a rolling regression), we see that ${}_hw_{\tau-T} \cong h^{-1/2}k^{-1}$ near T, and 0 far away from T, with $\Sigma w\,\Delta\tau = 1$. Here m is the number of leads and n is the number of lags. In a standard one-sided rolling regression, m is set equal to zero and ${}_hw_{(ij)t-T} = 1/nh$ for $T - nh \le t < T$ and zero otherwise.

When $m = 0$ and the weights are nonnegative but otherwise unconstrained in (2.5), we have a special case of the multivariate GARCH model of Bollerslev *et al.* (1988). The method of treating conditional covariances as constant over blocks of time (e.g., Merton, 1980; Poterba and Summers, 1986; French *et al.*, 1987) is also easily accommodated: Here $w = 1/hK$ whenever $t - T$ is in the same time block as time T and equals zero otherwise. K is the number of observations within the block.

2.1. Assumptions

The first assumption requires the first few conditional moments of ${}_hX_t$ and ${}_h\Omega_t$ remain bounded with small changes over small time intervals as $h \to 0$. This assumption essentially allows us to apply the Central Limit Theorem locally in time.

Assumption A. *The following eight expressions are all $O_p(1)$:*

(*i*) $\sup_{s,t\in[T,T+h^{1/2}]} |\hat{\mu}_s - \mu_t|$

(*ii*) $\sup_{t\in[T,T+h^{1/2}]} |{}_h\lambda_{(i)t} - {}_h\lambda_{(i)T}|$

(*iii*) ${}_h\lambda_T$

(*iv*) ${}_h\Omega_T$

(*v*) ${}_h\theta_T$

(*vi*) ${}_h\Lambda_T$

(*vii*) *For some* $\epsilon > 0$, $\mathrm{E}(|h^{-1/2}{}_h\Delta M_T^*|^{2+\epsilon} \mid \mathscr{F}_{T-2h})$

(*viii*) *For some* $\epsilon > 0$, $\mathrm{E}[|h^{-1/2}{}_h\Delta B_T|^{2+\epsilon} \mid \mathscr{F}_{T-h})$.

Assumption A is not as formidable as its eight parts appear. For example, if all these processes are actually continuous semi-martingales, then Assumption A will hold with only non-explosiveness conditions. This is made precise in the following definition and following restatement of Assumption A.

DEFINITION. We will call ${}_hX_\tau$ a "discretized continuous semi-martingale" if there exists a process ${}_0X_\tau$, such that ${}_hX_{ih} = {}_0X_{ih}$ and ${}_0X_\tau$ is a continuous semi-martingale with differential representation of $\mathrm{d}_0X_\tau = {}_0\mu_\tau\,\mathrm{d}\tau + {}_0\Omega_\tau^{1/2}\,\mathrm{d}_0W_\tau$, where both ${}_0\mu_\tau$ and ${}_0\Omega_\tau$ are continuous semi-martingales with Ω positive definite (a.s.). Further, $\mathrm{d}_0\Omega_\tau = {}_0\lambda_\tau\,\mathrm{d}\tau + {}_0\Lambda_\tau\,\mathrm{d}_0W_\tau'$, where both ${}_0\lambda_\tau$ and ${}_0\Lambda_\tau$ are continuous semi-martingales, and W_τ and W_τ' are multivariate Brownian motions.

ASSUMPTION A′. *Here* ${}_hX_\tau$ *is a discretized continuous semi-martingale for which there exists a random variable M with finite mean such that for all* $\tau \leq K$ (*K finite*) *the following holds* (*almost surely*): $|{}_0\mu_\tau| + |{}_0\Omega_\tau| + |{}_0\theta_\tau| + |{}_0\lambda_\tau| + |{}_0\Lambda_\tau| \leq M$. *Also assume* ${}_h\hat{\mu}_\tau \equiv 0$.

From standard arguments Assumption A′ can be shown to imply Assumption A. Thus, we see that Assumption A is more of a regularity condition rather than a restrictive assumption.

ASSUMPTION B. *The* ${}_h\theta_\tau$, ${}_h\Lambda_\tau$, *and* ${}_h\rho_\tau$ *change slowly over time. That is to say,*

$$\sup_{T \leq \tau \leq T+h^{1/2}} |{}_h\theta_\tau - {}_h\theta_T| = o_p(1),$$

$$\sup_{T \leq \tau \leq T+h^{1/2}} |{}_h\Lambda_\tau - {}_h\Lambda_T| = o_p(1), \qquad \textit{and}$$

$$\sup_{T \leq \tau \leq T+h^{1/2}} |{}_h\rho_{(ijkl)\tau} - {}_h\rho_{(ijkl)T}| = o_p(1).$$

Assumption B tells us that the "hyperparameters" are regular enough that they can be estimated. Again this is not a very restrictive assumption in the sense that these terms would naturally be $O_p(h^{1/2})$ if θ, Λ, and ρ followed stochastic differential equations (SDEs).

ASSUMPTION C. *The diagonal elements of* ${}_h\theta_\tau$ *and* ${}_h\Lambda_\tau$ *are nonvanishing. That is to say,* $\forall i\,\forall j$: $1/{}_h\theta_{(ijij)T} = O_p(1)$, *and* $1/{}_h\Lambda_{(ijij)T} = O_p(1)$.

Assumption C tells us that we can get a nondegenerate asymptotic distribution at the natural rate of convergence. If Assumption C were dropped, our asymptotic variance calculation would still hold. But the results might be trivial in the sense that we get an asymptotic normal with zero variance. Assumption C avoids this.

The ${}_h\mu_t$, ${}_h\hat{\mu}_t$, and ${}_h\lambda_t$ drop out of the asymptotic distribution of the measurement error in the conditional covariance estimate produced by the rolling regression—i.e., these terms are of only second-order importance in determining the measurement error. In fact, if we explode ${}_h\mu_t, {}_h\hat{\mu}_t$, and ${}_h\lambda_t$ to infinity as $h \to 0$ at a sufficiently slow rate, these conditional moments *still* drop out of the asymptotic distribution of the measurement error.

DEFINITION. The ${}_hT_*$ and ${}_hT^*$ are the "start" and "end" times of the rolling regression. That means ${}_hw_{\tau-T} = 0$ for $\tau <{}_hT_*$ or $\tau >{}_hT^*$.

Note ${}_hw_{\tau-T}$ is not required to be nonzero between T_* and T^*. This will be useful when considering two different weights. T_* then typically will be the earlier of the starting times and T^* the later of the ending times. The next assumption restricts the behavior of the weights ${}_hw_{\tau-T}$.

ASSUMPTION D.

$$ {}_hT^* - {}_hT_* = O(h^{1/2}), $$

$$ \sum_{\tau=T_*, T_*+h, \dots}^{T^*} {}_hw_{(ij)\tau-T}\,\Delta\tau = 1, \qquad \textit{and} $$

$$ \sup_{\tau}\left(|{}_hw_{(ij)\tau T}|\right) = O(h^{-1/2}). $$

Assumption D requires that the total number of lags and leads used in the rolling regression is going to infinity at rate $h^{-1/2}$, although the time interval over which the weights are nonzero is shrinking to 0 at rate $h^{1/2}$. Assumption A guarantees that changes in ${}_h\Omega_t$ are small over small time intervals: As in the illustration at the beginning of this section, as $h \to 0$ the rolling regression generates its conditional covariance estimate ${}_h\Omega_t$, using a *growing* number of residuals generated over a *shrinking* period of time. Unfortunately, however, Assumption D also *requires* that the number of residuals assigned nonzero weights be bounded for each h. This accommodates the ARCH(p) model of Engle (1982) with p growing at rate $h^{-1/2}$ as $h \to 0$, but formally excludes the GARCH(p, q) model of Bollerslev (1986). We can, however, *approximate* GARCH models to arbitrary accuracy by considering ARCH(p) models for arbitrarily large but finite (for each h) order.

Typically $w_{(ij)\tau-T} \geq 0$, but this is not required. Assumption D also requires $\sum w_{(ij)\tau-T}\,\Delta\tau = 1$. Interpreting the rolling regression as a multivariate GARCH model, this corresponds to an IGARCH (integrated GARCH) model (see Engle and Bollerslev, 1986). For theorems we can relax this condition to assume only that $\sum w_{(ij)\tau - T}\,\Delta\tau = 1 + o(h^{1/4})$. For intuition on why IGARCH is approached as $h \to 0$, see Nelson (1992).

DEFINITION.

$$
{}_h\Psi_{(ij)x} \equiv \begin{cases} \displaystyle\sum_{\tau = x+h,\, x+2h \ldots}^{\infty} {}_h w_{(ij)\tau}\,\Delta\tau & \text{if } x \geq 0 \\ \displaystyle -\sum_{\tau = -\infty}^{x} {}_h w_{(ij)\tau}\,\Delta\tau & \text{if } x < 0. \end{cases}
$$

Note ${}_h\Psi_x$ is defined only if x/h is an integer. This is like an integral of ${}_h w_\tau$ in the sense that $\Delta\Psi(x)/\Delta x = -{}_h w_x$. For example, in the case of the flat-weight rolling regression for a univariate process,

$$
{}_h\Psi_{s-T} = \frac{({}_hT^* - s)I_{s \geq T} - (s - {}_hT_*)I_{s<T}}{{}_hT^* - {}_hT_*},
$$

where ${}_hT^*$ is the right end point of the rolling regression and ${}_hT_*$ is the left end point. Define the following sums,

$$
{}_hS_{ww} \equiv h^{1/2} \sum_\tau {}_h w_\tau^2\,\Delta\tau
$$

$$
{}_hS_{\Psi\Psi} \equiv h^{-1/2} \sum_\tau {}_h\Psi_\tau^2\,\Delta\tau
$$

$$
{}_hS_{w\Psi} \equiv \sum_\tau {}_h w_\tau \cdot {}_h\Psi_\tau\,\Delta_\tau.
$$

In the multivariate case, w_τ is replaced by $w_{\tau(ij)}$. So, ${}_hS_{ww}$ is $k \times k \times k \times k$, these sums are actually tensors. For example, ${}_hS_{w_{ij}\Psi_{kl}} \equiv \sum_\tau {}_h w_{\tau(ij)} \cdot {}_h\Psi_{\tau(kl)}\,\Delta\tau$.

Finally, define the normalized measurement error process

$$
{}_hQ_t \equiv h^{-1/4}\big({}_h\hat{\Omega}_t - {}_h\Omega_t\big).
$$

Its conditional covariances are asymptotically the $k \times k \times k \times k$ tensor process ${}_hC_\tau$ with elements given by

$$
\begin{aligned}
{}_hC_{(ijkl)t} \equiv {}& {}_hS_{w_{ij}w_{kl}} \cdot {}_h\theta_{(ijkl)t} + {}_hS_{\Psi_{ij}\Psi_{kl}} \cdot {}_h\Lambda_{(ijkl)t} \\
& + {}_hS_{w_{ij}\Psi_{kl}} \cdot {}_h\rho_{(ijkl)t} \sqrt{{}_h\theta_{(ijij)t} \cdot {}_h\Lambda_{(klkl)t}} \\
& + {}_hS_{w_{kl}\Psi_{ij}} \cdot {}_h\rho_{(klij)t} \sqrt{{}_h\theta_{(klkl)t} \cdot {}_h\Lambda_{(ijij)t}}\,.
\end{aligned}
$$

Which in the scalar case is just (where the h has been deleted from C, θ, Λ, ρ, Ψ, and S)

$$
C_t \equiv S_{ww}\theta_t + 2S_{w\Psi}\rho_t\sqrt{\theta_t\Lambda_t} + S_{\Psi\Psi}\Lambda_t. \tag{2.7}
$$

2.2 Main Convergence Theorems

THEOREM 1 (REPRESENTATION). *If Assumptions A and D hold, then*

$$
\begin{aligned}
{}_hQ_{(ij)T} &\equiv h^{-1/4}\left({}_h\hat{\Omega}_{(ij)T} - {}_h\Omega_{(ij)T}\right) \\
&= h^{1/4}\sum_{\tau} {}_hw_{\tau-Th}\,\Delta B_{(ij)\tau} + h^{-1/4}\sum_{\tau} {}_h\Psi_{\tau-Th}\Delta M^{*}_{(ij)\tau} + o_p(1).
\end{aligned}
$$

THEOREM 2 (ASYMPTOTIC DISTRIBUTION). *If Assumptions A–D hold, then*

$$
{}_hQ_T \mid \mathscr{F}_{T_*} \text{ is asymptotically distributed } N(0, {}_hC_{T_*}). \tag{2.8}
$$

For proofs, see the Appendix.

The matrix normal distribution in Theorem 2 has the obvious interpretation: The asymptotic covariance of ${}_hQ_{(ij)t}$ and ${}_hQ_{(kl)t}$ given $\mathscr{F}_{T_*}$ is $C_{(ijkl)T_*}$. Alternatively, using an appropriate sense of a tensor square root, equation (2.8) says $C^{-1/2}Q \xrightarrow{d} N(0, 1)$, where 1 is the tensor identity.

To illustrate the application of Theorem 2, consider a multivariate rolling regression with flat weights. Assume that ${}_hn_{ij} = n_0h^{-1/2}$, and ${}_hm_{ij} = m_0h^{-1/2}$. This is a restricted form of a rolling regression in which all of the windows are the same size. For all i and j the same weighting is then used. In other words,

$$
{}_hw_\tau = h^{1/2}(n_0 + m_0)^{-1} I\{\tau \in [-n_0h^{1/2}, m_0h^{1/2}]\}.
$$

Thus, Assumption D is satisfied. So, in this case (the following approximations are easy to see if one thinks of each sum as being approximated by an integral),

$$
\begin{aligned}
h^{-1/2}\Psi_{(h^{1/2}s)} &= \frac{(m_0 - s)I_{s \geq T} + (s - n_0)I_{s<T}}{m_0 + n_0} \\
S_{ww} &= \sum w_s^2 h \cong \frac{1}{m_0 + n_0} \\
S_{\Psi\Psi} &= \sum \Psi_s^2 h \cong \frac{m_0^3 + n_0^3}{3(m_0 + n_0)^2} \\
S_{w\Psi} &= \sum w_s\Psi_s h \cong \frac{m_0 - n_0}{2(m_0 + n_0)}
\end{aligned}
$$

Because of our assumption that all w_{ij} are the same, we need not distinguish between $S_{w_{ij}w_{kl}}$ and just call all of them S_{ww}. Likewise for $S_{w\Psi}$ and $S_{\Psi\Psi}$. We can now compute the variance of $\hat{\Omega}_{ij}$. Then (where to simplify the equations we have taken: ${}_h\theta_{(ijij)T_*} = \theta$, ${}_h\Lambda_{(ijij)T_*} =$

$\Lambda, {}_h\rho_{(ijij)T_*} = \rho)$,

$$C_{(ijij)T_*} = \theta S_{ww} + 2\sqrt{\theta\Lambda}\,\rho S_{w\Psi} + \Lambda S_{\Psi\Psi}$$

$$= \frac{\theta}{m_0 + n_0} + \sqrt{\theta\Lambda}\,\rho\frac{m_0 - n_0}{n_0 + m_0} + \Lambda\frac{m_0^3 + n_0^3}{3(n_0 + m_0)^2} \quad (2.9)$$

Consider the three components of the asymptotic covariances in (2.9): The first term, θS_{ww}, would be present even in the i.i.d. case. This term reflects sampling error and can be made arbitrarily small by making $n_0 + m_0$ sufficiently large. Indeed, if the conditional covariance matrix ${}_h\Omega_t$ were constant, the other terms in $C_{(ijkl)T_*}$ would vanish, and letting $n_0 + m_0$ be infinite would be optimal. The third term, $\Lambda S_{\Psi\Psi}$, reflects the variability in ${}_h\Omega_t$. This term can be made arbitrarily small by making $n_0 + m_0$ sufficiently small: The smaller the window over which the rolling regression is conducted, the more like a constant ${}_h\Omega_t$ is within the window. As indicated in our discussion of ${}_h\rho$, the second term, $\sqrt{\theta\Lambda}\,\rho S_{w\Psi}$, comes from the covariance between the first and last terms. This term drops out when the data are generated by a diffusion but not, for example, when the data are generated by a GARCH model. This term also controls how much information about ${}_h\Omega_\tau$ is in the "past" residuals as opposed to the future residuals.

2.3 Consistent Estimation of Nuisance Parameters

To construct correct asymptotic confidence intervals, we must have consistent estimates of the components of the conditional covariance of the measurement error ${}_hQ_t$; namely, ${}_h\theta_t$, ${}_h\Lambda_t$, and ${}_h\rho_t$. Sometimes some of these are known a priori; for example, when $\{{}_hX_t, {}_h\Omega_t\}$ is generated by a diffusion process, ${}_h\rho_{(ijkl)t} \to 0$, ${}_h\theta_{(iiii)t}/{}_h\Omega^2_{(ii)t} \to 2$, and $\theta_{ijkl} \mapsto 0$ otherwise as $h \to 0$, thus leaving only Λ_t to estimate. In more general circumstances, however, they all must be estimated.

We next consider estimation of ${}_h\theta_r$ and ${}_h\Lambda_\tau$.

Since we have only the most indirect methods of obtaining information about these parameters, we will need to assume that the processes under consideration are "regular" over a slightly longer interval. To do this we will use the following uniform convergence idea. We will say that $X_T = o_p(1)$ holds uniformly over $T \in [T', T' + K_h h^{1/2}]$ if for all $\epsilon > 0$,

$$\sup_{T \in [T', T' + K_h h^{1/2}]} P(|X_T| > \epsilon) \to 0 \quad \text{as } h \to 0.$$

ASSUMPTION E. *Assume there exist a function K_h such that $K_h \to \infty$ as $h \to 0$ such that Assumption A holds uniformly over $T \in [T', T' + K_h h^{1/2}]$.*

By way of example, consider assumption A part iii. It tells us that λ_T is small: $|_h \lambda_T| = O_p(1)$. In other words, Assumption A part iii by itself says: $\forall \epsilon > 0$, $\exists M$ such that $P(|_h\lambda_T| > M) < \epsilon$ for sufficiently small h. Under Assumption E, we have the following stronger statement: $\forall \epsilon > 0$, $\exists M$ such that

$$\sup_{T \in [T', T' + K_h h^{1/2}]} P(|_h \lambda_T| > M) < \epsilon$$

for sufficiently small h. We now need to assume that our "targets" change very little over short time intervals. In other words, we need a stronger version of Assumption B.

ASSUMPTION F. *For the K_h in Assumption E,*

$$\tau \in \sup_{[T', T' + K_h h^{1/2}]} |_h \theta_\tau - {}_h \theta_T| = o_p(1)$$

$$\tau \in \sup_{[T', T' + K_h h^{1/2}]} |_h \Lambda_\tau - {}_h \Lambda_T| = o_p(1)$$

$$\tau \in \sup_{[T', T' + K_h h^{1/2}]} |_h \rho_\tau - {}_h \rho_T| = o_p(1)$$

ASSUMPTION (F′). *For the K_h in Assumption E (in the univariate case only),*

$$\sup_{\tau \in [T', T' + K_h h^{1/2}]} \left|_h \theta_\tau / {}_h \Omega_\tau^2 - {}_h \theta_T / {}_h \Omega_T^2\right| = o_p(1)$$

$$\sup_{\tau \in [T', T' + K_h h^{1/2}]} \left|_h \Lambda_\tau / {}_h \Omega_\tau^2 - {}_h \Lambda_T / {}_h \Omega_T^2\right| = o_p(1)$$

$$\sup_{\tau \in [T', T' + K_h h^{1/2}]} |_h \rho_\tau - {}_h \rho_T| = o_p(1)$$

Assumption F trivially implies Assumption B. That F′ implies Assumption B follows from the "near" constancy of Ω_T over intervals of length $h^{1/2}$. Assumption F is more natural for the proof of our convergence theorem and is easily understood in the multivariate setting.

In the univariate case, the advantage of using θ/Ω^2 and Λ/Ω^2 instead of θ and Λ, respectively, is that it may be more believable that the "shape" parameters are constant than the parameters themselves: Constant θ/Ω^2 is equivalent to constant conditional kurtosis of the increments in ${}_hX_t$. When ${}_hX_t$ is generated by a diffusion, for example, $\theta/\Omega^2 = 2$. Constant Λ/Ω^2 is equivalent to $\ln({}_h\Omega_t)$ being conditionally homoskedastic. Many ARCH and stochastic volatility models effectively assume this (see Nelson and Foster, 1994b) and, as we see in the empirical application below, this homoskedastic $\ln({}_h\Omega_t)$ seems a reasonable approximation for U.S. stock prices.

In the univariate case, these assumption are equivalent in the sense that a process that satisfies F for some K_h will satisfy F′ for some other K_h (and vice versa). But if one of these K_h's is significantly larger than the other, it will allow the use of more data in estimating θ and Λ.

We will now outline estimators for θ and Λ. First define a matrix $f_\epsilon(\tau)$, which can be thought of as being $\hat{\Omega}_\tau - \hat{\Omega}_{\tau-\epsilon}$ for an appropriately defined $\hat{\Omega}$:

$$f_\epsilon(\tau) = \sum_{s=\tau,\tau+h,\dots}^{\tau+\epsilon h^{1/2}} h^{-1/2}(\Delta_h X_s - {}_h\hat{\mu}_s \Delta s)(\Delta_h X_s - {}_h\hat{\mu}_s \Delta s)'/\epsilon$$

$$- \sum_{s=\tau,\tau-h,\tau-2h,\dots}^{\tau-\epsilon h^{1/2}} h^{-1/2}(\Delta_h X_s - {}_h\hat{\mu}_s \Delta s)(\Delta_h X_s - {}_h\hat{\mu}_s \Delta s)'/\epsilon.$$

$\hat{\theta}$ and $\hat{\Lambda}$ can now defined as

$$\hat{\theta}_{(ijkl)T} = \frac{\epsilon}{2K_h} \sum_{\tau=T,T+h,\dots}^{T+K_n h^{1/2}} f_{(ij)\epsilon}(\tau) f_{(kl)\epsilon}(\tau) - \hat{\Lambda}_{(ijkl)T}\epsilon^2/3 \quad (2.10)$$

$$\hat{\Lambda}_{(ijkl)T} = \frac{3}{2\delta K_h} \sum_{\tau=T,T+,\dots}^{T+K_n h^{1/2}} f_{(ij)\delta}(\tau) f_{(kl)\delta}(\tau) - 3\hat{\theta}_{(ijkl)T}/\delta^2. \quad (2.11)$$

In the case where $\epsilon \ll \delta$ these estimators are more intuitive because the "corrections" are small and only the sums themselves need to be considered. To actually get the estimators, we have to solve the simultaneous equations (2.10) and (2.11). These estimators are designed to work with Assumption F. The following theorem shows they achieve this goal.

THEOREM 3 (CONSISTENCY). *Under Assumption D, E, and F, both $\hat{\theta}_T$ and $\hat{\Lambda}_T$ are consistent pointwise in T. For proof, see the Appendix.*

For the scalar case, Assumption F′ should hold over a longer interval and so "better" estimates of θ and Λ should be available. Estimators appropriate for this situation will now be given. The definition of f_ϵ is notationally simpler in the scalar case:

$$f_\epsilon(\tau) = \sum_{s=\tau,\tau+h,\dots}^{\tau+\epsilon h^{1/2}} h^{-1/2}(\Delta_h X_s - {}_h\hat{\mu}_s \Delta s)^2/\epsilon$$

$$- \sum_{s=\tau,\tau-h,\dots}^{\tau-\epsilon h^{1/2}} h^{-1/2}(\Delta_h X_s - {}_h\hat{\mu}_s \Delta s)^2/\epsilon.$$

Now modify (2.10) and (2.11) as follows:

$$\hat{\theta}_T = \frac{\hat{\Omega}_T^2 \epsilon}{2K_h} \sum_{\tau=T,T+h,\dots}^{T+K_n h^{1/2}} f_\epsilon(\tau)^2/\hat{\Omega}_\tau^2 - \hat{\Lambda}_T \epsilon^2/3 \qquad (2.10')$$

$$\hat{\Lambda}_T = \frac{3\hat{\Omega}_T^2}{2\delta L_h} \sum_{\tau=T,T+h,\dots}^{T+K_n h^{1/2}} f_\delta(\tau)^2/\hat{\Omega}_\tau^2 - 3\hat{\theta}_T/\delta^2. \qquad (2.11')$$

These are the estimators that we actually use in the empirical example.

We will see from the simulations that the following estimator appears to do somewhat better for $\hat{\Lambda}$ in the scalar case:

$$\hat{\Lambda}_T = \frac{3\hat{\Omega}_T^2}{2\delta K_h} \sum_{\tau=T}^{T+K_n h^{1/2}} \log\left(\hat{\Omega}(T+\delta h^{1/2})/\hat{\Omega}_T\right)^2 - 3\hat{\theta}_T/\delta^2, \quad (2.12)$$

where $\hat{\Omega}$ is taken to be a one-sided rolling regression of length $\delta h^{1/2}$. Equation (2.12) can be seen to be close to (2.11′) if a one-term Taylor series for the log is used.

Without further assumptions on the processes X_t and Ω_t it is impossible to estimate ${}_h\rho_{(ijkl)t}$. In order to prove this, we have to find two models that have identical observable random variables (i.e., the distributions for the X_t's are the same) but have different values for the parameter ${}_h\rho_{(ijkl)t}$. Luckily, Shephard (1994) does exactly this. He starts out with a stochastic volatility model, which by construction has a ρ of zero. In other words, $\Omega_t = \text{Var}(X_t \mid \mathcal{F}_{t-h})$, where $\mathcal{F}$ is the σ-field generated by both the observed state variable X and the unobservable latent variable Ω. He integrates out this latent state variable and generates a GARCH model. This is equivalent to looking at $\Omega_t' = \text{Var}(X_t \mid \mathcal{G}_{t-h})$, where $\mathcal{G}$ is the σ-field generated only by the observable state variable X. In this new model, $\rho = 1$. Since only the process X is observed, it is impossible to distinguish between these two models. Thus, ρ is unidentifiable since it changes with the definition of the σ-field.

3. EFFICIENCY AND OPTIMALITY

Throughout this section, we will use various techniques of estimating a particular Ω_{ij}. Thus, we will think of i, j as fixed. We will call ${}_h\theta_{(ijij)T_*} = \theta$, ${}_h\Lambda_{(ijij)T_*} = \Lambda$, ${}_h\rho_{(ijij)T_*} = \rho$. Further, because we will want to compare windows of different lengths, we will take our conditioning time to be $T_* = T - kh^{-1/2}$ for some sufficiently large k.

3.1 Optimal Lead and Lag Lengths in Standard (Flat-Weight) Rolling Regressions

Consider again the example of the standard rolling regression in the previous section, in which, for some nonnegative n_0 and m_0, the weights are given by ${}_h w_t = h^{1/2}(n_0 + m_0).\ I(t \in [-n_0 h^{1/2}, m_0 h^{1/2}])$. This weighting scheme is of special interest, since it is most frequently encountered in practice. The asymptotic standard error for the ijth element of the measurement error in the conditional covariance matrix is given in (2.9). In other words,

$$\mathrm{SE}\left(\hat{\Omega}_{ij}\right)^2 = (\text{bias})^2 + \mathrm{Var}\left(\hat{\Omega}_{ij}\right) \cong 0^2 + C_{ijij}$$

$$\cong \frac{\theta}{m_0 + n_0} + \sqrt{\theta\Lambda}\,\rho\,\frac{m_0 - n_0}{n_0 + m_0} + \Lambda\,\frac{m_0^3 + n_0^3}{3(n_0 + m_0)^2}.$$

Where the first equality follows by definition of the mean squared error, and the approximation follows from our Theorem 2.

THEOREM 4 (FLAT WEIGHT).

- *The asymptotic variance-minimizing backward-looking flat-weight rolling regression (i.e., $m_0 = 0$) is given by setting $n_0 = \sqrt{3\theta/\Lambda}$. The asymptotic measurement error variance (see (2.9)) achieved by this choice of m_0 and n_0 is $(2 - \rho)\sqrt{\Lambda\theta/3}$.*
- *The asymptotic variance-minimizing forward-looking flat-weight rolling regression (i.e., $n_0 = 0$) is given by setting $m_0 = \sqrt{3\theta/\Lambda}$. The asymptotic measurement error variance achieved with this choice of m_0 and n_0 is $(2 + \rho[\Lambda\theta/3]^{1/2}$.*
- *When $\rho > (3/4)^{1/2}$, the one-sided backward-looking flat-weight rolling regression is asymptotically optimal in the class of flat-weight rolling regressions. When $\rho < -(3/4)^{1/2}$, the optimum is a one-sided forward-looking rolling regression. When $|\rho| \leq (3/4)^{1/2}$, the asymptotic optimum is a two-sided rolling regression with*

$$n_0 = \sqrt{3(1 - \rho^2)\theta/\Lambda} + \rho\sqrt{\theta/\Lambda}, \quad \textit{and} \tag{3.1}$$

$$m_0 = \sqrt{3(1 - \rho^2)\theta/\Lambda} - \rho\sqrt{\theta/\Lambda} \tag{3.2}$$

The minimized asymptotic variance when $|\rho| \leq (3/4)^{1/2}$ is $\sqrt{\Lambda\theta(1 - \rho^2)/3}$.

For proof, see the Appendix.

Note the role of ${}_h\rho$ in determining the optimal weighting scheme: When GARCH generates the data, ${}_h\rho = 1$ and all information used by the rolling regression about ${}_h\Omega_t$ is in the *lagged* residuals. The closer ${}_h\rho$ is to 1, therefore, the more weight is optimally put on lagged (as opposed to led) residuals.

The ${}_h\rho = 0$ case is also instructive: Here the optimal weighting scheme is two-sided with equal window lengths on each side. This cuts the asymptotic variance exactly in half compared with the optimal one-sided rolling regression.

3.2. Optimal Weighted Rolling Regressions

Although flat-weight rolling regressions are widely used, they are generally nonoptimal.

THEOREM 5 (OPTIMAL WEIGHTS). *Define θ and Λ as in (2.7) and let $\alpha \equiv \sqrt{\Lambda/\theta}$.*

- *The asymptotic variance-minimizing backward-looking (i.e., all the weight is on lagged residuals) weight function ${}_0w_t$ is given by $I_{\{t<0\}}\alpha e^{\alpha t}$. This achieves an asymptotic measurement error variance of $\sqrt{\Lambda\theta}(1-\rho)$.*
- *The asymptotic variance-minimizing forward-looking weight function ${}_0w_t$ is given by $I_{\{t>0\}}\alpha e^{-\alpha t}$. This achieves an asymptotic measurement error variance of $\sqrt{\Lambda\theta}(1+\rho)$.*
- *The asymptotic variance-minimizing weight function ${}_0w_t$ is given by*

$$
{}_0w_s = \begin{cases} p\alpha e^{-\alpha s} & \text{for } s \geq 0 \\ (1-p)\alpha e^{\alpha s} & \text{for } s < 0. \end{cases} \tag{3.3}
$$

where $p = (1-\rho)/2$. This achieves an asymptotic measurement error variance of $\frac{1}{2}\sqrt{\Lambda\theta}(1-\rho^2)$.

For proof, see the Appendix.

Others (Geno-Catalot *et al.*, 1992; Corradi and White, 1994; Banon, 1978; Dohnal, 1987; Florens-Zmirou, 1993) have estimated Ω_t by nonparametric methods. Their estimators often achieve better rates of convergence than we do since they assume that Ω_t is much smoother than we assume it to be. On the other hand, we can often handle a more general situation than they can. So, the choice of estimator and its resulting rate of convergence depends on which assumptions are appropriate.

Note that the estimators recommended Theorem 5 violate our assumptions in the sense that ${}_0w_s$ does not have compact support. Of course the recommended ${}_0w_s$ can be arbitrarily well approximated by a w that does have compact support.

Further notice that in terms of forecasting (i.e., backward looking) the optimal weighting is the same regardless of the value of ρ. Thus, even if ρ

cannot be estimated, optimal forecasts for Ω are still available. Of course, we would not know how accurate these forecasts in fact are!

Another popular strategy for estimating conditional covariances—chopping the data into short blocks and estimating covariances as if they were constant within the blocks (see, e.g., Merton, 1980; Poterba and Summers 1986; French *et al.*, 1987)—is a special case of the two-sided flat-weight rolling regression. Suppose the block is composed of a total of K observations. At the left (right) end point of the block, the covariance matrix estimate is a one-sided rolling regression using K led (lagged) residuals. Between the two end points, the estimate is a two-sided rolling regression. If we set $K \equiv h^{-1/2}k_0$, then the asymptotic measurement error variance at a point a fraction η through the block ($0 \leq \eta \leq 1$) is obtained from (2.9) by setting $n_0 = k_0\eta$ and $m_0 = k_0(1-\eta)$:

$$\hat{C} = \theta/k_0 + \rho\sqrt{\theta\Lambda}(1-2\eta) + (\Lambda k_0/3)\left[\eta^3 + (1-\eta)^3\right] \quad (3.4)$$

which, when $|\rho\sqrt{\theta\Lambda}\,k_0| \leq \frac{1}{2}$, is minimized when $\eta = \frac{1}{2} - \rho\sqrt{\theta/\Lambda}\,k_0$, lending a bow shape to the confidence intervals.

An obvious implication of Theorem 5 is that flat-weighting schemes such as one- or two-sided rolling regressions or block-constant estimators are inefficient. Unfortunately, however, constructing the asymptotically efficient weights requires consistent estimates of the nuisance parameter processes $\{\rho_t\}$, $\{\Lambda_t\}$, and $\{\theta_t\}$. Can we construct dominating weighting schemes without knowing $\{\rho_t\}$, $\{\Lambda_t\}$, and $\{\theta_t\}$? The answer, it turns out, is yes.

THEOREM 6 (DOMINATING FLAT WEIGHTS). *For every i and j, define the weights* ${}_hw_{(ij)\tau-T}$ *by* (2.6) (*i.e., we use* $n = n_0h^{-1/2}$ *lagged residuals and* $m = m_0h^{-1/2}$ *led residuals*). *Define an alternative set of weights* ${}_hw^*_{(ij)\tau-T}$ *by*

$$ {}_hw^*_{(ij)\tau-T} = \begin{cases} 3^{1/2}(n_0 - m_0)\exp\left[-3^{1/2}h^{1/2}(T-\tau)/m_0\right] & \text{if } \tau > T \\ 3^{1/2}(n_0 + m_0)\exp\left[-3^{1/2}h^{1/2}(T-\tau)/n_0\right] & \text{if } \tau < T. \end{cases} \quad (3.5)$$

Then the asymptotic variance obtained using ${}_hw^*_{(ij)\tau-T}$ *is lower than the asymptotic variance obtained by using* ${}_hw_{(ij)\tau-T}$ *for any* ρ, θ, *and* Λ, *with*

$$\hat{C} = \hat{C}^* = (1-\sqrt{3}/2)\Big(\psi(0-)^2(\theta/m_0 + \Lambda m_0/3) + \psi(0)^2(\theta/n_0 + \Lambda n_0/3)\Big) > 0. \quad (3.6)$$

The idea behind Theorem 6 is simple: We leave the total *share* of the weight put on led and lagged residuals unchanged, but alter the *shape* of the weights on each side of time T from a block shape to an exponential decline.

There is another natural way to dominate a block-constant estimation scheme, provided we are willing to consider average, rather than pointwise, measures of accuracy: Integrate the measurement error variance (3.4) across the block (i.e., integrate (3.4) over η from 0 to 1), yielding an average measurement error variance across the block of ("b.c." is for "block constant") $\hat{C}_{\text{b.c.}} = \theta/k_0 + (\Lambda k_0/6)$. Now consider a flat-weight, two-sided rolling regression using $K/2 = .5k_0 h^{-1/2}$ leads and the same number of lags. By (2.9), this achieves an average measurement error variance of ("t.s." is for "two-sided") $\hat{C}_{\text{t.s.}} = \theta/k_0 + (\Lambda k_0/12)$, which is strictly smaller whenever $\Lambda > 0$, regardless of the values of k_0, ρ, and Ω. Of course, this two-sided rolling regression is itself dominated by an exponentially weighted rolling regression constructed as in Theorem 6.

If we are willing to assume that $\rho = 0$, as it would be, for example, if the data are generated by a diffusion observed at discrete intervals, further dominance relations follow; in particular, a one-sided rolling regression using, say, n lags and no leads has exactly twice the asymptotic variance of a rolling regression using n lags and n leads. The resulting two-sided rolling regression is itself dominated by an exponential-weighted rolling regression constructed as in Theorem 6.

Several of the dominance relations are illustrated in Figure 1. Using numbers from the empirical applications in Section 5, Figure 1 plots the ratios of the standard deviation of measurement errors in S & P 500 volatility estimates using various estimation schemes to that obtained using the optimal two-sided exponentially weighted estimator. The graph was constructed under the assumption that $\rho = 0$. In switching from the optimal two-sided exponentially weighted estimator to the optimal flat-weight estimator, the standard deviation of the measurement error rises about 7%. In switching from the optimal two-sided to the optimal one-sided estimate, the standard deviation goes up by a factor of $\sqrt{2}$. The bow-shaped pattern attained by the block-constant scheme of French *et al.* (1987) and of Poterba and Summers (1986) is clear in Figure 1: When $\rho = 0$, this estimate does relatively well midmonth but poorly at the beginning and the end of the month. Switching from this block-constant scheme to using a two-sided rolling regression with the same number of residuals (as proposed above) achieves a standard error equal to the (minimized) midmonth standard error.

If standard errors are estimated for the variance estimate under the false assumption that the covariance matrix *truly* is constant within blocks, only the sampling error term θ/k_0 appears, giving an unrealistically optimistic picture of the accuracy of the estimated covariance matrix. This is illustrated in Figure 2.

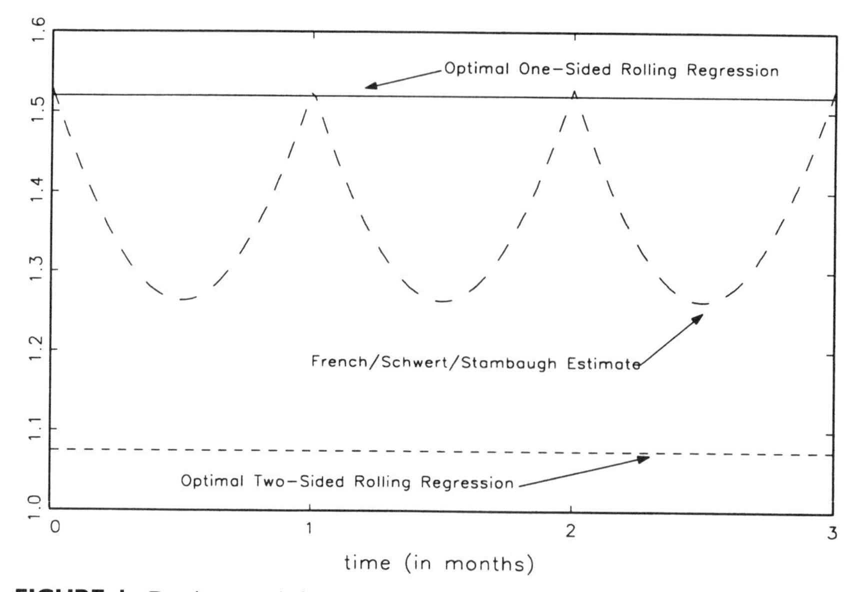

FIGURE 1 Dominance relations. This graph shows the accuracy of various estimates of the variance vs. time. The optimal two-sided exponentially weighted estimate is normalized to have a standard error of one. Relative to the accuracy of this estimator, the optimal two-sided flat weight has a standard error of 1.07. The French/Schwert/Stambaugh estimator assumes a fixed variance over the month so its performance changes over the month. The optimal one-sided flat-weight estimator uses only historical data and so does worse than the other estimators, which use both the past and the future to estimate Ω_t.

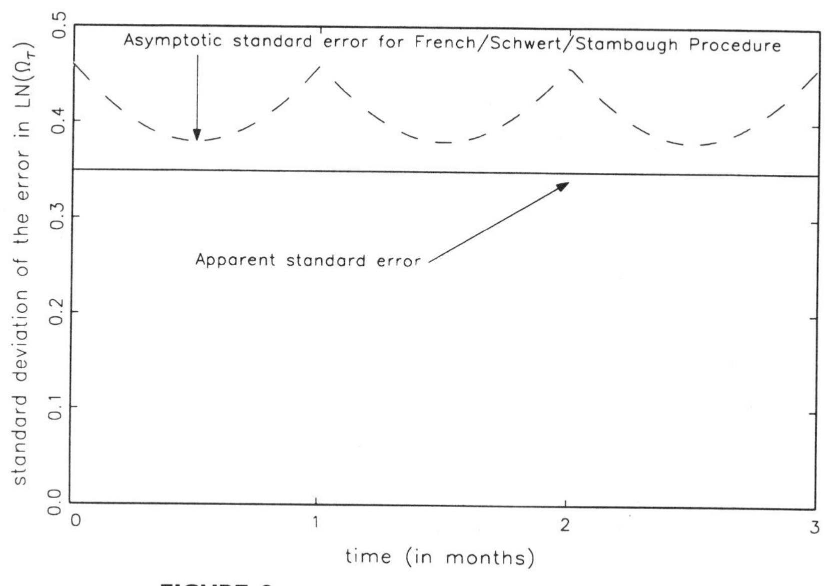

FIGURE 2 Real vs. apparent measurement accuracy.

3.3 The Relation between the Regularity Conditions and the Optimality Results

Clearly, there are relaxations in the regularity conditions which would invalidate the optimality results. For example, suppose that, within each month, volatility is constant, with each month's volatility an i.i.d. draw from some distribution. Presumably, in this case, the block-constant estimation scheme of Poterba and Summers (1986) and French *et al.* (1987) would dominate two-sided exponentially declining weights. This, however, would violate our regularity conditions, which (asymptotically) ruled out discrete jumps in ${}_h\Omega_t$.

A more subtle example was suggested to us by John Campbell: Suppose volatility follows a moving average process in which volatility shocks persist—with constant weight—for some period and then suddenly die out. In this case, a flat-weight rolling regression would presumably dominate an exponential weighting scheme. (This is obviously true, for example, if volatility follows Engle's (1982) ARCH(p) process with equal weights on p lagged residuals.) Here, discrete jumps are not the problem, since it is easy to show that such a moving-average scheme is consistent with a continuous sample path for volatility in the limit as $h \to 0$. For example, for some $\Delta > 0$, set $\Omega_t = \exp(W_t - W_{t-\Delta})$.

Though it may not be as obvious, this scheme is also ruled out by our regularity conditions, which not only assumed that the sample paths of the state variables were (asymptotically) continuous, but also that, over short time intervals, the *unpredictable component of changes in the state variables swamps the predictable component*[5]—i.e., the noise swamps the signal for sufficiently small h. In the moving average example just given, the noise and the signal are of the same stochastic order as $h \to 0$. Our regularity conditions effectively assume that shocks to the state variables decay either gradually or not at all. This means that, over very short time intervals, the movements in ${}_h\Omega_t$ and ${}_hX_t$ look like random walks.

Since our estimates of Ω_t are formed over short intervals, and since X_t and Ω_t behave asymptotically like random walks over such short intervals, it should not be too surprising that our optimal weighting scheme is two-sided exponential: This is the weighting scheme obtained in the literature on random coefficient models under the assumption of a Gaussian random walk (independent if the right-hand side variables) for the regression coefficients (see, for example, Fisher and Kamin, 1985).

If the regularity conditions asymptotically ruling out discrete jumps in ${}_hX_t$ are relaxed, our results are invalidated: Suppose, for example, that ${}_hX_t$ is generated by a jump process, say a Poisson, observed at discrete

[5] A continuous-time semi-martingale is decomposable (by definition) into the sum of a martingale (which may be of unbounded variation, and so very rapidly oscillating) and an instantaneously predictable component of bounded variation (which is much more slowly varying over short time intervals).

intervals of length h. For each T, the normalized residual $h^{-1/2}[_hX_t - {}_hX_{t-h}]$ converges in probability to zero as $h \to 0$, yet its conditional variance does not vanish to zero with h. Clearly a rolling regression using $O(h^{1/2})$ window widths cannot consistently extract this variance, since unless there is a jump within the window (which happens with vanishingly small probability as h goes to zero), the variance estimate produced by the rolling regression is 0! The problem here is that the normalized residuals $h^{-1/2}[_hX_t - {}_hX_{t-h}]$ are too thick-tailed (they are nearly always small but are occasionally enormous—i.e., $\theta = \infty$). This prevents us from applying a Law of Large Numbers and a Central Limit Theorem locally in time to extract ${}_h\Omega_t$ from the squared increments in ${}_hX_t$.

We have also assumed that our variance process, Ω, does not have jumps. In this case though, the problem becomes in some sense easier instead of harder. If the variability of Ω is contained in jumps, then "most of the time," Ω is relatively constant. So, long windows can be used for the rolling regression. Unfortunately, the asymptotic variance will still be infinite, but this is now due to a few large errors. In other words, most of the time, we will be getting very accurate estimates, but when a jump occurs, we get asymptotically an infinite error.

4. ESTIMATING CONDITIONAL BETAS

In many applications, especially in finance, conditional betas are of greater importance than conditional variances or covariances. Suppose that $\Delta_h X_{1,t}$ is the return on some market index, while $\Delta_h X_{j,t}$ is the return on some other asset or portfolio. The true and estimated conditional betas of asset j with respect to the market index are defined, respectively, as

$$ {}_h\beta_{j,t} \equiv {}_h\Omega_{1,j,t}/{}_h\Omega_{1,1,t}, \qquad \text{and} \quad {}_h\hat{\beta}_{j,t} \equiv {}_h\hat{\Omega}_{1,j,t}/{}_h\hat{\Omega}_{1,1,t}. $$

Since the estimated beta is a differentiable function of the asymptotically normal covariance and variance estimates ${}_h\hat{\Omega}_{1,j,t}$ and ${}_h\hat{\Omega}_{1,1,t}$, it is also asymptotically normal (see, e.g., Serfling, 1980, section 3.3, Theorem A), with mean zero and asymptotic variance

$$ {}_h\hat{\Omega}_{1,1,t}^{-2}\left[{}_hC_{(1j1j)t} + {}_h\beta_{j,t}^2\,{}_hC_{(1111)t} - 2\,{}_h\beta_{j,t}\,{}_hC_{(111j)t}\right]. \tag{4.1} $$

We next consider optimality, assuming, for simplicity, that the same weights are used in forming both ${}_h\hat{\Omega}_{1,j,t}$ and ${}_h\hat{\Omega}_{1,1,t}$. This corresponds to using weighted least squares (regressing $\Delta_h X_{j,\cdot}$, on $\Delta_h X_{1,\cdot}$) to estimate

${}_h\beta_{j,t}$. Substituting from (2.7) into (4.1) yields

$$\mathrm{AVAR}_t\left(h^{-1/4}\left[{}_h\hat{\beta}_{j,t} - {}_h\beta_{j,t}\right]\right) = \left[\theta_\beta S_{ww} + \Lambda_\beta S_{\Psi\Psi} + 2\rho_\beta \sqrt{\theta_\beta \Lambda_\beta}\, S_{w\Psi}\right] \tag{4.2}$$

where

$$\theta_\beta \equiv \left({}_h\theta_{(1j1j)t} + {}_h\beta_{j,t}^2\, {}_h\theta_{(1111)t} - 2\,{}_h\beta_{j,t}\, {}_h\theta_{(111j)t}\right) / {}_h\hat{\Omega}_{1,2,t}^2, \tag{4.3}$$

$$\Lambda_\beta \equiv \left({}_h\Lambda_{(1j1j)t} + {}_h\beta_{j,t}^2\, {}_h\Lambda_{(1111)t} - 2\,{}_h\beta_{j,t}\, {}_h\Lambda_{(111j)t}\right) / {}_h\Omega_{1,1,t}^2, \tag{4.4}$$

and (deleting the h and t subscripts to improve legibility)

$$\rho_\beta \equiv \frac{\rho_{1j1j}\sqrt{\theta_{1j1j}\Lambda_{1j1j}} + \beta^2\rho_{1111}\sqrt{\theta_{1111}\Lambda_{1111}} - \beta\rho_{111j}\sqrt{\theta_{1111}\Lambda_{1j1j}} - \beta\rho_{1j11}\sqrt{\theta_{1j1j}\Lambda_{1111}}}{\Omega_{11}^2\sqrt{\theta_\beta\Lambda_\beta}}$$

As in section 2, the three terms are easily interpreted: θ_β is the sampling error variance, Λ_β is the instantaneous conditional variance of the increments in ${}_h\beta_{j,t}$. The $2\rho_\beta\sqrt{\theta_\beta\Lambda_\beta}$ term arises from the covariance between the other two terms. Again, this term is zero for diffusion models and many stochastic volatility models. Note that (4.2) has the same form as (2.9) if we substitute θ_β, Λ_β, and ρ_β for θ, Λ, and ρ. Apart from these substitutions, the optimality and dominance results of section 3 are unaffected. In particular, the asymptotically optimal weights are two-sided and exponentially declining, just as derived in the random coefficients literature under the assumption that betas follow random walks independent of returns on the market index.

5. AN APPLICATION: VOLATILITY ON THE S & P 500

To illustrate the application of our results, we estimate the conditional variance of continuously compounded daily capital gains on the S & P 500. Our data extend from January 1928 through December 1990. Poterba and Summers (1986) and French *et al.* (1987) employed the same series (up to 1985) in their work. The series exhibits small but statistically significant serial correlation of about 6% at one lag, presumably caused by thin trading of the stocks in the underlying index (see, e.g., Scholes and Williams, 1977). There is little serial correlation at longer lags. Since this serial correlation is not of interest to our application, we prewhitened the series with an AR(1). Another "nuisance" aspect of this data is the contribution of nontrading days to variance: i.e., stock volatility is typically higher following weekends and holidays, since the information arriving

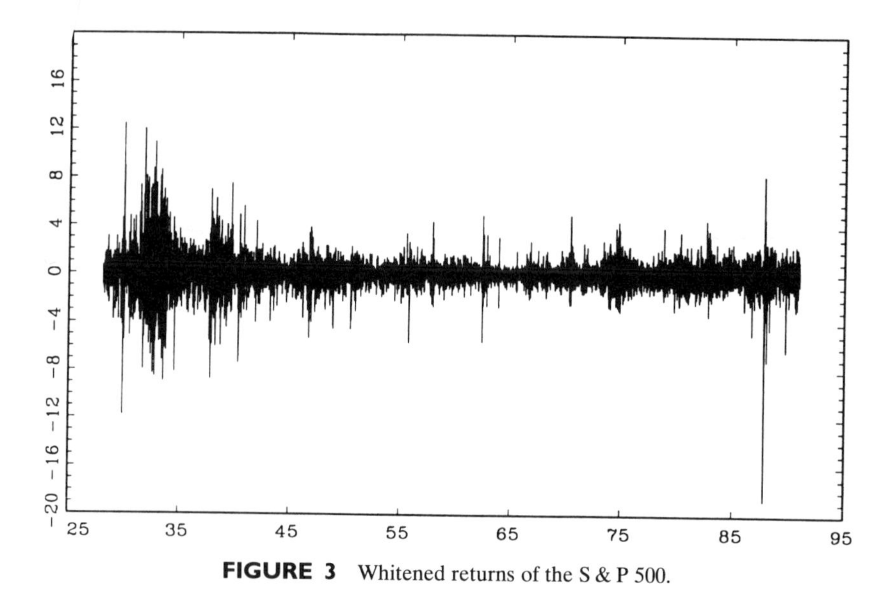

FIGURE 3 Whitened returns of the S & P 500.

during the period of market closure must be reflected in asset prices when the market reopens (see, e.g., French and Roll, 1986). Nelson (1989) estimated that each nontrading day adds 22.8% to the variance of the S & P 500 on the next trading day. Accordingly, we divide each of the prewhitened capital gains ξ_t by $(1 + .228 \cdot N_t)^{1/2}$, where N_t is the number of nontrading days preceding trading day t. The transformed series is plotted in Figure 3.

As noted earlier, French *et al.* (1987) employed a block-constant estimation strategy for the variance. They noted that the resulting $\hat{\Omega}_t$ series is skewed to the right and that the variance of the innovations in $\hat{\Omega}_t$ is an increasing function of $\hat{\Omega}_t$. French and colleagues took the log of $\hat{\Omega}_t$ and found that this transformation adequately stabilized the variance. This is apparent in Figure 4, which plots the log of a single flat-weight rolling regression with a window length of 25 days on each side. We, therefore, make the simplifying assumption that $\ln(\Omega_t)$ is conditionally homoskedastic (i.e., $\Lambda_t = \Lambda\Omega_t^2$). We also make the simplifying assumptions that conditional kurtosis is constant (i.e., $\theta_t = \theta\Omega_t^2$) and that $\rho_t = 0$; i.e., stochastic volatility or diffusion rather than GARCH in the data-generating process. These assumptions allow us to set $K_h = \infty$ in Theorem 3. We then formed initial conditional variance estimates using two-sided flat-weight rolling regressions. From these initial variance estimates, we created estimates of θ and Λ using the method of Theorem 3. These estimates in turn implied optimal n and m values $(n - m)$ for

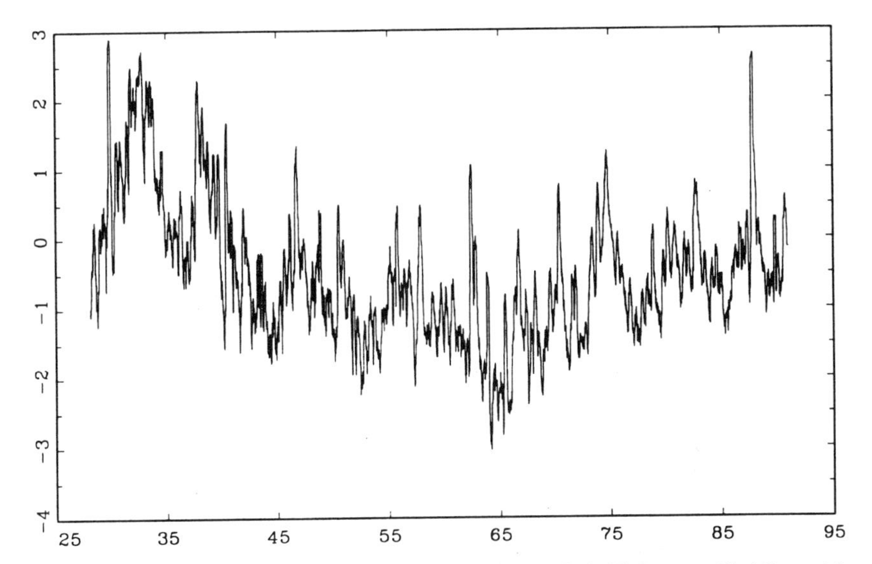

FIGURE 4 Estimated log of the variance of the S & P 500. A 25-day two-sided flat-weight rolling regression was used to estimate the variance of the S & P 500 (50 days total). There are 240 nonoverlapping estimates of the variance (60 years × 200 trading days per year/50 days per estimate). The graph shows that assuming conditional homoskedasticity is reasonable.

two-sided rolling regressions through formulas (3.1) and (3.2). We then iterated this procedure, at each stage using the "optimal" n and m suggested at the previous step until the procedure converged. (This occurred very rapidly, since for $m + n$ values below 52, a higher value was suggested; while for $n + m$ above 54, a lower value was suggested. We settled on a window length of 52.) The estimated θ and Λ values were 2.72 and .0120, respectively, implying through Theorem 5 an optimal exponential decay rate of $\alpha = .0665$ for a two-sided exponentially weighted rolling regression.[6]

To gauge the reliability of our asymptotic approximations, we performed 600 replications of the following experiment calibrated to the S & P 500 data: First, we generated 16,885 observations of $\ln(\Omega_t)$ and ΔM_t as

$$\ln(\Omega_t) = -.4246 + .9944 \cdot [\ln(\Omega_{t-1}) + .4246] + z_{2,t} \tag{5.1}$$

$$\Delta M_t = \Omega_t^{1/2} \cdot z_{1,t}, \tag{5.2}$$

[6] To gauge the importance of our prewhitening and nontrading days adjustment, we repeated the estimation procedure using the raw (i.e., unadjusted) capital gains data. The results changed very little: The estimated θ and Λ were, respectively, 2.668 and .0124, and the optimal $m + n$ and exponential decay rates were 51 and .068, respectively.

TABLE 1 Using Equation (2.12)

	Mean of estimated coefficient	Actual coefficient	Sample standard deviation
Λ	0.01051 (0.00005)	0.0120	0.0012
θ	2.480 (0.003)	2.75	0.082

where $z_{1,t}$ and $z_{2,t}$ are mutually independent and i.i.d., with $z_{1,t}$ distributed as a Student's t with 12 degrees of freedom, mean 0, and variance 1, and $z_{2,t}$ is $N(0, .0120)$. The degrees of freedom of the Student's t distribution were selected to match the estimated conditional kurtosis from the S & P 500 data. The variance of $z_{2,t}$ was selected to match the estimate of Λ for the S & P data. The population mean of $\ln(\Omega_t)$, which was $-.4246$, matched the sample mean of the fitted $\ln(\hat{\Omega}_t)$. The slow mean reversion (.9944) was selected to match the unconditional variance of $\ln(\Omega_t)$ to the sample variance of the fitted $\ln(\hat{\Omega}_t)$ plus the variance of $(\ln \Omega_\tau - \ln \hat{\Omega}_\tau)$.

For each replication, we repeated precisely the same estimation procedure we had applied to the S & P data. Tables 1 and 2 report means and standard deviations of the estimated parameters in the simulations. Standard errors (i.e., sample standard deviations divided by the square root of the number of simulations) are given in parenthesis.

The estimates for both Λ and θ are downward biased (by 1.2 and 3.3 standard deviations, respectively, in Table 1 and 5.1 and 2.5 standard deviations in Table 2). The width of the asymptotic confidence intervals, the optimal $m + n$, and the like are functions of $\sqrt{\Lambda/\theta}$. The bias in this ratio is quite small—for example the optimal $m + n$ for two-sided rolling regressions is given by (3.1) and (3.2) as $(12 \cdot \theta/\Lambda)^{1/2} = 52.4$ for the simulation. The mean estimated optimal $m + n$ was 53 with a standard

TABLE 2 Using Equation (2.11′)

	Mean of estimated coefficient	Actual coefficient	Sample standard deviation
Λ	0.007 (0.001)	0.0120	0.00088
θ	2.527 (0.004)	2.75	0.088

TABLE 3 Using Equation (2.12)

Standard deviations	Mean coverage in simulation	Asymptotic coverage
1	0.6393 (0.0007)	0.6827
2	0.9306 (0.0004)	0.9545
3	0.9929 (0.0001)	0.9973

deviation of 3.6 (using equation (2.12) it was 6.4 ± 4.3). Our estimates of $(\Lambda/\theta)^{1/2}$ were close despite the biases in both Λ and θ. Since Λ and θ are biased in the same directions, the biases partially offset in $(\Lambda/\theta)^{1/2}$. It is also worth noting that the asymptotic standard deviation of the measurement error achieved by the optimal flat-weight or exponentially weighted rolling regressions is proportional to $(\Lambda\theta)^{1/4}$. This means that measurement errors in Λ and θ must be quite large to have much effect on the accuracy of the confidence intervals. For example, getting θ wrong by a factor of 2 throws off the confidence intervals by only about 19%. Tables 3 and 4 compare the asymptotic versus actual coverages in the measurement error, giving the proportion of measurement errors falling between ±1, ±2, and ±3 estimated asymptotic standard deviations, along with the standard errors. The asymptotic confidence bands are slightly too narrow, but not drastically so.

Figure 5 plots 95% confidence bands. We used the delta method to transform our asymptotic distribution for $h^{1/4}(\hat{\Omega} - \Omega)$ into an asymptotic distribution for $h^{-1/4}(\ln\hat{\Omega} - \ln\Omega)$. This, combined with our assumption that $\theta_t = \theta \cdot \Omega_t^2$ and $\Lambda_t = \Lambda\Omega_t^2$, implies that the width of the confidence

TABLE 4 Using Equation (2.11′)

Standard deviations	Mean coverage in simulation	Asymptotic coverage
1	0.5915 (0.0008)	0.6827
2	0.8991 (0.0006)	0.9545
3	0.9848 (0.0002)	0.9973

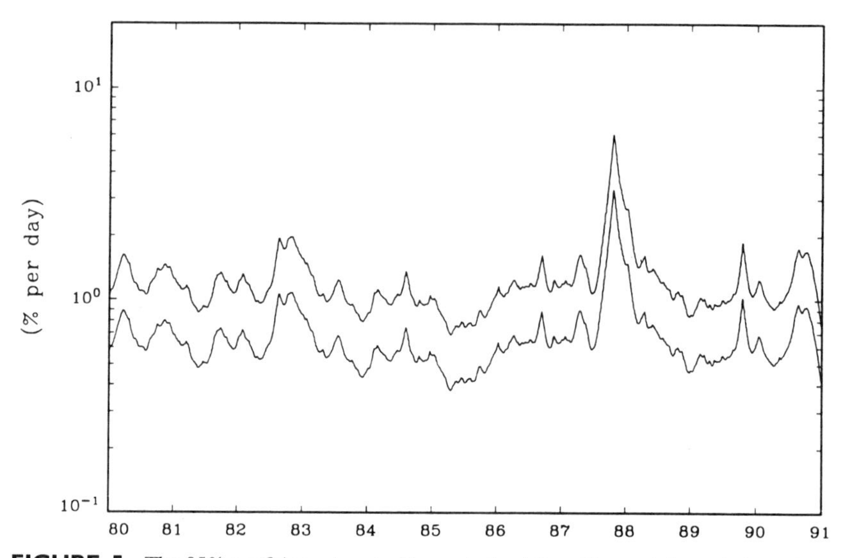

FIGURE 5 The 95% confidence bands. For each fixed time Figure 4 showed the estimate of Ω. This graph shows a 95% confidence interval around that estimate ($\hat{\Omega}_t \pm 2$ S. E.). The 95% holds pointwise, not uniformly, over the entire interval so we would expect that 5% of the time that the confidence interval does not cover the truth. Notice that this is a log scale plot of the $\Omega_t^{1/2}$.

bounds in a log plot is constant, so the extension from Figure 5 to confidence bounds for the whole sample is immediate.

Figure 6 is analogous to Figure 5, except that it uses simulated data and plots the true (simulated) $\Omega_t^{1/2}$ along with the ± 2 standard deviation confidence bounds. Overall, the asymptotic approximation performs tolerably well in the simulations using equation (2.11′) and extremely well using (2.12).

6. CONCLUSION

While this chapter, we believe, has shed new light on rolling regressions as conditional variance and covariance estimators, much work remains. For example, in tests of asset-pricing theories the link between conditional means and conditional covariance matrices is usually crucial. As we have seen, conditional covariances can be accurately measured using high-frequency data (i.e., taking h to zero). Unfortunately, estimating conditional means requires a long *span* of data as opposed to a high observation frequency (see, e.g., Merton, 1980). Since the asymptotic results developed

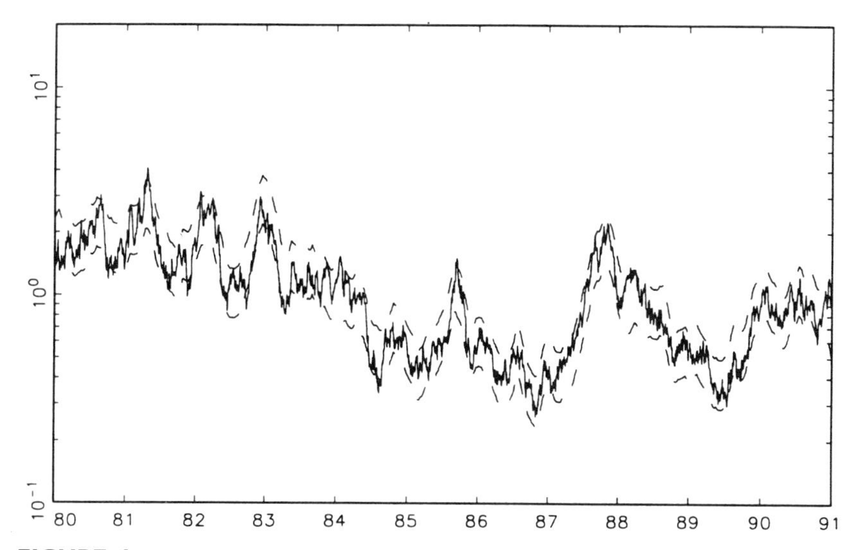

FIGURE 6 Simulated data with 95% confidence bands. The "truth" is seen to bounce around mostly between the confidence bounds—95% of the time lying between the bounds and 5% of the time bouncing outside them. Since these bands hold pointwise, the "truth" *should* be outside them 5% of the time.

in this chapter are *pointwise* in time, they do not adequately equip us to study the joint evolution of conditional means and covariances *over* time.

A second limitation is our consideration of only unconstrained linear regression to compute the estimated conditional covariance matrix. Constraints on the conditional covariance matrix (e.g., on the eigenvalues or eigenvectors) are likely to prove important in dynamic factor analysis or principle components.

Finally, as we have seen, conditionally thick-tailed processes reduce the efficiency of least squares based procedures such as rolling regression. It should be possible to adapt the methods for robust estimation of covariance matrices developed for the i.i.d. case (see, e.g., Huber, 1981) to the rolling regression framework.[7] Extending our results in these directions may prove quite challenging, but should be worth the effort.

[7] Robust conditional variance estimation methods have been employed in the ARCH literature. For example, Taylor (1986) and Schwert (1989) estimate the conditional standard deviation as a distributed lag of absolute residuals (rather than estimating the conditional variance as a distributed lag of squared residuals). Schwert was explicitly motivated by the robust variance estimation methods of Davidian and Carroll (1987). For a formal analysis of the robustness properties of these models, see Nelson and Foster (1995).

APPENDIX

We will drop the prefix h from our stochastic processes to conserve space in our proofs. Lemmas, theorems, and so forth will include the h's. All processes depend on h.

Proof of Theorem 1. We will first divide the problem into two pieces.

DEFINITION

$$ {}_h\overline{\Omega}_{(ij)T} = \sum_{\tau} {}_h\Omega_{(ij)\tau} {}_h w_{\tau-T} \, \Delta\tau. $$

LEMMA 1. *If Assumptions A and D hold, then*

$$ h^{-1/4}\left(\hat{\Omega}_{(ij)T} - \overline{\Omega}_{(ij)T}\right) = h^{1/4} \sum_{\tau} w_{\tau-T} \, \Delta B_{(ij)\tau} + o_p(1) $$

$$ h^{-1/4}\left(\overline{\Omega}_{(ij)T} - \Omega_{(ij)T}\right) = h^{-1/4} \sum_{\tau} \Psi_{\tau-T} \, \Delta M^{*}_{(ij)\tau} + o_p(1). $$

From Lemma 1, it is obvious that Theorem 1 holds. The proof of Lemma 1 relies on some other lemmas, which we will prove first.

LEMMA 2.

$$ \hat{\Omega}_{(ij)T} \equiv \overline{\Omega}_{(ij)T} + \sqrt{h} \sum_{\tau} w_{\tau-T} \, \Delta Q_{(ij)\tau} + h\left(\mathscr{B}_{ij} + \mathscr{B}_{ji} + \mathscr{D}\right), $$

where

$$ \mathscr{B}_{ji} \equiv \sum_{\tau} \left(\mu_{(j)\tau} - \hat{\mu}_{(j)\tau}\right) w_{\tau-T} \, \Delta M_{(i)\tau} $$

and

$$ \mathscr{D} \equiv \sum_{\tau} \left(\mu_{(j)\tau} - \hat{\mu}_{(j)\tau}\right)\left(\mu_{(i)\tau} - \hat{\mu}_{(i)\tau}\right) w_{\tau-T} \, \Delta\tau. $$

Proof of Lemma 2. First, note that

$$ \Delta X_{(j)\tau} - h\hat{\mu}_{(j)\tau} = \Delta M_{(j)\tau} + h\mu_{(j)} - h\hat{\mu}_{(j)\tau} = \Delta M_{(j)\tau} + h\left(\mu_{(j)\tau} - \hat{\mu}_{(j)\tau}\right). $$

So,

$$ \begin{aligned} \left[\Delta X_{(i)\tau} - h\hat{\mu}_{(i)\tau}\right]\left[\Delta X_{(j)\tau} - h\hat{\mu}_{(j)\tau}\right] &= \left(\Delta M_{(i)\tau} + h\left(\mu_{(i)\tau} - \hat{\mu}_{(i)\tau}\right)\right) \\ &\quad \times \left(\Delta M_{(j)\tau} + h\left(\mu_{(j)\tau} \hat{\mu}_{(j)\tau}\right)\right) \\ &= \Delta M_{(i)\tau} \, \Delta M_{(j)\tau} + h\left(\mu_{(j)\tau} - \hat{\mu}_{(j)\tau}\right) \Delta M_{(i)\tau} \\ &\quad + h\left(\mu_{(i)\tau} - \hat{\mu}_{(i)\tau}\right) \Delta M_{(j)\tau} + h^2\left(\mu_{(j)\tau} - \hat{\mu}_{(j)\tau}\right)\left(\mu_{(i)\tau} - \hat{\mu}_{(i)\tau}\right). \end{aligned} $$

Define $\mathscr{A} = \sum_\tau w_{\tau-T}\,\Delta M_{(i)\tau}\,\Delta M_{(j)\tau}$. Therefore, $\hat{\Omega}_{(ij)T} \equiv \mathscr{A} + h(\mathscr{B}_{ij} + \mathscr{B}_{ji} + \mathscr{D})$.

Now, analyzing $\mathscr{A}$,

$$\begin{aligned}
\mathscr{A} &= \sum_\tau w_{\tau-T}\,\Delta M_{(i)}\,\Delta M_{(j)\tau}\\
&= \sum_\tau \Omega_{(ij)\tau} w_{\tau-T}\,\Delta\tau + \sum_\tau w_{\tau-T}\big(\Delta M_{(i)\tau}\,\Delta M_{(j)\tau} - \Omega_{(ij)\tau}\,\Delta\tau\big)\\
&= \overline{\Omega}_{(ij)T} + \sqrt{h}\,\sum_\tau w_{\tau-T}\,\Delta Q_{(ij)}\tau.
\end{aligned}$$

LEMMA 3.

$$\overline{\Omega}_{(ij)T} - \Omega_{(ij)T}\sum w_{\tau-T}\,\Delta\tau = \sum_{s=0}^{\infty} \Psi_{s-T}\,\Delta M^*_{(ij)s} + \mathscr{E} + \mathscr{F},$$

where $\mathscr{E} \equiv \lambda(T)\Sigma_{s=0}^{\infty}\,\Psi_{s-T}\,\Delta s$ *and* $\mathscr{F} \equiv \Sigma_{s=0}^{\infty}\Psi_{s-T}(\lambda(s) - \lambda(T))\,\Delta s$.

Proof.

$$\begin{aligned}
\overline{\Omega}_{(ij)T} - \Omega_{(ij)T}\sum_{s=0}^{\infty} w_{\tau-T}\,\Delta\tau &= \sum_\tau \big(\Omega_{(ij)\tau} - \Omega_{(ij)T}\big) w_{\tau-T}\,\Delta\tau\\
&= \left(\sum_{\tau>T}\sum_{s=T}^{\tau-h} \Delta\Omega_{(ij)s} w_{\tau-T}\,\Delta\tau - \sum_{\tau<T}\sum_{s=\tau}^{T-h} \Delta\Omega_{(ij)s} w_{\tau-T}\,\Delta\tau\right)\\
&= \left(\sum_{T\le s<T} w_{\tau-T}\,\Delta\Omega_{(ij)s}\,\Delta\tau - \sum_{\tau\le s<T} w_{\tau-T}\,\Delta\Omega_{(ij)s}\,\Delta\tau\right)\\
&= \left(\sum_{s=T}^{\infty}\sum_{\tau=s+h}^{\infty} w_{\tau-T}\,\Delta\tau\,\Delta\Omega_{(ij)s} - \sum_{s=0}^{T-h}\sum_{\tau=0}^{s} w_{\tau-T}\,\Delta\tau\,\Delta\Omega_{(ij)s}\right)\\
&= \sum_{s=0}^{\infty}\left(I_{s\ge T}\sum_{\tau=s+h}^{\infty} w_{\tau-T}\,\Delta\tau - I_{s<T}\sum_{\tau=0}^{s} w_{\tau-T}\,\Delta\tau\right)\Delta\Omega_{(ij)s}\\
&= \sum_{s=0}^{\infty} \Psi_{s-T}\,\Delta\Omega_{(ij)s}.
\end{aligned}$$

Now use the Doob–Meyer decomposition of $\Delta\Omega$. We get

$$\begin{aligned}
\overline{\Omega}_{(ij)T} - \Omega_{(ij)T}\sum w_{\tau-T} &= \sum_{s=0}^{\infty} \Psi_{s-T}\big(\lambda(s)\Delta s + \Delta M^*_{(ij)s}\big)\\
&= \lambda(T)\sum_{s=0}^{\infty} \Psi_{s-T}\,\Delta s + \sum_{s=0}^{\infty} \Psi_{s-T}\big(\lambda(s) - \lambda(T)\big)\Delta s\\
&\quad + \sum_{s=0}^{\infty} \Psi_{s-T}\,\Delta M^*_{(ij)s}.
\end{aligned}$$

LEMMA 4. *Under Assumptions A and D, the following hold:*

$$\mathscr{B}_{ij} = o_p(h^{-3/4}) \tag{A.1}$$

$$\mathscr{D} = O_p(1) \tag{A.2}$$

$$\mathscr{E} = O_p(h^{1/2}) \tag{A.3}$$

$$\mathscr{F} = O_p(h^{1/2}). \tag{A.4}$$

Proof of (A.1): Because μ, $\hat{\mu}$, w are all predictable, and ΔM is a martingale different array,

$$\mathrm{E}(\mathscr{B}_{ij}) = 0$$

and

$$\mathrm{E}(\mathscr{B}_{ij}^2) = \mathrm{E}\left(\sum_{\tau} (\mu_{(i)\tau} - \hat{\mu}_{(i)\tau})^2 w_{\tau-T}^2 \Omega_{(jj)\tau} \Delta\tau\right).$$

But, by part i of Assumption (A), we know that

$$\sup_{T_* \le \tau \le T^*} (\mu_{(i)\tau} - \hat{\mu}_{(i)\tau})^2 = O_p(1).$$

By Assumption D, we know $\sup(w_{\tau-T}^2) = O_p(h^{-1})$. By parts iv and vii of Assumption A, we know that $\sup(\Omega_{(jj)\tau}) = O_p(1)$. By D, we know that there are $O(h^{-1/2})$ terms in our sum. And, by definition, $\Delta\tau = h$. Thus, $\mathrm{E}(\mathscr{B}_{ij}^2) = O_p(h^{-1/2})$. So, by Jensen's inequality,

$$\mathrm{P}(\mathscr{B}_{ij} > Mh^{-3/4}) < O_p(h^{-1/2})/M^2 h^{-3/2} = O_p(h^1) = o_p(1).$$

Proof of (A.2). Using part i of Assumption A and Assumption D we see that $\mathscr{D} = O_p(1)$.

Proof of (A.3). Using part iii of Assumption A, we see that $\lambda_T = O_p(1)$. By Assumption D and the definition of Ψ, we therefore can conclude that $\mathscr{E} = O_p(h^{1/2})$.

Proof of (A.4). Using part ii of Assumption A, the definition of ψ, and Assumption D, we see that $\mathscr{F} = O_p(h^{1/2})$.

Proof of Lemma A.1. Follows by substituting Lemma A.4 into Lemmas A.2 and A.3.

Thus, we have now completed the proof of Theorem 1.

Proof of Theorem 2: By Theorem 1, we need only analyze

$$h^{1/4} \sum_{\tau} w_{\tau-T} \Delta B_{(ij)\tau} + h^{-1/4} \sum_{\tau} \Psi_{\tau-T} \Delta M^*_{(ij)\tau}.$$

But since B and M^* are martingales, we know its mean to be zero and its covariance between terms ij and kl to be

$$h^{1/2} \sum_{\tau} w_{(ij)\tau-T} w_{(kl)\tau-T} \theta_{(ijkl)\tau} \, \Delta\tau + h^{-1/2} \sum_{\tau} \psi_{(ij)\tau-T} \psi_{(kl)\tau-T} \Lambda_{(ijkl)\tau} \, \Delta\tau$$
$$+ \sum_{\tau} w_{(ij)\tau-T} \psi_{(kl)\tau-T} \sqrt{\theta_{(ijij)\tau} \Lambda_{(klkl)\tau}} \, \rho_{(ijkl)\tau} \, \Delta\tau$$
$$+ \sum_{\tau} w_{(kl)\tau-T} \psi_{(ij)\tau-T} \sqrt{\theta_{(klkl)\tau} \Lambda_{(ijij)\tau}} \, \rho_{(klij)\tau} \, \Delta\tau,$$

which by Assumptions (A.v), (A.vi), and B is asymptotically equal to $C_{(ijkl)T}$. Now, applying the standard martingale central limit (which uses Assumptions C and A.vii and A.viii) (see, e.g., Liptser and Shiryayev, 1980), we get the desired result.

Proof of Theorem 3. Before we begin, we have to mention a detail about what we are going to prove. We will prove that trimmed-mean versions of (2.10) and (2.11) will work have the desired properties. Thus, we will replace the sum

$$\sum_{\tau} f_{\epsilon}(\tau)^2$$

by a trimmed version; namely,

$$\sum_{\tau} \min\left(f_{\epsilon}(\tau)^2, M\right).$$

(In the multivariate case, each element of the matrix $f_{\epsilon}(\tau) f_{\epsilon}(\tau)'$ should be trimmed by the constant M.)

First, we need to represent $f_{\epsilon}(\tau)$ as

$$f_{\epsilon}(\tau) = \sum_{s=\tau}^{\tau+\epsilon h^{1/2}} h^{-1/2} (\Delta X_s - \hat{\mu}_s \, \Delta s)^2 / \epsilon$$
$$- \sum_{s=\tau-\epsilon h^{1/2}}^{\tau} h^{-1/2} (\Delta X_s - \hat{\mu}_s \, \Delta s)^2 / \epsilon$$
$$= \sum_{s} h^{-1/2} w_{\epsilon}((s-\tau) h^{1/2}) (\Delta X_s - \hat{\mu}_s \, \Delta s)^2,$$

where $w_e(x)$ is defined as $h^{-1/2}(1/\epsilon)\text{sgn}(x) I_{[-\epsilon, \epsilon]}(x)$, where $\text{sgn}(x)$ is the sign of x. That is, $\text{sgn}(x) = 1$ if $x > 0$, and $\text{sgn}(x) = -1$ if $x < 0$. Thus, we have written f_{ϵ} is the form of equation (2.5). If $\sum_s w_e(s) h = 1$, then Assumption D would hold and we could apply Lemmas 2–4. But, looking at the proofs of Lemmas 2–4 we see that this is not used. Thus, from Lemmas 2–4 we have an asymptotic representation for $f_{\epsilon}(\tau)$ in terms of martingales. Using the same Central Limit Theorem as before, we can find the asymptotic distribution for $f_\epsilon(\tau)$. In particular $f_\epsilon(\tau)$ converges to a normal with mean zero and variance of $2\theta_\tau/\epsilon + 2\epsilon\Lambda_\tau/3$. Thus, asymptotically,

$$\lim_{M\to\infty}\lim_{h\to 0} \mathrm{E}\Big(\min\big(f_\epsilon(\tau)^2, M\big)\Big) \to 2\theta_\tau/\epsilon + 2\epsilon\Lambda_\tau/3.$$

Now, applying the Law of Large Numbers,

$$\lim_{M\to\infty}\lim_{h\to 0} 1/K \sum_\tau \min\big(f_\epsilon(\tau)^2, M\big)/\Omega(\tau) \to 2\theta_\tau/\epsilon + 2\epsilon\Lambda_\tau/3.$$

Substituting this into equations (2.10) and (2.11) we get the desired result.

Note: This proof also works for the multivariate problem.

Note: The importance of the truncation is that convergence in distribution will imply convergence in mean only for bounded random variables. So, we must make the $f_\epsilon(\tau)^2$ bounded to use in the Law of Large Numbers.

Proof of Theorem 4. This theorem consists of three different optimizations of equation (2.9). Part (a) forces m_0 to be zero, part (b) forces n_0 to be zero, and part (c) only constrains n_0 and m_0 to be nonnegative. Parts (a) and (b) follow from taking derivatives and setting them equal to zero. By the form of equation (2.9), it is obvious that there is a unique minimum. Part (c) is solved by using partial derivatives. The side constraints of nonnegativity for n_0 and m_0 come into play for extreme values of ρ. Thus, we get the three-part solution.

Lemmas 5 through 8 set up Theorem 5.

LEMMA 5 (*Some Calculations for Exponential Weights*). *Let* $c_i = \beta^i(1-\beta)$, *for* $i = 0, 1, 2, \ldots$. *Define*

$$C_i \equiv \sum_{j=i}^{\infty} c_j = \beta^i.$$

Then,

$$\sum_{i=0}^{\infty} c_i^2 = (1-\beta)^2/(1-\beta^2)$$

and

$$\sum_{i=0}^{\infty} C_i^2 = 1/(1-\beta^2).$$

The minimum of

$$\sum_{i=0}^{\infty} c_i^2 + A\sum_{i=0}^{\infty} C_i^2 \tag{A.5}$$

occurs at $\beta = 1 - \sqrt{A} + o(\sqrt{A})$, *and the minimum value obtained is* $\sqrt{A} + o(\sqrt{A})$.

Proof. Note that formula (A.5) is equivalent to

$$(1-\beta)^2/(1-\beta^2) + A/(1-\beta^2) \tag{A.6}$$

The following algebra minimizes (A.6) to generate our result ($\epsilon = 1 - \beta$).

$$\min_{\epsilon}(\epsilon^2 + A)\left(1 - (1-\epsilon)^2\right)$$

$$\min_{\epsilon}(\epsilon + A/\epsilon)/(2-\epsilon),$$

for which the minimum occurs at $\epsilon^2 = A(1-\epsilon)$, which is $\epsilon = \sqrt{A} + o(\sqrt{A})$, and the value of (A.6) at this point is $\sqrt{A} + o(\sqrt{A})$.

LEMMA 6 (*Discrete Approximately Equals Continuous*). *Any* c_i*'s that sum to one have the property that the value of equation* (*A*.5) *is at least* $\sqrt{A}$. *In particular, let* $\mathscr{D}^+ = \{f(\cdot) \mid \int_0^\infty f(t)\,dt = 1\}$, *then*

$$\sum_{i=1}^{n} c_i^2 + A \sum_{i=1}^{n} C_i^2 \geq \min_{f \in \mathscr{D}} \int_0^\infty f(t)^2\,dt + A \int_0^\infty \left(\int_0^\infty f(s)\,ds\right)^2 dt \tag{A.7}$$

$$\leq \sqrt{A}.$$

Proof. Taking

$$f(t) = w_i \qquad \text{for } i \leq t < i+1,$$

which is in $\mathscr{D}$, and its value is exactly the left-hand side of (A.7). This proves the inequality part. Write

$$f(t) = \alpha e^{-\alpha t} + \eta(t),$$

with $\alpha = \sqrt{A}$. Then

$$\int_0^\infty \eta(t)\,dt = 0. \tag{A.8}$$

Because $\int f(t)\,dt = 1$, and $\int \alpha e^{-\alpha t}\,dt = 1$. The following follows by an interchange of integrals and the definition of $\eta(\cdot)$:

$$\int_0^\infty f(t)^2\,dt = \alpha/2 + \int_0^\infty \eta(t)\,\alpha e^{-\alpha t}\,dt + \int_0^\infty \eta(t)^2\,dt. \tag{A.9}$$

Obviously,

$$\int_t^\infty f(s)\,ds = e^{-\alpha t} + \int_t^\infty n(s)\,ds.$$

Some more calculus yields

$$\int_0^\infty \left(\int_t^\infty f(s)\, ds \right)^2 dt$$

$$= \int_0^\infty \eta(t)\, dt - \int_0^\infty \eta(t) e^{-\alpha t}\, dt + \int_0^\infty \left(\int_t^\infty \eta(s)\, ds \right)^2 dt. \quad \text{(A.10)}$$

Substituting (A.8) into (A.10) yields

$$\int_0^\infty \left(\int_t^\infty f(s)\, ds \right)^2 dt$$

$$= -\int_0^\infty \eta(t) e^{-\alpha t}\, dt + \int_0^\infty \left(\int_0^\infty \eta(s)\, ds \right)^2 dt \qquad \text{(A.11)}$$

Our desired result is now A times equation (A.11) plus equation (A.9). Putting these together yields (recall $\alpha = \sqrt{A}$)

$$\text{goal} = \alpha + \int_0^\infty \eta(t)^2\, dt + \int_0^\infty \left(\int_0^\infty \eta(s)\, ds \right)^2 dt \geq \alpha,$$

with equality holding if $f(\cdot) = \alpha e^{-\alpha t}$.

LEMMA 7. *If* $\sum_{i=0}^\infty c_i = p$, *then* $c_i = p\beta^i(1 - \beta)$ *is asymptotically* (*as* $A \to 0$), *the minimizer of equation* (*A*.5) *with an asymptotic value of* $p^2 \sqrt{A}$.

Proof. Lemma 5 shows the value of (A.5) for these c_i's and Lemma 6 shows that they cannot be improved upon.

LEMMA 8. *Restrict* w_s *such that* $\int_0^\infty {}_0 w_s\, ds = p$. *Then the optimum* w_s *is*

$${}_0 w_s = \begin{cases} p\alpha e^{-\alpha s} & \textit{for } s \geq 0, \\ (1 - p)\alpha e^{-\alpha s} & \textit{for } s < 0, \end{cases}$$

(*where* $\alpha = \sqrt{\Lambda/\theta}$), *which yields an asymptotic variance of*

$$\sqrt{\Lambda\theta}\left((2p - 1)^2/2 + 1/2 - (2p - 1)\right)\rho.$$

Proof. We will break the problem into two pieces, the positive part ($s \geq 0$) and the negative part ($s < 0$). Each will be separately minimized for each value of $p = \int_0^\infty {}_0 w_s\, ds$. First consider

$$\int_0^\infty w_s \Psi_s\, ds = \int_0^\infty w_s \int_s^\infty w_t\, dt\, ds$$

$$= (1/2)\left(\int_0^\infty w_s \int_s^\infty w_t\, dt\, ds + \int_0^\infty w_s \int_s^\infty w_s\, ds\, dt \right)$$

$$= (1/2) \int_0^\infty \int_0^\infty w_s w_t\, dt\, ds = p^2/2.$$

Therefore, for fixed p, minimizing the w's is the same as minimizing equation (A.7) with $A = \Lambda/\theta$. Thus, the parameter of the exponential function is identical regardless of p and regardless of which side of zero we are on. So, $\alpha = \sqrt{A}$ is optimal

Proof of Theorem 5. Lemmas 5 through 8 prove everything except picking the value of p. For parts (A) and (B), the value of p is determined, so we are done. For part (C) we need to minimize the variance with respect to p. The variance is

$$\sqrt{\Lambda\theta}\left((2p-1)^2/2 + 1/2 - (2p-1)\rho\right),$$

which is minimized at $(2p - 2) = \rho$. Thus, the minimum occurs at $p = (1-\rho)/2$, so the optimum variance is

$$= (1/2)\sqrt{\Lambda\theta}\left(1-\rho^2\right).$$

Proof of Theorem 6: Equation (3.5) follows from (2.9) by substitution. Equation (3.6) is obviously positive, which proves our result.

ACKNOWLEDGMENTS

We thank Phillip Braun, John Cochrane, Gene Fama, Wayne Ferson, Ken French, Gerard Gennotte, Lars Hansen, Robin Lumsdaine, two referees, and the seminar participants at the University of Chicago, Northwestern, the 1991 NBER Program on Asset Pricing Meetings, and at the 1991 NBER/NSF Time Series Meeting for helpful comments. This material is based on work supported by the National Science Foundation under grants #SES 9110131 and #SES 9310683. We thank the University of Chicago Graduate School of Business, the Center for Research in Security Prices, and William S. Fishman Research Scholarship for additional support.

REFERENCES

Banerjee, A., Lumsdaine, R. L., and Stock, J. H. (1992). Recursive and sequential tests of the unit root and trend break hypothesis. *Journal of Business and Economic Statistics* **10**, 271–288.

Banon, G. (1978). Nonparametric identification for diffusion processes. *SIAM Journal Control and Optimization* **16**, 380–395.

Bollerslev, T. (1986). Generalized autoregressive conditional heteroskedasticity. *Journal of Econometrics* **31**, 307–327.

Bollerslev, T., Engle, R. F., and Wooldridge, J. M. (1988). A capital asset pricing model with time varying covariances. *Journal of Political Economy* **96**, 116–131.

Bollerslev, T., Chou, R. Y., and Kroner, K. (1992). ARCH modeling in finance: A review of the theory and empirical evidence. *Journal of Econometrics* **52**, 5–60.

Chow, G. C. (1984). Random and changing coefficient models. *In* "The Handbook of Econometrics" (Z. Griliches and M. D. Intriligator, eds.), *Vol. 2*. North-Holland Publ., Amsterdam.

Corradi, V., and White, H. (1994). "Consistent Nonparametric Estimation and Testing for the Variance of a Diffusion from Discretely Sampled Observations." University of California at San Diego.

Davidian, M., and Carroll, R. J. (1987). Variance function estimation. *Journal of the American Statistical Association* **82**, 1079–1091.

Dohnal, G. (1987). On estimating the diffusion coefficient. *Journal of Applied Probability* **24**, 105–114.

Engle, R. F. (1982). Autoregressive conditional heteroskedasticity with estimates of the variance of United Kingdom inflation. *Econometrica* **50**, 987–1008.

Engle, R. F., and Bollerslev, T. (1986). Modelling the persistence of conditional variances. *Econometric Reviews* **5**, 1–50.

Fama, E. F., and Macbeth, J. D. (1973). Risk, return, and equilibrium: Empirical tests. *Journal of Political Economy* **81**, 607–636.

Fisher, L. (1970). "The Estimation of Systematic Risk: Some New Findings," Proceedings of the Seminar on the Analysis of Security Prices. University of Chicago, Chicago.

Fisher, L., and Kamin, J. H. (1985). Forecasting systematic risk: Estimates of 'raw' beta that take account of the tendency of beta to change and the heteroskedasticity of residual returns. *Journal of Financial and Quantitative Analysis* **20**, 127–149.

Florens-Zmirou, D. (1993). On estimating the diffusion coefficient from discrete observations. *Journal of Applied Probability* **30**, 790–804.

French, K. R., and Roll, R. (1986). Stock return variances: The arrival of information and the reaction of traders. *Journal of Financial Economics* **17**, 5–26.

French, K. R., Schwert, G. W., and Stambaugh, R. F. (1987). Expected stock returns and volatility. *Journal of Financial Economics* **19**, 3–29.

Geno-Catalot, V., Laredo, C., and Picard, D. (1992). Non-parametric estimation of the diffusion coefficient by wavelets methods. *Scandinavian Journal of Statistics* **19**, 317–335.

Gonedes, N. (1973). Evidence on the information content of accounting numbers: Accounting-based and market based estimates of systematic risk. *Journal of Financial and Quantitative Analysis* **8**, 407–444.

Huber, P. J. (1981). "*Robust Statistics*." Wiley, New York.

Hull, J., and White, A. (1987). The pricing of options on assets with stochastic volatilities. *Journal of Finance* **42**, 281–300.

Jacod, J., and Shiryaev, A. N. (1987). "*Limit Theorems for Stochastic Processes*." Springer-Verlag, Berlin.

Liptser, R. S., and Shiryayev, A. N. (1980): A functional central limit theorem for semimartingales. *Theory of Probability and Its Applications* **25**, 667–688

Melino, A., and Turnbull, S. (1990). Pricing foreign currency options with stochastic volatility. *Journal of Econometrics* **45**, 239–266.

Merrill Lynch, Pierce, Fenner, and Smith, Inc. (1986). "*Security Risk Evaluation*." Merrill Lynch, New York.

Merton, R. C. (1980). "On estimating the expected return on the market. *Journal of Financial Economics* **8**, 323–361.

Nelson, D. B. (1989). "Commentary: Price volatility, international market links, and their implications for regulatory policies. *Journal of Financial Services Research* **3**, 247–254.

Nelson, D. B. (1992). Filtering and forecasting with misspecified ARCH models I: Getting the right variance with the wrong model. *Journal of Econometrics* **52**, 61–90.

Nelson, D. B. (1994). "Asymptotic Filtering and Smoothing Theory for multivariate ARCH Models." University of Chicago, Chicago.

Nelson, D. B., and Foster, D. P. (1994). Asymptotic filtering theory for univariate ARCH models. *Econometrica* **62**, 1–41.

Nelson, D. B., and Foster, D. P. (1995). Filtering and forecasting with misspecified ARCH models II: Making the right forecast with the wrong model. *Journal of Econometrics* **67**, 303–335.

Officer, R. R. (1973). The variability of the market factor of the New York Stock Exchange. *Journal of Business* **46**, 434–453.

Poterba, J. M., and Summers, L. H. (1986). The persistence of volatility and stock market fluctuations. *American Economic Review* **76**, 1142–1151.

Scholes, M., and Williams, J. (1977). Estimating betas from nonsynchronous data. *Journal of Financial Economics* **5**, 309–327.

Schwert, G. W. (1989). Why does stock market volatility change over time? *Journal of Finance* **44**, 1115–1154.

Serfling, R. J. (1980). "*Approximation Theorems of Mathematical Statistics*." Wiley, New York.

Shephard, N. (1994). Local scale of models: State space alternative to integrated GARCH processes. *Journal of Econometrics* **60**, 181–202.

Taylor, S. (1986): "*Modeling Financial Time Series*." Wiley, New York.

Wiggins, J. B. (1987). Option values under stochastic volatility: Theory and empirical estimates. *Journal of Financial Economics* **19**, 351–372.

PART III

SPECIFICATION AND ESTIMATION OF CONTINUOUS TIME PROCESSES

11

ESTIMATING DIFFUSION MODELS OF STOCHASTIC VOLATILITY

ROBERT F. ENGLE*
GARY G. J. LEE†

*Department of Economics
University of California, San Diego
La Jolla, California 92093

†Treasury Quantitative Research
ABN AMRO Bank, N.A.
Chicago, Illinois 60602

I. INTRODUCTION

Diffusion models have provided the basic statistical models for financial research over the last 25 years. These models provide analytical convenience and inherent plausibility. Probably the greatest single success is the Black–Scholes model of option pricing, which is a standard for practitioners and the starting point for a wide range of academic research. Black and Scholes postulate a diffusion process for the underlying asset and then derive the fair price for a European call or put option. The two difficulties with this strategy is, first, that the postulated distribution of the underlying returns may not correspond to the actual distribution and second, that unknown parameters remain in this distribution, which must be estimated before the options prices can be calculated. Much of the vast research output following from Black and Scholes addresses these issues. Alternative and more elaborate processes for underlying assets are postulated and methods to imply volatilities from options prices and from historical data are widely studied.

Modelling Stock Market Volatility
Copyright © 1996 by Academic Press, Inc. All rights of reproduction in any form reserved.

An attractive extension of the Black–Scholes model treats unknown and time varying parameters as random processes themselves and then seeks revised formulae for option values. In particular, many treat volatility as a state variable which also follows a diffusion. These extensions imply different option pricing formulae, although in many cases, these are no longer arbitrage-based relations.

This chapter will focus attention on the formulation and estimation of diffusion models for volatilities. While it may be easy to think intuitively about the best class of models for asset returns, it is not clear that the same logic leads to volatility diffusions with sensible properties. It will be necessary to compare the behavior of various volatility diffusions with the properties of data. The chapter adopts an approach which allows specification tests of diffusion models using discrete data. Furthermore, this approach will provide parameter estimates of the underlying diffusion parameters which can be used for options pricing and hedging.

The novel technique used in this chapter is called "indirect inference" by Gourieroux *et al.* (1993) and "which moments to match" by Gallant and Tauchen (1992). The idea is simply to find data generating processes which produce simulated data with the same properties as observed discrete data. This approach is described in the second section of this chapter.

An alternative mapping is available, if the time interval is short enough that it approximates a diffusion. This mapping, called the "diffusion limit" by Nelson (1990c), is a major innovation in the analysis of financial data. It is unfortunate that Dan cannot be with us to see the future developments of his seminal ideas. Nevertheless, his research will live on and his memory will be carried not only by his friendships but by his mark on the profession.

In the third and fourth sections, characteristics of the volatility diffusions and discrete time volatility models are discussed and in sections 5 to 8 the estimation results are presented for equity and exchange rate data. Section 9 discusses the case with interest rate models and section 10 concludes.

2. INDIRECT INFERENCE FOR ESTIMATING VOLATILITY DIFFUSIONS

To develop the statistical foundation of indirect inference, consider the following underlying statistical model, which is assumed to describe the price of an asset, p, in terms of an unobservable volatility, σ^2, and two possibly correlated Brownian motions, dw and dy.

$$\begin{aligned} dp &= \mu(\sigma, p, t : \theta)\, dt + \sigma(p, t : \theta)\, dw \\ d\sigma &= g(\sigma, p, t; \theta)\, dt + h(\sigma, p, t; \theta)\, dv, \end{aligned} \tag{2.1}$$

where $\mu(\cdot)$ is the drift of the price movement; the functions $g(\cdot)$ and $h(\cdot)$ describe how the mean and variance of the volatility process, $\sigma(\cdot)$, depend upon the state variables. Within each of these functions there are unknown parameters, θ. The task is to specify these functions, test their adequacy for various data sets, and then estimate the parameters.

A closely related parametric model is called the "stochastic volatility" model, which is a discrete-time model of time-varying variances motivated by the diffusion in (2.1). See, for example, Harvey *et al.* (1994), Ruiz and Shephard (1994) and Jacquier *et al.* (1994) among others. In its simplest form this specifies the model as

$$\begin{aligned} \Delta \log(p_t) &= \mu + \sigma_t \cdot \varepsilon_{1t} \\ \log(\sigma_t) &= \omega + \beta \cdot \log(\sigma_{t-1}) + \psi \cdot \varepsilon_{2t}, \end{aligned} \tag{2.2}$$

where ε_{1t} and ε_{2t} are independent standard normals. Estimation and forecasting of this model is a statistical challenge. It is amenable to the same techniques as the diffusion models and will be discussed below.

The equations in (2.1) and (2.2) are examples of data-generating processes (DGP) in the sense that they completely define the joint and conditional densities of p and σ, but in both cases, they do not lead to closed form likelihood functions. Just as nature can use such equations to generate actual data, the researcher can use them to generate artificial data on asset prices. For any specification of the functions and parameters, stimulated values of the continuous time process of prices can be generated by using a random number generator for the shocks and sampling each process at a very high frequency in order to approximate continuous time. When such a process is sampled discretely, such as at a daily or weekly frequency, it can be compared with actual data. If the parameters and the functional forms are correct, then the simulated data should have the same properties as the actual data.

The method of indirect inference is a systematic approach to matching the stochastic process of the stimulated data to that of the real data. Because this problem involves no exogenous variables and has unimportant initialization assumptions, the full power of the Gourieroux *et al.* (1993) and Gallant and Tauchen (1992) methods are not needed. This exposition therefore will be simpler and focus on the continuous time problem.

With a discrete time data set on asset returns $\{y_1, \ldots, y_T\}$, it is standard to specify an objective function of the form

$$Q_T(\beta, y_T^1) = \frac{1}{T} \sum_{t=1}^{T} \log[q_t(\beta, y_t^1)], \tag{2.3}$$

where β is a set of unknown parameters and the notation $y_t^1 \equiv \{y_1, \ldots, y_t\}$. Commonly, the investigator assumes and ideally tests the assumption that this is a satisfactory approximation to the data-generating process so that Q is the log likelihood. The estimator is

$$\hat{\beta} = \arg\max Q_T(\beta, y_T^1), \tag{2.4}$$

and if the Q is the log likelihood, then this is the MLE and under some regularity conditions, the standard properties of MLE apply. Examples are GARCH and EGARCH estimates of variance processes, so the vector β would be the parameters of these processes.

When this objective function does not correspond to the DGP, then the properties of the estimator must be reassessed. In some cases, the estimator in (2.4) remains consistent and is called a "quasi" or "pseudo maximum likelihood"; however, in general, one would not even expect it to be consistent. Nevertheless, it may be possible to learn about the true parameters from the estimated betas. This is the contribution of the "indirect inference" procedure.

Assuming that the DGP is a set of equations such as (2.1) or (2.2), the parameters of interest are θ, which are potentially different in number and implication from those in β. If a data set of length HT (a number H times T) is generated from a set of parameters θ and a sequence of random numbers, it can be denoted by $x(\theta)_{HT}^1 \equiv \{x(\theta)_1, \dots, x(\theta)_{HT}\}$.

We assume that the observed data is drawn from one member of the class of DGP's. This can be called the "simulation of nature."

ASSUMPTION 1. *There is a value $\theta^0 \in \Theta$, such that the joint density of y is the same as $x(\theta^0)$.*

The criterion in (2.3) becomes

$$Q_T\left(\beta, x(\theta)_{HT}^1\right) = \frac{1}{HT} \sum_{t=1}^{HT} \log\left[q_t\left(\beta, x(\theta)_t^1\right)\right]. \tag{2.5}$$

For well-behaved objective functions, it is natural to assume the following.

ASSUMPTION 2. *Q_T converges uniformly and almost surely to a differentiable function $Q(\beta, \theta)$, which has a unique maximizing $\beta \in B$ for each $\theta \in \Theta$ and which is positive definite in θ in a neighborhood of θ^0.*

The function that maps θ into β asymptotically, called the "binding function," is written as

$$\beta = b(\theta) = \arg\max_{\beta \in B} Q(\beta, \theta). \tag{2.6}$$

The binding function defines β^0, the "pseudo true value" in Gourieroux and Monfort (1994) by

$$\beta^0 = b(\theta^0). \tag{2.7}$$

From Assumptions 1 and 2, it is clear that

$$p\lim \hat{\beta} = \beta^0. \tag{2.8}$$

In some respects, the problem now appears solved. A consistent estimator of the pseudo true value can be found by conventional analysis of the data set, and then if the binding function is invertible, this can be used to solve for θ^0. This is exactly the approach taken by Drost and Werker (1994), who define a class of diffusion processes as "GARCH diffusions" if all the discrete time aggregates are weak GARCH as in Drost and Nijman (1993). They then calculate the binding function for all aggregates, which enables one to map back from an aggregate estimate to the underlying diffusion parameter.

An approximate solution is also available from Nelson (1990c) who proves that discrete time GARCH models can be viewed as approximations to underlying volatility diffusions if the parameters of the discrete time models bear a simple relation to the sampling frequency. If the observed frequency is high enough relative to the dynamics of the underlying process, then this approximation again yields a binding function.

Neither of these approaches is capable of detecting misspecification in the sense that the discrete-time data is inconsistent with an underlying diffusion process. In fact, Nelson's approach can give a variety of diffusion limits for a single discrete-time model. Furthermore, the Drost and Werker procedure is limited to simple models.

The method of "indirect inference" essentially proposes calculating the binding function by simulating data from the assumed true DGP. The method allows overidentification in the sense that more parameters can be estimated in the discrete-time model than in the continuous-time model, and consequently, evidence against the hypothesized diffusion can be examined.

Consider the derivative of the objective in (2.5) as a moment condition:

$$\frac{\partial Q_T(\beta, x(\theta))}{\partial \beta} = \frac{1}{HT} \sum \frac{\partial \log\left[q\left(\hat{\beta}, x(\theta)_{HT}^{1}\right)\right]}{\partial \beta} \equiv m_T, \tag{2.9}$$

which will converge to zero as T goes to infinity when $\theta = \theta^0$. This occurs because (2.8) implies that the estimated betas converge to the pseudo true value, and this pseudo true value maximizes the quasi likelihood function. For other values of θ, the maximizer of the asymptotic quasi likelihood function is different by Assumption 2, and therefore the score in (2.9) will not converge to zero. Hence, this is a moment which can be used to estimate the parameter θ.

DEFINITION. The indirect estimator is now defined

$$\hat{\theta} = \arg\min_{\theta \in \Theta} \left\{ \frac{\partial Q_T\left(\hat{\beta}, x(\theta)_{HT}^{1}\right)}{\partial \beta} \Gamma \frac{\partial Q_T\left(\hat{\beta}, x(\theta)_{HT}^{1}\right)}{\partial \beta} \right\}$$

for a positive definite matrix Γ.

To establish the asymptotic properties and the optimal choice of Γ, assume that the following regularity conditions hold.

ASSUMPTION 3. *The random vector*

$$\xi_T = \sqrt{T} \frac{\partial Q_T(\beta^0, y_T^1)}{\partial \beta} \xrightarrow{\text{d}} N(0, W),$$

where $W = \lim_{T \to \infty} V(\xi_T)$.

Furthermore, the following regularity conditions are also needed.

ASSUMPTION 4. *The hessian converges to a positive definite matrix* J_0 *in a neighborhood of* β^0:

$$p \lim_{T \to \infty} \frac{\partial^2 Q_T(\beta^0, y_T^1)}{\partial \beta \, \partial \beta'} = J_0,$$

and the partial derivatives of the score converge to a $p \times q$ *matrix* M *given by*

$$p \lim_{T \to \infty} \frac{\partial^2 Q_T\left(\beta^0, x(\theta^0)_{HT}^{1}\right)}{\partial \beta \, \partial \theta'} = M$$

in a neighborhood of β^0 *where* $p = \dim(\beta)$, $q = \dim(\theta)$.

From Assumptions 3 and 4, it can be shown by Taylor series expansion that the simulated scores also converge to a multiple of the same covariance matrix:

$$\sqrt{T} \frac{\partial Q_T\left(\hat{\beta}, x(\theta)_{HT}^{1}\right)}{\partial \beta} \xrightarrow{\text{d}} N\left(0, \left(1 + \frac{1}{H}\right) \cdot W\right), \tag{2.10}$$

as $T \to \infty$ with H fixed, since $\hat{\beta} \xrightarrow{\text{P}} \beta^0$ and errors in estimates of β are independent of errors in the simulated data. From (2.10) it is clear that the optimal choice of Γ is W^{-1} so that the optimal indirect estimator can be defined by

$$\theta^* = \arg\min_{\theta \in \Theta} R_T, \qquad \text{where } R_T = \left(\frac{H}{H+1}\right) \cdot m_T' \cdot W^{-1} \cdot m_T, \tag{2.11}$$

and from GMR, proposition 5, this estimator has the asymptotic distribution

$$\sqrt{T}(\theta^* - \theta^0) \xrightarrow{d} N\left[0, \left(1 + \frac{1}{H}\right)(M'W^{-1}M)^{-1}\right]. \tag{2.12}$$

This result provides Wald tests of the parameters of interest. A general specification test is also immediately available. From GMR Appendix 3, R_T has an asymptotic chi-squared distribution with degrees of freedom equal to the number of overidentifying restrictions.

$$R_T \xrightarrow{d} \chi^2_{p-q} \tag{2.13}$$

It remains only to develop an estimate of W, the variance of the scores of the data. Following Newey and West (1987), for example, this can be estimated by

$$\hat{W} = V_0 + \sum_{k=1}^{K} (V_k + V_{-k}) \cdot \left(1 - \frac{k}{K+1}\right), \tag{2.14}$$

where K is determined by an optimal bandwidth assumption or estimation, and

$$V_k = \frac{1}{T} \sum_{t=k+1}^{T} \frac{\partial \log q(\beta^0, y_t^1)}{\partial \beta} \frac{\partial \log q(\beta^0, y_{t+k}^1)}{\partial \beta'} \tag{2.15}$$

are the cross-products of the scores. For all of the applications in this chapter, the scores are nearly uncorrelated over time, as will be pointed out below. Hence, the optimal K is zero so that $\hat{W}$ is simply the outer product of the gradient estimate of the information matrix.

3. VOLATILITY DIFFUSIONS AND STOCHASTIC VOLATILITY MODELS

Although data are only observed discretely, it will be assumed that the underlying data-generating process is a continuous-time diffusion. A variety of processes will be assumed to give ever-increasing realism and complexity. Throughout, it will be assumed that prices follow a geometric Brownian motion, such as

$$dp_t = \mu p_t \, dt + v_t^{1/2} p_t \, dw, \tag{3.1}$$

where dw is a Wiener process. So, stock returns $r_t = dp_t/p_t$ follow a simple arithmetic Brownian motion. The instantaneous variance of stock returns v_t is often assumed to be constant over time in the literature. There is much empirical evidence against this assumption, and the goal of this chapter is to determine promising classes of volatility processes.

Letting volatility itself follow a diffusion, the joint data-generating process for stock prices becomes

$$dp_t/p_t = \mu\, dt + v_t^{1/2}\, dw_1 \tag{3.2}$$

$$dv_t = \phi(\theta - v_t)\, dt + \xi v_t^{\delta}\, dw_2 \tag{3.3}$$

$$\mathrm{Corr}(dw_1, dw_2) = \rho, \tag{3.4}$$

where dw_1 and dw_2 are correlated Wiener processes. Now volatility is assumed to follow a mean-reverting Ornstein–Uhlenbeck process with the innovation proportional to the volatility level raised to the power δ. The mean level of the volatility is represented by θ. If $\xi = 0$, then volatility is constant; and if $\phi > 0$, then it is mean-reverting. When $\delta = 0$, the process is an arithmetic Ornstein–Uhlenbeck process; whereas for $\delta = 1$, it is geometric. An interesting intermediate case is $\delta = \frac{1}{2}$, as in Heston (1993) following Cox *et al.* (1985). Processes similar to (3.2), (3.3), and (3.4) have been recently studied in Johnson and Shanno (1987), Wiggins (1987), Hull and White (1987), and Stein and Stein (1991) for option pricing, generally with $\rho = 0$.

To estimate members of this class of volatility diffusions, it is necessary to simulate the process. This generally is accomplished by computing a discrete-time simulation for finely spaced time intervals. In this chapter a very simple Euler approximation is used. For the equations (3.2)–(3.4) with sampling interval τ, these equations become

$$r_t = \mu\tau + (v_t\tau)^{1/2}\eta_{1t} \tag{3.5}$$

$$v_t = \phi\theta\tau + (1 - \phi\tau)v_{t-\tau} + \xi\tau^{1/2}v_{t-\tau}^{\delta}\eta_{2t}, \tag{3.6}$$

where η_{1t} and η_{2t} are correlated standard Gaussian noises with correlation coefficient ρ. Data thus easily can be simulated from these models for any choice of τ. Then the approximately continuous series of returns must be aggregated over time, giving a discrete simulated return series, x_t,

$$x_t = \sum_{s=1}^{1/\tau} r_{t+\tau\cdot s}. \tag{3.7}$$

The discrete time series x_t alone is used in the moment conditions just discussed.

The Euler approximation has interest directly as a stochastic volatility model. When $\tau = 1$, the variance follows a first-order autoregression with a shock that is possibly independent of the return shock. Because (3.3) can be rewritten in terms of $\log(v_t)$, (3.6) is a close approximation to a first-order log autoregression of volatility. When the probability of negative volatilities is negligible and $\delta = 1$, equation (3.6) can be thought of as a stochastic volatility model, just as if it were written in logs.

4. ARCH-CLASS MODELS FOR TIME-VARYING VOLATILITY OF ASSET RETURNS

A natural class of discrete time volatility models are available for financial time series. The general ARCH class of models are relatively easy to estimate and give substantial flexibility in modelling asymmetries and extreme returns. These models, proposed initially by Engle (1982), are found in many different forms, as shown by the surveys in Bollerslev *et al.* (1992, 1994) and Engle and Mezrich (1995).

Define h_t as the conditional variance of r_t so that $h_t \equiv \mathrm{Var}(r_t \mid F_{t-1})$ with F_{t-1} being the information set containing the relevant history at time $t-1$. The GARCH(1, 1) process of Bollerslev (1986) can be written in a form close to the diffusion models as

$$r_t = m_t + h_t^{1/2}\eta_t \tag{4.1}$$

$$h_t = \omega(1 - \alpha - \beta) + (\alpha + \beta)h_{t-1} + \alpha h_{t-1}\left(\eta_{t-2}^2 - 1\right) \tag{4.2}$$

where η_t is a martingale difference with mean zero and variance one. The parameter ω represents the unconditional mean of volatility. Various extensions based on this basic autoregressive relationship in (4.2) have been seen in the literature. Three will be used in this chapter. The first is to allow the disturbance η_t to be generated by a Student's t distribution with mean zero, variance one, and degrees of freedom ψ following Bollerslev (1987). A second extension is to allow an asymmetric or leverage term in (4.2) to account for the fact that large upward moves in equity markets typically have smaller volatility impacts than large downward moves. The version to be used was suggested by Glosten, Jagannathan, and Runkle (GJR) (1993) and Engle and Ng (1993). The third extension is to allow ω itself to evolve slowly over time. This model is called a "component" model by Engle and Lee (1993), since it models volatility as having both a permanent component and a transitory component.

There are obvious links between the discrete-time ARCH class models and the diffusion processes discussed. Drost and Werker (1994) derive exact formulae to calculate the limiting parameters from the GARCH(1, 1) estimates on data of different frequencies of observation. Nelson (1990c) proves that a diffusion in the form of (3.3) serves as the limiting process of the GARCH(1, 1) model when the time interval is allowed to be arbitrarily short. Nelson's results are a useful starting point. Essentially, he argues that, when the time interval becomes small, the first and second moments of volatility must match the theoretical moments from the diffusion. Since this is a Gaussian diffusion, matching these moments at all points of time is sufficient to justify the diffusion limit argument. Although there is much more math, this is sufficient for the purposes here.

Equating moments of r_t in equations (3.5) and (4.1) gives $h_t = v_t\tau$. This approximation is very good if τ, the discrete time sampling interval, is small enough for the Euler approximation to be close to the diffusion process. From matching the mean and variance of equations (3.6) and (4.2),

$$(1 - \alpha - \beta) \cdot \omega + (\alpha + \beta) \cdot h_t = (\phi \cdot \theta \cdot \tau + (1 - \phi \cdot \tau) \cdot v_t) \cdot \tau$$
$$\alpha^2 \cdot (\kappa - 1) \cdot h_t^2 = (\xi^2 \cdot \tau \cdot v_t^{2 \cdot \delta}) \cdot \tau^2, \tag{4.3}$$

where the conditional kurtosis of the shocks is κ. From estimates of the discrete-time process, diffusion parameter estimates can be computed from the following equations:

$$\begin{aligned} \hat{\theta} &= \hat{\omega}/\tau \\ \hat{\phi} &= (1 - \hat{\alpha} - \hat{\beta})/\tau \\ \hat{\xi} &= \hat{\alpha} \cdot \sqrt{(\kappa - 1) \cdot \tau} \\ \hat{\delta} &= 1. \end{aligned} \tag{4.4}$$

If the diffusion parameters are interpreted in terms of daily returns, then $\tau = 1$, but in a more common interpretation where annualized values are used, $\tau = 1/252$ if there are considered to be 252 trading days in a year. When τ is small, then ξ becomes small and ϕ becomes large unless $\alpha + \beta$ is very close to one. The conditional kurtosis of the standardized residuals can be simply estimated from the data or can be inferred from the degrees of freedom of the student's t model if that is being used.

5. ESTIMATION OF STOCK PRICE DIFFUSIONS WITH STOCHASTIC VOLATILITY

The data used are the daily returns on the S & P 500 index from 1971.1 to 1990.9, altogether 5000 observations. The daily returns are constructed as the log difference of the indices on two consecutive trading days. The mean of daily returns on the S & P 500 index is often trivial and not statistically significantly different from zero so no attempt is made to estimate the mean process. Instead, the returns are normalized to be mean zero and variance one in order to be compared with the simulated data.

5.1. Estimates of the GARCH(1, 1) Model

The first estimates are for the simple GARCH(1, 1) model, assuming normality and using the quasi maximum-likelihood method. To account the excess kurtosis of the standardized residuals, the conditional t distribution model is also estimated. The results are presented in Table 1.

TABLE 1 Estimates of GARCH(1, 1) for Daily Returns on the S & P 500 Index 1971.1 – 1990.9[a]

	ω	α	β	$1/\psi$[b]	χ^2 Test[c]
Gaussian	1.1486	0.0799	0.9050		322.1[d]
	(8.57)[f]	(41.4)	(202.6)		7.86[e]
t distribution	0.8690	0.0493	0.9381	0.1322	322.1[d]
	(7.35)	(8.40)	(122.32)	(1951.95)	16.1[e]

$$r_t = \varepsilon_t$$

$$h_t = \omega(1 - \alpha - \beta) + \alpha\varepsilon_{t-1}^2 + \beta h_{t-1}$$

[a] The daily return is constructed by log difference of two consecutive daily indices, $R_r = \log(p_t/p_{t-1})$.
[b] ψ is the degree of freedom of the t distribution.
[c] The Ljung–Box test with fifteen lags follows $\chi^2(15)$ with the .05 critical value being 25.
[d] The Ljung–Box test for squared residual ε_t^2.
[e] The Ljung–Box test for squared standard residual ε_t^2/h_t.
[f] The t statistics are in parentheses.

There is clear evidence of time varying volatility in the Ljung–Box statistic for squares of the data. This takes the value 322 but would be a chi square with 15 degrees of freedom if variances were constant. The GARCH(1, 1) estimates dramatically reduce this statistic, leaving a value for the standardized data of only 7.9. Volatility is persistent as forecasts decay with powers of .985. The t distribution gives somewhat similar results with a persistence parameter .987. The degrees of freedom of the t are 7.6, which is much more leptokurtic than the normal.

5.2. Estimates of the Diffusion Processes

These estimates form the basis for estimating volatility diffusions. Using the scores and these parameter estimates, simulated data is examined to find underlying parameterizations that set the scores as close as possible to zero. The underlying DGP is assumed to be (3.6) with $\text{corr}(\eta_{1t}, \eta_{2t}) = 0$ and $\tau = 1$ or $\frac{1}{10}$. In each case the length of the simulation in days is taken to be the same as the underlying data so that $H = 1$.

The GARCH(1, 1) with a Gaussian likelihood has only three unknown parameters and therefore there are only three moment conditions. Thus, it should be possible to find parameters that set the criterion exactly to zero as long as there is a value $\theta \in \Theta$ that corresponds to the estimated GARCH parameters. This appears not to be the case for the square root model listed first in Table 2. The parameter ξ must be positive and yet the estimation insists on making it as small as possible. Presumably this is to ensure the positivity of variances. Even when sampling at $\tau = .1$, the

TABLE 2 Estimates of Volatility Diffusion for the S & P 500 Index 1970.1 – 1991.9[a]

	ϕ	θ	ξ	R_T	Kurtosis	χ^2 Test[b]
$\tau = 1$						
Normal distribution	0.4385	0.7850	0.000001	2.24	3.0[f]	23.73[c]
$\delta = 1/2$	(*)[e]	(3.07)	(*)		3.1[g]	67.37[d]
Normal distribution	0.0114	1.3055	0.1439	10^{-19}	10.4[f]	4173.93[c]
$\delta = 1$	(1.80)	(2.13)	(3.35)		3.6[g]	23.1[d]
t distribution	0.0127	0.6947	0.0938	10^{-5}	3.8[f]	1616.2[c]
$\delta = 1$	(2.80)	(7.47)	(5.93)		3.1[g]	23.5[d]
$\tau = 0.1$						
Normal distribution	0.0120	1.0268	0.1065	10^{-19}	15.1[f]	3976.61[c]
$\delta = 1$	(2.14)	(3.17)	(2.79)		3.6[g]	18.45[d]
t distribution	0.0115	0.6000	0.0583	.025	4.7[f]	1025.7[c]
$\delta = 1$	(2.14)	(9.92)	(4.18)	—	3.1[g]	17.09[d]

$$r_t = v_t^{1/2}\tau^{1/2}\eta_{1t}$$

$$v_t = v_{t=\tau} + \phi(\theta - v_{t-\tau})\tau + \xi\tau^{1/2}v_{t-\tau}^{\delta}\eta_{2t}$$

$$\text{corr}(\eta_{1t}, \eta_{1t}) = 0$$

[a] The simulation has 5000 observations as does the real sample.
[b] The Ljung–Box test with 15 lags follows $\chi^2(15)$ with the .05 critical value being 25.
[c] The Ljung–Box test for squared residuals of the simulated data.
[d] The Ljung–Box test for squared standard residuals.
[e] The t statistics are in parentheses. Since $H = 1$, they should be divided by $\sqrt{2}$. Those with asterisk indicate no statistical significance.
[f] Kurtosis of the simulated data set.
[g] Kurtosis of the standardized simulated data set.

optimizer is not able to find a set of parameters for the diffusion. The simulated data from the best parameter values appears not even to have any ARCH. When the parameter δ is estimated, it comes out very close to the value one as suggested by the GARCH models.

Turning to $\tau = 0.1$ in the lower panel of Table 2, the estimation procedure involves temporal aggregation of the simulated data. The estimates are fairly similar to those with $\tau = 1$ but the estimate of the volatility mean is closer to unity, which is the discrete sample value. Since we do not know whether a discrete-time stochastic volatility model or a continuous-time volatility diffusion is the DGP, it is not clear which set of estimates should be superior. At least one set should be inconsistent.

If one day is close enough to a diffusion process, and if the diffusion is the true process, then the formulae in (4.4) should give good estimates of both the diffusion and the stochastic volatility model. These formulae are quite close approximations to the observed estimates from "indirect estimation." For example, for the Gaussian model the estimate of θ in the stochastic volatility and diffusion models would be 1.14 rather than the observed 1.30 and 1.03, respectively. The estimates of ϕ would be .0151 rather than .0114 and .0115, respectively. Because the standardized kurtosis is 8.4, the estimate of ξ would be .22 rather than .14 and .11, respectively. These estimates are all rather close and would probably be closer still if a fully adequate model were estimated.

Table 2 also lists the diffusion estimates with the conditional t distribution used to measure the scores. The estimates are also close to those under the conditional normality assumption except for the long-run volatility estimate, which is well below that under normality. With the addition of the degree-of-freedom parameter in the Student's t, there is now an overidentifying restriction. This appears satisfied.

5.3. Comparison of the Simulated Data with the Real Data

Comparing the statistical characteristics between the simulated data and the real data gives us another measure to evaluate the diffusion estimates. First, the simulated data does exhibit the serial correlation in squares observed in the real data. The Ljung–Box statistics for the simulated data from the successful models are around 4000 with the normal likelihood, and more than 1000 under the t distribution. Both easily reject the null hypothesis of no serial correlation, and the correlation is corrected by the volatility diffusions successfully, as seen from the Ljung–Box statistics for the squared standardized residuals. In comparison with the real data, however, the diffusions seem to exaggerate serial correlation of the volatility, especially with the conditional normality assumption.

Asset returns exhibit high estimates of kurtosis and index returns, which include October 1987 are particularly dramatic. In this data set, the kurtosis of the data is 63.8, which is far above the normal value of 3. When standardized by the Student's t GARCH(1, 1), this drops to 10.7, and by the Gaussian GARCH(1, 1), it is 8.4. Since the t distribution is estimated to have 7.7 degrees of freedom, the implied conditional kurtosis is $3(\psi - 2)/(\psi - 4)$, which is 4.6. Thus, in neither case does the estimated model have sufficient kurtosis to be consistent with the observed data. The t distribution reduces the kurtosis less because it is less sensitive to extremes in the data and therefore does not match these extremes as much as the normal model.

While it is possible that the simulated data from the diffusion or the stochastic volatility model would have kurtosis which matches the data, this did not occur. The raw kurtosis of the simulated data is under 5 for both t models and reaches 15 for the Gaussian diffusion estimate. In each case, the model correctly reduces this kurtosis so that the standardized residuals are nearly normal. Essentially, the data has a thick-tailed distribution which is too extreme for both the t distribution with GARCH and for the simulated model. However, much of this result would be different if the 1987 crash were excluded from the sample. The bulk of this kurtosis is due to less than 5 data points out of 5000 and suggests that the sampling properties of these estimates are probably very uncertain.

6. ESTIMATION OF DIFFUSIONS FOR S & P 500 INDEX WITH ASYMMETRIC VOLATILITY

It has been observed since Nelson (1990b) and Pagan and Schwert (1990) that stock return volatility has an asymmetric structure which can be incorporated into volatility models. Typically market declines forecast more volatility in the future than an equal market increase. ARCH-type models, by focusing on squared returns, missed this information, which is often referred to as the "leverage effect" in the finance literature, due to the consequent changes in firm debt–equity ratios (see Black, 1976, Christie, 1982).

Substantial research has been focused upon the appropriate approach to incorporating this effect into volatility models. Engle and Ng (1993) recommend a model proposed by Glosten *et al.* (1993), which essentially allows positive and negative market moves to have different quadratic impacts upon future volatility. This can be formulated as

$$h_t = (1 - \alpha - \beta - 0.5\gamma)\,\omega + (\alpha + \gamma D_{t-1})\,\varepsilon_{t-1}^2 + \beta h_{t-1}, \quad (6.1)$$

where D_t is defined to indicate the shock direction with $D_t = 1$ if $\varepsilon_t < 0$ and $D_t = 0$ if $\varepsilon_t > 0$. So a positive γ reflects the "leverage effect." Table 3 contains the estimation results of the asymmetric GARCH(1, 1) model in (6.1).

To model this asymmetry in a diffusion or a stochastic volatility model, it is natural to suppose that there is some correlation between the shock to the mean and the shock to the variance. Thus, the correlation coefficient, which has been assumed to be zero, will be needed to model this effect. The results are in Table 4. Only $\delta = 1$ is considered.

The positive estimate of γ indicates the significance of the "leverage effect" in the S & P. The "leverage effect" becomes larger with the conditional normality assumption, which may be due to the better ability of the conditional t distribution to model large market shocks since the 1987 crash is an important contributor to the leverage effect. The estimate of the degrees of freedom of the t distribution is 8.1.

The correlation between the innovations is found to be negative as expected under both distributional assumptions and for both values of τ. The correlation is larger under the conditional normality assumption, around 40 to 50%, negative. Again, this may be due to the robustness of

TABLE 3 Estimates of GARCH(1, 1) with Leverage Effect for Returns on the S & P 500 Index 1971.1 – 1990.9[a]

	ω	α	β	γ	$1/\psi$[b]	χ^2 **Test**[c]
Gaussian	1.0035	0.0332	0.9122	0.0725	—	322.1[d]
	(12.8)[f]	(6.05)	(209.2)	(12.6)	—	5.51[e]
t distribution	0.8597	0.0306	0.9358	0.0399	0.1237	322.1[d]
	(7.76)	(4.44)	(125.4)	(4.33)	(12.92)	10.04[e]

$$r_t = \varepsilon_t, \qquad h_t = \mathrm{Var}_{t-1}(\varepsilon_t)$$

$$h_t = \omega(1 - \alpha - \beta - 0.5\gamma) + \alpha\varepsilon_{t-1}^2 + \gamma D_{t-1}\varepsilon_{t-1}^2 + \beta h_{t-1}$$

[a] The daily return is constructed by log difference of two consecutive daily indices, $R_t = \log(p_t/p_{t-1})$.
[b] ψ is the degree of freedom of the t distribution.
[c] The Ljung–Box test with 15 lags follows $\chi^2(15)$ with the .05 critical value being 25.
[d] The Ljung–Box test for squared residual ε_t^2.
[e] The Ljung–Box test squared standard residual ε_t^2/h_t.
[f] The t statistics are in parentheses.

TABLE 4 Estimates of Diffusion Processes for Returns on the S & P 500 Index 1971.1 – 1990.9[a]

	ϕ	θ	ζ	ρ	R_T	χ^2 **Test**[b]
$\tau = 1$						
Gaussian	0.0141	1.2353	0.1462	−0.4196	10^{-11}	2652.8[c]
	(1.66)[e]	(2.38)	(4.11)	(−2.69)	—	14.4[d]
t distribution	0.0138	0.7899	0.1119	−0.2850	8.3	1762.21
	(2.84)	(6.53)	(7.81)	(−2.83)	—	16.83
$\tau = 0.1$						
Gaussian	0.0103	1.1088	0.1046	−0.5121	10^{-10}	5187.8
	(2.09)	(2.98)	(3.16)	(−4.31)	—	17.33
t distribution	0.0109	0.6637	0.0685	−0.2346	10.0	1951.4
	(2.76)	(8.90)	(5.60)	(−1.61)	—	18.5

$$r_t = (v_t\tau)^{1/2}\eta_{1t}$$

$$v_t = v_{t-\tau} + \phi(\theta - v_{t-\tau})\tau + \xi\tau^{1/2}v_{t-\tau}\eta_{2t}$$

$$\mathrm{corr}(\eta_{1t}, \eta_{2t}) = \rho$$

[a] The simulation has 5000 observations as in the real sample.
[b] The Ljung–Box test with 15 lags follows $\chi^2(15)$ with the .05 critical value being 25.
[c] The Ljung–Box test for squared residuals of the simulated data.
[d] The Ljung–Box test for squared standard residuals.
[e] The t statistics are in parentheses but should be divided by $\sqrt{2}$ to adjust for the simulation uncertainty.

the t distribution to large market shocks, which in this sample period correspond to negative market movements. The other aspects of the estimates are similar to those with no correlation between the innovations.

To compare these results with Nelson's diffusion limit, it is necessary to compute the theoretical correlation between the variance surprises and the mean surprises for the GARCH model in (6.1). Substituting $\varepsilon_t = \eta_t h_t^{1/2}$ and assuming that η_t is a standard normal, the variances and covariances of the GJR GARCH model can be computed to be

$$
\begin{aligned}
V_{t-1}(h_{t+1}) &= h_t^2\left(2\alpha^2 + 2\alpha\gamma + \tfrac{5}{4}\gamma^2\right) \\
V_{t-1}(r_t) &= h_t \\
\mathrm{Cov}_{t-1}(r_t, h_{t+1}) &= -\gamma h_t^{3/2}\sqrt{\frac{2}{\pi}} \qquad (6.2)\\
\mathrm{corr}_{t-1}(r_t, h_{t+1}) &= -\frac{\gamma}{\sqrt{\left(2\alpha^2 + 2\alpha\gamma + \tfrac{5}{4}\gamma^2\right)2/\pi}}.
\end{aligned}
$$

Equating moments with those of the diffusion, the expression in the parentheses in the first line is ξ^2 and the last moment is simply ρ. These expressions assume that the underlying process is Gaussian. Equating the last expression to ρ gives a diffusion limit estimate of $-.14$ using the Gaussian estimates in Table 3. This is somewhat below the "indirect" estimates.

The estimates using the t distribution have one additional parameter, hence the criterion function has one degree of freedom. Thus, $R/2$ should be chi-squared with one degree of freedom. In both cases, these values are slightly above the 5% critical value indicating that the simulated data does not have sufficient kurtosis to match the empirical data.

7. ESTIMATION OF DIFFUSION PROCESS FOR STOCK PRICE WITH PERSISTENT VOLATILITY

From the estimates of Tables 1 and 3, the volatility of stock returns is found to be highly persistent. Actually this is a common finding in the ARCH literature. Engle and Lee (1993) proposed a component volatility model which decomposes the conditional variance into a permanent component (trend) and a transitory component. Defining q_t as the trend of h_t, the difference between the conditional variance and its trend, $(h_t - q_t)$, thus is naturally interpreted as the transitory component. The components are assumed to have the following dynamics:

$$(h_t - q_t) = (\alpha + \beta)(h_{t-1} - q_{t-1}) + \alpha(\varepsilon_{t-1}^2 - h_{t-1}) \tag{7.1}$$

$$q_t = \omega + \lambda q_{t-1} + \phi(\varepsilon_{t-1}^2 - h_{t-1}), \tag{7.2}$$

with an autoregressive root λ for the trend, and an autoregressive root $(\alpha + \beta)$ for the transitory component. The empirical findings generally exhibit an extremely close-to-one λ and a stationary root $(\alpha + \beta) < 1$, especially for high-frequency financial data. This indicates that the trend is almost a unit-root process and the long-run forecast of volatility is governed by the trend only, regardless the transitory component. Readers can refer to the original paper for more detailed discussion.

To match the behavior of the component model it is necessary to formulate a more elaborate diffusion model which has differing speeds of adjustment.

$$dp_t/p_t = \mu \, dt + v_t^{1/2} \, dw_1 \tag{7.3}$$

$$dv_{Lt} = \gamma_L(\theta_L - v_{Lt}) \, dt + \xi_L v_t \, dw_2 \tag{7.4}$$

$$dv_{St} = -\gamma_S v_{St} \, dt + \xi_S v_t \, dw_2 \tag{7.5}$$

$$v_t = v_{St} + v_{Lt} \tag{7.6}$$

where the subscripts S and L denote the short-run (transitory) and the long-run (permanent) components of the instantaneous volatility.

The long-run part of the instantaneous volatility in (7.4) could be assumed to be a random walk in response to the empirical finding of nearly integrated volatility with high-frequency data. On the other hand, the finding of persistent volatility on high-frequency financial data is actually consistent with the mean-reverting volatility diffusion in (7.4) since the autocorrelation increases as the interval of observation shrinks.

Table 5 contains the estimation results of the volatility component model with the S & P 500 index. It can be seen that the trend is highly persistent with the autoregressive root very close to one, $\lambda = 0.9943$, while the autoregressive root of the transitory component is also pretty high $(\alpha + \beta) = 0.9628$ but decays much faster than the trend. The shock effects reflected by the parameters α and ϕ indicate that market shocks have larger effects on the transitory component than on the trend.

Table 6 contains the estimation results of diffusions of (7.3) to (7.6). First, the mean-reverting property of the instantaneous volatility is found as suggested by the discrete-time estimates; the long-run part has a slower rate of mean-reversion than the short-run part, $\lambda_L = 0.0045$ and $\gamma_S = 0.0452$ for $\tau = 1$; $\gamma_L = 0.0022$ and $\gamma_S = 0.0799$ for $\tau = 0.1$. Second, the

TABLE 5 Estimates of the Volatility Component Model for Returns on the S & P 500 Index 1971.1 – 1986.10[a]

	ω	α	β	ϕ	λ	χ^2 Test[b]
Gaussian	0.006585	0.0459	0.9169	0.0377	0.9943	292.3[c]
	(0.002)[d]	(0.01)	(0.02)	(0.01)	(0.002)	11.15[e]

$$r_t = \varepsilon_t$$

$$(h_t - q_t) = (\alpha + \beta)(h_{t-1} - q_{t-1}) + \alpha(\varepsilon_{t-1}^2 - h_{t-1})$$

$$q_t = \omega + \lambda q_{t-1} + \phi(\varepsilon_{t-1}^2 - h_{t-1})$$

[a] The daily return is constructed by log difference of two consecutive daily indices, $R_t = \log(p_t/p_{t-1})$.
[b] The Ljung–Box test with 15 lags follows $\chi^2(15)$ with the .05 critical value being 25.
[c] The Ljung–Box test for squared residual ε_t^2.
[d] The Ljung–Box test for squared standard residual ε_t^2/h_t.
[e] The standard deviations are in parentheses.

shock effects also show that the short-run part of the instantaneous volatility is more volatile than the long-run part, $\xi_L = 0.0734$ and $\xi_S = 0.1006$ for $\tau = 1$; $\xi_L = 0.0503$ and $\xi_S = 0.1401$ for $\tau = 0.1$. Therefore, the volatility diffusion estimates are generally consistent with the discrete-time model estimates.

TABLE 6 Estimates of Diffusion Processes for Returns on the S & P Index 1971.1 – 1986.10[a]

λ_L	θ_L	ξ_L	γ_S	ξ_S	χ^2 Test[b]
$\tau = 1$					
0.0045	1.2994	0.0735	0.0452	0.1006	3261.8[c]
(0.005)[e]	(0.89)	(0.04)	(0.03)	(0.06)	20.31[d]
$\tau = 0.1$					
0.0022	1.2162	0.0503	0.0799	0.1401	2980.0
(0.002)	(1.04)	(0.03)	(0.07)	(0.14)	11.20

$$r_t = (v_t \tau)^{1/2} \eta_{it}$$

$$v_{Lt} = v_{Lt-\tau} + \gamma_L(\theta_L - v_{Lt-\tau})\tau + \xi_L \tau^{1/2} v_{t-\tau} \eta_{2t}$$

$$v_{St} = (1 - \gamma_S \tau) v_{St-\tau} + \xi_S \tau^{1/2} v_{t-\tau} \eta_{2t}$$

$$v_t = v_{St} + v_{Lt}$$

[a] The simulation has 500 observations in consistence with the real sample size.
[b] The Ljung–Box test with 15 lags follows $\chi^2(15)$ with the .05 critical value being 25.
[c] The Ljung–Box test for squared residuals of the simulated data.
[d] The Ljung–Box test for squared standard residuals.
[e] The t statistics are in parenthesis. Since $H = 1$, the t statistics should be divided by $\sqrt{2}$.

8. RESULTS FOR DM/$US EXCHANGE RATE WITH VOLATILITY DIFFUSIONS AND STOCHASTIC VOLATILITY

The theoretical framework for estimating diffusions for exchange rates is much the same as that for stock prices. The data used here are the daily changes on the Deutsche mark versus U.S. dollar exchange rate during the period of 1985.1 to 1988.9, with a total of 1000 observations. The daily changes are constructed from the log difference of the exchange rates on two consecutive trading days. Again, the mean of daily changes of the DM/$US rate is trivial and ignored. The daily changes are also normalized before estimation.

The QMLE estimation results for the GARCH(1, 1) model and the "indirect inference" estimates are listed in Tables 7 and 8, respectively. The exchange rate data usually does not have excess kurtosis so only the conditional normality distribution is estimated.

9. DIFFUSION PROCESSES FOR INTEREST RATES WITH STOCHASTIC VOLATILITY

Diffusion processes describing interest rate changes are quite different from the diffusions discussed. Cox, Ingersoll, and Ross (1985) propose a well-known model (CIR) for interest rates,

$$dr_t = \gamma(\theta - r_t)\,dt + \xi r_t\,dw_1, \tag{9.1}$$

TABLE 7 Estimates of GARCH(1, 1) for Daily Changes on the DM/$US Rate 1985.1 – 1988.10[a]

	ω	α	β	χ^2 Test[b]
Gaussian	0.0176 (2.55)[e]	0.1073 (6.34)	0.8816 (47.2)	75.48[c] 20.73[d]

$$r_t = \varepsilon_t$$

$$h_t = \omega + \alpha\varepsilon_{t-1}^2 + \beta h_{t-1}$$

[a] The daily return is constructed by log difference of two consecutive exchange rates.

[b] The Ljung–Box test with 15 lags follows $\chi^2(15)$ with the .05 critical value being 25.

[c] The Ljung–Box test for squared residual ε_t^2.

[d] The Ljung–Box test for squared standard residual ε_t^2/h_t.

[e] The t statistics are in parentheses.

TABLE 8 Estimates of Diffusion Processes for Daily Returns on the DM / $US Rate 1985.1 – 1988.10[a]

	ϕ	θ	ξ	χ^2 Test[b]
$\delta = 1$	0.0139	1.2161	0.1653	270.1[c]
	(44.1)[d]	(*)	(522.9)	12.4

$$r_t = (v_t \tau)^{1/2} \eta_{it}$$

$$v_t = v_{t-\tau} + \phi(\theta - v_{t-\tau})\tau + \xi \tau^{1/2} v_{t-\tau} \eta_{2t}$$

[a] The simulation has one thousand observations as in the real sample.
[b] The Ljung–Box test with fifteen lags follows $\chi^2(15)$ with 5 percent critical value being 25.
[c] The Ljung–Box test for squared standard residuals.
[d] The t statistics are in parentheses and should be divided by $\sqrt{2}$. Those with an asterisk indicate no statistical significance.

which assumes the volatility of interest rates is proportional to the interest rate level, the so-called level effect. This model was generalized in the literature by

$$dr_t = \gamma(\theta - r_t)\, dt + \xi r_t^{\pi}\, dw_1, \tag{9.2}$$

as summarized by Chan *et al.* (1992). The parameter π is important since it measures how sensitive the volatility is to the level of interest rates. Many researchers have found that π is bigger than one, and the results in Chan *et al.* (1992) suggest models with $\pi > 1$ outperform those with $\pi < 1$.

This model, however, cannot account for the autocorrelation of interest rate volatilities, as observed by Fama (1976) and modeled with ARCH by Weiss (1984) and Engle *et al.* (1987). Two recent studies focus on the appropriate model for interest rates. Pagan *et al.* (1994) find that there are significant ARCH effects in the volatility of monthly data on one-, three-, six-, and nine-month zero coupon bond yields over the period from 1946.12 to 1987.2; however, the volatilities of these monthly yields show no significant "level effect" after adjusted by the "ARCH" volatility. Brenner *et al.* (1994) pursue the issue from a different angle. They found that the volatility of weekly observations on 13-week Treasury-Bill yields from February 9, 1973, to July 6, 1990, show significant "ARCH" effects after adjusting by the "level effect." It seems that the answer is still not clear.

A more potentially fruitful way to investigate the "ARCH" effect as well as the "level effect" may lie on estimating the diffusion processes:

$$dr_t = \gamma(\theta_r - r_t)\, dt + \xi_1 r_t^{\pi} v_t^{1/2}\, dw_1 \tag{9.3}$$

$$dv_t = \phi(\theta_\sigma - v_t)\, dt + \xi_2 v_t^{\delta}\, dw_2, \tag{9.4}$$

following the discrete-time model in Brenner *et al.* (1994). This provides an alternative way to look into the data-generating process with stochastic volatility. The diffusions can even be extended by the diffusions in (7.4), (7.5), and (7.6) when accounting for persistent volatility. Estimation of the proposed models above, however, is beyond the current context of the chapter.

10. CONCLUSIONS

Simple diffusion models and stochastic volatility models have been estimated for various models and data sets using the method of "indirect inference." In general, the approach is successful in finding a best-fitting diffusion, although in many cases, a very similar model could have been inferred from the discrete-time estimates using Nelson's diffusion limit argument. Estimates of discrete-time stochastic volatility models and continuous time diffusions are typically not very different. A tentative conclusion from this study is that a satisfactory way to estimate diffusion parameters is to estimate the best daily discrete time model and then extrapolate to the diffusion.

The treatment of the extremes in the distribution remain the biggest problem. The discrete-time methods are not fully successful in fitting the tails of the distribution, even when a *t* distribution or semi-parametric methods are used as in Engle and González-Rivera (1991). Continuous-time solutions from volatility diffusions are also not successful in producing infrequent extreme values. Presumably, beginning with a class of diffusions which allow the possibility of jumps would go far toward solving this problem.

REFERENCES

Black, F. (1976). Studies of stock price volatility changes. *Proceedings of the American Statistical Association, Business and Economics Statistics Section*, pp. 177–181.

Bollerslev, T. (1986). Generalized autoregressive conditional heteroskedasticity. *Journal of Econometrics* **31**, 307–327.

Bollerslev, T. (1987). A conditional heteroskedastic time series model for speculative prices and rates of return. *Review of Economics and Statistics* **69**, 542–547.

Bollerslev, T., Chou, R. Y., and Kroner, K. F. (1992). ARCH modeling in finance: A review of the theory and empirical evidence. *Journal of Econometrics* **52**, 5–59.

Bollerslev, T., Engle, R. and Nelson, D. (1994). ARCH models. *In* "Handbook of Econometics" (R. F. Engle and D. McFadden, eds.), Vol. 4, Chapter 49, pp. 2959–3038. North-Holland Publ., Amsterdam.

Brenner, R., Harjes, R. H. and Kroner, K. (1994). "Another Look at Alternative Models of the Short-Term Interest Rate." Discussion Paper. University of Arizona, Tuscon.

Chan, K.-C., Karolyi, A., Longstaff, F. A., and Sanders, A. B. (1992). An empirical comparison of alternative models of the short-term interest rate. *Journal of Finance* **47**(3), 1209–1227.

Chou, R. Y. (1988). Volatility persistence and stock valuations: Some empirical evidence using GARCH. *Journal of Applied Econometrics* **3**, 279–294.

Christie, A. A. (1982). The stochastic behavior of common stock variances: Value, leverage and interest rate effects. *Journal of Financial Economics* **10**, 407–432.

Cox, J. C., Ingersoll,, J. E., Jr., and Ross, S. A. (1985). A theory of the term structure of interest rates. *Econometrica* **53**(2), 385–407.

Drost, F. C., and Nijman, T. E. (1993). Temporal aggregation of GARCH processes. *Econometrica* **61**(4), 909–927.

Drost, F., and Werker, B. J. M. (1994). "Closing the GARCH Gap: Continuous Time GARCH Modeling," Discussion Paper. Tilburg University, Center for Economic Research, Tilburg, The Netherlands.

Engle, R. F. (1982). Autoregressive conditional heteroscedasticity with estimates of the variance of U.K. inflation. *Econometrica* **50**, 987–1008.

Engle, R. F., and Bollerslev,, T. (1986). Modeling the persistence of conditional variances. *Econometric Reviews*, **5**, 1–50, 81–87.

Engle, R. F., and González-Rivera, G. (1991). Semiparametric ARCH models. *Journal of Business and Economic Statistics* **9**, 345—359.

Engle, R. F., and Lee, G. (1993). "A Permanent and Transitory Component Model of Stock Return Volatility," Discussion Paper 92-44R. University of California at San Diego.

Engle, R. F., and Mezrich, J. (1995). Grappling with GARCH. *RISK Magazine*, September, pp. 112–117.

Engle, R. F., and Ng, V. K. (1993). Measuring and testing the impact of news on volatility. *Journal of Finance* **48**(5), 1749–1778.

Engle, R. F., Lilien, D. M., and Robins, R. P. (1987). Estimating time varying risk premia in the term structure: The ARCH-M model. *Econometrica* **55**, 391–407.

Fama, E. F. (1965). The behavior of stock market prices. *Journal of Business* **38**, 34–105.

Gallant, A. R., and Tauchen, G. (1992). "Which Moments to Match," Discussion Paper. Duke University, Durham, NC.

Glosten, L. R., Jagannathan, R., and Runkle, D. E. (1993). On the relation between the expected value and the volatility of the nominal excess return on stocks. *Journal of Finance* **48**(5), 1779–1801.

Gourieroux, C., and Monfort, A. (1994). Testing non-nested hypotheses. *In* "Handbook of Econometrics" (R. F. Engle and D. McFadded, eds.) Vol. 4, Chapter 44, pp. 2583–2637. North-Holland Publ., Amsterdam.

Gourieroux, C., Monfort, A., and Renault, E. (1993). Indirect inference. *Journal of Applied Econometrics* **8**, Suppl. and S85–S118.

Harvey, A., Ruiz, E., and Shephard, N. (1994). Multivariate stochastic variance models. *Review of Economic Studies* **61**(2), 247–264.

Heston, S. L. (1993). A closed-form solution for options with stochastic volatility with applications to bond and currency options. *Review of Financial Studies* **6**(2), 327–343.

Hull, J., and White, A. (1987). The pricing of options on assets with stochastic volatilities. *Journal of Finance* **42**, 281–301.

Jacquier, E, Polson, N. G., and Rossi, P. E. (1994). Bayesian analysis of stochastic volatility models. *Journal of Business and Economic Statistics* **12**(4), 371–389.

Johnson, H., and Shanno, D. (1987). Option pricing when the variance is changing. *Journal of Financial and Quantitative Analysis* **22**, 143–151.

Merville, L. J., and Pieptea, D. R. (1989). Stock-price volatility, mean-reverting diffusion, and noise. *Journal of Financial Economics* **24**, 193–214.

Nelson, D. B. (1990a). Stationarity and persistence in the GARCH(1, 1) model. *Econometric Theorey* **6**, 318–334.

Nelson, D. B. (1990b). Conditional heteroskedasticity in asset returns: A new approach. *Econometrica* **59** 347–370.

Nelson, D. B. (1990c). ARCH models as diffusion approximations. *Journal of Econometrics* **45**, 7–38.

Newey, W. K., and West, K. D. (1987). A simple, positive semi-definite, heteroskedasticity and autocorrelation consistent covariance matrix. *Econometrica* **55**(3), 703–708.

Pagan, A. R., and Schwert, G. W. (1990). Alternative models for conditional stock volatility. *Journal of Econometrics* **45**, 267–290.

Pagan, A. R., Hall, A. D., and Martin, V. (1994). "Exploring the Relationship between the Finance and Econometrics Literatures on the Term Structure," Discussion Paper. Australian National University and University of Rochester, Rochester, NY.

Schwert, G. W. (1989). Why does stock volatility change over time. *Jounral of Finance* **44**, 1115–1153.

Schwert, G. W. (1990). Stock volatility and the crash of '87. *Review of Financial Studies* **3**, 77–102.

Stein, E. M., and Stein, J. C. (1991). Stock price distributions with stochastic volatility—an analytic approach. *Review of Financial Studies* **4**, 727–752.

Weiss, A. A. (1984). ARMA models with ARCH errors. *Journal of Time Series Analysis* **5**, 129–143.

Wiggins, J. B. (1987). Option values under stochastic volatility: Theory and empirical estimates. *Journal of Financial Economics* **19**, 351–372.

12

SPECIFICATION ANALYSIS OF CONTINUOUS TIME MODELS IN FINANCE

A. RONALD GALLANT*
GEORGE TAUCHEN[†]

**Department of Economics*
University of North Carolina
Chapel Hill, North Carolina 27599
[†]Department of Economics
Duke University
Durham, North Carolina 27708

I. INTRODUCTION

This is an expository chapter that describes how the efficient method of moments (EMM) estimator proposed by Gallant and Tauchen (1996) can be used for estimation and specification analysis of a system of stochastic differential equations (SDEs). The chapter is a summary of previous work. Ideas are illustrated by a yield curve application, and sources of software are indicated.

The EMM estimator is a generalized method of moments (GMM) estimator in which the moment function that enters the GMM criterion is the expectation of the score function of an auxiliary model for the data. We refer to the auxiliary model as the score generator hereafter. The estimator follows the line work on dynamic simulation estimators initiated by Ingram and Lee (1991) and Duffie and Singleton (1993). These estimators use long simulations to compute expectations given a candidate value of the parameter vector. An important feature of simulation estimators is

Copyright © 1996 by Academic Press, Inc. All rights of reproduction in any form reserved.

that they are applicable when the state vector is partially observed. The distinguishing characteristic of the EMM method is the recommendation of specific moment functions that guarantee efficiency and facilitate specification analysis. EMM is less computationally demanding than the simulation estimator proposed by Gourieroux *et al.* (1993) and Broze *et al.* (1995) because a binding function does not need to be computed at each candidate parameter value nor does a hessian need to be computed.

Alternative estimation strategies for scalar stochastic differential equations are due to Aït-Sahalia (1996a, b) and Hansen and Scheinkman (1995). These strategies rely on moment functions computed directly from the data, rather than moment functions computed by simulation. They require the state to be fully observed in order to estimate all of the parameters. Lo (1988) discusses maximum-likelihood methods. Maximum-likelihood methods are difficult to implement when the state is partially observed and do not provide statistics for specification analysis.

As noted above, appropriate choice of the score generator in EMM estimation guarantees efficiency and protects against specification error (Gallant and Long, 1995; Tauchen, 1995). Specifically, the score generator must provide a complete statistical description of the observed series. Here we use the semi-nonparametric (SNP) density proposed by Gallant and Tauchen (1989). The SNP density has three identifiable components: a VAR-type location function, an ARCH-type scale function, and a Hermite polynomial. The Hermite polynomial accommodates general features of the data beyond those embodied in the location and scale functions which justifies a nonparametric interpretation of SNP estimates. The value of the optimized EMM objective function is chi-squared and thus can be used for forming confidence intervals and testing system adequacy. If a fitted stochastic differential equation is rejected, then the studentized scores associated with the parameters of the three components of the SNP density suggest how the system can be modified to improve the fit.

The chapter is organized as follows. In section 2, we describe the SNP nonparametric time-series estimator. In section 3, we describe the EMM estimator. In section 4, we describe application to SDEs and describe SDE simulation schemes. In section 5, we illustrate by application to continuous-time models of the term structure of interest rates based on the yield factor model of Duffie and Kan (1993). Section 6 contains concluding remarks.

2. SNP NONPARAMETRIC TIME SERIES ANALYSIS

Let $\{y_t\}_{t=-\infty}^{\infty}$, $y_t \in \mathscr{R}^M$, be a stationary, multiple, discrete time series that is adapted to a filtration $\{\mathscr{F}_t\}_{t=-\infty}^{\infty}$ (Karatzas and Shreve, 1991, p. 4) and that is Markovian in L lags in the sense that conditional expectation satisfies $\mathscr{E}(y_t \mid \mathscr{F}_t) = \mathscr{E}(y_t \mid x_{t-1})$, where $x_{t-1} = (y_{t-L}, \dots, y_{t-1})$.

The stationary distribution of the process is presumed to have a density $p(y_{-L}, \ldots, y_0)$ which is defined over $\mathscr{R}^l$, $l = M(L+1)$. Put $y = y_0$, $x = x_{-1} = (y_{-L}, \ldots, y_{-1})$, and write the stationary, marginal, and conditional densities of the process $\{y_t\}_{t=-\infty}^{\infty}$ as $p(x, y) = p(y_{-L}, \ldots, y_0)$, $p(x) = \int p(y_{-L}, \ldots, y_0) dy_0$, $p(y \mid x) = p(x, y)/p(x)$, respectively. Let $\{\tilde{y}_t\}_{t=-L}^{n}$ denote realization from the process $\{y_t\}_{t=-\infty}^{\infty}$; that is, $\{\tilde{y}_t\}_{t=-L}^{n}$ denotes data and $\{y_t\}_{t=-\infty}^{\infty}$ denotes the random variables to which the data correspond. We require estimates of the conditional density $p(y \mid x)$ rather than the stationary density $p(x, y)$ and therefore propose to estimate $p(y \mid x)$ directly rather than indirectly via an estimate of $p(x, y)$.

We shall describe a class of conditional densities

$$\mathscr{H}_K = \left\{ f_K(y \mid x, \theta) : \theta = \left(\theta_1, \theta_2, \ldots, \theta_{p_K}\right) \right\}$$

proposed by Gallant and Tauchen (1989), which they termed "SNP" for semi-nonparametric, which has two properties: (1) The union $\mathscr{H} = \bigcup_{K=1}^{\infty} \mathscr{H}_K$ is quite rich and it is reasonable to assume that the true density $p(y \mid x)$ of stationary data from a financial market is contained in $\mathscr{H}$. (2) If θ is estimated by quasi maximum likelihood; namely,

$$\tilde{\theta}_n = \arg \max_{\theta \in \mathscr{R}^{p_K}} \frac{1}{n} \sum_{t=0}^{n} \log\left[f_K\left(\tilde{y}_t \mid \tilde{y}_{t-L}, \ldots, \tilde{y}_{t-1}, \theta\right)\right],$$

and if K grows with sample size n (either adaptively as a random variable $\tilde{K}_n$ or deterministically as a function $K(n)$), then

$$\tilde{p}_n(y \mid x) = f_K\left(y \mid x, \tilde{\theta}_n\right)$$

is a consistent (Gallant and Nychka, 1987) and efficient (Fenton and Gallant, 1996a; Gallant and Long, 1995) nonparametric estimator of $p(y \mid x)$ with desirable qualitative features (Fenton and Gallant, 1996b).

A standard method of describing a conditional density $f(y \mid x, \theta)$ is to set forth a location function μ_x and a scale function R_x that reduces the process $\{y_t\}_{t=-\infty}^{\infty}$ to an innovation process $\{z_t\}_{t=-\infty}^{\infty}$ via the transformation

$$z_t = R_{x_{t-1}}^{-1}\left(y_t - \mu_{x_{t-1}}\right).$$

The description is completed by setting forth a conditional density $h(z \mid x)$ for the innovation process. We follow this recipe in describing $f_K(y \mid x, \theta) \in \mathscr{H}_K$.

The location function μ_x is affine in x

$$\mu_{x_{t-1}} = b_0 + Bx_{t-1}. \tag{2.1}$$

It is presumed to depend on $L_\mu \leq L$ lags, which is accomplished by putting leading columns of B to zero as required. Note that were we to put R_x to a constant matrix and eliminate the dependence of the innovation density on x by writing $h(z)$ instead of $h(z \mid x)$ then $\{y_t\}_{t=-\infty}^{\infty}$ would be a vector autoregression (VAR).

The scale function R_x is affine in the absolute value of the difference between x and the location function μ_x:

$$\operatorname{vech}(R_{x_{t-1}}) = \rho_0 + P\,|x_{t-1} - \mu_{x_{t-2}}|, \tag{2.2}$$

where vech(R) denotes a vector of length $M(M+1)/2$ containing the elements of the upper triangle of R and $|x|$ denotes elementwise absolute value. It is presumed to depend on $L_R + L_\mu \leq L$ lags, which is accomplished by putting leading columns of P to zero as required. For $M > 1$ we recommend setting the elements of P that correspond to off-diagonal elements of R_x to zero. Note that were we to eliminate the dependence of the innovation density on x by writing $h(z)$ instead of $h(z \mid x)$ then $\{y_t\}_{t=-\infty}^{\infty}$ would be an ARCH-type process akin to that proposed by Nelson (1991). Since the absolute value function is not differentiable, $|u|$ is approximated in the formula for R_x above by the twice continuously differentiable function

$$a(u) = \begin{cases} (|100u| - \pi/2 + 1)/100 & |100u| \geq \pi/2 \\ (1 - \cos(100u))/100 & |100u| < \pi/2 \end{cases}.$$

Let $z^\alpha = z_1^{\alpha_1} \ldots z_M^{\alpha_M}, |\alpha| = \Sigma_{k=1}^M \alpha_k$, similarly for x^β, and consider the density

$$h_K(z \mid x) = \frac{[P_K(z,x)]^2 \phi(z)}{\int [P_K(u,x)]^2 \phi(u)\,du} \tag{2.3}$$

formed from the polynomial

$$P_K(z,x) = \sum_{\alpha=0}^{K_z} \left(\sum_{\beta=0}^{K_x} a_{\beta\alpha} x^\beta \right) z^\alpha,$$

where $\phi(z) = (2\pi)^{-M/2} e^{-z'z/2}$. $P_K(z,x)$ is a polynomial of degree K_z in z whose coefficients are, in turn, polynomials of degree K_x in x. The product $[P_K(z,x)]^2\phi(z)$ is a Hermite polynomial in z with positivity enforced whose coefficients depend on x. The shape of the innovation density $h_K(z_t \mid x_{t-1})$ varies with x_{t-1} which permits $h_K(z_t \mid x_{t-1})$ to exhibit general, conditional shape heterogeneity. By putting selected elements of the matrix $A = [a_{\beta\alpha}]$ to zero, $P_K(z,x)$ can be made to depend on only $L_p \leq L$ lags from x. Also, when M is large, coefficients $a_{\beta\alpha}$ corresponding to monomials z^α that represent high-order interactions can be set to zero with little effect on the adequacy of approximations. Let $I_z = 0$ indicate that no interaction coefficients are set to zero, $I_z = 1$ indicate that coefficients corresponding to interactions z^α of order larger than $K_z - 1$ are set to zero, and so on; similarly for x^{β} and I_x. One may note that if K_z is put to zero, then the innovation density $h_K(z \mid x)$ is Gaussian. If $K_z > 0$ and $K_x = 0$, then the density can assume arbitrary shape but innovations are homogeneous.

The change of variables $y_t = R_{x_{t-1}} z_t + \mu_{x_{t-1}}$ to obtain the density

$$f_K(y_t \mid x_{t-1}, \theta) = \frac{\left\{P_K\left[R_{x_{t-1}}^{-1}(y_t - \mu_{x_{t-1}}), x_{t-1}\right]\right\}^2 \phi\left[R_{x_{t-1}}^{-1}(y_t - \mu_{x_{t-1}})\right]}{\left|\det(R_{x_{t-1}})\right|^{1/2} \int [P_K(u, x_{t-1})]^2 \phi(u)\,du}$$

completes the description of the SNP density. The vector θ contains the coefficients $A = [a_{\beta\alpha}]$ of the Hermite polynomial, the coefficients $[b_0, B]$ of the location function, and the coefficients $[\rho_0, P]$ of the scale function. To achieve identification, the coefficient $a_{0,0}$ is set to 1. The tuning parameters are L_u, L_r, L_p, K_z, I_z, K_z, and I_z, which determine the dimension p_K of θ.

Some structural characteristics of $f_K(y_t \mid x_{t-1}, \theta)$ may be noted. If K_z, K_x, and L_r are put to zero, then $f_K(y_t \mid x_{t-1}, \theta)$ is a Gaussian vector autoregression. If K_x and L_r are put to zero, then $f_K(y_t \mid x_{t-1}, \theta)$ is a non-Gaussian vector autoregression with homogeneous innovations. If K_z and K_x are put to zero, then $f_K(y_t \mid x_{t-1}\ \theta)$ is a Gaussian ARCH model. If K_x is put to zero, then $f_K(y_t \mid x_{t-1}, \theta)$ is a non-Gaussian ARCH model with homogeneous innovations. If $K_z > 0$, $K_x > 0$, $L_p > 0$, $L_\mu > 0$, and $L_r > 0$, then $f_K(y_t \mid x_{t-1}, \theta)$ is a general nonlinear process with heterogeneous innovations.

How best to select the tuning parameters L_u, L_r, L_p, K_z, I_z, K_z, and I_z is an open question. At present we recommend moving along an upward expansion path using the BIC criterion (Schwarz, 1978)

$$\text{BIC} = s_n(\tilde{\theta}) + (1/2)(p_K/n)\ln(n),$$

$$s_n(\theta) = -\frac{1}{n}\sum_{t=0}^{n} \log\left[f_K(\tilde{y}_t \mid \tilde{y}_{t-L}, \ldots, \tilde{y}_{t-1}, \theta)\right],$$

to guide the search, models with small values of BIC being preferred. An alternative to BIC is the Hannan and Quinn (Hannan, 1987) criterion

$$\text{HQ} = s_n(\hat{\theta}) + (p_\theta/n)\ln[\ln(n)].$$

The expansion path has a tree structure. Rather than examining the full tree, our strategy is to expand first in L_u with $L_r = L_p = K_z = K_z = 0$ until BIC turns upward. We then expand in L_r with $L_p = K_z = K_z = 0$.

Next we expand K_z with $K_x = 0$ and lastly L_p and K_x. We also expand K_z, L_p, and K_x at a few intermediate values of L_r because it sometimes happens that the smallest value of BIC lies elsewhere within the tree. Frequently, this process terminates with $K_x = 0$. As a precaution we compute the residuals

$$\tilde{e}_t = \left[\mathrm{Var}(y_t \mid x_{t-1})\right]^{-1/2}\left[y_t - \mathscr{E}(y_t \mid x_{t-1})\right],$$

with $\mathscr{E}(\cdot \mid x)$ and Var $(\cdot \mid x)$ computed under $f_K(y \mid x, \tilde{\theta})$, and test for the significance of a regression of $\tilde{e}_t$ on a polynomial of degree three in x_{t-1}. This regression often detects a need to increase L_p and K_x and also suggests the appropriated value for L_p. The tuning parameter selection strategy just described has given reasonable results in much applied work (see Fenton and Gallant, 1996b, and the references therein).

Fortran code, a user's guide in PostScript (Gallant and Tauchen, 1995a), and PC executables for quasi-maximum-likelihood estimation of the parameters of the SNP density are available by anonymous ftp at site ftp.econ.duke.edu in directory pub/arg/snp. The Fortran code and user's guide are also available from the Carnegie-Mellon University e-mail server by sending the one-line e-mail message "send snp from general" to statib@lib.stat.cmu.edu.

3. EFFICIENT METHOD OF MOMENTS

Let $\{y_t\}_{t=-\infty}^{\infty}$, $y_t \in \mathscr{R}^M$, be a stationary, multiple, discrete time series. The stationary distribution of a stretch $y_{t-L}, \ldots, y_t$ from the process $\{y_t\}_{t=-\infty}^{\infty}$ is hypothesized to have density $p(y_{-L}, \ldots, y_0 \mid \rho)$, where ρ is a vector of unknown parameters that must be estimated. Put $y = y_0$, $x = x_{-1} = (y_{-L}, \ldots, y_{-1})$, and write $p(x, y \mid \rho) = p(y_{-L}, \ldots, y_0 \mid \rho)$, $p(x \mid \rho) = \int p(y_{-L}, \ldots, y_0 \mid \rho) dy_0$, $p(y \mid x, \rho) = p(x, y \mid \rho)/p(x \mid \rho)$. Let $\{\tilde{y}_t\}_{t=-L}^{n}$ denote the data from which ρ is to be estimated.

We are interested in the situation where an analytic expression for the maintained model $p(y_{-L}, \ldots, y_0 \mid \rho)$ is not available yet expectations of the form

$$\mathscr{E}_\rho(g) = \int \cdots \int g(y_{-L}, \ldots, y_0) p(y_{-L}, \ldots, y_0 \mid \rho) dy_{-L} \cdots dy_0$$

can be computed by simulation, quadrature, or other numerical means for given ρ. We focus on the case where simulation is used to compute $\mathscr{E}_\rho(g)$.

That is, for given a ρ, one generates the simulation $\{\hat{y}_t\}_{t=-L}^{N}$ from $p(y \mid x, \rho)$ and puts

$$\mathscr{E}_\rho(g) = \frac{1}{N} \sum_{t=0}^{N} g(\hat{y}_{t-L}, \ldots, \hat{y}_t)$$

with N large enough that Monte Carlo error is negligible. Examples of this situation include representative agent asset-pricing models, general equilibrium rational expectations models, auction models, and of particular interest here, systems of stochastic differential equations.

Our objective is threefold: (1) estimate ρ; (2) test the hypothesis that $p(y_{-L}, \ldots, y_0 \mid \rho)$ is the stationary distribution the process $\{y_t\}_{t=-\infty}^{\infty}$; and, (3) provide diagnostics that indicate how a rejected model should be modified to better describe the distribution of the process.

Gallant and Tauchen (1996) proposed an estimator for estimating ρ in the situation above. Termed the "efficient method of moments" estimator, it would be regarded as a minimum chi-square estimator in the statistics literature and as a generalized method of moments estimator in the econometrics literature. Being a minimum chi-square, the optimized chi-square criterion can be used to test model adequacy. Also, as seen below, the moments that enter the criterion provide diagnostics that indicate how the maintained model $p(y_{-L}, \ldots, y_0 \mid \rho)$ should be modified if it is rejected by the test of model adequacy.

The moment equations for the EMM estimator are obtained from the score vector $(\partial/\partial\theta)\ln f(y \mid x, \theta)$ of an auxiliary model $f(y \mid x, \theta)$ that is termed the "score generator." Gallant and Tauchen (1996) show that if the score generator $f(y \mid x, \theta)$ encompasses the maintained model $p(y \mid x, \rho)$, then their estimator is as efficient as maximum likelihood. Tauchen (1995) sets forth formulae that would lead one to expect that the EMM estimator will be nearly as efficient as maximum likelihood when the score generator $f(y \mid x, \theta)$ is a good statistical approximation to the process $\{y_t\}_{t=-\infty}^{\infty}$ in the sense of passing diagnostic tests and the like. Gallant and Long (1996) support this conjecture by showing that if the score generator is the SNP density $f_K(y \mid x, \theta)$ described in section 2, then the efficiency of the EMM estimator can be made as close to that of maximum likelihood as desired by taking K large enough. Moreover, they show that if $\{y_t\}_{t=-\infty}^{\infty}$ is stationary but not Markovian, then the same is true for large L.

The Gallant–Tauchen EMM estimator $\hat{\rho}_n$ is computed as follows. Use the score generator

$$f(y_t \mid y_{t-L}, \ldots, y_{t-1}, \theta) \qquad \theta \in R^{p_\theta}$$

and the data $\{\tilde{y}_t\}_{t=-L}^{n}$ to compute the quasi-maximum-likelihood estimate

$$\tilde{\theta}_n = \arg\max_{\theta \in \Theta} \frac{1}{n} \sum_{t=0}^{n} \log\left[f\left(\tilde{y}_t \mid \tilde{y}_{t-L}, \ldots, \tilde{y}_{t-1}, \theta\right)\right]$$

and the corresponding estimate of the information matrix

$$\tilde{\mathscr{I}}_n = \frac{1}{n} \sum_{t=0}^{n} \left[\frac{\partial}{\partial \theta} \log f\left(\tilde{y}_t \mid \tilde{x}_{t-1} \tilde{\theta}_n\right)\right]\left[\frac{\partial}{\partial \theta} \log f\left(\tilde{y}_t \mid \tilde{x}_{t-1}, \tilde{\theta}_n\right)\right]'.$$

If the score generator $f(y \mid x, \theta)$ is not an adequate statistical approximation to the data, then one of the more complicated expressions for $\tilde{\mathscr{I}}_n$ set forth in Gallant and Tauchen (1996) must be used. If the SNP density $f_K(y \mid x, \theta)$ of section 2 is used as the score generator, and the model selection protocol of section 2 is used to determine (L_u, L_r, L_p, K_z, I_z, K_x, I_x), then one may expect $\tilde{\mathscr{I}}_n$ as above to be appropriate (Gallant and Long, 1996).

Define

$$m(\rho, \theta) = \mathscr{E}_\rho\left\{\frac{\partial}{\partial \theta} \log[f(y_0 \mid y_{-L}, \ldots, y_{-1}, \theta)]\right\},$$

which is computed by averaging over a long simulation

$$m(\rho, \theta) \doteq \frac{1}{N} \sum_{t=0}^{N} \frac{\partial}{\partial \theta} \log\left[f\left(\hat{y}_t \mid \hat{y}_{t-L}, \ldots, \hat{y}_{t-1}, \theta\right)\right].$$

The estimator is

$$\hat{\rho}_n = \arg\min_{\rho \in R} m'\left(\rho, \tilde{\theta}_n\right)\left(\tilde{\mathscr{I}}_n\right)^{-1} m\left(\rho, \tilde{\theta}_n\right).$$

The asymptotics of the estimator, which are derived in Gallant and Tauchen (1996), are as follows. If ρ^o denotes the true value of ρ and θ^o is an isolated solution of the moment equations $m(\rho^o, \theta) = 0$, then

$$\lim_{n \to \infty} \hat{\rho}_n = \rho^o \text{ a.s.(almost surely)}$$

$$\sqrt{n}\,(\hat{\rho}_n - \rho^o) \xrightarrow{\mathscr{L}} N\left\{0, \left[(M^o)'(\mathscr{I}^o)^{-1}(M^o)\right]^{-1}\right\}$$

$$\lim_{n \to \infty} \hat{M}_n = M^o \text{ a.s.}$$

$$\lim_{n \to \infty} \tilde{\mathscr{I}}_n = \mathscr{I}^o \text{ a.s.},$$

where $\hat{M}_n = M(\hat{\rho}_n, \tilde{\theta}_n)$, $M^o = M(\rho^o, \theta^o)$, $M(\rho, \theta) = (\partial/\partial\rho')m(\rho, \theta)$, and

$$\mathscr{I}^o = \mathscr{E}_{\rho^o}\left[\frac{\partial}{\partial\theta}\log f(y_0 \mid x_{-1}, \theta^o)\right]\left[\frac{\partial}{\partial\theta}\log f(y_0 \mid x_{-1}, \theta^o)\right]'.$$

Under the null hypothesis that $p(y_{-L}, \ldots, y_0 \mid \rho)$ is the correct model,

$$L_0 = nm'\left(\hat{\rho}_n, \tilde{\theta}_n\right)\left(\tilde{\mathscr{I}}_n\right)^{-1} m\left(\hat{\rho}_n, \tilde{\theta}_n\right) \tag{3.1}$$

is asymptotic chi-squared on $p_\theta - p_\rho$ degrees freedom. Under the null hypothesis that $h(\rho^o) = 0$, where h maps $\mathscr{R}^{p_\rho}$ into $\mathscr{R}^q$,

$$L_h = n\left[m'\left(\hat{\hat{\rho}}_n, \tilde{\theta}_n\right)\left(\tilde{\mathscr{I}}_n\right)^{-1} m\left(\hat{\hat{\rho}}_n, \tilde{\theta}_n\right) - m'\left(\hat{\rho}_n, \tilde{\theta}_n\right)\left(\tilde{\mathscr{I}}_n\right)^{-1} m\left(\hat{\rho}_n, \tilde{\theta}_n\right)\right]$$

is asymptotic chi-squared on q degrees freedom, where

$$\hat{\hat{\rho}}_n = \arg \min_{h(\rho)=0} m'\left(\rho, \tilde{\theta}_n\right)\left(\tilde{\mathscr{I}}_n\right)^{-1} m\left(\rho, \tilde{\theta}_n\right).$$

A Wald confidence interval on an element ρ_i of ρ can be constructed in the usual way from an asymptotic standard error $\sqrt{\hat{\sigma}_{ii}}$. A standard error may be obtained by computing the Jacobian $M_n(\rho, \theta)$ numerically and taking the estimated asymptotic variance $\hat{\sigma}_{ii}$ to be the ith diagonal element of $\hat{\Sigma} = (1/n)[(\hat{M}_n)'(\tilde{\mathscr{I}}_n)^{-1}(\hat{M}_n)]^{-1}$. These intervals, which are symmetric, are somewhat misleading because they do not reflect the rapid increase in the EMM objective function $s_n(\rho) = m'(\rho, \tilde{\theta}_n)(\tilde{\mathscr{I}}_n)^{-1}m(\rho, \tilde{\theta}_n)$ when ρ_i approaches a value for which simulations from $p(y_{-L}, \ldots, y_0 \mid \rho)$ become explosive. Confidence intervals obtained by inverting the criterion difference test L_h do reflect this phenomenon and therefore are more useful. To invert the test, one puts in the interval those ρ_i^* for which L_h for the hypothesis $\rho_i^o = \rho_i^*$ is less than the critical point of a chi-square on one degree freedom. To avoid reoptimization one may use the approximation

$$\hat{\hat{\rho}}_n = \hat{\rho}_n + \frac{\rho_i^* - \hat{\rho}_{in}}{\hat{\sigma}_{ii}}\hat{\Sigma}_{(i)}$$

in the formula for L_h where $\hat{\Sigma}_{(i)}$ is the ith column of $\hat{\Sigma}$.

When $L_0 = nm'(\hat{\rho}_n, \tilde{\theta}_n)(\tilde{\mathscr{I}}_n)^{-1}m(\hat{\rho}_n, \tilde{\theta}_n)$ exceeds the chi-square critical point, diagnostics that suggest improvements to the model are desirable. Because

$$\sqrt{n}\, m\left(\hat{\rho}_n, \tilde{\theta}_n\right) \xrightarrow{\mathscr{L}} N\left\{0, \mathscr{I}^o - (M^o)\left[(M^o)'(\mathscr{I}^o)^{-1}(M^o)\right]^{-1}(M^o)'\right\}$$

inspection of the t ratios

$$T_n = S_n^{-1} \sqrt{n}\, m(\hat{\rho}_n, \tilde{\theta}_n), \tag{3.2}$$

where $S_n = (\text{diag}\{\tilde{\mathscr{I}}_n - (\hat{M}_n)[(\hat{M}_n)'(\tilde{\mathscr{I}}_n)^{-1}(\hat{M}_n)]^{-1}(\hat{M}_n)'\})^{1/2}$ can suggest reasons for failure. Different elements of the score correspond to different characteristics of the data. Large t ratios reveal the characteristics that are not well approximated. For this purpose, the quasi t ratios

$$\hat{T}_n = \left[\left(\text{diag}\, \tilde{\mathscr{I}}_n\right)^{1/2}\right]^{-1} \sqrt{n}\, m(\hat{\rho}_n, \tilde{\theta}_n), \tag{3.3}$$

which are underestimates, may be adequate and are cheaper to compute because they avoid numerical approximation to $\hat{M}_n$. We investigate the adequacy of quasi-t-ratios in Section 5.

Fortran code and a user's guide in PostScript (Gallant and Tauchen, 1995b) that implement the EMM estimator using $f_K(y_t \mid x_{t-1}, \theta)$ as described in section 2 as the score generator are available by anonymous ftp at site ftp.econ.duke.edu in directory pub/get/emm.

4. STOCHASTIC DIFFERENTIAL EQUATIONS

Consider application of the EMM estimation and inference methods described in section 3 to a system of stochastic differential equations

$$dU_t = A(U_t, \rho)dt + B(U_t, \rho)dW_t \qquad 0 < t < \infty.$$

The parameter ρ has dimension p_ρ, the state vector U_t has dimension d, W_t is a k-dimensional vector of independent Wiener processes, $A(\cdot, \rho)$ maps $\mathscr{R}^d$ into $\mathscr{R}^d$, and $B(\cdot, \rho)$ is a $d \times k$ matrix composed of the column vectors $B_1(\cdot, \rho), \ldots, B_k(\cdot, \rho)$, each of which maps $\mathscr{R}^d$ into $\mathscr{R}^d$. U_t is interpreted as the solution of the integral equations

$$U_t = U_0 + \int_0^t A(U_s, \rho)ds + \sum_{i=1}^{k} \int_0^t B_i(U_s, \rho)dW_{is},$$

where U_0 is the initial condition at time $t = 0$, and $\int_0^t B_i(U_s, \rho)dW_{is}$ denotes the Ito stochastic integral (Karatzas and Schreve, 1991).

The system is observed at equally spaced time intervals $t = 0, 1, \ldots$ and selected characteristics

$$y_t = T(U_{t+L}, \rho) \qquad t = -L, -L+1, \ldots$$

of the state are recorded, where y_t is an M-dimensional vector and $L > 0$ is the number of lagged variables that enter the formulas of section 3.

An example is the continuous time version of the stochastic volatility model that has been proposed by Clark (1973), Tauchen and Pitts (1983), and others, as a description of speculative markets. For daily price observations on two securities the model is

$$\left.\begin{aligned} dU_{1t} &= (\rho_1 - \rho_2 U_{1t})\,dt + \rho_3\,dW_{1t} \\ dU_{2t} &= (\rho_4 - \rho_5 U_{2t})\,dt + \rho_6 \exp(U_{1t})\,dW_{2t} \\ dU_{3t} &= (\rho_7 - \rho_8 U_{3t})\,dt + \rho_9 \exp(U_{1t})\,dW_{3t} \end{aligned}\right\} \quad 0 \le t < \infty$$

$$\left.\begin{aligned} y_{1t} &= U_{2,t+L} \\ y_{2t} &= U_{3,t+L} \end{aligned}\right\} \quad t = -L, L+1, \ldots, n.$$

U_{1t} represents an unobserved flow of new information to the market that influences the volatility of asset prices U_{2t} and U_{3t} by changing the instantaneous conditional variances of U_{2t} and U_{3t}. The observed data $\tilde{y}_{1t}$ and $\tilde{y}_{2t}$ are prices at the end of each trading day. To achieve identification in estimation, a normalization rule such as $\rho_9 = 1$ should be imposed.

We assume that U_t, and hence y_t, is stationary and ergodic. We further assume that the stationary distribution of y_t is absolutely continuous. Thus, for each setting of parameter ρ and lag length L, there exists a time-invariant density $p(y_{-L}, \ldots, y_0 \mid \rho)$ of the form described in section 3 such that

$$\lim_{N\to\infty} \frac{1}{N} \sum_{t=0}^{N} g(\hat{y}_{t-L}, \ldots, \hat{y}_t)$$

$$= \int \cdots \int g(y_{-L}, \ldots, y_0)\, p(y_{-L}, \ldots, y_0 \mid \rho)\, dy_{-L} \cdots dy_0$$

where $\{\hat{y}_t\}_{t=-L}^{N}$ is realization of length $N + L + 1$ from the system. This assumes that g is integrable and that either U_0 is a sample from the stationary distribution of U_t or that a longer realization was observed and enough initial observations were discarded for transients to have dissipated. Therefore, all that is needed for implementation of the EMM methodology is a practical means for generating the simulation $\{\hat{y}_t\}_{t=-L}^{N}$.

A simulation $\{\hat{y}_t\}_{t=-L}^{N}$ for computing

$$m(\rho, \theta) \doteq \frac{1}{N} \sum_{t=0}^{N} \frac{\partial}{\partial\theta} \log\left[f(\hat{y}_t \mid \hat{y}_{t-L}, \ldots, \hat{y}_{t-1}, \theta)\right]$$

may be obtained by first generating the sequence

$$\hat{U}_0, \hat{U}_\Delta, \hat{U}_{2\Delta}, \hat{U}_{3\Delta}, \ldots, \hat{U}_{N+L-\Delta}, \hat{U}_{N+L}$$

with step-size $\Delta = 1/n_0$ from the system

$$dU_t = A(U_t, \rho)\,dt + B(U_t, \rho)\,dW_t \qquad 0 \le t < N + L$$

by means of an explicit order 2 weak scheme as described below and then transforming and retaining every n_0th element to get

$$\hat{y}_t = T\left(\hat{U}_{\Delta n_0(t+L)}\right) = T\left(\hat{U}_{t+L}\right) \qquad t = -L, -L+1, \ldots, N.$$

A simulation $\{\hat{y}_t\}_{t=-L}^{N}$ for graphical display or for computation of statistics such as density estimates may be generated similarly using an explicit order 1 strong scheme, which is described below. For the results in subsection 5.3 time t is in weeks and $n_0 = 10$, which implies that the simulation step-size is $\Delta = 1/10$.

The two schemes used here follow. They represent certain choices of tuning parameters on our part to schemes from Kloeden and Platen (1992, p. 347, 376, 486). The weak scheme is valid for autonomous systems $dU_t = A(U_t, \rho)dt + B(U_t, \rho)dW_t$ only. The strong scheme is valid for nonautonomous systems $dU_t = A(t, U_t, \rho)dt + B(t, U_t, \rho)dW_t$ as well. Fortran code implementing them are available by anonymous ftp at site ftp.econ.duke.edu in directory pub/arg/libf as files stng1.f and weak2.f.

4.1. Explicit Order 2 Weak Scheme

Recursion:

$$\begin{aligned}
\hat{U}_{t+\Delta} = \hat{U}_t &+ \frac{1}{2}\left[A(\Upsilon, \rho) + A\left(\hat{U}_t, \rho\right)\right]\Delta \\
&+ \frac{1}{4}\sum_{j=1}^{k}\Bigg\{\left[B_j\left(R_j^{+}, \rho\right) + B_j\left(R_j^{-}, \rho\right) + 2B_j\left(\hat{U}_t, \rho\right)\right]\Delta W_j \\
&\qquad + \sum_{\substack{r=1 \\ r\neq j}}^{k}\left[B_j\left(Y_r^{+}, \rho\right) + B_j\left(Y_r^{-}, \rho\right) - 2B_j\left(\hat{U}_t, \rho\right)\right]\Delta W_j \Delta^{-1/2}\Bigg\} \\
&+ \frac{1}{2}\sum_{j=1}^{k}\Bigg\{\left[B_j\left(R_j^{+}, \rho\right) - B_j\left(R_j^{-}, \rho\right)\right] I_{jj} \\
&\qquad + \sum_{\substack{r=1 \\ r\neq j}}^{k}\left[B_j\left(Y_r^{+}, \rho\right) - B_j\left(Y_r^{-}, \rho\right)\right] I_{rj}\Bigg\}\Delta^{-1/2}.
\end{aligned}$$

Supporting values:

$$\begin{aligned}
\Upsilon &= \hat{U}_t + A\left(\hat{U}_t, \rho\right)\Delta + \sum_{j=1}^{k} B_j\left(\hat{U}_t, \rho\right)\Delta W_j \\
R_j^{\pm} &= \hat{U}_t + A\left(\hat{U}_t, \rho\right)\Delta \pm B_j\left(\hat{U}_t, \rho\right)\Delta^{1/2} \\
Y_j^{\pm} &= \hat{U}_t \pm B_j\left(\hat{U}_t, \rho\right)\Delta^{1/2}.
\end{aligned}$$

Integral approximation:

$$I_{rj} = (1/2)\left[\Delta W_j \Delta W_r + V_{rj}\right]$$

$$V_{rj} = -\Delta I_{(0,\frac{1}{2}]}(U_{rj}) + \Delta I_{(\frac{1}{2},1]}(U_{rj}) \qquad r < j$$

$$V_{rj} = -\Delta \qquad r = j$$

$$V_{rj} = -V_{jr} \qquad r > j.$$

Independent random variables:

$$\Delta W_j \sim N(0, \Delta) \qquad j = 1, \ldots, k$$

$$U_{rj} \sim U(0, 1] \qquad r = 1, \ldots, j-1, \quad j = 1, \ldots, k.$$

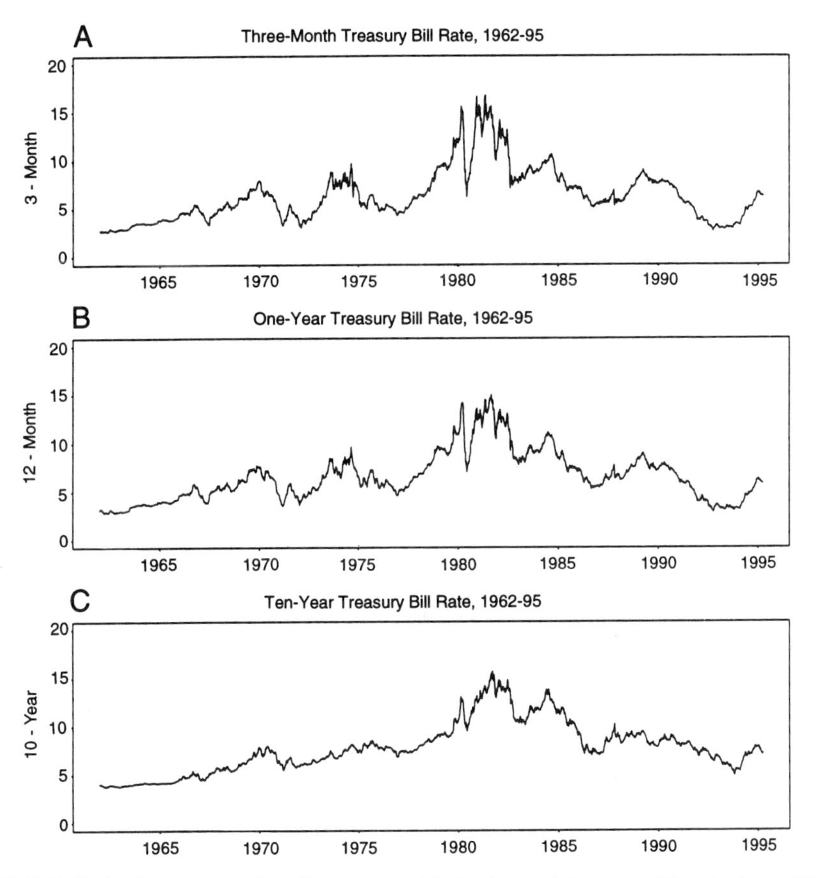

FIGURE 1 Interest rate data. The top panel is the 3-month Treasury-Bill rate, the middle is the 12-month Treasury-Bill rate, and the bottom is the 10-year constant maturity Treasury-Bond rate; weekly, January 5, 1962–March 31, 1995.

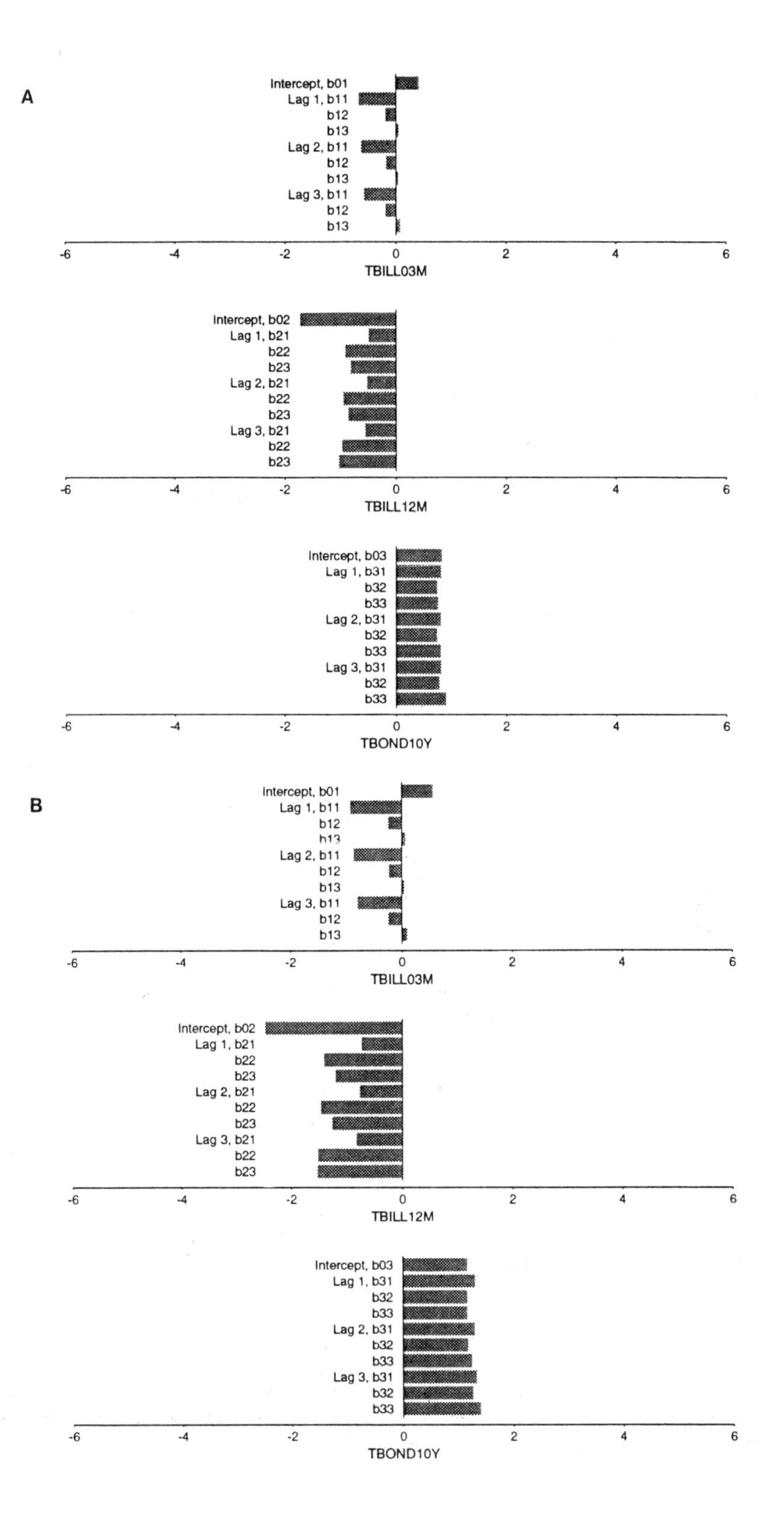
A
Intercept, b01
Lag 1, b11
b12
b13
Lag 2, b11
b12
b13
Lag 3, b11
b12
b13
-6
-4
-2
0
2
4
6
TBILL03M
Intercept, b02
Lag 1, b21
b22
b23
Lag 2, b21
b22
b23
Lag 3, b21
b22
b23
TBILL12M
Intercept, b03
Lag 1, b31
b32
b33
Lag 2, b31
b32
b33
Lag 3, b31
b32
b33
TBOND10Y
B
Intercept, b01
Lag 1, b11
b12
b13
Lag 2, b11
b12
b13
Lag 3, b11
b12
b13
TBILL03M
Intercept, b02
Lag 1, b21
b22
b23
Lag 2, b21
b22
b23
Lag 3, b21
b22
b23
TBILL12M
Intercept, b03
Lag 1, b31
b32
b33
Lag 2, b31
b32
b33
Lag 3, b31
b32
b33
TBOND10Y

4.2. Explicit Order 1 Strong Scheme

Recursion:

$$\hat{U}_{t+\Delta} = \hat{U}_t + A(t, \hat{U}_t, \rho)\Delta + \sum_{j=1}^{k} B_j(t, \hat{U}_t, \rho)\Delta W_j$$

$$+ \frac{1}{\sqrt{\Delta}} \sum_{j=1}^{k} \sum_{r=1}^{k} \left[B_j(t, \Upsilon_r, \rho) - B_j(t, \hat{U}_t, \rho) \right] I_{rj}.$$

Supporting values:

$$\Upsilon_j = \hat{U}_t + A(t, \hat{U}_t, \rho)\Delta + B_j(t, \hat{U}_t, \rho)\sqrt{\Delta}.$$

Integral approximation:

$$I_{rj} = (1/2)\left[(\Delta W_j)^2 - \Delta \right] \qquad r = j$$

$$I_{rj} = (1/2)\Delta W_r \Delta W_j + (\Delta C)^{1/2}(\mu_r \Delta W_j - \mu_j \Delta W_r)$$

$$+ \frac{\Delta}{2\pi} \sum_{l=1}^{p} \frac{1}{l} \left[\zeta_{rl}\left(\frac{\Delta W_j}{\sqrt{\Delta/2}} + \eta_{jl} \right) - \zeta_{jl}\left(\frac{\Delta W_r}{\sqrt{\Delta/2}} + \eta_{rl} \right) \right] \qquad r \neq j$$

$$C = \frac{1}{12} - \frac{1}{2\pi^2} \sum_{l=1}^{p} \frac{1}{l^2} \qquad p = 50.$$

Independent random variables

$$\Delta W_j \sim N(0, \Delta) \qquad j = 1, \ldots, k$$

$$u_j \sim N(0, 1) \qquad j = 1, \ldots, k$$

$$n_{jl} \sim N(0, 1) \qquad j = 1, \ldots, k, \quad l = 1, \ldots, p$$

$$\zeta_{jl} \sim N(0, 1) \qquad j = 1, \ldots, k, \quad l = 1, \ldots, p.$$

5. AN APPLICATION TO THE TERM STRUCTURE

5.1. Data and Score Generator

The data are 1735 weekly observations, January 5, 1962–March 31, 1995, on three interest rates: the 3-month Treasury-Bill rate from the secondary market (TBILL03M), the 12-month Treasury-Bill rate from the secondary

FIGURE 2 Location function diagnostics for exponent = .5. The t ratio diagnostics on SNP scores corresponding to the parameters of the location function defined in (2.1) for the estimate of the SDE defined in (5.1)–(5.2) with $\gamma = (0.50\,0.50\,0.50)$. Top: quasi t ratios defined in (3.3). Bottom: adjusted t ratios defined in (3.2).

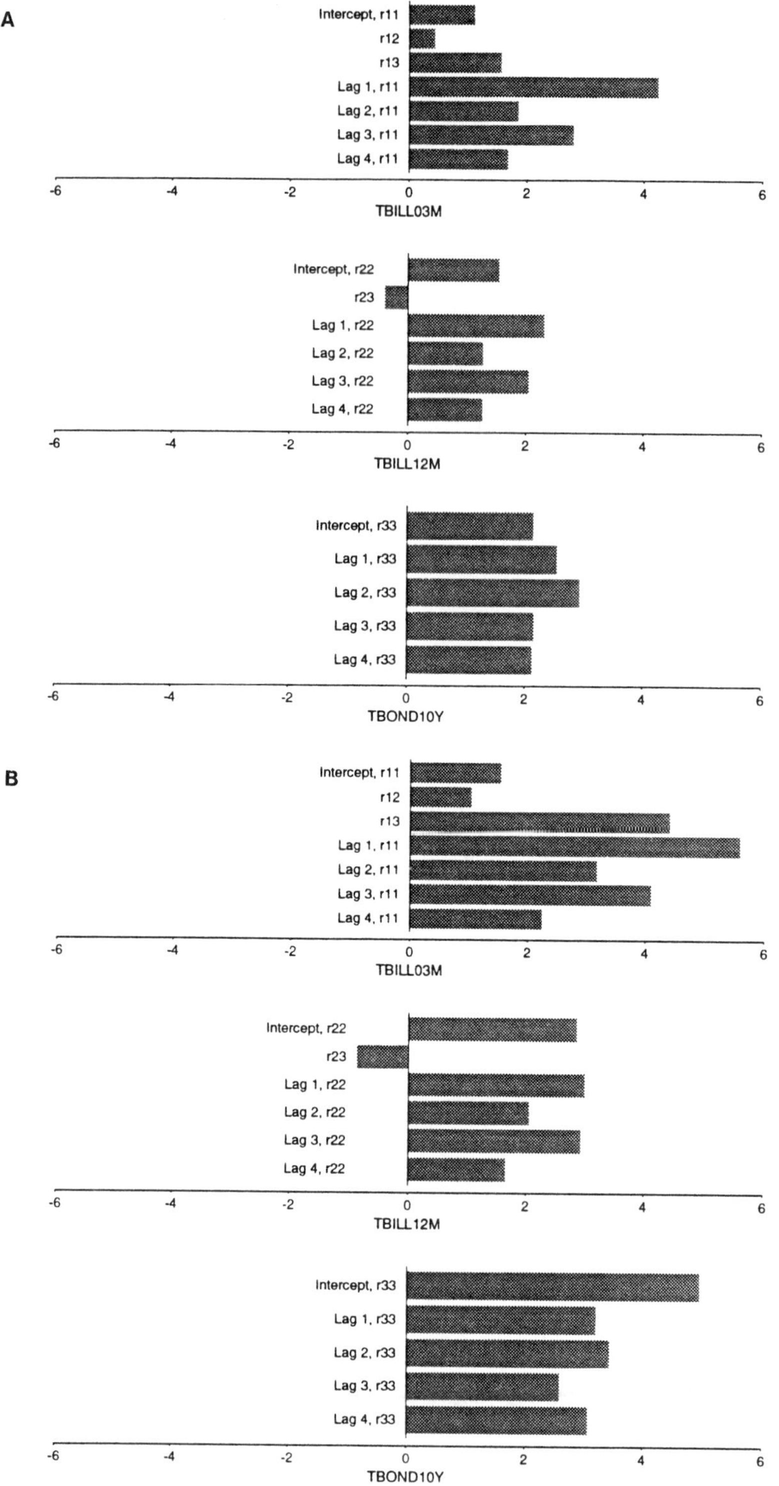

A
Intercept, r11
r12
r13
Lag 1, r11
Lag 2, r11
Lag 3, r11
Lag 4, r11
-6 -4 -2 0 2 4 6
TBILL03M
Intercept, r22
r23
Lag 1, r22
Lag 2, r22
Lag 3, r22
Lag 4, r22
-6 -4 -2 0 2 4 6
TBILL12M
Intercept, r33
Lag 1, r33
Lag 2, r33
Lag 3, r33
Lag 4, r33
-6 -4 -2 0 2 4 6
TBOND10Y
B
Intercept, r11
r12
r13
Lag 1, r11
Lag 2, r11
Lag 3, r11
Lag 4, r11
-6 -4 -2 0 2 4 6
TBILL03M
Intercept, r22
r23
Lag 1, r22
Lag 2, r22
Lag 3, r22
Lag 4, r22
-6 -4 -2 0 2 4 6
TBILL12M
Intercept, r33
Lag 1, r33
Lag 2, r33
Lag 3, r33
Lag 4, r33
-6 -4 -2 0 2 4 6
TBOND10Y

market (TBILL12M), and a 10-year constant maturity Treasury-Bond rate (TBOND10Y). Friday rates are used except when unavailable due to a holiday, in which case the Thursday rate is used. We put $\tilde{y}_t = (\text{TBILL03M}_t, \text{TBILL12M}_t, \text{TBOND10Y}_t)' \in \mathscr{R}^3$ as the observed variable. Figure 1 shows plots of the three series.

On this data set we estimated by quasi maximum-likelihood the parameters θ of the SNP density $f_K(y \mid x, \theta)$ under the restriction $K_x = 0$. Following the model selection protocol described in section 2, we obtain $L_u = 3$, $L_r = 4$, $K_z = 4$, $I_z = 3$ for the tuning parameters. This model takes the form of an ARCH model with conditionally homogeneous non-Gaussian errors. The error density is a modified Hermite density, as discussed in section 2. The polynomial $P(z)$ is a quartic in $z \in \mathscr{R}^3$ with all interactions suppressed. We call the score from this fit the "Semi-parametric ARCH score," which is of length $p_\theta = 60$.

Below, we use this score generator for EMM estimation of continuous time models of the term structure. EMM applied to a score from an SNP density constrained so that $K_x = 0$ is consistent and asymptotically normal, so long as the underlying continuous time model is correctly specified (Gallant and Tauchen, 1996), albeit with a possible efficiency loss (Gallant and Long, 1996). Failure to fit this score, of course, can be construed as sharp evidence against the specification of the continuous time model. However, for reasons discussed in Tauchen (1996), successful fitting of this score does not necessarily signal correct specification of the continuous-time model, as this score need not incorporate all salient features of the data.

5.2. Generalized Power Specifications

Let $\{U_t\}$, $t \geq 0$, denote the continuous record on the three interest rates for which $\tilde{y}_{t-L} = U_t$, $t = 0, 1, 2, \ldots$. We assume U_t is generated by the autonomous system

$$dU_t = A(U_t, \rho)dt + B(U_t, \rho)dW_t,$$

where $A(U, \rho)$ is a 3×1 vector-valued function and $B(U, \rho)$ is 3×3 matrix-valued function, with $U \in \mathscr{R}^3$ and $\rho \in \mathscr{R}^{l_\rho}$.

FIGURE 3 Scale function diagnostics for exponent = .5. The t ratio diagnostics on SNP scores corresponding to the parameters of the scale function defined in (2.2) for the estimate of the SDE defined in (5.1)–(5.2) with $\gamma = (0.50\ 0.50\ 0.50)$. Top: quasi t ratios defined in (3.3). Bottom: adjusted t ratios defined in (3.2).

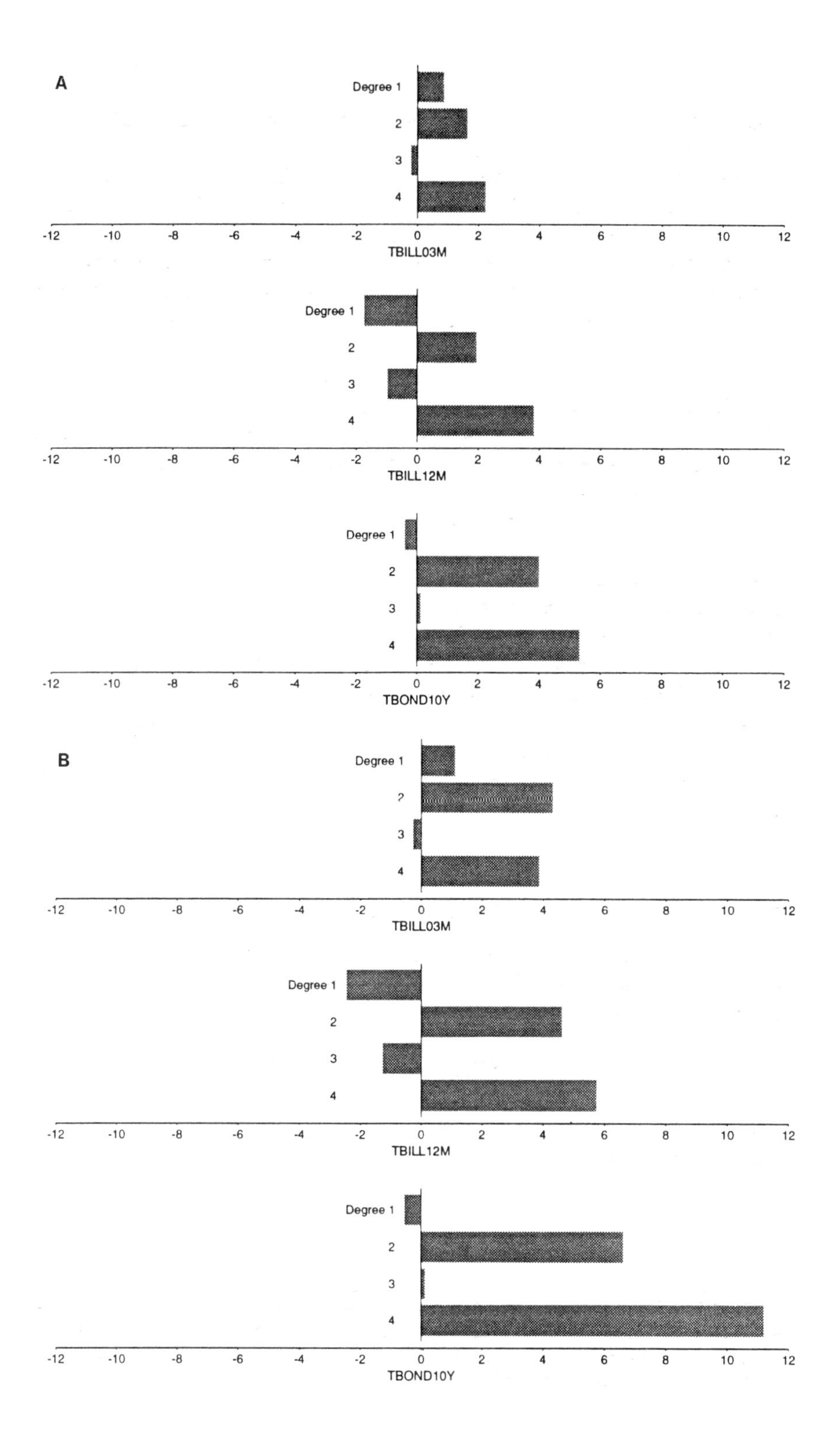

A
Degree 1
2
3
4
-12 -10 -8 -6 -4 -2 0 2 4 6 8 10 12
TBILL03M
Degree 1
2
3
4
-12 -10 -8 -6 -4 -2 0 2 4 6 8 10 12
TBILL12M
Degree 1
2
3
4
-12 -10 -8 -6 -4 -2 0 2 4 6 8 10 12
TBOND10Y
B
Degree 1
2
3
4
-12 -10 -8 -6 -4 -2 0 2 4 6 8 10 12
TBILL03M
Degree 1
2
3
4
-12 -10 -8 -6 -4 -2 0 2 4 6 8 10 12
TBILL12M
Degree 1
2
3
4
-12 -10 -8 -6 -4 -2 0 2 4 6 8 10 12
TBOND10Y

We examine empirical characteristics of a multivariate power specification in which the drift and diffusion functions take the form

$$A(U, \rho) = \alpha_0 + \alpha_1 U \tag{5.1}$$

$$B(U, \rho) = \beta_0 \operatorname{diag}\left[(\beta_1 + \beta_2 U)^{\gamma}\right], \tag{5.2}$$

where α_0 is 3×1, α_1 is 3×3, β_0 is 3×3, β_1 is 3×1, β_2 is 3×3, γ is 3×1, ρ contains the free elements of α_0, α_1, β_0, β_1, β_2, γ, and

$$\operatorname{diag}(V^{\gamma}) = \begin{bmatrix} V_1^{\gamma_1} & 0 & 0 \\ 0 & V_2^{\gamma_2} & 0 \\ 0 & 0 & V_3^{\gamma_3} \end{bmatrix}$$

for $V = \beta_1 + \beta_2 U \in \mathscr{R}^3$. For $v, \psi \in \mathscr{R}_1$ we interpret $v^{\psi} = \operatorname{sign}(v)\,|v|^{\psi}$ as the signed power function. Using the signed power function makes the diffusion function globally defined, which is tremendously convenient in empirical work, though it does mean some parameters are identified only up to sign.

The multivariate power specification (5.1)–(5.2) subsumes the yield-factor model of Duffie and Kan (1993), which is obtained by taking $\gamma = (0.50\,0.50\,0.50)$. The yield-factor model is the multivariate generalization of the square-root model of Cox *et al.* (1985). It is analytically convenient, because the local variance function $B(U, \rho)B(U, \rho)'$ is affine in U, which generates closed-form expressions for bond prices. If U_t itself is viewed as the vector of fundamental factors, then the bond prices are log-linear functions of U_t. This simplicity is lost, though, if $\gamma \neq (0.50\,0.50\,0.50)$. In this case, the multivariate power specification is a natural multivariate generalization of the univariate constant elasticity of variance specification developed by Chan *et al.* (1992).

5.3. Estimation

We employ the EMM method described in section 3 to estimate the multivariate power specification. In computation, we take

$$m(\rho, \theta) \doteq \frac{1}{N} \sum_{t=0}^{N} \frac{\partial}{\partial \theta} \log\left[f_K(\hat{y}_t \mid \hat{y}_{t-L}, \ldots, \hat{y}_{t-1}, \theta)\right]$$

FIGURE 4 Hermite polynomial diagnostics for exponent = .5. The t ratio diagnostics on SNP scores corresponding to the parameters of the Hermite polynomial defined in (2.3) for the estimate of the SDE defined in (5.1)–(5.2) with $\gamma = (0.50\,0.50\,0.50)$. Top: quasi t ratios defined in (3.3). Bottom: adjusted t ratios defined in (3.2).

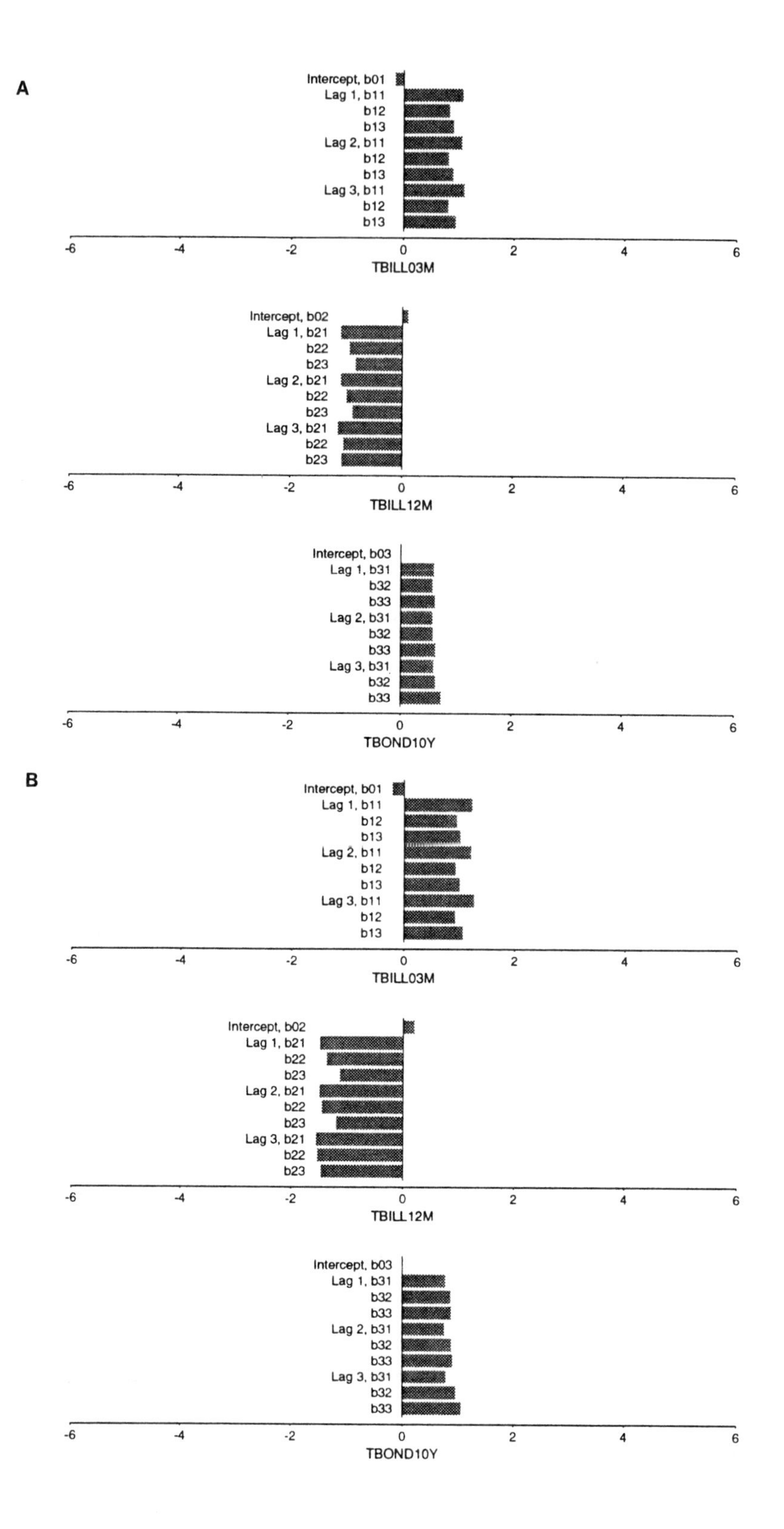

A
Intercept, b01
Lag 1, b11
b12
b13
Lag 2, b11
b12
b13
Lag 3, b11
b12
b13
-6 -4 -2 0 2 4 6
TBILL03M
Intercept, b02
Lag 1, b21
b22
b23
Lag 2, b21
b22
b23
Lag 3, b21
b22
b23
-6 -4 -2 0 2 4 6
TBILL12M
Intercept, b03
Lag 1, b31
b32
b33
Lag 2, b31
b32
b33
Lag 3, b31
b32
b33
-6 -4 -2 0 2 4 6
TBOND10Y
B
Intercept, b01
Lag 1, b11
b12
b13
Lag 2, b11
b12
b13
Lag 3, b11
b12
b13
-6 -4 -2 0 2 4 6
TBILL03M
Intercept, b02
Lag 1, b21
b22
b23
Lag 2, b21
b22
b23
Lag 3, b21
b22
b23
-6 -4 -2 0 2 4 6
TBILL12M
Intercept, b03
Lag 1, b31
b32
b33
Lag 2, b31
b32
b33
Lag 3, b31
b32
b33
-6 -4 -2 0 2 4 6
TBOND10Y

and use an explicit order 2 weak scheme as described in subsection 4.1. For this work, time t is scaled so the interval $[t, t+1]$ is one week and $n_0 = 10$, which implies the simulation step size is $\Delta = 0.10$. Based on Gallant and Tauchen (1995c) and Tauchen (1996), we use $N = 50{,}000$ throughout. With the elements of γ pinned ρ contains $p_\rho = 30$ parameters. Since $p_\theta = 60$, then the scaled objective function L_0, defined in (3.1), follows a $\chi^2(30)$ under the null hypothesis of correct specification of the SDE.

Estimating the multivariate power specification (5.1)–(5.2) with $\gamma = (0.50\,0.50\,0.50)$ gives $L_0 = 133.563$, p- value $< 10^{-12}$. Evidently, an affine local variance function is sharply at odds with the data.

Figures 2–4 reveal the empirical shortcomings of the specification with $\gamma = (0.50\,0.50\,0.50)$. The figures show the t diagnostics presented in section 3 for the three groups of parameters of the SNP density. These diagnostics are the studentized elements of the score vector $m(\hat{\rho}_n, \tilde{\theta}_n)$. The figures contain both forms of the diagnostics in order to assess robustness. The quasi t ratios, defined in (3.3) and shown in the top panels of the figures, are very easy to compute but biased downward in magnitude relative to 2.0. The t ratios, defined in (3.2) and shown in the bottom panels, are asymptotically unbiased indicators of misspecification but more difficult to compute. Numerical first derivatives were used to compute the Jacobian matrix estimate $\hat{M}_n$.

Figure 2 shows the t ratio diagnostics on the elements of the score vector for the parameters of the location $\mu_{x_{t-1}}$ defined in (2.1). The t ratio on the intercept in the TBILL12M equation suggests this specification has trouble fitting the unconditional mean of the TBILL12M series, though the order t ratios would be considered statistically insignificant.

Figure 3 shows the t ratio diagnostics on the elements of the score vector corresponding to the scale function $R_{x_{t-1}}$ defined in (2.2). Nearly all t ratios on the scale function scores are above 2.0 in magnitude, indicating that this specification has trouble tracking the conditional second moments of all three interest rates. The signs of the t ratios are the same because of correlations across the scores.

Figure 4 shows the t ratio diagnostics on the elements corresponding to the Hermite polynomial parameters. From this figure it is seen that the specification fits the scores on the odd powers but fails to fit those on the even powers of the polynomial. From extensive previous experience with

FIGURE 5 Location function diagnostics for exponent = .8. The t ratio diagnostics on SNP scores corresponding to the parameters of the location function defined in (2.1) for the estimate of the SDE defined in (5.1)–(5.2) with $\gamma = (0.80\,0.80\,0.80)$. Top: quasi t ratios defined in (3.3). Bottom: adjusted t ratios defined in (3.2).

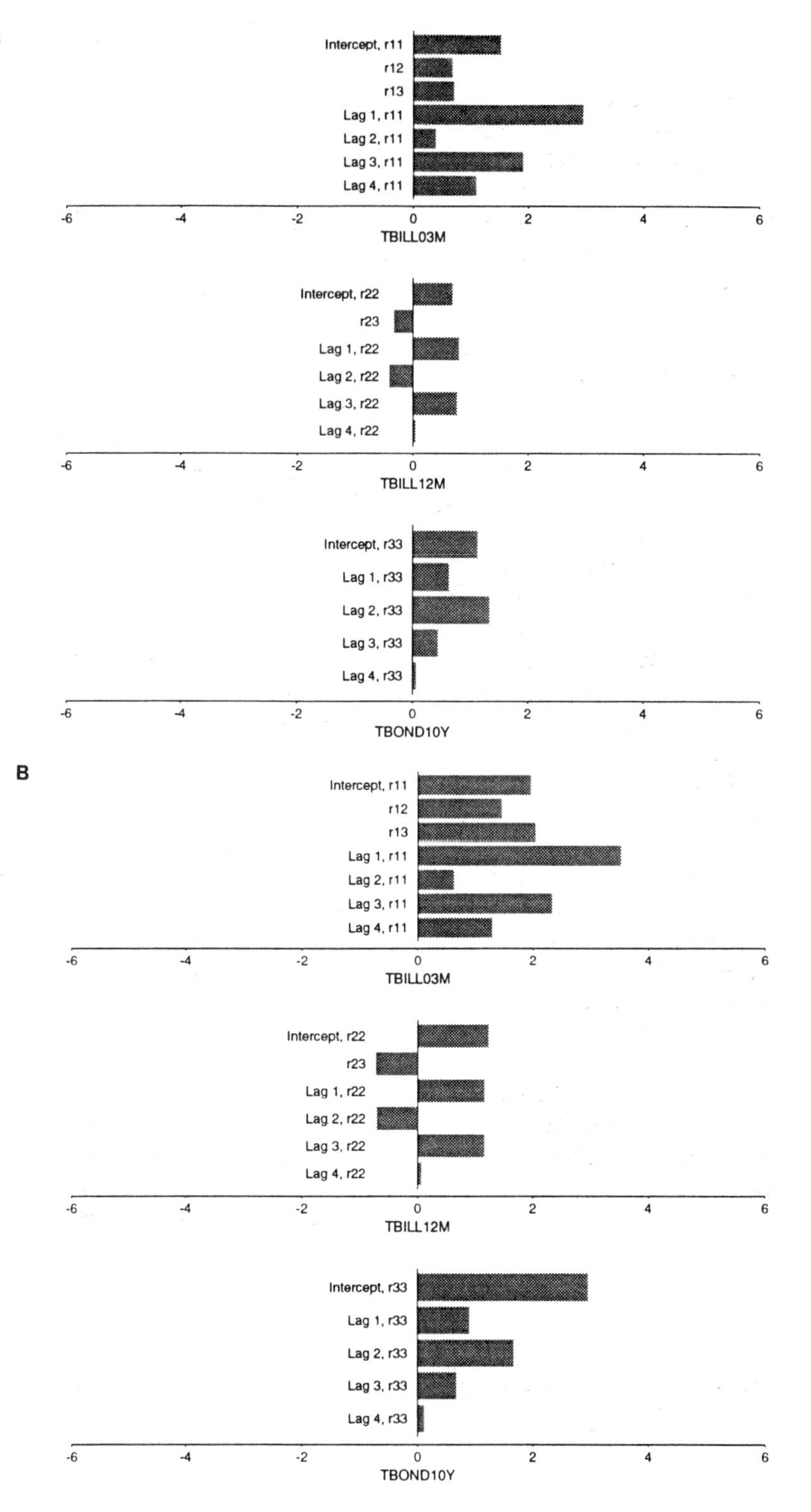
A
Intercept, r11
r12
r13
Lag 1, r11
Lag 2, r11
Lag 3, r11
Lag 4, r11
-6
-4
-2
0
2
4
6
TBILL03M
Intercept, r22
r23
Lag 1, r22
Lag 2, r22
Lag 3, r22
Lag 4, r22
TBILL12M
Intercept, r33
Lag 1, r33
Lag 2, r33
Lag 3, r33
Lag 4, r33
TBOND10Y
B
Intercept, r11
r12
r13
Lag 1, r11
Lag 2, r11
Lag 3, r11
Lag 4, r11
TBILL03M
Intercept, r22
r23
Lag 1, r22
Lag 2, r22
Lag 3, r22
Lag 4, r22
TBILL12M
Intercept, r33
Lag 1, r33
Lag 2, r33
Lag 3, r33
Lag 4, r33
TBOND10Y

SNP, we know that the even powers of the Hermite tend to control the tail thickness, while odd powers control asymmetries, although the association between degree and these features of the density is not quite exact because the polynomial is squared in (2.3). The evidence in Figure 4 suggests that the predicted discrete-time innovation density from the fitted SDE is not sufficiently leptokurtic relative to the Gaussian density.

From findings reported in Conley *et al.* (1995), Aït-Sahalia (1996a), and Tauchen (1996), who estimate continuous-time models on univariate short-rate data, one expects a specification with exponents above 0.50 to do better. A very detailed examination of the objective is beyond the scope of this chapter, and some preliminary findings are reported in Gallant and Tauchen (1995c). Here we re-estimate the SDE specification (5.1)–(5.2) with $\gamma = (0.80\,0.80\,0.80)$, which gives $L_0 = 57.089$, p value $= 0.00169$. Although this specification does not quite accommodate the semi-parametric ARCH score, the improvement in fit indicates that a local variance function that is convex in linear combinations of the interest rates is needed to fit the data.

Figures 5–7 show the t ratio diagnostics on this specification. Comparing these three figures to Figures 2–4 shows how the increase in the common value of the exponent diminishes many of the problems with the fit. The scores for the parameters associated with $\mu_{x_{t-1}}$ are reduced to insignificance as are many of those for the parameters of $R_{x_{t-1}}$. Perhaps the most noticeable improvement is in the Hermite scores (Figure 7 versus Figure 4), indicating that higher value of the exponent generates higher order non-Gaussianity in the data.

Interestingly, the top versus the bottom panels of Figures 6 and 7 show that the adjustments to the standard errors of the scores matter. Several scores have quasi t ratios below 2.0 in magnitude in the top panels but would be judged statistically significant on the basis of the bottom panels. If, for reasons of computational limitations, the quasi t ratios are used, then caution is needed in evaluating quasi t ratios just below 2.0, and the overall chi-square goodness of test should be used to protect the inference.

6. CONCLUSION

Our purpose in this chapter has been to explore the use of EMM diagnostics to guide the determination of the specification of a continuous time model for financial data. In the empirical work, we show how the

FIGURE 6 Scale function diagnostics for exponent = .8. The t ratio diagnostics on SNP scores corresponding to the parameters of the scale function defined in (2.2) for the estimate of the SDE defined in (5.1)–(5.2) with $\gamma = (0.80\,0.80\,0.80)$. Top: quasi t ratios defined in (3.3). Bottom: adjusted t ratios defined in (3.2).

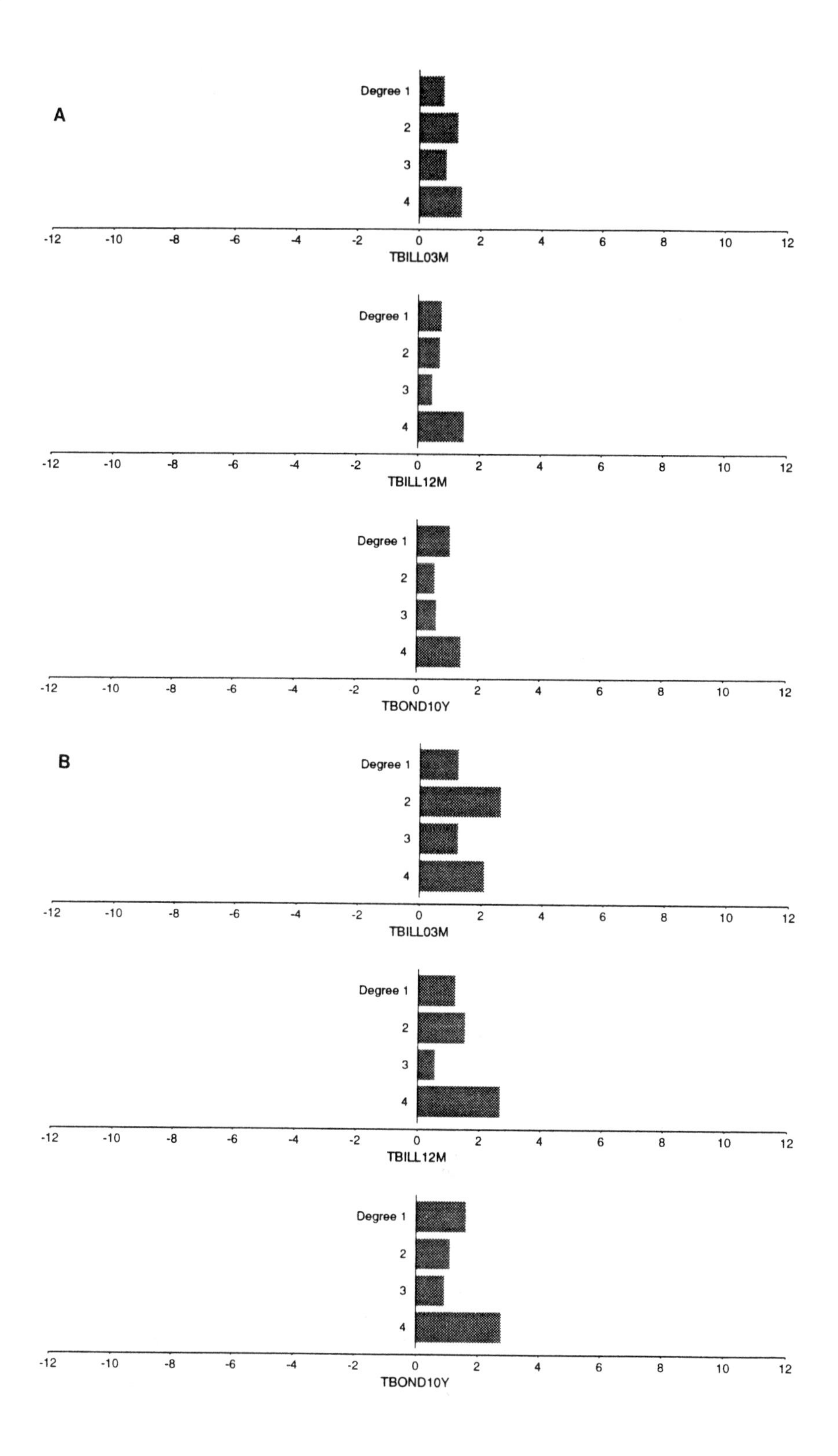

A
Degree 1
2
3
4
-12
-10
-8
-6
-4
-2
0
2
4
6
8
10
12
TBILL03M
Degree 1
2
3
4
TBILL12M
Degree 1
2
3
4
TBOND10Y
B
Degree 1
2
3
4
TBILL03M
Degree 1
2
3
4
TBILL12M
Degree 1
2
3
4
TBOND10Y

scores of the score generator provide specific guidance on which features of the data the fitted SDE accommodates well and which features it misses. This information can be quite useful in determining the directions along which to expand the model. There is one key caveat. We have shown how to adjust the specification of the SDE to accommodate better the score function of a particular score generator, which takes the form of a semi-parametric ARCH model. Although this score generator is a reasonable data descriptor, it would not emerge as the best fitting model from a complete and exhaustive search. Such a search is well beyond the scope of this chapter, but preliminary evidence (not reported here) indicates that it would detect additional features of the data beyond those captured by semi-parametric ARCH. As discussed at length in Tauchen (1996), it is imperative to ultimately confront the SDE specification to the score function of this larger model to avoid specification error. That effort would entail using the EMM diagnostics in much the same way as considered in this chapter.

ACKNOWLEDGMENTS

Supported by the National Science Foundation. The corresponding author is A. Ronald Gallant, Department of Economics, University of North Carolina, CB# 3305, 6F Gardner Hall, Chapel Hill NC 27599-3305 USA; phone: 1-919-966-5338; fax: 1-919-966-4986; e-mail: ron-gallant@unc.edu.

REFERENCES

Aït-Sahalia, Y. (1996a). Testing continuous-time models of the spot interest rate. *Review of Financial Studies* **9**, 385–426.

Aït-Sahalia, Y. (1996b). Nonparametric pricing of interest rate derivatives. *Econometrica* **64**, 527–560.

Broze, L., Scaillet, O., Zakoian, J.-M. (1995). "Quasi Indirect Inference for Diffusion Processes," Working Paper. C.O.R.E., Université Catholique de Louvain, Louvain, Belgium.

Chan, K. C., Karolyi, G. A., Longstaff, F. A., and Sanders, A. B. (1992). An empirical comparison of alternative models of the short-term interest rate. *Journal of Finance* **47**, 1209–1227.

Clark, P. K. (1973). A subordinated stochastic process model with finite variance for speculative prices. *Econometrica* **41**, 135–156.

FIGURE 7 Hermite polynomial diagnostics for exponent = .8. The t ratio diagnostics on SNP scores corresponding to the parameters of the Hermite polynomial defined in (2.3) for the estimate of the SDE defined in (5.1)–(5.2) with $\gamma = (0.80\,0.80\,0.80)$. Top: quasi t ratios defined in (3.3). Bottom: adjusted t ratios defined in (3.2).

Conley, T., Hansen, L. P., and Scheinkman, J. (1995). "Estimating Subordinated Diffusions From Discrete Time Data," Working Paper. University of Chicago, Chicago.

Cox, J. C., Ingersoll, J. E., Jr., and Ross, S. A. (1985). A theory of the term structure of interest rates. *Econometrica* **53**, 385–407.

Duffie, D., and Kan, R. (1993). "A Yield-Factor Model of Interest Rates," Working Paper. Stanford University, Stanford, CA.

Duffie, D., and Singleton, K. J. (1993). Simulated moments estimation of Markov models of asset prices. *Econometrica* **61**, 929–952.

Fenton, V. M., and Gallant, A. R. (1996a). Convergence rates of SNP density estimators. *Econometrica* **64**, 719–727.

Fenton, V. M., and Gallant, A. R. (1996b). Qualitative and asymptotic performance of SNP density estimators. *Journal of Econometrics*.

Gallant, A. R., and Long, J. R. (1996). "Estimating Stochastic Differential Equations Efficiently by Minimum Chi-Square," *Biometrika*, forthcoming.

Gallant, A. R., and Nychka, D. W. (1987). Seminonparametric maximum likelihood estimation. *Econometrica* **55**, 363–390.

Gallant, A. R., and Tauchen, G. (1989). Seminonparametric estimation of conditionally constrained heterogeneous processes: Asset pricing applications. *Econometrica* **57**, 1091–1120.

Gallant, A. R., and Tauchen, G. (1995a). "SNP: A Program for Nonparametric Time Series Analysis, A User's Guide." University of North Carolina, Chapel Hill. (The Guide and Fortran code are available by anonymous ftp at site ftp.econ.duke.edu in directory pub/arg/snp or by sending the one-line e-mail message "send snp from general" to statlib@lib.stat.cmu.edu.)

Gallant, A. R., and Tauchen G. (1995b). "EMM: A Program for Efficient Method of Moments Estimation, A User's Guide." Duke University, Durham, NC. (The Guide and Fortran code are available by anonymous ftp at site ftp.econ.duke.edu in directory pub/get/emm.)

Gallant, A. R., and Tauchen, G. (1995c). "Estimation of Continuous Time Models for Stock Returns and Interest Rates," Working Paper. Duke University, Durham, NC.

Gallant, A. R., and Tauchen, G. (1996). Which moments to match? *Econometric Theory*.

Gourieroux, C., Monfort, A., and Renault, E. (1993). Indirect inference. *Journal of Applied Econometrics* **8**, S85–S118.

Hannan, E. J. (1987). Rational transfer function approximation. *Statistical Science* **2**, 1029–1054.

Hansen, L. P., and Scheinkman, J. A. (1995). Back to the future: Generating moment implications for continuous time Markov process. *Econometrica* **50**, 1029–1054.

Ingram, B. F., and Lee, B. S. (1991). Simulation estimation of time series models. *Journal of Econometrics* **47**, 197–250.

Karatzas, I., and Schreve, S. E. (1991). "*Brownian Motion and Stochastic Calculus*," 2nd ed. Springer-Verlag, Berlin.

Kloeden, P. E., and Platen, E. (1992). "*Numerical Solution of Stochastic Differential Equations*." Springer-Verlag, Berlin.

Lo, A. (1988). Maximum likelihood estimation of generalized Ito process with discretely sampled data. *Econometric Theory* **4**, 231–247.

Nelson, D. (1991). Conditional heteroskedasticity in asset returns: A new approach. *Econometrica* **59**, 347–370.

Schwarz, G. (1978). Estimating the dimension of a model. *Annals of Statistics* **6**, 461–464.

Tauchen, G. (1996). New minimum chi-square methods in empirical finance. In *Advances in Econometrics* (Kenneth Wallace and David Kreps, eds.). Cambridge University Press, Cambridge, UK.

Tauchen, G., and Pitts, M. (1983). The price variability-volume relationship on speculative markets. *Econometrica* **51**, 485–505.

13

BACK TO THE FUTURE: GENERATING MOMENT IMPLICATIONS FOR CONTINUOUS-TIME MARKOV PROCESSES*

LARS PETER HANSEN
JOSÉ ALEXANDRE SCHEINKMAN

Department of Economics
University of Chicago
Chicago, Illinois 60637

1. INTRODUCTION

In this chapter we derive moment conditions for estimating and testing continuous-time Markov models using discrete-time data. An extensive literature exists on estimating continuous time *linear* models from discrete-time data emanating from the work of A. W. Phillips (1959). This literature includes treatments of identification (e.g., see P. C. B. Phillips, 1973; Hansen and Sargent, 1983) as well as estimation (e.g., see P. C. B. Phillips, 1973; Robinson, 1976; Harvey and Stock, 1985). Our aim is to develop new methods for estimation and inference that can be applied to continuous-time *nonlinear* Markov models, again from the vantage point of discrete-time sampling. Recently there has been a considerable interest among economists in understanding the role of nonlinearities in dynamic

* Reprinted from *Econometrica* **63** (1995), 767–804.

models (see Scheinkman, 1990, for a survey of this literature.) Furthermore, several particular nonlinear continuous-time models have been proposed for the term structure of interest rates (e.g., see Cox *et al.*, 1985; Heath *et al.*, 1992) and for exchange rates (e.g., see Froot and Obstfeld, 1991; Krugman, 1991). Among other things, we develop tools for assessing the empirical plausibility of these models.

Likelihood-based methods of estimation and inference for nonlinear continuous-time models can be very difficult to implement due to the computational costs associated with evaluating the likelihood function (e.g., see Lo, 1988). This is true even when the Markov state vector is completely observable at any point in time, as we assume here. The reason for this difficulty is that a continuous-time Markov process is typically specified in terms of its local evolution. Evaluating the discrete-time transition density can be quite costly because it may require solving numerically a partial differential equation as in the case of a diffusion. The computational costs can become excessive because these calculations must be repeated for each hypothetical parameter value and each observed state.[1]

In this chapter we adopt a more pragmatic approach. We begin by considering a Markov process specified in terms of its *infinitesimal generator*. Formally, this generator is defined as an operator on a function space, and, in effect, this operator stipulates the local evolution of the process. For instance, for a diffusion model specified as a solution to a stochastic differential equation, the generator can be constructed from the coefficients of the differential equation and the associated boundary conditions.

Given the infinitesimal generator, we show how to construct two sets of moment conditions that are often easy to compute in practice. As a consequence, both sets of moment conditions can be used to construct generalized-method-of-moments estimators of an unknown parameter vector and diagnostic tests. One set involves only the contemporaneous Markov state vector and hence uses only the marginal distribution of that vector. The second set includes functions of the state vector in two adjacent time periods, and hence, like the score vector from a likelihood function, exploits properties of the conditional distribution of the current state vector given the past.

Moment conditions in the second set are most conveniently represented in terms of the original generator as well as the *reverse-time*

[1] Exceptions to this are the nonlinear Markov process assumed by Cox *et al.* (1985) in their analysis of the term structure of interest rates and the reflecting barrier model of exchange rates assumed by Krugman (1991). The transition densities for this process have been fully characterized (e.g., see Feller, 1951; Wong, 1964; Levy, 1993). However, small departures in the form of the nonlinearities make likelihood-based methods much more numerically intensive to implement.

generator for a process running *backwards* in time. Although there exist general characterizations of reverse-time generators, the second set of moment conditions may be easiest to apply when the underlying Markov processes under consideration are (time) reversible. As a consequence, we use results in the probability theory literature to show that a potentially rich collection of models are reversible, including many multi-factor models of the term structure of interest rates that have been suggested in the literature.

Another strategy that has been proposed for estimating continuous-time Markov processes is to use numerical methods to approximate moments. Duffie and Singleton (1993) and Gourieroux *et al.* (1992) suggested the use of simulation while He (1990) proposed the use of binomial approximations. An attractive feature of the simulation approach is that the Markov state vector does not have to be fully observed. However, for both of these approaches it may be difficult to account for the magnitude of the approximation error, and for some applications it may be numerically costly to ensure that the approximation error is small.

When alternative parametric estimation methods are computationally feasible, the moment conditions we derive can be used to construct diagnostic tests of the model which are easy to implement in practice. Moreover, they permit the estimated model to be nested within a larger class of alternative continuous-time models including ones whose transition probabilities may be much more difficult to compute in practice. Our interest in this approach extends beyond consideration of computational costs vis-a-vis other methods of estimating parametric models, however. For instance, we also study nonparametric identification via these moment conditions. It is well-known that identifying continuous-time models from discrete-time data can be problematic because of what is known as the *aliasing phenomenon*: Distinct continuous-time processes may look identical when sampled at regular time intervals. Using spectral representation theory for self-adjoint operators, we show that there is no aliasing problem when it is known that the continuous-time process is reversible. Although the generalized-method-of-moments approach described in our chapter is designed to be computationally tractable, complete (nonparametric) identification of reversible processes is not possible using the moment conditions we derive. However, we do show that in the context of scalar diffusion models, the local mean (drift) and local variance (diffusion coefficient) can be identified up to a common scale factor.

The focus of this chapter is on deriving moment conditions implied by infinitesimal generators and on characterizing the extent to which these moment conditions can discriminate among the members of a class of infinitesimal generators. In addition, we provide restrictions on infinitesimal generators that ensure that the Law of Large Numbers and Central Limit Theorem apply to a discrete-sampled process. Armed with these approximation results, we can apply the results in Hansen (1982) to justify formally estimation and inference using generalized method of moments. On the other hand, we do not address formally issues of statistical efficiency and nonparametric estimation and inference using the moment conditions we derive. Such issues are deferred to future work.

This chapter is organized as follows. In section 2 we review the mathematical construction of an infinitesimal generator for a continuous time Markov process and describe properties of the generator that are important for our analysis. Since leading examples of continuous-time Markov processes are scalar diffusion, we use these processes to illustrate many of the results in this chapter. Consequently, in this section we recall the well-known connection between the generator of a scalar diffusion and its local mean and variance. In section 3 we show how to use the infinitesimal generator to construct two families of moment conditions expressed in terms of the Markov state vector. Other examples of Markov processes together with their infinitesimal generators are presented in section 4. We study the observable implications of each family of moment conditions in section 5. For instance, we show that our first family of moment conditions can be used to distinguish alternative candidate generators that imply distinct marginal distributions for the Markov state vector. We also provide a characterization of the additional informational content provided by the second family of moment conditions obtained by reversing calendar time. Since the domains of the generators are sometimes difficult to characterize fully, in section 6 we show how to reduce the family of test functions used in the moment conditions to include only ones for which the generator can be represented in a convenient manner. For the moment conditions to be of use in practice, we must be able to approximate expectations of functions of the state vector using a discrete-time moment analog. In section 7 we present sufficient conditions for these approximations to be valid. To facilitate verification, these conditions are expressed as restrictions on the infinitesimal generator. These large-sample approximations can also be used to justify other estimation methods than the one described in this chapter.

2. INFINITESIMAL GENERATORS

In this section we give the mathematical basis for our analysis. The focal point is on the construction of the *infinitesimal generator* of a strictly stationary, continuous-time, n-dimensional, vector Markov process $\{x_t\}$ defined on a probability space $(\Omega, \mathscr{F}, \mathscr{P}r)$.

Let $\mathscr{Q}$ be the probability measure induced on $\mathbf{R}^n$ by x_t (for any t), $\mathscr{L}^2(\mathscr{Q})$ be the space of all Borel measurable functions $\phi : \mathbf{R}^n \to \mathbf{R}$ such that

$$\int_{R^n} \phi^2 \, d\mathscr{Q} < \infty,$$

and let $\langle \cdot \mid \cdot \rangle$ and $\| \cdot \|$ be the usual inner product and norm on $\mathscr{L}^2(\mathscr{Q})$. Associated with the Markov process is a family of operators $\{\mathscr{T}_t : t \geq 0\}$ where for each $t \geq 0$, $\mathscr{T}_t$ is defined by[2]

$$\mathscr{T}_t \phi(y) \equiv \mathrm{E}[\phi(x_t) \mid x_0 = y]. \tag{2.1}$$

This family is known to satisfy several properties. For instance, from Nelson (1958, Theorem 3.1) it follows that for each $t \geq 0$, the following properties hold.

PROPERTY P1. *$\mathscr{T}_t : \mathscr{L}^2(\mathscr{Q}) \to \mathscr{L}^2(\mathscr{Q})$ is well defined, i.e., if $\phi = \psi$ with $\mathscr{Q}$ probability one, then $\mathscr{T}_t \phi = \mathscr{T}_t \psi$ with $\mathscr{Q}$ probability one and for each $\phi \in \mathscr{L}^2(\mathscr{Q})$, $\mathscr{T}_t \phi \in \mathscr{L}^2(\mathscr{Q})$.*

PROPERTY P2. *$\|\mathscr{T}_t \phi\| \leq \|\phi\|$ for all $\phi \in \mathscr{L}^2(\mathscr{Q})$, i.e., $\mathscr{T}_t$ is a (weak) contraction.*

PROPERTY P3. *For any $s, t \geq 0$, $\mathscr{T}_{t+s} = \mathscr{T}_t \mathscr{T}_s$, i.e., $\{\mathscr{T}_t : t \geq 0\}$ is a (one-parameter) semi-group.*

Property P2 is the familiar result that the conditional expectation operator can only reduce the second moment of a random variable. Property P3 is implied by the Law of Iterated Expectations since the expectation of $\phi(x_{t+s})$ given x_0 can be computed by first conditioning on information available at time s.

Our approach to exploring the implications of continuous-time Markov models for discrete time data is to study limits of expectations over small increments of time. In order for this approach to work, we impose a *mild* restriction on the smoothness properties of $\{\mathscr{T}_t\}$.

ASSUMPTION A1. *For each $\phi \in \mathscr{L}^2(\mathscr{Q})$, $\{\mathscr{T}_t \phi : t \geq 0\}$ converges [in $\mathrm{L}^2(\mathscr{Q})$] to ϕ as $t \downarrow 0$.*

Assumption A1 is implied by measurability properties of the underlying stochastic process $\{x_t\}$. Recall that $\{x_t\}$ can always be viewed as a mapping from $\mathbf{R} \times \Omega$ into $\mathbf{R}^n$. Form the product σ-algebra using the Borelians of $\mathbf{R}$ and the events of Ω. A sufficient condition for Assumption

[2] Throughout this chapter we follow the usual convention of not distinguishing between an equivalence class and the functions in the equivalence class. Moreover, in (2.1) we are abusing notation in a familiar way (e.g., see Chung, 1974, pp. 299, 230).

A1 is that the mapping defined by the stochastic process be Borel measurable with respect to the product σ-algebra.[3]

For some choices of $\phi \in \mathscr{L}^2(\mathscr{Q})$, the family of operators is *differentiable* at zero, i.e. $\{[\mathscr{T}_t\phi - \phi]/t : t > 0\}$ has an $\mathscr{L}^2(\mathscr{Q})$ limit as t goes to 0. Whenever this limit exists, we denote the limit point $\mathscr{A}\phi$. We refer to $\mathscr{A}$ as the *infinitesimal generator*. The domain $\mathscr{D}$ of this generator is the family of functions ϕ in $\mathscr{L}^2(\mathscr{Q})$ for which $\mathscr{A}\phi$ is well-defined. Typically, $\mathscr{D}$ is a proper subset of $\mathscr{L}^2(\mathscr{Q})$. Since $\mathscr{A}$ is the *derivative* of $\{\mathscr{T}_t : t \geq 0\}$ at $t = 0$, and $\{\mathscr{T}_t : t \geq 0\}$ is a semigroup, $\mathscr{A}$ and $\mathscr{T}_t$ commute on $\mathscr{D}$. Moreover the following properties are satisfied (e.g., see Pazy, 1983, Theorem 2.4, p. 4).

PROPERTY P4. *For any* $\phi \in \mathscr{L}^2(\mathscr{Q})$, $\int_0^t \mathscr{T}_s\phi\, ds \in \mathscr{D}$ *and* $\mathscr{A}\int_0^t \mathscr{T}_s\phi\, ds = \mathscr{T}_t\phi - \phi$.

PROPERTY P5. *For any* $\phi \in \mathscr{D}$, $\mathscr{T}_t\phi - \phi = \int_0^t \mathscr{A}[\mathscr{T}_s\phi]\, ds = \int_0^t \mathscr{T}_s[\mathscr{A}\phi]\, ds$.

Property P4 gives the operator counterpart to the familiar relation between derivatives and integrals. Property P5 extends the formula by interchanging the order of integration and applications of the operators $\mathscr{T}_s$ and $\mathscr{A}$.

There are three additional well-known properties of infinitesimal generators of continuous-time Markov operators that we will use in our analysis.

PROPERTY P6. $\mathscr{D}$ *is dense in* $\mathscr{L}^2(\mathscr{Q})$.

PROPERTY P7. $\mathscr{A}$ *is a closed linear operator, i.e., if* $\{\phi_n\}$ *in* $\mathscr{D}$ *converges to* ϕ_0 *and* $\{\mathscr{A}\phi_n\}$ *converges to* ψ_0, *then* ϕ_0 *is in* $\mathscr{D}$ *and* $\mathscr{A}\phi_0 = \psi_0$.

PROPERTY P8. *For every* $\lambda > 0$, $\lambda I - \mathscr{A}$ *is onto.*

(A reference for P6 and P7 and P8 is Pazy, 1983, ch. 1, Corollary 2.5 and Theorem 4.3).

A final property of $\mathscr{A}$ that will be of value to us is P9.

PROPERTY P9. $\langle \phi \mid \mathscr{A}\phi \rangle \leq 0$ *for all* $\phi \in \mathscr{D}$, *i.e.,* $\mathscr{A}$ *is quasi negative semi-definite.*[4]

This property follows from the (weak) contraction property of $\{\mathscr{T}_t\}$ and the Cauchy–Schwarz Inequality since for any $t > 0$,

$$\langle \phi \mid \mathscr{T}_t\phi - \phi \rangle \leq \|\phi\| \left(\|\mathscr{T}_t\phi\| - \|\phi\|\right) \leq 0. \tag{2.2}$$

[3] It follows from Halmos (1974, p. 148) that $\mathscr{T}_t\phi : t \geq 0$ is weakly measurable for any $\phi \in \mathscr{L}^2(\mathscr{Q})$ and hence from Theorem 1.5 of Dynkin (1965, p. 35) that A1 is satisfied.

[4] When a generator of a semi-group defined on an arbitrary Banach space satisfies the appropriate generalization of P9, it is referred to as being *dissipative* (e.g., see Pazy, 1983, p. 13).

In modelling Markov processes one may start with a candidate infinitesimal generator satisfying a particular set of properties, and then show that there exists an associated Markov process (e.g., see Ethier and Kurtz, 1986, Corollary 2.8, p. 170). Furthermore, combinations of the above properties are sufficient for $\mathscr{A}$ to be the infinitesimal generator of a semi-group of contractions satisfying A1. For instance, it suffices that P6 and P9 hold as well as P8 for *some* $\lambda > 0$. This is part of the Lumer–Phillips Theorem (e.g., see Pazy, 1983, Chapter 1, Theorem 4.3).

In many economic and finance applications it is common to start with a stochastic differential equation, instead of specifying the infinitesimal generator directly. As the example below shows, there are well-known connections between the coefficients of the stochastic differential equation and the infinitesimal generator.

EXAMPLE 1 (SCALAR DIFFUSION PROCESS). Suppose that $\{x_t\}$ satisfies the stochastic differential equation:

$$dx_t = \mu(x_t)\, dt + \sigma(x_t)\, dW_t \tag{2.3}$$

where $\{W_t : t \geq 0\}$ is a scalar Brownian motion. As we elaborate in the discussion below, under some additional regularity conditions the generator for this process can be represented as

$$\mathscr{A}\phi = u\phi' + (1/2)\sigma^2\phi''$$

on a large subset of its domain.

Prior to justifying this representation, we consider sufficient conditions for the existence of a unique stationary solution to the stochastic differential equation. There are many results in the literature that establish the existence and uniqueness of a Markov process $\{x_t\}$ satisfying (2.3) for a prespecified initialization. One set of sufficient conditions, that will be exploited in section 6, requires that the *diffusion coefficient* σ^2 be strictly positive with a bounded and continuous second derivative and that the *local mean* μ have a bounded and continuous first derivative. These conditions also imply that $\{x_t\}$ is a *Feller* process, that is, for any continuous function ϕ, $\mathscr{T}\phi$ is continuous.[5]

Of course, for the process to be stationary, we must initialize appropriately. An additional restriction is required for there to exist such an initialization. We follow Karlin and Taylor (1981) and others by introducing a *scale function S* and its derivative:

$$s(y) \equiv \exp\left\{-\int^y 2[\mu(z)/\sigma^2(z)]\, dz\right\};$$

[5] For alternative Lipschitz and growth conditions, see the hypotheses in Has'minskii (1980, Theorem 3.2, p. 79) or the weaker local hypotheses of Theorem 4.1, p. 84. Alternatively, the Yamada–Watanabe Theorem in Karatzas and Shreve (1988, p. 291) can be applied.

and a *speed density* $1/s\sigma^2$, which we assume to be integrable.[6] If we use the measure $\mathscr{Q}$ with density

$$q(y) = \frac{1}{s(y)\sigma^2(y)\int\left[s(z)\sigma^2(z)\right]^{-1} dz} \tag{2.4}$$

to initialize the process, then the Markov process $\{x_t\}$ generated via the stochastic differential equation (2.3) will be stationary. In fact, under these hypotheses $\mathscr{Q}$ is the unique stationary distribution that can be associated with a solution of (2.3).

The infinitesimal generator is defined on a subspace of $\mathscr{L}^2(\mathscr{Q})$ that contains at least the subset of functions ϕ for which ϕ' and ϕ'' are continuous, $\sigma\phi' \in \mathscr{L}^2(\mathscr{Q})$, and such that the second-order differential operator:

$$L\phi(y) \equiv \mu(y)\phi'(y) + (1/2)\sigma^2(y)\phi''(y) \tag{2.5}$$

yields an $\mathscr{L}^2(\mathscr{Q})$ function. Ito's Lemma implies that $\{\phi(x_t) - \phi(x_0) - \int_0^t L\phi(x_s)\,ds\}$ is a continuous martingale. Take expectations conditional on x_0, to obtain

$$\mathrm{E}[\phi(x_t) \mid x_0] - \phi(x_0) = \mathrm{E}\left[\int_0^t L\phi(x_s)\,ds \mid x_0\right]. \tag{2.6}$$

Or by Fubini's Theorem,

$$[\mathscr{T}_t\phi(x_0) - \phi(x_0)]/t = (1/t)\int_0^t \mathscr{T}_s L\phi(x_0)\,ds.$$

Using the continuity property of $\mathscr{T}_s$ and applying the Triangle Inequality we conclude that $\mathscr{A}\phi = L\phi$.

3. MOMENT CONDITIONS

In this section we characterize two sets of moment conditions that will be of central interest for our analysis. The basic idea underlying both sets of moment conditions is to exploit the fact that expectations of time invariant functions of say x_{t+1} and x_t do not depend on calendar time t. We connect this invariance to the local specification of the underlying Markov process, or more precisely to the infinitesimal generator. Formally, these moment conditions are derived from two important (and well-known) relations. The first relation links the stationary distribution $\mathscr{Q}$ and the

[6] It suffices for integrability for some strictly positive C and K and $|y| \geq K$, $\mu y + \sigma^2/2 \leq -C$. This can be verified by using this inequality to construct an integrable upper bound for the speed density or by appealing to more general results in Has'minskii (1980).

infinitesimal generator $\mathscr{A}$, and the second one exploits the fact that $\mathscr{A}$ and $\mathscr{T}_t$ commute. Much of our analysis will focus on a fixed sampling interval which we take to be one. From now on we write $\mathscr{T}$ instead of $\mathscr{T}_1$.

Since the process $\{x_t\}$ is stationary, $\mathrm{E}[\phi(x_t)]$ is independent of t, implying that its derivative with respect to t is zero. To see how this logic can be translated into a set of moment conditions, note that by the Law of Iterated Expectations, the expectation of $\phi(x_t)$ can either be computed directly or by first conditioning on x_0:

$$\int_{\mathbf{R}^n} \phi \, d\mathscr{Q} = \int_{\mathbf{R}^n} \mathrm{T}_t \phi \, d\mathscr{Q} \quad \text{for all } \phi \in \mathscr{L}^2(\mathscr{Q}). \tag{3.1}$$

Hence for any $\phi \in \mathscr{D}$, we have that

$$\int_{\mathbf{R}^n} \mathscr{A}\phi \, d\mathscr{Q} = \lim_{t \downarrow 0} (1/t) \int_{\mathbf{R}^n} [\mathscr{T}_t \phi - \phi] \, d\mathscr{Q} = 0 \tag{3.2}$$

since (3.1) holds for all positive t and $\{(1/t)[\mathscr{T}_t \phi - \phi] : t > 0\}$ converges in $\mathscr{L}^2(\mathscr{Q})$ to $\mathscr{A}\phi$. Relation (3.2) shows the well-known link between the generator $\mathscr{A}$ and the stationary distribution $\mathscr{Q}$ (e.g., see Ethier and Kurtz, 1986, Proposition 9.2, p. 239). Hence our first set of moment conditions is as follows.

C1. $\mathrm{E}[\mathscr{A}\phi(x_t)] = 0$ for all $\phi \in \mathscr{D}$.

The stationarity of $\{x_t\}$ also implies that $\mathrm{E}[\phi(x_{t+1})\phi^*(x_t)]$ does not depend on calendar time. Rather than deriving moment conditions by exploiting this invariance directly, a more convenient derivation is to start by using the fact that $\mathscr{A}$ and $\mathscr{T}$ commute. That is,

$$\mathrm{E}[\mathscr{A}\phi(x_{t+1}) \mid x_t = y] = \mathscr{A}\{\mathrm{E}[\phi(x_{t+1}) \mid x_t = y]\} \qquad \text{for all } \phi \in \mathscr{D}. \tag{3.3}$$

It may be difficult to evaluate the right side of (3.3) in practice because it entails computing the conditional expectation of $\phi(x_{t+1})$ prior to the application of $\mathscr{A}$. For this reason, we also derive an equivalent set of unconditional moment restrictions which are often easier to use. These moment restrictions are representable using the semi-group and generator associated with the reverse-time Markov process.

The semi-group $\{\mathscr{T}_t^*\}$ associated with the reverse-time process is defined via

$$\mathscr{T}_t^* \phi^*(y) \equiv \mathrm{E}[\phi^*(x_0) \mid x_t = y]. \tag{3.4}$$

The family $\{\mathscr{T}_t^* : t \geq 0\}$ is also a contraction semi-group of operators satisfying continuity restriction A1. Let $\mathscr{A}^*$ denote the infinitesimal generator for this semi-group with domain $\mathscr{D}^*$. The operator $\mathscr{T}_t^*$ is the adjoint of $\mathscr{T}_t$ and $\mathscr{A}^*$ is the operator adjoint of $\mathscr{A}$. To verify these results, note that it follows from (2.1), (3.4), and the Law of Iterated Expectations that

$$\begin{aligned}\langle \phi^* \mid \mathscr{T}_t \phi \rangle &= \mathrm{E}\{\phi^*(x_0)\mathrm{E}[\phi(x_t) \mid x_0]\} \\ &= \mathrm{E}[\phi^*(x_0)\phi(x_t)] \\ &= \mathrm{E}\{\phi(x_t)\mathrm{E}[\phi^*(x_0) \mid x_t]\} \\ &= \langle \phi \mid \mathscr{T}_t^* \phi^* \rangle. \end{aligned} \tag{3.5}$$

We can now use Corollary 10.6 of Pazy (1983, p. 41) to show that $\mathscr{A}^*$ is the operator adjoint of $\mathscr{A}$ and vice versa. A process is reversible if the generator $\mathscr{A}$ is self-adjoint, i.e., $\mathscr{A} = \mathscr{A}^*$.

For any ϕ in the domain $\mathscr{D}$ of $\mathscr{A}$ and ϕ^* in $\mathscr{L}^2(\mathscr{Q})$, the fact that $\mathscr{A}$ and $\mathscr{T}$ commute implies that

$$\langle \mathscr{T}\mathscr{A}\phi \mid \phi^* \rangle - \langle \mathscr{A}\mathscr{T}\phi \mid \phi^* \rangle = 0. \tag{3.6}$$

When ϕ^* is restricted to be in the domain $\mathscr{D}^*$ of $\mathscr{A}^*$,

$$\langle \mathscr{A}\mathscr{T}\phi \mid \phi^* \rangle = \langle \phi \mid \mathscr{T}^*\mathscr{A}^*\phi^* \rangle. \tag{3.7}$$

Substituting (3.6) into (3.7) and applying the Law of Iterated Expectations we find the following.

C2. $\mathrm{E}[\mathscr{A}\phi(x_{t+1})\phi^*(x_t) - \phi(x_{t+1})\mathscr{A}^*\phi^*(x_t)] = 0$ for all $\phi \in \mathscr{D}$ and $\phi^* \in \mathscr{D}^*$.

Since $\mathscr{A}^*$ enters C2, these moment conditions exploit both the forward- and reverse-time characterization of the Markov process. As we hinted above, these moment conditions can be interpreted as resulting from equating the time derivative of $\mathrm{E}[\phi(x_{t+1})\phi^*(x_t)]$ to zero.

When ϕ^* is a constant function, $\mathscr{A}^*\phi^*$ is identically zero and moment condition C2 collapses to C1. In our subsequent analysis, we will find it convenient to look first at moment conditions in the class C1, and then to ascertain the incremental contribution of the class C2. Moreover, for Markov processes that are not reversible ($\mathscr{A} \neq \mathscr{A}^*$), the additional moment conditions in the class C2 can be more difficult to use in practice because they use the reverse-time generator. The more extensive set of moment conditions are still of interest, however, because sometimes it can be shown that the processes of interest are reversible. Even when they are not, it is typically easier to compute the reverse-time generator than the one-period transition distribution.

The stationary scalar diffusions described previously in Example 1 are reversible. This is easy to verify by following the usual approach of introducing the integrating factor $1/q$, and writing

$$\mathscr{A}\phi = (1/q)(\sigma^2 q \phi'/2)'. \tag{3.8}$$

Let $\mathscr{C}_K^2$ denote the space of functions that are twice continuously differentiable with a compact support. Integration by parts implies that for functions ψ and ϕ in $\mathscr{C}_K^2$,

$$\langle \mathscr{A}\phi \mid \psi \rangle = \langle \phi \mid \mathscr{A}\psi \rangle. \tag{3.9}$$

As we will discuss in section 6, if (3.9) holds for ψ and ϕ in $\mathscr{C}_K^2$, it actually holds for each ψ and ϕ in $\mathscr{D}$; that is, $\mathscr{A}$ is symmetric. Since $\mathscr{A}$ is the generator of a contraction semigroup in a Hilbert space, it is self-adjoint and $\{x_t\}$ is reversible (see Brezis, 1983, Proposition VII.6, p. 113).

One strategy for using these moment conditions for estimation and inference is as follows. Suppose the problem confronting an econometrician is to determine which, if any, among a parameterized set of infinitesimal generators is compatible with a discrete-time sample of the process $\{x_t\}$. For instance, imagine that the aim is to estimate the "true parameter value," say β_o, associated with a parameterized family of generators $\mathscr{A}_\beta$ for β in some admissible parameter space. Vehicles for accomplishing this task are the sample counterparts to moment conditions C1 and C2. By selecting a finite number of test functions ϕ, the unknown parameter vector β_o can be estimated using generalized method of moments (e.g., see Hansen, 1982) and the remaining over-identifying moment conditions can be tested. In section 5 of this chapter we characterize the information content of each of these two sets of moment conditions for discriminating among alternative sets of infinitesimal generators; and in section 7 we supply some supporting analysis for generalized method of moments estimation and inference by deriving some sufficient conditions on the infinitesimal generators for the Law of Large Numbers and Central Limit Theorem to apply.

For particular families of scalar diffusions and test functions, moment conditions in the class C1 have been used previously, albeit in other guises. For instance Wong (1964) showed that first-order polynomial specifications of μ, and second-order polynomial specifications of σ^2 are sufficient to generate processes with stationary densities in the Pearson class. Pearson's method of moment estimation of these densities can be interpreted, except for its assumption that the data generation is i.i.d., as appropriately parameterizing the polynomials defining the drift and diffusion coefficients and using polynomial test functions in C1. This approach has been extended to a broader class of densities by Cobb *et al.* (1983). These authors suggested estimating higher order polynomial specifications of μ for a prespecified (second-order) polynomial specification of σ^2. Their estimation can again be interpreted as using moment conditions C1 with polynomial test functions. Of course, moment conditions C1 and C2 are easy to apply for other test functions ϕ and ϕ^* whose first and second derivatives can be computed explicitly. Although Cobb *et al.* (1983) prespecify the local variance to facilitate identification, this assumption is not convenient for many applications in economics and finance. As we will see in section 5, moment conditions in the class C2 can be used to help identify and estimate unknown parameters of the local variance.

4. EXAMPLES

In this section we give additional illustrations of infinitesimal generators for continuous-time Markov processes.

EXAMPLE 2 (SCALAR DIFFUSIONS WITH BOUNDARY CONDITIONS). Many economic examples deal with processes that are restricted to a finite interval or to the nonnegative reals. The reasoning in Example 1 applies immediately to processes defined on an interval when both end points are entrance boundaries. Moreover, this analysis also can be extended to processes with reflecting boundaries as the ones assumed in the literature on exchange rate bands. Typically, a process with reflecting boundaries in an interval (ℓ, u) is constructed by changing equation (2.3) to include an appropriate additional term that is "activated" at the boundary points:

$$dx_t = \mu(x_t)\, dt + \sigma(x_t)\, dW_t + \theta(x_t)\, d\kappa_t$$

where $\{\kappa_t\}$ is a particular nondecreasing process that increases only when x_t is at the boundary, and

$$\theta(\ell) = 1 \quad \text{and} \quad \theta(u) = -1.$$

In addition we assume that the original μ and σ^2 define a *regular* diffusion, that is the functions s and $1/\sigma^2 s$ are integrable. We will also assume that s is bounded away from zero.[7] Ito's Lemma applies to such processes with an additional term $\phi'(x_t)\theta(x_t)\, d\kappa_t$. This term vanishes if $\phi'(\ell) = \phi'(u) = 0$, and $\mathscr{A}\phi$ is again given as (2.5) or (3.8) for ϕ's that satisfy these additional restrictions. To check that $\mathscr{A}$ is self-adjoint, it is now sufficient to verify (3.9) for ψ and ϕ that are twice continuously differentiable with first derivatives that vanish at the boundaries. When the drift coefficient has continuous first derivatives and diffusion coefficient continuous second derivatives, the right-hand side of (2.4) again defines the unique stationary distribution associated with the reflexive barrier process. The case with one reflecting barrier and the other one nonattracting can be handled in an analogous fashion.

EXAMPLE 3 (MULTIVARIATE FACTOR MODELS). We have just seen that many scalar diffusions are reversible. Reversibility carries over to multivariate diffusions built up as time invariant functions of a vector of

[7] If σ^2 is bounded away from zero, all these assumptions hold but they also hold for some processes where σ^2 vanishes at the boundary.

independent scalar diffusions. More precisely, let $\{f_t\}$ be a vector diffusion with component processes that are independent, stationary, and reversible. Think of the components of $\{f_t\}$ as independent unobservable "factors." Suppose that the observed process is a time invariant function of the factors:

$$x_t = F(f_t)$$

for some function F. Factor models of bond prices like those proposed by Cox *et al.* (1985), Longstaff and Schwartz (1992), and Frachot *et al.* (1992) all have this representation. The resulting $\{x_t\}$ process will clearly be stationary as long as the factors are stationary. Moreover, since $\{f_t\}$ is reversible, so is $\{x_t\}$. To ensure that $\{x_t\}$ is a Markov process, we require that the factors can be recovered from the observed process. That is, F must be one-to-one. Finally, to guarantee that $\{x_t\}$ satisfies a stochastic differential equation, we restrict F to have continuous second derivatives. For a more general characterization of multivariable reversible diffusions, see Kent (1978).

EXAMPLE 4 (MULTIVARIATE DIFFUSION MODELS). More generally, suppose that $\{x_t\}$ satisfies the stochastic differential equation

$$dw_t = \mu(x_t)\,dt + \Sigma(x_t)^{1/2}\,dW_t$$

where $\{W_t : t \geq 0\}$ is an n-dimensional Brownian motion. Entry i of the *local mean* μ, denoted μ_i, and entry (i, j) of the *diffusion matrix* Σ, denoted σ_{ij}, are functions from $\mathbf{R}^n$ into $\mathbf{R}$. The functions μ_i and the entries of $\Sigma(x_t)^{1/2}$ are assumed to satisfy Lipschitz conditions (see Has'minskii, 1980, Theorem 3.2, p. 79; the weaker local hypothesis of Theorem 4.1, p. 84; or Ethier and Kurtz, 1986, Theorem 3.7, p. 297, Theorem 3.11, p. 300, and Remark 2.7, p. 374). The existence of a unique stationary distribution is assured if there exists a K such that $\Sigma(y)$ is positive definite for $|y| \leq K$ and for $|y| \geq K$,

$$\mu(y) \cdot V_y + (1/2)\text{trace}[\Sigma(y)V] < -1$$

for some positive definite matrix V. Moreover this guarantees that the process is mean recurrent (see Has'minskii, 1980, Corollary 1, p. 99; Example 1, p. 103; and Corollary 2, p. 123). If $\Sigma(y)$ is positive definite for all $y \in \mathbf{R}^n$, then the density q is given by the unique nonnegative bounded solution q to the partial differential equation:

$$-\sum_{j=1}^{n} \partial/\partial y_j[\mu_j(y)q(y)]$$
$$+(1/2)\sum_{i=1}^{n}\sum_{j=1}^{n}\left(\partial^2/\partial y_i\,\partial y_j\right)[\sigma_{ij}(y)q(y)] = 0$$

that integrates to one (see Has'minskii, 1980, Lemma 9.4, p. 138). Again by initializing the process using the measure $\mathscr{Q}$ with density q, we construct a stationary Markov process $\{x_t\}$.

The infinitesimal generator is defined on a subspace of $\mathscr{L}^2(\mathscr{Q})$ that contains at least the space of functions ϕ for which $\partial\phi/\partial y$ and $\partial^2\phi/\partial y\,\partial y'$ are continuous and the entries of $\Sigma^{1/2}\,\partial\phi/\partial y \in \mathscr{L}^2(\mathscr{Q})$ via

$$\mathscr{A}\phi(y) = \mu(y)\cdot\partial\phi(y)/\partial y + (1/2)\text{trace}\left[\Sigma(y)\,\partial^2\phi(y)/\partial y\,\partial y'\right], \quad (4.1)$$

if the right-hand side of (4.1) is in $\mathscr{L}^2(\mathscr{Q})$. Under suitable regularity conditions, a time-reversed diffusion is still a diffusion and the adjoint $\mathscr{A}^*$ can be represented as[8]

$$\mathscr{A}^*\phi^*(y) = \mu^*(y)\cdot\partial\phi^*(y)/\partial y + (1/2)\text{trace}\left[\Sigma^*(y)\,\partial^2\phi^*(y)/\partial y\,\partial y'\right]. \quad (4.2)$$

In (4.2) the diffusion matrix Σ^* turns out to be equal to Σ; however, the local mean μ^* may be distinct from μ. Let Σ_j denote column j of Σ. It follows from Nelson (1958), Anderson (1982), Hausmann and Pardoux (1985), or Millet *et al.* (1989) that

$$\mu^*(y) = -\mu(y) + [1/q(y)]\sum_j \partial/\partial y_j\left[q(y)\Sigma_j(y)\right].$$

Therefore, on a dense subset of its domain, the adjoint $\mathscr{A}^*$ can be constructed from knowledge of μ, Σ, and the stationary density q.

Moment conditions in the class C1 remain easy to apply to test functions whose two derivatives can be readily computed. For multivariate diffusions that are not time reversible, it is, in general, much more difficult to calculate the reverse-time generator used in C2. An alternative approach is to approximate the reverse-time generator using a nonparametric estimator of the logarithmic derivative of the density. With this approach, nonparametric estimators of the density and its derivative appear in the constructed moment conditions even though the underlying estimation problem is fully parametric. The nonparametric estimator is used only as a device for simplifying calculations. Rosenblatt (1969) and Roussas (1969) described and justified nonparametric estimators of densities for stationary Markov processes. Moreover, estimation using moment conditions constructed with nonparametric estimates of functions such as $\partial[\log q(y)]/\partial y$ has been studied in the econometrics literature (e.g., see Gallant and Nychka, 1987; Powell *et al.*, 1989; Robinson, 1989; Chamberlain, 1992; Newey, 1993; Lewbel, 1994). Presumably results from these literatures could be extended to apply to our estimation problem.

[8] For example, Millet *et al.* (1989) showed that it suffices for coefficients μ and Σ of the diffusion to be twice continuously differentiable, to satisfy Lipschitz conditions, and for the matrix Σ to be uniformly nonsingular (see Theorem 2.3 and Proposition 4.2).

EXAMPLE 5 (MARKOV JUMP PROCESS). Let η be a nonnegative bounded function mapping $\mathbf{R}^n$ into $\mathbf{R}$ and let $\pi(y, \Gamma)$ denote a transition function in the Cartesian product of $\mathbf{R}^n$ and the Borelians of $\mathbf{R}^n$. Imagine the following stochastic process $\{x_t\}$. Dates at which changes in states occur are determined by a Poisson process with parameter $\eta(y)$ if the current state is y. Given that a change occurs, the transition probabilities are given by $\pi(y, \cdot)$. Hence we can think of a jump process as a specification of an underlying discrete time Markov chain with transition probabilities.

Additional restrictions must be imposed for this process to be stationary. First, suppose there exists a nonzero Borel measure $\mathscr{Q}$ satisfying the equation

$$\tilde{\mathscr{Q}}(\Gamma) = \int \pi(y, \Gamma)\, d\tilde{\mathscr{Q}}(y) \quad \text{for any Borelian } \Gamma.$$

In other words $\tilde{\mathscr{Q}}$ is the stationary distribution of the underlying discrete time Markov chain. If the Poisson intensity parameter η is constant, then the Markov jump process also has $\tilde{\mathscr{Q}}$ as a stationary distribution because the randomization of the jump times is independent of the current Markov state. In the more general case in which η depends on the state, there is an observationally equivalent representation with a constant Poisson parameter and a different underlying Markov chain (e.g., see Ethier and Kurtz, 1986, p. 163). The alternative Poisson intensity parameter is $\hat{\eta} \equiv \sup \eta$ and the transition probabilities for the alternative Markov chain are

$$\hat{\pi}(y, \Gamma) \equiv (1 - \eta/\hat{\eta})\,\delta_y(\Gamma) + (\eta/\hat{\eta})\,\pi(y, \Gamma)$$

where $\delta_y(\Gamma)$ is one if y is in Γ and zero otherwise. Thus under this alternative representation jumps are more frequent, but once a jump occurs there is state dependent positive probability that the process will stay put. Elementary computations show that if we assume that

$$\int [1/\eta(y)]\, d\tilde{\mathscr{Q}}(y) < \infty,$$

and construct a probability measure $\mathscr{Q}$ to be

$$d\mathscr{Q} = \frac{d\tilde{\mathscr{Q}}}{\eta \int (1/\eta)\, d\tilde{\mathscr{Q}}}, \tag{4.3}$$

then $\mathscr{Q}$ is a stationary distribution for the alternative Markov chain. Therefore, the Markov jump process $\{x_t\}$ will be stationary so long as it is initialized at $\mathscr{Q}$.

Define the conditional expectation operator $\tilde{\mathcal{T}}$ associated with the underlying Markov chain:

$$\tilde{\mathcal{T}}\phi \equiv \int \phi(y')\pi(y, dy').$$

Analogous to the operator $\mathcal{T}, \tilde{\mathcal{T}}$ maps $\mathcal{L}^2(\tilde{\mathcal{Q}})$ into itself. Using the fact that

$$\mathcal{T}_t\phi - \phi = t\eta\tilde{\mathcal{T}} - t\eta\phi + o(t),$$

one can show that the generator for the continuous time jump process can be represented as

$$\mathcal{A}\phi = \eta\left[\tilde{\mathcal{T}}\phi - \phi\right] \tag{4.4}$$

(e.g., see Ethier and Kurtz, 1986, pp. 162–163). It is easy to verify that the generator $\mathcal{A}$ is a bounded operator on all of $\mathcal{L}^2(\mathcal{Q})$. Since η is bounded, any function ϕ in $\mathcal{L}^2(\mathcal{Q})$ must also be in $\mathcal{L}^2(\tilde{\mathcal{Q}})$.

It is also of interest to characterize the adjoint $\mathcal{A}^*$ of $\mathcal{A}$. To do this we study the transition probabilities for the reverse-time process. Our candidate for $\mathcal{A}^*$ uses the adjoint $\tilde{\mathcal{T}}^*$ in place of $\tilde{\mathcal{T}}$:

$$\mathcal{A}^*\phi^* = \eta\left[\tilde{\mathcal{T}}^*\phi^* - \phi^*\right].$$

To verify that $\mathcal{A}^*$ is the adjoint of $\mathcal{A}$, first note that

$$\langle \mathcal{A}\phi \mid \phi^* \rangle = \int \eta\left[\tilde{\mathcal{T}}\phi - \phi\right]\phi^* \, d\mathcal{Q} = \int \tilde{\mathcal{T}}\phi\phi^*\eta \, d\mathcal{Q} - \int \eta\phi\phi^* \, d\mathcal{Q}.$$

By construction $\tilde{\mathcal{T}}^*$ is the $\mathcal{L}^2(\tilde{\mathcal{Q}})$ adjoint of $\tilde{\mathcal{T}}$. It follows from (4.3) that $\eta \, d\mathcal{Q}$ is proportional to $d\tilde{\mathcal{Q}}$. Consequently,

$$\begin{aligned} \int \tilde{\mathcal{T}}\phi\phi^* \, d\mathcal{Q} - \int \eta\phi\phi^* \, d\mathcal{Q} &= \int \phi\tilde{\mathcal{T}}^*\phi^*\eta \, d\mathcal{Q} - \int \eta\phi\phi^* \, d\mathcal{Q} \\ &= \langle \phi \mid \mathcal{A}^*\phi^* \rangle. \end{aligned}$$

To use moment conditions C1 and C2 for this Markov jump process requires that we compute $\tilde{\mathcal{T}}\phi$ and $\tilde{\mathcal{T}}^*\phi$ for test functions ϕ. Suppose that the Markov chain is a discrete-time Gaussian process. It is then straightforward to evaluate $\tilde{\mathcal{T}}\phi$ and $\tilde{\mathcal{T}}^*\phi$ for polynomial test functions. Nonlinearities in the continuous-time process could still be captured by nonlinear specification of the function η. On the other hand, when nonlinearities are introduced into the specification of the Markov chain, it may be difficult to compute $\tilde{\mathcal{T}}\phi$ and $\tilde{\mathcal{T}}^*\phi$. In these cases our approach may not be any more tractable than, say, the method of maximum likelihood.

Recall that the moment conditions C1 and C2 can be used in situations in which the sampling interval is fixed and hence where the econometrician does not know the number of jumps that occurred between observations. This should be contrasted with econometric methods designed to exploit the duration time in each state.

5. OBSERVABLE IMPLICATIONS

Recall that in section 3 we derived two sets of moment conditions to be used in discriminating among a family of candidate generators. In this section we study the informational content of these two sets of moment conditions. Formally, there is a true generator $\mathscr{A}$ underlying the discrete-time observations. We then characterize the class of observationally equivalent generators from the vantage point of each of the sets of moment conditions. In the subsequent discussion, when we refer to a *candidate generator* we will presume that it is a generator for a Markov process.

We will establish the following identification results. First we will show that if a candidate generator $\hat{\mathscr{A}}$ satisfies the moment conditions in the set C1, it has a stationary distribution in common with the true process. Second we will show that if $\hat{\mathscr{A}}$ also satisfies the moment conditions in the set C2, it must commute with the true conditional expectation operator $\mathscr{T}$. If in addition $\hat{\mathscr{A}}$ is self-adjoint, and the true process is reversible (i.e., $\mathscr{A}$ is self-adjoint), then $\hat{\mathscr{A}}$ and $\mathscr{A}$ commute. One implication of this last result is that the drift and diffusion coefficients of stationary scalar diffusions can be identified up to a common scale factor using C1 and C2. A by-product of our analysis is that using the conditional expectation operator allows one to identify fully the generator of reversible processes.

Consider first moment conditions C1. Since one of the goals of the econometric analysis is to ascertain whether a candidate generator $\hat{\mathscr{A}}$ has $\mathscr{Q}$ as a stationary distribution, it is preferable to begin with a specification of $\hat{\mathscr{A}}$ without reference to $\mathscr{L}^2(\mathscr{Q})$. Instead let $\mathscr{B}$ denote the space of bounded functions on $\mathbf{R}^n$ endowed with the sup norm. Suppose that $\hat{\mathscr{A}}$ is the infinitesimal generator for a strongly continuous contraction semi-group $\{\hat{\mathscr{T}}_t : t \geq 0\}$ defined on a closed subspace $\mathscr{L}$ of $\mathscr{B}$ containing at least all of the continuous functions with compact support. In this setting, strong continuity is the sup-norm counterpart to Assumption A1, i.e., $\{\hat{\mathscr{T}}_t : t \geq 0\}$ converges uniformly to ϕ as t declines to zero for all ϕ in $\mathscr{L}$. Let $\hat{\mathscr{D}}$ denote the domain of $\hat{\mathscr{A}}$. We say that a candidate $\hat{\mathscr{A}}$ has $\mathscr{Q}$ as its stationary distribution if $\mathscr{Q}$ is the stationary distribution for a Markov process associated with this candidate. The following result is very similar to part of Proposition 9.2 of Ethier and Kurtz (1986, p. 239).

PROPOSITION 1. *Let $\hat{\mathscr{A}}$ be a candidate generator defined on $\hat{\mathscr{D}} \subseteq \mathscr{L}$. Then $\hat{\mathscr{A}}$ satisfies C1 for all $\phi \in \hat{\mathscr{D}}$ if and only if $\mathscr{Q}$ is a stationary distribution of $\hat{\mathscr{A}}$.*

Proof. Since convergence in the sup norm implies $\mathscr{L}^2(\mathscr{Q})$ convergence, our original derivation of C1 still applies. Conversely, note that the analog of Property P4 implies that for any $\phi \in \mathscr{L}$: $\int_0^t \hat{\mathscr{T}}_s(\phi)\, ds \in \hat{\mathscr{D}}$ and

$$\hat{\mathscr{A}} \int_0^t \hat{\mathscr{T}}_s \phi \, ds = \hat{\mathscr{T}}_t(\phi) - \phi. \tag{5.1}$$

Hence integrating both sides of (5.1) with respect to $\mathscr{Q}$ and using the fact that $\hat{\mathscr{A}}$ satisfies C1, we have that for all $\phi \in \mathscr{L}$:

$$\int_{\mathbf{R}^n} \left(\hat{\mathscr{T}}_t \phi - \phi\right) d\mathscr{Q} = 0. \tag{5.2}$$

Relation (5.2) can be shown to hold for all indicator functions of Borelians of $\mathbf{R}^n$ because $\mathscr{L}$ contains all continuous functions with a compact support. Q.E.D.

In light of Proposition 1, any infinitesimal generator $\hat{\mathscr{A}}$ satisfying C1 has $\mathscr{Q}$ as a stationary distribution. In other words, moment conditions in the family C1 cannot be used to discriminate among models with stationary distribution $\mathscr{Q}$. On the other hand, if $\hat{\mathscr{A}}$ does not have $\mathscr{Q}$ as its stationary distribution, then there exists a test function ϕ in $\hat{\mathscr{D}}$ such that $E\hat{\mathscr{A}}\phi$ is different from zero.

EXAMPLES 1 AND 2 (Continued). Suppose $\{x_t\}$ is a scalar diffusion that satisfies equation (2.3). If the stationary density q is given by the right-hand side of (2.4), then

$$\mu = (1/2)\left[(\sigma^2)' + \sigma^2 q'/q\right]. \tag{5.3}$$

For a fixed q, this equation relates the diffusion coefficient σ^2 with the local drift μ. This equation gives us sets of observationally equivalent pairs (μ, σ^2) from the vantage point of C1. In fact Banon (1978) and Cobb *et al.* (1983) used equation (5.3) as a basis to construct flexible (nonparametric) estimators of μ for prespecified σ^2. As is evident from (5.3) parameterizing (μ, σ^2) is equivalent, modulo some invertibility and regularity conditions, to parameterizing $(q'/q, \sigma^2)$. Sometimes the latter parameterization is simpler and more natural. If we start by describing our candidate models using this parameterization, moment conditions C1 yield no information about the diffusion coefficient. On the other hand, as we will see later in this section, moment conditions C2 provide a considerable amount of information about the diffusion coefficient.

To illustrate these points, as in Cobb *et al.* (1983), consider a family of diffusions defined on the nonnegative reals parameterized by a "truncated" Laurent series:

$$(q'/q)(y, \alpha) = \sum_{j=-k}^{l} \alpha_j y^j,$$

where $\alpha = (\alpha_{-k}, \ldots, \alpha_l)$ is a vector of unknown parameters that must satisfy certain restrictions for q to be nonnegative and integrable. We also assume that $\sigma^2 = \rho y^\gamma$ where $\rho > 0$ and $\gamma > 0$. This parameterization is sufficiently rich to encompass the familiar "square-root" process used in the bond pricing literature as well as other processes that exhibit other volatility elasticities. The implicit parameterization of μ can be deduced from (5.3).[9] Moment conditions C1 will suffice for the identification of α. Since for fixed α, variations in γ leave invariant the stationary distribution q, γ cannot be inferred from moment conditions C1. However, as we will see in our subsequent analysis, moment conditions C2 will allow us to identify γ, but not ρ.

To assess the incremental informational content of the set of moment conditions C2, we focus only on generators that satisfy C1. In light of Proposition 1, all of these candidates have $\mathscr{Q}$ as a stationary distribution. Strong continuity of the semi-group $\{\hat{\mathscr{T}}_t : t \geq 0\}$ in $\mathscr{L}$ implies Assumption A1. Thus we are now free to use $\mathscr{L}^2(\mathscr{Q})$ (instead of the more restrictive domain $\mathscr{L}$) as the common domain of the semi-groups associated with the candidate generators. To avoid introducing new notation, for a candidate generator $\hat{\mathscr{A}}$ satisfying C1, we will still denote by $\hat{\mathscr{A}}$ the generator of the semi-group defined on $\mathscr{L}^2(\mathscr{Q})$ and by $\hat{\mathscr{D}}$ its domain.

Recall that C2 was derived using the fact that $\mathscr{A}$ and $\mathscr{T}$ commute. In fact, if a candidate $\hat{\mathscr{A}}$ satisfies C2, then $\hat{\mathscr{A}}$ must commute with $\mathscr{T}$, i.e., for any ϕ in $\hat{\mathscr{D}}$, $\mathscr{T}\phi$ is in $\hat{\mathscr{D}}$ and $\hat{\mathscr{A}}\mathscr{T}\phi = \mathscr{T}\hat{\mathscr{A}}\phi$.

PROPOSITION 2. *Suppose $\hat{\mathscr{A}}$ satisfies C1. Then $\hat{\mathscr{A}}$ satisfies C2 for $\phi \in \hat{\mathscr{D}}$ and $\phi^* \in \hat{\mathscr{D}}^*$ if, and only if, $\hat{\mathscr{A}}\mathscr{T} = \mathscr{T}\hat{\mathscr{A}}$.*

Proof. By mimicking the reasoning in section 3 one shows that $\hat{\mathscr{A}}\mathscr{T} = \mathscr{T}\hat{\mathscr{A}}$ is sufficient for C2. To prove necessity, note that for any ϕ in $\hat{\mathscr{D}}$ and any ϕ^* in $\hat{\mathscr{D}}^*$, it follows from C2 and the Law of Iterated Expectations that

$$\langle \mathscr{T}\hat{\mathscr{A}}\phi \mid \phi^* \rangle = \langle \phi \mid \mathscr{T}^*\hat{\mathscr{A}}^*\phi^* \rangle. \tag{5.4}$$

Since $\mathscr{T}$ is the adjoint of $\mathscr{T}^*$,

$$\langle \phi \mid \mathscr{T}^*\hat{\mathscr{A}}^*\phi^* \rangle = \langle \mathscr{T}\phi \mid \hat{\mathscr{A}}^*\phi^* \rangle. \tag{5.5}$$

It follows that $\hat{\mathscr{A}}\mathscr{T}\phi = \mathscr{T}\hat{\mathscr{A}}\phi$ for all ϕ in $\hat{\mathscr{D}}$, because the adjoint of $\hat{\mathscr{A}}^*$ is $\hat{\mathscr{A}}$. Q.E.D.

[9] Additional restrictions must be imposed on the parameters to guarantee that there is a solution to the associated stochastic differential equation. As mentioned in footnote 5, there are a variety of alternative sufficient conditions that can be employed to ensure that a solution exists. Alternatively, we can work directly with the implied infinitesimal generators and verify that there exist associated Markov processes for the admissible parameter values.

Since it is typically hard to compute $\mathcal{T}\phi$ for an arbitrary function ϕ in $\mathcal{L}^2(\mathcal{Q})$, it may be difficult to establish directly that $\hat{\mathcal{A}}$ commutes with $\mathcal{T}$. As an alternative, it is often informative to check whether $\hat{\mathcal{A}}$ commutes with $\mathcal{A}$. To motivate this exercise, we investigate moment conditions C2 for arbitrarily small sampling intervals. By Proposition 2, this is equivalent to studying whether $\mathcal{T}_t$ and $\hat{\mathcal{A}}$ commute for arbitrarily small t.

PROPOSITION 3. *Suppose $\hat{\mathcal{A}}$ commutes with $\mathcal{T}_t$ for all sufficiently small t. Then $\hat{\mathcal{A}}\mathcal{A}\phi = \mathcal{A}\hat{\mathcal{A}}\phi$ for all ϕ in $\hat{\mathcal{D}}$ with $\hat{\mathcal{A}}\phi$ in $\mathcal{D}$.*

Proof. Note that

$$\mathcal{A}\hat{\mathcal{A}}\phi = \lim_{t \downarrow 0} [(\mathcal{T}_t - I)/t]\hat{\mathcal{A}}\phi = \lim_{t \downarrow 0} \hat{\mathcal{A}}[(\mathcal{T}_t - I)/t]\phi. \tag{5.6}$$

Since $\hat{\mathcal{A}}$ has a closed graph, the right side of (5.6) must converge to $\hat{\mathcal{A}}\mathcal{A}\phi$. Q.E.D.

In light of this result, we know that any $\hat{\mathcal{A}}$ satisfying the *small interval* counterpart to C2 must commute with $\mathcal{A}$. Thus, from Propositions 2 and 3, if there exists an admissible test function ϕ such that

$$\hat{\mathcal{A}}\mathcal{A}\phi \neq \mathcal{A}\hat{\mathcal{A}}\phi,$$

then there is a sampling interval for which $\hat{\mathcal{A}}$ fails to satisfy some of the moment conditions in the collection C2. While we know there exists such a sampling interval, the conclusion is not necessarily applicable to all sampling intervals and hence may not be applicable to one corresponding to the observed data. As we will now see, this limitation can be overcome when additional restrictions are placed on the candidate and true generators. Conversely, suppose that $\hat{\mathcal{A}}$ commutes with $\mathcal{A}$ in a subspace of $\mathcal{L}^2(\mathcal{Q})$. Bequillard (1989) gave sufficient conditions on that subspace to ensure that $\hat{\mathcal{A}}$ commutes with $\{\mathcal{T}_t : t \geq 0\}$. Hence such a candidate generator can never be distinguished from $\mathcal{A}$ using moment conditions C2.

When both the candidate and true processes are reversible, requiring that $\hat{\mathcal{A}}$ commute with $\mathcal{T}$ is equivalent to requiring that $\hat{\mathcal{A}}$ commute with $\mathcal{A}$. To verify this equivalence, recall that reversible Markov processes have infinitesimal generators that are self adjoint. Such operators have unique spectral representations of the following form:

$$\mathcal{A}\phi = \int_{(-\infty,\,0]} \lambda \, d\mathcal{E}(\lambda)\phi \tag{5.7}$$

where $\mathcal{E}$ is a "resolution of the identity":

DEFINITION. $\mathcal{E}$ is a *resolution of the identity* if

(i) $\mathcal{E}(\Lambda)$ is a self-adjoint projection operator on $\mathcal{L}^2(\mathcal{Q})$ for any Borelian $\Lambda \subset \mathbf{R}$;

(ii) $\mathscr{E}(\Phi) = 0$, $\mathscr{E}(R) = I$;

(iii) for any two Borelians Λ_1 and Λ_2, $\mathscr{E}(\Lambda_1 \cap \Lambda_2) = \mathscr{E}(\Lambda_1)\mathscr{E}(\Lambda_2)$;

(iv) for any two disjoint Borelians Λ_1 and Λ_2, $\mathscr{E}(\Lambda_1 \cup \Lambda_2) = \mathscr{E}(\Lambda_1) + \mathscr{E}(\Lambda_2)$;

(v) for $\phi \in \mathscr{D}$ and $\psi \in \mathscr{L}^2(\mathscr{Q})$, $\langle \mathscr{E}\phi \mid \psi \rangle$ defines a measure on the Borelians.

Spectral representation (5.7) is the operator counterpart to the spectral representation of symmetric matrices. It gives an orthogonal decomposition of the operator $\mathscr{A}$ in the sense that if Λ_1 and Λ_2 are disjoint, $\mathscr{E}(\Lambda_1)$ and $\mathscr{E}(\Lambda_2)$ project onto orthogonal subspaces (see condition (iv) characterizing the resolution of the identity). When the spectral measure of $\mathscr{E}$ has a mass point at a particular point λ, then λ is an eigenvalue of the operator and $\mathscr{E}(\lambda)$ projects onto the linear space of eigenfunctions associated with that eigenvalue. In light of condition (v), the integration in (5.7) can be defined formally in terms of inner products. The integration can be confined to the interval $(-\infty, 0]$ instead of all of $\mathbf{R}$ because $\mathscr{A}$ is negative semi-definite (e.g., Rudin, 1973, see Theorems 13.30 and 13.31, p. 349). Finally, spectral decomposition (5.7) of $\mathscr{A}$ permits us to represent the semi-group $\{\mathscr{T}_t : t \geq 0\}$ via the exponential formula

$$\mathscr{T}_t\phi = \int_{(-\infty\ ,\ 0]} \exp(\lambda t)\, d\mathscr{E}(\lambda)\,\phi \tag{5.8}$$

(e.g., see Rudin, 1973, Theorem 13.37, p. 360).

Consider now a candidate generator $\hat{\mathscr{A}}$ that is reversible and satisfies C1. Then

$$\hat{\mathscr{A}}\phi = \int_{(-\infty\ ,\ 0]} \lambda\, d\hat{\mathscr{E}}(\lambda)\,\phi, \tag{5.9}$$

where $\hat{\mathscr{E}}$ is also resolution of the identity. Suppose that $\hat{\mathscr{A}}$ commutes with $\mathscr{T}$. Then $\mathscr{T}$ must commute with $\hat{\mathscr{E}}(\Lambda)$ for every Borelian Λ (see Rudin, 1973, Theorem 13.33, p. 351). Let $\{\hat{\mathscr{T}}_t\}$ be the semi-group generated by $\hat{\mathscr{A}}$. Since $\hat{\mathscr{T}}_\tau$ is a bounded operator and can be constructed from $\hat{\mathscr{A}}$ via the exponential formula analogous to (5.8), $\hat{\mathscr{T}}_\tau$ must commute with $\mathscr{T}$ for every nonnegative τ. Moreover, $\hat{\mathscr{T}}_\tau$ must commute with $\mathscr{E}(\Lambda)$ for any Borelian Λ and hence with $\mathscr{T}_t$ for any nonnegative t. We have thus established the following proposition.

PROPOSITION 4. *Suppose $\hat{\mathscr{A}}$ satisfies C1 and $\mathscr{A}$ and $\hat{\mathscr{A}}$ are self-adjoint. If $\hat{\mathscr{A}}\mathscr{T} = \mathscr{T}\hat{\mathscr{A}}$, then $\hat{\mathscr{A}}\mathscr{T}_t = \mathscr{T}_t\hat{\mathscr{A}}$ for all $t \geq 0$.*

Propositions 1–4 support the following approach to identification. Suppose one begins with a parameterization of a family of candidate generators. First partition this family into collections of generators with the same stationary distributions. Then partition further the family into groups of generators that commute (on a sufficiently rich collection of test functions). Two elements in the same subset of the original partition cannot be distinguished on the basis of moment conditions in the set C1, and two elements in the same subset of the finest partition cannot be distinguished on the basis of C2.

Moment conditions C1 and C2 do not capture all of the information from discrete-time data pertinent to discriminating among generators. To see this, we consider identification results based instead on knowledge of the discrete-time conditional expectation operator $\mathscr{T}$. In the case of reversible Markov processes, any candidate $\hat{\mathscr{A}}$ that implies $\mathscr{T}$ as a conditional expectation operator, must satisfy the exponential formula:

$$\mathscr{T}\phi = \int_{-\infty}^{0} \exp(\lambda t)\, d\hat{\mathscr{E}}(\lambda)\,\phi. \tag{5.10}$$

Moreover, the spectral decomposition for self-adjoint operators is unique (see Rudin, 1973, Theorem 13.30, p. 348). Since the exponential function is one-to-one, it follows that $\hat{\mathscr{E}}$ and $\mathscr{E}$ and hence $\hat{\mathscr{A}}$ and $\mathscr{A}$ must coincide. Thus we have shown the following proposition.

PROPOSITION 5. *Suppose that the generators $\mathscr{A}$ and $\hat{\mathscr{A}}$ are self-adjoint and imply the same conditional expectation operator $\mathscr{T}$. Then $\mathscr{A} = \hat{\mathscr{A}}$.*

Proposition 5 implies that there is no *aliasing* problem when the Markov processes are known to be reversible. Aliasing problems in Markov processes arise because of the presence of complex eigenvalues of the generators. For reversible processes all of the eigenvalues are real and negative (the corresponding resolutions of the identity are concentrated on the nonpositive real numbers). Similarly, the eigenvalues of $\mathscr{T}$ must be in the interval $(0, 1]$. As we will see in our examples, moment conditions C1 and C2 fail to achieve complete identification of $\mathscr{A}$ for reversible processes.

More generally, even if we do not impose reversibility, $\hat{\mathscr{A}}$ and $\mathscr{T}$ are connected through the following alternative exponential formula:

$$\lim_{n\to\infty} \left(I - \hat{\mathscr{A}}/n\right)^{-n} \phi = \mathscr{T}\phi \qquad \text{for all } \phi \in \mathscr{L}^2(\mathscr{Q}) \tag{5.11}$$

(e.g., see Pazy, 1983, p. 33). This exponential formula is the generalization of a formula used to study *aliasing* and *embedability* for finite state Markov chains (e.g., see Johansen, 1973; Singer and Spillerman, 1976). Since $\hat{\mathscr{A}}$ is a continuous operator in this case, the exponential formulas simplify to $\mathscr{T} = \exp(\hat{\mathscr{A}})$. Relation (5.11) also encompasses the exponential formulas derived by Phillips (1973) and Hansen and Sargent (1983) in their analysis of aliasing in the class of multivariate Gaussian diffusion models.

We now apply and illustrate our results in the context of three examples.

EXAMPLES 1 AND 2 (Continued). As we saw earlier, using moment conditions in the set C1 may still leave a large class of observationally equivalent Markov processes. It is also clear that the two families of moment conditions cannot distinguish between two diffusions with generators that are equal up to a scale factor. We now show that using moment conditions C2 we can in fact identify the diffusion up to scale.

Suppose that both the true and the candidate processes are reversible scalar diffusions that share a stationary distribution $\mathscr{Q}$. Let μ and σ^2 denote the local mean and diffusion coefficients associated with the true process, and let $\hat{\mu}$ and $\hat{\sigma}^2$ denote their counterparts for the candidate process. We maintain the assumptions made in Example 1 of section 2. Write L for the second-order differential operator associated with the pair (μ, σ^2) and $\hat{L}$ for the corresponding operator associated with $(\hat{\mu}, \hat{\sigma}^2)$.

We start by examining the case of diffusions on a compact interval $[\ell, u]$ with two reflective barriers and a strictly positive diffusion coefficient, to which the standard Sturm–Liouville theory of second-order differential equations applies. Consider the following eigenvalue problem associated with $\hat{L}$:

$$\hat{L}\phi = \hat{\lambda}\phi, \qquad \phi'(\ell) = \phi'(u) = 0. \tag{5.12}$$

In light of the boundary conditions imposed in (5.12), we know that if ϕ is a twice continuously differentiable ($\mathscr{C}^2$) solution to this eigenvalue problem, then ϕ is an eigenvector for $\hat{\mathscr{A}}$, i.e., $\hat{\mathscr{A}}\phi = \hat{\lambda}\phi$. From Sturm–Liouville theory there exists an infinite sequence of negative numbers $\hat{\lambda}_0 > \hat{\lambda}_1 > \hat{\lambda}_2 > \cdots$ with $\lim_{n\to\infty} \hat{\lambda}_n = -\infty$ and corresponding unique, up to constant factors, $\mathscr{C}^2$ functions ϕ_n such that the pair $(\hat{\lambda}_n, \phi_n)$ solves the eigenvalue problem (5.12). Furthermore, the sequence of eigenfunctions $\{\phi_n\}$ are mutually orthogonal and form a basis for $\mathscr{L}^2(\mathscr{Q})$. Choose a negative $\hat{\lambda}_n$. Suppose that all of the moment conditions in the class C2 are satisfied by $\hat{\mathscr{A}}$. By Proposition 4, $\mathscr{T}_t$ and $\hat{\mathscr{A}}$ commute, and hence $\mathscr{T}\phi_n$ is also an eigenvector of $\hat{\mathscr{A}}$ associated with $\hat{\lambda}_n$. Consequently, $\mathscr{T}\phi_n$ is proportional to ϕ_n, implying that $\mathscr{T}$ and hence $\mathscr{A}$ must have ϕ_n as an eigenvector. Let λ_n denote the corresponding eigenvalue of $\mathscr{A}$. Since ϕ_n is $\mathscr{C}^2$ and satisfies the appropriate boundary conditions, $\mathscr{A}\phi_n$ coincides with $L\phi_n$ and hence ϕ_n satisfies the counterpart to (5.12)

$$L\phi_n = \lambda_n \phi_n. \tag{5.13}$$

Multiply this equation by q and substitute $(1/2)(\sigma^2 q)'$ for μq, to obtain

$$(1/2)(\sigma^2 q)'\phi_n' + (1/2)(\sigma^2 q)\phi_n'' = \lambda_n \phi_n q,$$

or

$$(1/2)(\sigma^2 q \phi_n')' = \lambda \phi_n q.$$

Therefore the eigenvector ϕ_n satisfies

$$\sigma^2(y)\phi_n'(y)q(y) = 2\lambda_n \int_\ell^y \phi_n(x)q(x)\,dx + C. \tag{5.14}$$

The boundary conditions on ϕ_n and condition C1 assure us that the constant C in (5.14) is in fact zero. Similarly, we conclude that

$$\hat{\sigma}^2(y)\phi_n'(y)q(y) = 2\hat{\lambda} \int_\ell^y \phi_n(x)q(x)\,dx. \tag{5.15}$$

It follows from (5.14) and (5.15) that σ^2 and $\hat{\sigma}^2$ are proportional and hence from formula (5.3) the μ and $\hat{\mu}$ are also proportional with the same proportionality factor given by the ratio of the eigenvalues. In other words, moment condition C2 permits us to identify the infinitesimal generator $\mathscr{A}$ up to scale.

We now show how this identification result can be extended to processes defined on the whole real line, even though in this case the generator may fail to have a nonzero eigenvalue. We will proceed by considering reflexive barrier processes that approximate the original process. We assume that both μ and σ^2 are $\mathscr{C}^2$ functions and with $\sigma^2 > 0$. When $\hat{\mathscr{A}}$ satisfies moment conditions C1, by Proposition 1 the candidate Markov process shares a stationary density q with the true process. Suppose $\hat{\mathscr{A}}$ also satisfies moment conditions C2. For a given $k > 0$ consider the processes created by adding to the original candidate and actual processes reflexive barriers at $-k$ and k. The reflexive barrier processes will share a common stationary distribution $\mathscr{Q}_k$ with a density q_k that, in the interval $[-k, k]$, is proportional to q. We write $\mathscr{A}_k$ and $\hat{\mathscr{A}}_k$ for the infinitesimal generators associated with the reflexive barrier processes in $\mathscr{L}^2(\mathscr{Q}_k)$. As before we can use Sturm–Liouville theory to establish the existence of a negative $\hat{\lambda}$ and a $\mathscr{C}^2$ function ϕ such that

$$\hat{L}\phi = \hat{\lambda}\phi, \qquad \phi'(-k) = \phi'(k) = 0. \tag{5.16}$$

Since μ and σ^2 are $\mathscr{C}^2$ functions, and $\sigma^2 > 0$, the eigenvector $\phi \in \mathscr{C}^4$ on $(-k, k)$ and all derivatives up to fourth order have well-defined limits as $y \to \pm k$. We write $f^{(n)}$ for the nth derivative of f. Construct ψ_+ to be a $\mathscr{C}^4$ function defined on (k, ∞) with $\psi_+(y) = 0$ when $y \geq k + 1$ and $\lim_{y \to k} \psi_+^{(n)}(y) = \lim_{y \to k} \phi^{(n)}(y)$ for $0 \leq n \leq 4$. Similarly construct ψ_- to be a $\mathscr{C}^4$ function defined on $(-\infty, -k)$ with $\psi_-(y) = 0$ when $y \leq -k - 1$ and $\lim_{y \to -k} \psi_-^{(n)}(y) = \lim_{y \to -k} \phi^{(n)}(y)$ for $0 \leq n \leq 4$. Finally let $\psi(y) \equiv \phi(y)$ if $-k \leq y \leq k$, and $\psi(y) \equiv \psi_+(\psi_-)$ if $y > k$ (resp. $y < -k$).

Notice ψ is a $\mathscr{C}^4$ function with support in $[-k-1, k+1]$ and that $\hat{L}\psi$ is $\mathscr{C}^2$ and has a compact support. Hence $\mathscr{A}\psi = \hat{L}\psi$ and $\mathscr{A}\hat{\mathscr{A}}\psi = L\hat{L}\psi$. Also, $L\psi$ is a $\mathscr{C}^2$ function with compact support and hence $\hat{\mathscr{A}}\mathscr{A}\psi = \hat{L}L\psi$. Since by Propositions 3 and 4, $\mathscr{A}$ and $\hat{\mathscr{A}}$ must commute, $\hat{L}L\psi = L\hat{L}\psi$, and hence $\hat{\mathscr{A}}_k \mathscr{A}_k \phi = \mathscr{A}_k \hat{\mathscr{A}}_k \phi$. The result for the compact support case with reflexive barriers implies that the drift and diffusion coefficient are identified up to scale in an interval $(-k, k)$ for arbitrary k.

Finally, we observe that if an eigenvector ϕ of $\mathscr{A}$ can be explicitly calculated, then the corresponding (real) eigenvalue and hence scale constant can also be identified by using the fact that $\mathscr{T}\phi = e^{\lambda}\phi$. Alternatively, suppose it is known that the spectral measure of $\mathscr{A}$ is concentrated at mass points with zero as an isolated mass point with multiplicity one.[10] Further suppose that $\hat{\mathscr{A}}$ is a candidate generator that is known to be proportional to $\mathscr{A}$. The smallest (in absolute value) nonzero eigenvalue $\hat{\lambda}$ of $\hat{\mathscr{A}}$ is given by

$$\hat{\lambda} \equiv \max_{\int \phi \, d\mathscr{Q} = 0, \, \|\phi\| = 1} \langle \phi \mid \hat{\mathscr{A}}\phi \rangle .$$

Under ergodicity restrictions (see section 7), the criterion for the maximization problem and constraints can be approximated by forming time series averages of the appropriate functions of the data. Similarly, the smallest (in absolute value) nonzero of $\mathscr{A}$ is the logarithm of the largest eigenvalue different from one of $\mathscr{T}$ and hence can be represented as

$$\lambda \equiv \log \max_{\int \phi \, d\mathscr{Q} = 0, \, \|\phi\| = 1} \langle \phi \mid \mathscr{T}\phi \rangle .$$

Using the Law of Iterated Expectations, the criterion for this maximization problem can be expressed as $E\phi(x_t)\phi(x_{t+1})$. Again the criterion and the constraints can be approximated by time series averages. The true generator can be identified via

$$\mathscr{A} = \frac{\lambda}{\hat{\lambda}} \hat{\mathscr{A}} .$$

Since the constraint set for the two maximization problems of interest is infinite-dimensional, it is not trivial to go from this identification scheme to a formal justification of an estimation method. While such an exercise is beyond the scope of this chapter, some of the results in section 7 should be helpful.

EXAMPLE 3 (Continued). In section 4 we introduced a class of factor models in which the process $\{x_t\}$ is a time invariant function of a vector of independent scalar diffusions. Suppose that both the candidate process

[10] In section 7 we will present sufficient conditions for zero to be an isolated point with multiplicity one.

and the true process satisfy these factor restrictions. Then it follows from Propositions 1–4 that $\mathscr{A}$ and $\hat{\mathscr{A}}$ can be distinguished on the basis of our moment conditions if they imply different stationary distributions or if they fail to commute. By extending the identification discussion of Examples 1 and 2, we can construct $\hat{\mathscr{A}}$'s that are not distinguishable from $\mathscr{A}$ as follows. Use the same function F mapping the factors into the observable processes as is used for $\mathscr{A}$. Then form an $\hat{\mathscr{A}}$ by multiplying the coefficients of each scalar factor diffusion by a possibly distinct scale factor. While such an $\hat{\mathscr{A}}$ cannot be distinguished using moment conditions C1 and C2, they can be distinguished based on knowledge of the conditional expectation operator $\mathscr{T}$ (see Proposition 5).

In concluding this section, we illustrate an additional identification problem for our moment conditions. Starting from an arbitrary infinitesimal generator $\mathscr{A}$, we can always construct a two-parameter family of candidate generators that always satisfy C1 and C2. Let $\{x_t\}$ be a continuous time stationary Markov process with infinitesimal generator $\mathscr{A}$ on a domain $\mathscr{D}$. For any $\phi \in \mathscr{D}$, construct $\hat{\mathscr{A}}\phi = \eta_1 \mathscr{A}\phi + \eta_2[\int \phi \, d\mathscr{Q} - \phi]$ where η_1 and η_2 are two positive real numbers. Notice that we formed our candidate $\hat{\mathscr{A}}$ by changing the *speed* of the original process by multiplying $\mathscr{A}$ by η_1 and adding to that the generator for a particular Markov jump process (see (4.4)). It is easy to check that $\hat{\mathscr{A}}$ has $\mathscr{Q}$ as its stationary distribution and commutes with $\mathscr{A}$. Therefore $\hat{\mathscr{A}}$ cannot be distinguished from $\mathscr{A}$ using C1 and C2. Moreover, when $\mathscr{A}$ is self-adjoint so is $\hat{\mathscr{A}}$. Consequently it follows from Proposition 5 that in this case the generators can be distinguished based on knowledge of the conditional expectations operator.

6. CORES

So far, we have analyzed observable implications by assuming all of the moment conditions in C1 and C2 could be checked. To perform such a check would require both knowledge of the domain $\hat{\mathscr{D}}$ and the ability to compute $\hat{\mathscr{A}}$. It is often difficult to characterize the domain $\hat{\mathscr{D}}$ and to evaluate a candidate generator applied to an arbitrary element of that domain. For instance, in the scalar diffusion example (Example 1) we only characterized the infinitesimal generator on a subset of the domain. Furthermore, it is desirable to have a *common* set of test functions to use for a parameterized family of generators.

Since the operator $\hat{\mathscr{A}}$ is not necessarily continuous, a dense subset of $\hat{\mathscr{D}}$ does not need to have a dense image under $\hat{\mathscr{A}}$. Consequently, in examining moment conditions C1 and C2, if we replace $\hat{\mathscr{D}}$ and $\hat{\mathscr{D}}^*$ by arbitrary dense subsets, we may weaken their implications. In addition, recall that the moment conditions in C2 require looking at means of random variables of the form $\hat{\mathscr{A}}\phi(x_{t+1})\phi^*(x_t) - \phi(x_{t+1})\hat{\mathscr{A}}^*\phi^*(x_t)$, which are differences of products of random variables with finite second moments. To ensure that standard central limit approximations work, it is convenient to restrict ϕ and ϕ^* so that $\hat{\mathscr{A}}\phi(x_{t+1})\phi^*(x_t) - \phi(x_{t+1})\hat{\mathscr{A}}^*\phi^*(x_t)$ has a finite second moment. For instance, this will be true when, in addition, both ϕ and ϕ^* are bounded.

To deal with these matters, we now describe a strategy for reducing the sets $\hat{\mathscr{D}}$ and $\hat{\mathscr{D}}^*$ that avoids losing information and results in random variables with finite second moments. The approach is based on the concept of a *core* for a generator $\mathscr{A}$. Recall that the graph of $\mathscr{A}$ restricted to a set $\mathscr{N}$ is $\{(\phi, \mathscr{A}\phi) : \phi \in \mathscr{N}\}$.

DEFINITION. A subspace $\mathscr{N}$ of $\mathscr{D}$ is a *core* for $\mathscr{A}$ if the graph of $\mathscr{A}$ restricted to $\mathscr{N}$ is dense in the graph of $\mathscr{A}$.

Clearly, if $\mathscr{N}$ is a core for $\mathscr{A}$, $\mathscr{A}(\mathscr{N})$ is dense in $\mathscr{A}(\mathscr{D})$. As we will argue below, in checking moment conditions C1 and C2 it suffices to look at sets whose linear spans are cores for $\hat{\mathscr{A}}$ and $\hat{\mathscr{A}}^*$.

For a Markov jump model (Example 5), a candidate generator $\hat{\mathscr{A}}$ is a bounded operator. In this case, it suffices to look at a countable collection of bounded functions whose linear span is dense in $\mathscr{L}^2(\hat{\mathscr{Q}})$ for all probability measures $\hat{\mathscr{Q}}$.

For a Markov diffusion model (Examples 1 to 4), $\hat{\mathscr{A}}$ is no longer a bounded operator. In order to apply Proposition 1, it is important to characterize a core for the candidate generator defined on $\mathscr{L}$. Recall that we assumed that $\mathscr{L}$ contained all the continuous functions with a compact support. For concreteness we now assume that $\mathscr{L} = \{\phi : \mathbf{R}^n \to \mathbf{R}$ such that ϕ is continuous, and $\lim_{|x| \to \infty} \phi(x) = 0\}$ with the sup norm. Notice that since convergence in the sup norm implies convergence of the mean, it suffices to verify moment conditions C1 on a core. Ethier and Kurtz (1986, Theorem 2.1, p. 371) showed that under the conditions stated in Example 1, the space of all infinitely differentiable functions with a compact support ($\mathscr{C}_K^\infty$) forms a core for the infinitesimal generator associated with the scalar stochastic differential equation. In this case since $\mathscr{C}_K^2 \subset \mathscr{D}$, it is also a core for $\mathscr{A}$. Since, in analogy to the result we presented in Example 1, for these functions the infinitesimal generator is given by the second-order differential operator L, we can easily perform the calculations needed to apply moment condition C1. The Ethier and Kurtz result also covers certain cases with finite support and inaccessible or reflexive boundary conditions. Extensions to the multi-dimensional case, that typically require stronger smoothness conditions, are given in Ethier and Kurtz (1986, Theorem 2.6, Remark 2.7, p. 374).

To apply Propositions 2–4 we need to consider candidate generators defined on $\mathscr{L}^2(\mathscr{Q})$. If $\mathscr{N}$ is a core for both $\mathscr{A}$ and $\mathscr{A}^*$ and our second moment condition holds for any $\phi \in \mathscr{N}$ and $\phi^* \in \mathscr{N}$, then it must hold for any pair $(\phi, \phi^*) \in \mathscr{D} \times \mathscr{D}^*$. This follows since, if $\phi_n \to \phi$, $\hat{\mathscr{A}}\phi_n \to \hat{\mathscr{A}}\phi$, $\phi_n^* \to \phi^*$, and $\hat{\mathscr{A}}^*\phi_n^* \to \hat{\mathscr{A}}\phi^*$, then since $\mathscr{T}$ is continuous,

$$\langle \mathscr{T}\hat{\mathscr{A}}\phi_n \mid \phi_n^* \rangle - \langle \hat{\mathscr{A}}^*\phi_n^* \mid \mathscr{T}\phi_n \rangle \to \langle \mathscr{T}\hat{\mathscr{A}}\phi \mid \phi^* \rangle - \langle \hat{\mathscr{A}}^*\phi^* \mid \mathscr{T}\phi \rangle.$$

Hence it suffices to characterize cores for $\mathscr{A}$ and $\mathscr{A}^*$. We can readily extend the results available in the literature on cores of infinitesimal generators for semi-groups on subspaces of the space of continuous bounded functions with the sup norm. For instance consider an extension of Ethier and Kurtz's Theorem 2.1 cited above. We use the fact that $\mathscr{N}$ is a core for $\mathscr{A}$ if and only if both $\mathscr{N}$ and the image of $\mathscr{N}$ under $\lambda I - \mathscr{A}$ for some $\lambda > 0$ are dense on the domain of the contraction semi-group.[11] For diffusions with continuous coefficients the infinitesimal generator coincides on $\mathscr{C}_K^2$ with the second-order differential operator. This is true whether the semi-group is defined on $\mathscr{L}$ or on $\mathscr{L}^2(\mathscr{Q})$. Since $\mathscr{C}_K^\infty$ forms a core for the candidate infinitesimal generator $\hat{\mathscr{A}}_{\mathscr{L}}$ of the semi-group defined on $\mathscr{L}$, there must exist a $\lambda > 0$ such that the image of $\mathscr{C}_K^\infty$ under $\lambda I - \hat{L}$ is dense in $\mathscr{L}$. It follows that the image of $\mathscr{C}_K^\infty$ under $\lambda I - \hat{L}$ is dense in $\mathscr{L}^2(\mathscr{Q})$, because $\mathscr{C}_K^\infty$ is dense in $\mathscr{L}^2(\mathscr{Q})$ and sup norm convergence implies $\mathscr{L}^2(\mathscr{Q})$ convergence. Hence $\mathscr{C}_K^\infty$ is a core for $\hat{\mathscr{A}}$, the infinitesimal generator for the semi-group defined in $\mathscr{L}^2(\mathscr{Q})$. Notice that, since sup norm convergence implies $\mathscr{L}^2(\mathscr{Q})$ convergence, we may apply exactly the same reasoning whenever we have a core $\mathscr{N}$ for $\hat{\mathscr{A}}_{\mathscr{L}}$ and we know that $\hat{\mathscr{A}}_{|\mathscr{N}} = (\hat{\mathscr{A}}_{\mathscr{L}})_{|\mathscr{N}}$.[12]

Even when reduced to a core the observable implications of conditions C1 and C2 presuppose the use of a large set of test functions. Of course, for a finite data set, only a small (relative to the sample size) number of test functions will be used. We now obtain a reduction that can be used to support theoretical investigations in which the number of test functions can increase with the sample size such as Bierens (1990), who suggested a way of testing an infinite number of moment conditions using penalty functions, and Newey (1990) who derived results for efficient estimation by expanding the number of moment conditions as an explicit function of the sample size. Their analysis could be potentially adapted to the framework of this chapter once we construct a countable collection of test functions whose *span* is a core. For the Markov jump process this reduction is easy

[11] See Ethier and Kurtz (1986, Proposition 3.1, p. 17).

[12] If the assumptions concerning the bounds on the derivatives of the coefficients made in Example 2.1 are replaced by weaker polynomial growth conditions, it is still possible to show directly that $\mathscr{C}_K^2$ is a core for $\hat{\mathscr{A}}$ if the stationary distribution possesses moments of sufficiently high order.

since it suffices to choose any collection of functions with a dense span in $\mathscr{L}^2(\mathscr{Q})$. For Markov diffusions that have $\mathscr{C}_K^2$ as a core we may proceed as follows. Fix a positive integer N and consider the subspace of $\mathscr{C}_K^2$ of functions with support on $\{y : |y| \leq N\}$. This subspace is separable if we use the norm given by the maximum of the sup norm of a function and of its first two derivatives. Choose a countable dense collection for each N and take the union over positive integers N. Since $\mathscr{C}_K^2$ is a core, it is straightforward to show that the linear span of this union is also a core for $\mathscr{A}$.

7. ERGODICITY AND MARTINGALE APPROXIMATION

To use the moment conditions derived in section 2 in econometric analyses, we must have some way of approximating expectations of functions of the Markov state vector x_t in the case of C1 and of functions of both x_{t+1} and x_t in the case of C2. As usual, we approximate these expectations by calculating the corresponding time-series averages from a discrete-time sample of finite length. To justify these approximations via a Law of Large Numbers, we need some form of ergodicity of the discrete-sampled process. In the first subsection, we investigate properties of the infinitesimal generator that are sufficient for ergodicity of both the continuous-time and the discrete-time processes.

It is also of interest to assess the magnitude of the sampling error induced by making such approximations. This assessment is important for determining the statistical efficiency of the resulting econometric estimators and in making statistical inferences about the plausibility of candidate infinitesimal generators. The vehicle for making this assessment is a central limit theorem. In the second subsection we derive central limit approximations via the usual martingale approach. Again we study this problem from the vantage point of both continuous and discrete records, and we derive sufficient conditions for these martingale approximations to apply that are based directly on properties of infinitesimal generators.

7.1. Law of Large Numbers

Stationary processes in either discrete time or continuous time obey a Law of Large Numbers. However, the limit points of time-series averages may not equal the corresponding expectations under the measure $\mathscr{Q}$. Instead these limit points are expectations conditioned on an appropriately constructed set of *invariant events* for the Markov process. Furthermore, the conditioning set for the continuous-time process may be a proper subset of the conditioning set for a fixed-interval discrete-sampled process. Therefore, the limit points for the discrete record Law of Large Numbers may be different than the limit points for the continuous record Law of Large Numbers.

Invariant events for Markov processes (in continuous or discrete time) turn out to have a very simple structure. They are measurable functions of the initial state vector x_0. Hence, associated with the invariant events for the continuous-time process $\{x_t\}$, there is a σ-algebra $\mathscr{G}$, contained in the Borelians of $\mathbf{R}^n$, such that any invariant event is of the form $A = \{x_0 \in B\}$ for some $B \in \mathscr{G}$. We can construct a conditional probability distribution $\mathscr{Q}_y$ indexed by the initial state $x_0 = y$ such that expectations conditioned on $\mathscr{G}$ (as a function of y) can be evaluated by integrating with respect to $\mathscr{Q}_y$. Imagine initializing the Markov process using $\mathscr{Q}_y$ in place of $\mathscr{Q}$ in our analysis where y is selected to be the observed value of x_0. Then under this new initial distribution the process $\{x_t\}$ remains stationary, but it is now ergodic for ($\mathscr{Q}$) almost all y. Therefore, by an appropriate initialization we can convert any stationary process into one that is also ergodic. Alternatively, it can be verified that the moment conditions C1 and C2 hold conditioned on the invariant events for the continuous-time process $\{x_t\}$. Consequently, imposition of ergodicity for the continuous-time process is made as a matter of convenience.

Since ergodicity of $\{x_t\}$ is connected directly to the initial distribution imposed on x_0, it is of interest to have a criterion for checking whether an appropriate initial distribution has been selected. As we will see, such a criterion can be obtained by examining the zeros of the operator $\mathscr{A}$. Note that constant functions are always zeros of the operator $\mathscr{A}$. We omit all such functions from consideration except for the zero function by focusing attention on the following closed linear subspace:

$$\mathscr{Z}(\mathscr{Q}) \equiv \left\{\phi \in \mathscr{L}^2(\mathscr{Q}) : \int_{\mathbf{R}^n} \phi \, \mathrm{d}\mathscr{Q} = 0\right\}.$$

For any t, $\mathscr{T}_t$ maps $\mathscr{Z}(\mathscr{Q})$ into itself. Hence we may consider $\{\mathscr{T}_t : t \geq 0\}$ as having the domain $\mathscr{Z}(\mathscr{Q})$. Furthermore, since $\mathscr{D}$ is a linear subspace of $\mathscr{L}^2(\mathscr{Q})$, it follows from P6 that $\mathscr{D} \cap \mathscr{Z}(\mathscr{Q})$ is a linear subspace and is dense in $\mathscr{Z}(\mathscr{Q})$.

Our first result relates the ergodicity of the continuous-time process $\{x_t\}$ to the uniqueness of the zeros of $\mathscr{A}$ on $\mathscr{Z}(\mathscr{Q})$ and is due to Bhattacharya (1982, Proposition 2.2).

Proposition 6. *The process $\{x_t\}$ is ergodic if and only if $\mathscr{A}\phi = 0$ for $\phi \in \mathscr{D} \cap \mathscr{Z}(\mathscr{Q})$ implies that $\phi = 0$.*

Next we consider ergodicity of the discrete-sampled process. For notational simplicity, we set the sampling interval to one. The discrete-time counterpart to Proposition 6 can be proven by modifying appropriately the argument of Bhattacharya (1982).

PROPOSITION 7. *The sampled process* $\{x_t : t = 0, 1, \ldots\}$ *is ergodic if and only if* $\mathcal{T}\phi = \phi$ *and* $\phi \in \mathcal{Z}(\mathcal{Q})$ *imply* $\phi = 0$.

Ergodicity of the continuous-time process $\{x_t\}$ does not necessarily imply ergodicity of this process sampled at integer points in time. From equation (5.11) $\mathcal{T}_t$ can be interpreted as the operator exponential of $t\mathcal{A}$ even though, strictly speaking, the exponential of $t\mathcal{A}$ may not be well defined. The Spectral Mapping Theorem for infinitesimal generators (e.g., see Pazy, 1983, Theorem 2.4, p. 46) states that the nonzero eigenvalues of $\mathcal{T}_t$ are exponentials of the eigenvalues of $t\mathcal{A}$. Therefore, the only way in which $\mathcal{T}$ can have a unit eigenvalue on $\mathcal{L}(\mathcal{Q})$ is for $\mathcal{A}$ to have $2\pi ki$ as an eigenvalue on $\mathcal{L}(\mathcal{Q})$, where k is an integer. In other words, there must exist a pair of functions ϕ_r and ϕ_i in $\mathcal{Z}(\mathcal{Q})$, at least one of which is different from zero, such that

$$\mathcal{A}(\phi_r + \mathrm{i}\phi_i) = 2\pi k\mathrm{i}(\phi_r + \mathrm{i}\phi_i).$$

When $\mathcal{A}$ has a purely imaginary eigenvalue there will always be a nondegenerate function ϕ in $\mathcal{L}(\mathcal{Q})$ such that $\phi(x_t)$ is perfectly predictable given x_0. In other words, there exists a nondegenerate function ϕ such that $\phi(x_t) = \mathcal{T}_t\phi(x_0)$ almost surely ($\mathcal{P}_r$) for all $t \geq 0$.[13] To ensure ergodicity for *all* sampling intervals, we must rule out *all* purely imaginary eigenvalues.

Generators of stationary, ergodic (in continuous time) Markov jump processes (Example 5) do not have purely imaginary eigenvalues. Suppose to the contrary that the pair (ϕ_r, ϕ_i) solves the eigenvector problem for eigenvalue $\mathrm{i}\theta$ different from zero. Then

$$\tilde{\mathcal{T}}(\phi_r + \mathrm{i}\phi_i) = [(\mathrm{i}\theta/\eta) + 1](\phi_r + \mathrm{i}\phi_i).$$

This leads to a contradiction because $|(\mathrm{i}\theta/\eta) + 1| > 1$ and $\tilde{\mathcal{T}}$ is a weak contraction.

Recall that generators of a reversible process, including those discussed in Examples 1, 2, and 3, have only real eigenvalues. Many other multivariate stationary, ergodic (in continuous time) Markov diffusion processes (Example 4) do not have purely imaginary eigenvalues. For instance, under the conditions given in section 4 for the existence of a unique stationary distribution, for sufficiently large t, $\phi(x_t) = \mathcal{T}_t\phi(x_0)$ almost surely ($\mathcal{P}r$) implies $\phi(x_0) = 0$ almost surely ($\mathcal{P}r$) (see Has'minskii,

[13] Let the pair of functions (ϕ_r, ϕ_i) in $\mathcal{D} \cap \mathcal{L}(\mathcal{Q})$ satisfy the eigenvalue problem for purely imaginary eigenvalue $\mathrm{i}\theta$. Thus (ϕ_r, ϕ_t) solves the analogous eigenvalue problem with eigenvalue $\exp(\mathrm{i}\theta t)$ for $\mathcal{T}_t$ and consequently, $\|\mathcal{T}_t\phi_r\|^2 + \|\mathcal{T}_t\phi_t\|^2 = \|\phi_r\|^2 + \|\phi_i\|^2$. Since $\|\phi_r\|^2 = \mathrm{E}[|\phi_r(x_t) - \mathcal{T}_t\phi_r(x_0)|^2] + \|\mathcal{T}_t\phi_r\|^2 = 0$, by the contraction property $\|\phi_r\|^2 = \|\mathcal{T}_t\phi_r\|^2 = 0$, and similarly for ϕ_i.

1980, Lemma 6.5, pp. 128, 129). Hence $\mathscr{A}$ cannot have a purely imaginary eigenvalue in this case.

We now consider the moment conditions derived in section 3. Recall that moment conditions C1 imply that $\mathscr{A}(\mathscr{D}) \subseteq \mathscr{Z}(\mathscr{Q})$. Consequently, the approximation results we obtained for functions in $\mathscr{Z}(\mathscr{Q})$ can be applied directly to justify forming finite sample approximations to the moment conditions C1.

Consider next moment conditions in the set C2. Recall from section 3 that these conditions are of the form

$$\mathrm{E}[\nu(x_{t+1}, x_t)] = 0 \tag{7.1}$$

for functions $\nu : \mathrm{R}^{2n} \to \mathbf{R}$. The function ν was not necessarily restricted so that the random variable $\nu(x_{t+1}, x_t)$ has a finite second moment, although this random variable will always have a finite first moment. Ergodicity of the sampled process is sufficient for the sample averages to converge to zero almost surely ($\mathscr{P}r$). Under the additional restriction that the random variable $\nu(x_{t+1}, x_t)$ has a finite second moment, it follows from the Mean Ergodic Theorem that the sample averages will also converge to zero in $\mathscr{L}^2(\mathscr{P}r)$, the space of all random variables with finite second moments on the original probability space.

7.2. Central Limit Theorem

To obtain central limit approximations for discrete time Markov processes, Rosenblatt (1971) suggested restricting the operator $\mathscr{T}$ to be a strong contraction on $\mathscr{Z}(\mathscr{Q})$. Among other things, this limits the temporal dependence sufficiently for the discrete-sampled process to be strongly mixing (see Rosenblatt, 1971, Lemma 3, p. 200). The way in which the central limit approximations are typically obtained is through martingale approximations. Explicit characterization of these martingale approximations is of independent value for investigating the statistical efficiency of classes of generalized method of moments estimators constructed from infinite-dimensional families of moment conditions (e.g., see Hansen, 1985) such as those derived in section 3.

Since our goal in this subsection is to deduce restrictions on $\mathscr{A}$ that are sufficient for martingale approximations to apply, we begin by investigating martingale approximations for the original continuous-time process. This will help to motivate restrictions on $\mathscr{A}$ that are sufficient for $\mathscr{T}$ to be a strong contraction. In studying moment conditions C1 using a continuous record, we use the standard argument of approximating the integral $\int_0^T \mathscr{A}\phi(x_t)\,dt$ by a martingale m_T, and applying a central limit theorem for martingales.[14] For each $T \geq 0$, define

$$m_T = -\phi(x_T) + \phi(x_0) + \int_0^T \mathscr{A}\phi(x_t)\,dt. \tag{7.2}$$

Then $\{m_T : T \geq 0\}$ is a martingale, relative to the filtration generated by the continuous-time process $\{x_t\}$ (e.g., see Ethier and Kurtz, 1986, Proposition 1.7, p. 162). The error in approximating $\int_0^T \mathscr{A}\phi(x_t)\,dt$ by m_T is just $-\phi(x_T) + \phi(x_0)$ which is bounded by $2\,\|\phi\|$. When scaled by $(1/\sqrt{T})$, this error clearly converges in $\mathscr{L}^2(\mathscr{P}r)$ to zero. Consequently, a central limit theorem for $\{(1/\sqrt{T})\int_0^T \mathscr{A}\phi(x_t)\,dt\}$ can be deduced from the central limit theorem for a scaled sequence of martingales $\{(1/\sqrt{T})m_T\}$ (see Billingsley, 1961).

The random variable m_T has mean zero and variance

$$\begin{aligned} \mathrm{E}\left[(m_T)^2\right] &= \sum_{j=1}^{J} \mathrm{E}\left[(m_{T(J-j+1)/J} - m_{T(J-j)/J})^2\right] \\ &= J\mathrm{E}\left[(m_{T/J})^2\right] \\ &= T\mathrm{E}\left[(J/T)\left(-\phi(x_{T/J}) + \phi(x_0) + \int_0^{T/J} \mathscr{A}\phi(x_t)\,dt\right)^2\right] \end{aligned} \tag{7.3}$$

for any positive integer J since the increments of the martingale are stationary and orthogonal. Thus we are led to investigate the limit

$$\lim_{\epsilon \to 0} \mathrm{E}\left[(1/\epsilon)\left(-\phi(x_\epsilon) + \phi(x_0) + \int_0^\epsilon \mathscr{A}\phi(x_t)\,dt\right)^2\right]. \tag{7.4}$$

By the Triangle Inequality

$$\begin{aligned} \left\{\mathrm{E}\left[(1/\epsilon)^{1/2}\int_0^\epsilon \mathscr{A}\phi(x_t)\,dt\right]^2\right\}^{1/2} &\leq \int_0^\epsilon \mathrm{E}\left[(1/\epsilon)\mathscr{A}\phi(x_t)^2\right]^{1/2} dt \\ &= \epsilon^{1/2}\,\|\mathscr{A}\phi\|. \end{aligned} \tag{7.5}$$

Thus the limit in (7.4) can be written as

$$\begin{aligned} \lim_{\epsilon \to 0} (1/\epsilon)\mathrm{E}\left\{[\phi(x_\epsilon) - \phi(x_0)]^2\right\} &= \lim_{\epsilon \to 0} (1/\epsilon)2[\langle \phi \mid \phi \rangle - \langle \mathscr{T}_\epsilon \phi \mid \phi \rangle] \\ &= -2\langle \phi \mid \mathscr{A}\phi \rangle. \end{aligned} \tag{7.6}$$

[14] A central limit theorem for diffusions on the line appears in Mandl (1968). Florens-Zmirou (1984) gave an alternative proof to Mandl's and derived a counterpart for the discretized version of some diffusions on the line. Our exposition of the central limit approximation for the continuous time record follows Bhattacharya (1982).

Taking limits on the right side of (7.3) as J gets large and substituting from (7.6), we find that

$$\mathrm{E}\left[(m_T)^2\right] = -2T\langle \phi \mid \mathscr{A}\phi \rangle. \tag{7.7}$$

Therefore, the asymptotic variance for the central limit approximation is given by $-2\langle \phi \mid \mathscr{A}\phi \rangle$, which is always nonnegative due to the fact that $\mathscr{A}$ is quasi-negative semi-definite (Property P9).

The fact that the right side of (7.2) is a martingale guarantees that the continuous-time central limit theorem can always be applied to functions $\psi = \mathscr{A}\phi$. Note, however, that the limiting distribution is nondegenerate only when $\langle \phi \mid \mathscr{A}\phi \rangle < 0$.

We now investigate the discrete-time counterpart to this martingale approximation under the restriction that $\mathscr{T}$ is a *strong contraction* on $\mathscr{Z}(\mathscr{Q})$, i.e., there exists a constant $C < 1$ such that $\|\mathscr{T}\phi\| \le C\,\|\phi\|$ for each $\phi \in \mathscr{Z}(\mathscr{Q})$. Later we will discuss conditions on the generator $\mathscr{A}$ that are sufficient for this property to hold. Using the strong contraction property, we will show that a martingale M_N approximates $\sum_{t=1}^{N} \psi(x_t)$ for ψ in $\mathscr{Z}(\mathscr{Q})$. Of course, the ψ's we are interested in are the ones constructed by applying $\mathscr{A}$ to an element of its domain $\mathscr{D}$.

The strong contraction property of $\mathscr{T}$ guarantees that $(I - \mathscr{T})$ has a bounded inverse on $\mathscr{Z}(\mathscr{Q})$. Note that since

$$\mathrm{E}[\psi(x_{t+1}) - \mathscr{T}\psi(x_t) \mid x_t] = 0,$$

the discrete-time process $\{M_N : N = 1, 2, \ldots\}$, defined by

$$M_N \equiv \sum_{t=1}^{N} \left[(I - \mathscr{T})^{-1}\psi(x_t) - \mathscr{T}(I - \mathscr{T})^{-1}\psi(x_{t-1})\right], \tag{7.8}$$

is a martingale adapted to the filtration generated by the discrete-sampled Markov process. Equivalently, we may write

$$M_N = \mathscr{T}(I - \mathscr{T})^{-1}[\psi(x_N) - \psi(x_0)] + \sum_{t=1}^{N} \psi(x_t),$$

which is the discrete-time counterpart to (7.2) and agrees with a martingale approximation suggested by Gordin (1969). Note that the $\mathscr{L}^2(\mathscr{P}r)$ norm of the error in approximating the partial sum $\sum_{t=1}^{N} \psi(x_t)$ by the martingale M_N has a bound independent of N. This bound is uniform for ψ in the unit ball of $\mathscr{Z}(\mathscr{Q})$ since

$$\left(\mathrm{E}\left\{|\mathscr{T}(I - \mathscr{T})^{-1}[\psi(x_N) - \psi(x_0)]|^2\right\}\right)^{1/2} \le 2\,\|\mathscr{T}(I - \mathscr{T})^{-1}(\psi)\|$$

and $\mathscr{T}$ and $(I - \mathscr{T})^{-1}$ are bounded operators on $\mathscr{Z}(\mathscr{Q})$. Scaling by $(1/\sqrt{N})$ makes the approximation error arbitrarily small as N goes to infinity, implying that central limit approximations for $\{(1/\sqrt{N})\Sigma_{t=1}^{N}\psi(x_t)\}$ can be deduced from central limit approximations for a scaled sequence of martingales $\{(1/\sqrt{N})M_N\}$ (see Billingsley, 1961). Finally, it follows from (7.8) that

$$\begin{aligned}(1/N)\mathrm{E}(M_N^2) &= \langle (I-\mathcal{T})^{-1}\psi \mid (I-\mathcal{T})^{-1}\psi\rangle \\ &\quad - \langle \mathcal{T}(I-\mathcal{T})^{-1}\psi \mid \mathcal{T}(I-\mathcal{T})^{-1}\psi\rangle \\ &= \langle \psi \mid \psi\rangle + 2\langle \psi \mid \mathcal{T}(I-\mathcal{T})^{-1}\psi\rangle, \end{aligned} \tag{7.9}$$

which gives the asymptotic variance for the discrete-time central time theorem. It can be shown that for $\psi = \mathcal{A}\phi$, the expression on the right side of (7.9) is greater than or equal to the corresponding expression $-2\langle \phi \mid \mathcal{A}\phi\rangle$ for the continuous-time martingale approximation. This reflects the loss of information due to sampling in discrete time.

The discrete-time martingale approximation given by equation (7.8) can also be applied to moment conditions in the class C2. As we argued previously, these moment conditions can be represented as in (7.1). Suppose that the random variable $\nu(x_{t+1}, x_t)$ has a finite second moment. Then there exists a ψ in $\mathcal{Z}(\mathcal{Q})$ such that

$$\mathrm{E}[\nu(x_{t+1}, x_t) \mid x_t] = \psi(x_t).$$

Then a martingale approximator for $\Sigma_{t=1}^{N}\nu(x_{t+1}, x_t)$ is given by the sum of the martingale $\Sigma_{t=1}^{N}[\nu(x_{t+1}, x_t) - \psi(x_t)]$ and the martingale approximator for $\Sigma_{t=1}^{N}\psi(x_t)$.

We now consider restrictions on $\mathcal{A}$ that are sufficient for $\mathcal{T}$ to be a strong contraction. We write Var (ϕ) for the variance of the random variable $\phi(x_t)$.

Condition G. There exists a subspace $\mathcal{N}$, a core for $\mathcal{A}$, and a $\delta > 0$ such that $-\langle \mathcal{A}\phi \mid \phi\rangle \geq \delta\ \mathrm{Var}(\phi)$ for all $\phi \in \mathcal{N}$.

Notice that since $\mathcal{N}$ is a core, Condition G can be extended to any $\phi \in \mathcal{D}$. Also, since $\phi - \int \phi\, d\mathcal{Q} \in \mathcal{Z}(\mathcal{Q})$, Condition G is equivalent to requiring

$$-\langle \mathcal{A}\phi \mid \phi\rangle \geq \delta\langle \phi \mid \phi\rangle \qquad \text{for all } \phi \text{ with mean zero.}$$

PROPOSITION 8. *The operator $\mathcal{T}$ is a strong contraction if and only if the generator $\mathcal{A}$ satisfies Condition G.*

Proof. As shown by Banon (1977, Lemma 3.11, p. 79), if $\mathcal{T}_t$ is a strong contraction for any fixed $t > 0$, then the semi-group $\{\mathcal{T}_t\}$ must satisfy the *exponential* inequality for some strictly positive δ:

$$\|\mathcal{T}_t(\phi)\| \leq \exp(-\delta t)\|\phi\| \qquad \text{for all } \phi \in \mathcal{Z}(\mathcal{Q}).$$

Hence $\mathscr{S}_t \equiv e^{\delta t}\mathscr{T}_t$ defines a contraction semi-group in $\mathscr{Z}(\mathscr{Q})$ with a generator $\mathscr{A} + \delta I$. By P9, Condition G holds. Conversely if Condition G holds, $\mathscr{A} + \delta I$ is a quasi-negative semi-definite operator with domain $\mathscr{D} \cap \mathscr{Z}(\mathscr{Q})$, and such that for any $\lambda > 0$, $\lambda I - (\mathscr{A} + \delta I)$ is onto. Hence, by the Lumer–Phillips Theorem, $\mathscr{A} + \delta I$ is the generator of a contraction semi-group, $\{\mathscr{S}_t\}$, in $\mathscr{Z}(\mathscr{Q})$ and since $\mathscr{T}_t = e^{-\delta t}\mathscr{S}_t$, it satisfies the exponential inequality. Q.E.D.

For reversible generators, Condition G is equivalent to zero being an isolated point in the support of $\mathscr{E}$, the resolution of the identity used in section 5. In particular, for diffusions with reflecting boundaries on a compact interval with strictly positive diffusion coefficient, Condition G holds. Condition G requires that the variances of the continuous-time martingale approximators (and hence the discrete-time approximators) be bounded away from zero for test functions ϕ with unit variances. In particular, when Condition G is satisfied, the central limit approximation will be nondegenerate whenever ϕ in $\mathscr{D}$ is not constant. Also, Condition G ensures that $\|\mathscr{A}(\phi)\|$ is bounded away from zero on the unit sphere and hence $\mathscr{A}^{-1}$ is a bounded operator on $\mathscr{A}(\mathscr{D})$.

We now study restriction on $\mathscr{A}$ that imply Condition G for the examples. In the case of the Markov jump process (Example 5), a sufficient condition is that η be bounded away from zero and the conditional expectation operator $\tilde{\mathscr{T}}$ on the associated chain be a strong contraction on $\mathscr{Z}(\mathscr{Q})$. To see this, first note that for $K_1 \equiv \int \phi \, d\tilde{\mathscr{Q}}$ and $K_2 \equiv \int (1/\eta)\, d\tilde{\mathscr{Q}}$,

$$\begin{aligned}\langle \phi \mid \mathscr{A}\phi \rangle &= \int \phi\big[\tilde{\mathscr{T}}(\phi) - \phi\big]\eta \, d\mathscr{Q} \\ &= (1/K_2)\int \phi\big[\tilde{\mathscr{T}}(\phi) - \phi\big]\, d\tilde{\mathscr{Q}} \\ &= (1/K_2)\int (\phi - K_1)\big[\tilde{\mathscr{T}}(\phi - K_1) - (\phi - K_1)\big]\, d\tilde{\mathscr{Q}} \\ &\le (1/K_2)(\gamma - 1)\int (\phi - K_1)^2 \, d\tilde{\mathscr{Q}} \qquad (7.10)\end{aligned}$$

for some $0 < \gamma < 1$ since $\tilde{\mathscr{T}}$ is a strong contraction on $\mathscr{Z}(\tilde{\mathscr{Q}})$. Next observe that for $\eta_\ell \equiv \inf \eta > 0$:

$$\begin{aligned}(1/K_2)(\gamma - 1)\int (\phi - K_1)^2 \, d\tilde{\mathscr{Q}} &\le (\gamma - 1)\int (\phi - K_1)^2 \eta \, d\mathscr{Q} \\ &\le (\gamma - 1)\eta_\ell \langle \phi \mid \phi \rangle, \qquad (7.11)\end{aligned}$$

as long as $\phi \in \mathscr{Z}(\mathscr{Q})$. The operator $\mathscr{A}$ in conjunction with any positive δ less than $(1 - \gamma)\eta_\ell$ will satisfy Condition G.

We now turn to Markov diffusion processes (Examples 1 to 4). Bouc and Pardoux (1984) provided sufficient conditions for G that include uniformly bounding above and below the diffusion matrix Σ as well as a pointing inward condition for the drift μ. To accommodate certain examples in finance where the diffusion coefficient may vanish at the boundary, it is necessary to relax the assumptions on the diffusion coefficient. For this reason we derive here an alternative set of sufficient conditions on the line.

Consider a scalar diffusion on an interval $[\ell, \mu]$, allowing for $\ell = -\infty$ and $u = +\infty$. We restrict σ^2 to be positive in the open interval (ℓ, u) and require the speed density to be integrable. If a finite boundary is attainable, we assume that it is reflexive. We start by presenting an alternative characterization of Condition G for diffusions. Let $\mathscr{M}$ be the space of twice continuously differentiable functions $\phi \in \mathscr{L}^2(\mathscr{Q})$ such that $\sigma\phi'$ and $L\phi$ are also in $\mathscr{L}^2(\mathscr{Q})$. In the case of a reflexive boundary we also add the restriction that ϕ' vanishes at the boundary.

For $\phi \in \mathscr{M}$ it follows from the characterization of the infinitesimal generator established in Example 1 and equation (2.4), that

$$-2\langle \phi - \int \phi \, d\mathscr{Q} \,|\, \mathscr{A}(\phi)\rangle = \int (\phi')^2 \sigma^2 \, d\mathscr{Q}.$$

Therefore, Condition G requires for some $\varepsilon > 0$:

$$\int (\phi')^2 \sigma^2 \, d\mathscr{Q} \geq \varepsilon \, Var(\phi) \qquad \text{for all } \phi \in \mathscr{M}. \tag{7.12}$$

When $\mathscr{M}$ is a core condition, (7.12) is equivalent to Condition G. Recall that in section 6 we discussed sufficient conditions for $\mathscr{C}_K^2$ and hence $\mathscr{M}$ to be a core.[15]

We now derive sufficient conditions for inequality (7.12). Let $z \in (\ell, u)$ and notice that, for any ϕ in $\mathscr{M}$,

$$\int [\phi - \phi(z)]^2 \, d\mathscr{Q} \geq \mathrm{Var}(\phi).$$

It follows from an inequality in Muckenhoupt (1972, Theorem 2) or Talenti (1969, p. 174) that there is a finite K_1 such that

$$\int_z^u [\phi - \phi(z)]^2 \, d\mathscr{Q} \leq K_1 \int_z^u |\phi'|^2 \, \sigma^2 \, d\mathscr{Q}$$

if, and only if,

$$\sup_{z<r<u} \mathscr{Q}\{[r, u)\} \int_z^r (\sigma^2 q)^{-1} \, dy < \infty. \tag{7.13}$$

[15] Inequality (7.12) has the obvious multivariate extension.

Hence (7.12) will hold provided (7.13) and the analogous condition

$$\sup_{\ell<r<z} \mathscr{Q}\{(\ell, r]\} \int_r^z (\sigma^2 q)^{-1} dy < \infty \tag{7.14}$$

are satisfied.

Applying L'Hospital's Rule, if

$$\lim_{r\to u} \mathscr{Q}\{[r, u)\} / \sigma(r) q(r) < \infty, \tag{7.15}$$

then (7.13) holds. In particular inequality (7.15) holds if

$$\liminf_{r\to u} \sigma(r) q(r) > 0.$$

If $\lim_{r\to u} \sigma(r)q(r) = 0$, we may apply L'Hospital's Rule again to obtain as a sufficient condition

$$\lim_{r\to u} \frac{1}{\sigma'(r) + \sigma(r)q'(r)/q(r)} \quad \text{exists and is finite.} \tag{7.16}$$

Using equation (5.3) that relates the logarithmic derivative of the stationary density to the drift and diffusion coefficients, we may rewrite (7.16) as

$$\lim_{r\to u} \frac{\sigma(r)}{2\mu(r) - \sigma(r)\sigma'(r)} \quad \text{exists and is finite.} \tag{7.17}$$

The conditions at the lower boundary are exactly analogous. We summarize this discussion in a proposition.

PROPOSITION 9. *Suppose* $\{x_t\}$ *solves* $dx_t = \mu(x_t)\,dt + \sigma(x_t)\,dW_t$ *in an interval* (ℓ, u) *with possibly* $\ell = -\infty$ *and/or* $u = +\infty$; σ *is* $\mathscr{C}^1$ *and positive on* (ℓ, u); *and the speed density* $1/s\sigma^2$ *is integrable. Then the following conditions are sufficient for* (7.12) *to hold.*

(*a*) *The right boundary satisfies either* $\liminf_{r\to u} \sigma(r)q(r) > 0$; *or* $\liminf_{r\to u} \sigma(r)q(r) = 0$ *and* $\lim_{r\to u} \sigma(r)/[2\mu(r) - \sigma(r)\sigma'(r)]$ *exists and is finite.*

(*b*) *The left boundary satisfies either* $\liminf_{r\to \ell} \sigma(r)q(r) > 0$; *or* $\liminf_{r\to \ell} \sigma(r)q(r) = 0$ *and* $\lim_{r\to \ell} \sigma(r)/[2\mu(r) - \sigma(r)\sigma'(r)]$ *exists and is finite.*

Notice that for models that are parameterized in terms q and σ^2, condition (7.16) is easier to verify than the equivalent condition (7.17) mentioned in the proposition. Also one may, in some cases, verify directly (7.15).

The square root process $dx_t = \kappa(x_t - \bar{x})\,dt + \sqrt{x_t}\,dW_t$, with $\kappa > 0$ and $\bar{x} > 0$, is an example of a process that satisfies the sufficient conditions of Proposition 9 even though σ^2 is not bounded away from zero. A process $\{x_t\}$ that solves a stochastic differential equation in $\mathbf{R}_+$, with

$\sigma(y) \equiv 1$ and $\mu(y) = -(\sqrt{y})^{-1}$ for large y, is an example where the sufficient conditions of Proposition 9 do not hold. However in this case every nonnegative real number is in the spectrum of $\mathscr{A}$ and hence C3 fails, even though $\{x_t\}$ is mean recurrent (Bouc and Pardoux, 1984, p. 378).

The conclusion of Proposition 9 also holds for the multivariate factor models described in Example 3 when the individual factor processes satisfy the specified conditions.

8. CONCLUSIONS

The analysis in this chapter is intended as support of empirical work aimed at assessing the empirical plausibility of particular continuous-time Markov models that arise in a variety of areas of economics. In many instances, such models are attractive because of conceptual and computational simplifications obtained by taking continuous time limits. The approach advanced in this chapter can, by design, be used to study these models empirically even when it is not possible for an econometrician to approximate a continuous data record. For this reason, we focused our analysis on fixed-interval sampling although our moment conditions are also applicable more generally. For instance, we could accommodate systematic patterns to the sampling, or the sampling procedure itself could be modelled as an exogenous stationary process. Both moment conditions are still satisfied, with moment condition C2 applied to adjacent observations. The observable implications we obtained extend in the obvious way. This means that our moment conditions can easily handle missing observations that occur in financial data sets due to weekends and holidays. On the other hand, such sampling schemes may alter the central limit approximations reported in section 7.

One of the many questions left unanswered here is that of the selection of test functions in practice. For finite-dimensional parameter models one could compare the asymptotic efficiency of estimators constructed with alternative configurations of test functions along the lines of Hansen (1985). Finite sample comparisons are likely to require Monte Carlo investigations.

ACKNOWLEDGMENTS

Conversations with Buz Brock, Henri Berestycki, Darrell Duffie, Ivar Ekeland, Hedi Kallal, Carlos Kening, Jean Michel Lasry, Pierre Louis Lions, Andy Lo, Erzo Luttmer, Ellen McGratten, Jesus Santos, Nizard Touzi, and Arnold Zellner are gratefully acknowledged. We received helpful comments from an editor and three anonymous referees on an earlier version of this chapter. Thanks to Andy Lo and Steven Spielberg for suggesting the title. Portions of this research were funded by grants from the National Science Foundation.

REFERENCES

Anderson, B. D. O. (1982). Reverse-time diffusion equation-models. *Stochastic Processes and Their Applications* **12**, 313–326.

Banon, G. (1977). "Estimation non Parametrique de densité de Probabilité pour les processus de Markov. Thesis, Université Paul Sabatier de Toulouse, France.

Banon, G. (1978). Nonparametric identification for diffusion processes. *SIAM Journal of Control and Optimization* **16**, 380–395.

Bequillard, A. L. (1989). A sufficient condition for two Markov semigroups to commute. *Annals of Probability* **17**, 1478–1482.

Bhattacharya, R. N. (1982). On the functional central limit theorem and the law of the iterated logarithm for Markov processes. *Zeitschrift fur Wahrscheinlichkeits-theorie und Verwandte Gebiete* **60**, 185–201.

Bierens, H. J. (1990). A consistent conditional moment test of functional form. *Econometrica* **58**, 1443–1458.

Billingsley, P. (1961). The Lindeberg-Levy theorem for martingales. *American Mathematical Society* **12**, 788–792.

Bouc, R., and Pardoux, E. (1984). Asymptotic analysis of P.D.E.s with wide-band noise disturbances and expansion of the moments. *Stochastic Analysis and Applications* **2**, 369–422.

Brezis, H. (1983). "Analyse fonctionnelle, Théorie et Applications." Masson, Paris.

Chamberlain, G. (1992). Efficiency bounds for semiparametric regression. *Econometrica* **60**, 567–596.

Chung, K. L. (1974). A Course in Probability Theory, 2nd ed. Academic Press, New York.

Cobb, L., Koopstein, P., and Chen, N. Y. (1983). Estimation and moment recursion relations for multimodal distributions of the exponential family. *Journal of the American Statistical Association* **78**, 124–130.

Cox, J., Ingersoll, J., and Ross, S. (1985). A theory of the term structure of interest rates. *Econometrica* **53**, 385–408.

Duffie, D., and Singleton, K. J. (1993). Simulated moments estimation of Markov models of asset prices. *Econometrica* **61**, 929–952.

Dunford, N., and Schwartz, J. T. (1988). "Linear Operators Part One: General Theory." Wiley, New York.

Dynkin, E. B. (1965). "Markov Processes." Springer-Verlag, Berlin.

Ethier, S. N., and Kurtz, T. G. (1986). "Markov Processes: Characterization and Convergence." Wiley, New York.

Feller, W. (1951). Two singular diffusion problems. *Annals of Mathematics* **54**, 173–182.

Florens-Zmirou, D. (1984). Théoreme de limite central pour une diffusion et pour sa discrétisée. *C.R. Sciences Acad. Sci. Ser. I* **299**(19), 995–998.

Frachot, A., Janci, D., and Lacoste, V. (1992). Factor analysis of the term structure: A probabilistic approach. NER No. 21. Banque de France, Paris, France.

Froot, K. A. and Obstfeld, M. (1991). Exchange-rate dynamics under stochastic regime shifts: A unified approach. *Journal of International Economics* **31**, 203–229.

Gallant, A. R., and Nychka, D. W. (1987). Semi-nonparametric maximum likelihood estimation. *Econometrica* **55**, 363–390.

Gordin, M. I. (1969). The central limit theorem for stationary processes, *Soviet Mathematics Doklady* **10**, 1174–1176.

Gourieroux, C., Monfort, A., and Renault, E. (1992). Indirect Inference, No. 9215, CREST-Departement de la Recherche INSEE, D.P., Paris, France.

Halmos, P. R. (1974). "Measure Theory." Springer-Verlag, New York.

Hansen, L. P. (1982). Large sample properties of generalized method of moments estimators. *Econometrica* **50**, 1029–1054.

Hansen, L. P. (1985). A method for calculating bounds on the asymptotic covariance matrices of generalized method of moments estimators. *Journal of Econometrics* **30**, 203–238.

Hansen, L. P., and Sargent, T. J. (1983). The dimensionality of the aliasing problem in models with rational spectral densities. *Econometrica* **51**, 377–387.

Harvey, A. C., and Stock, J. H. (1985). The estimation of higher order continuous time autoregressive models. *Journal of Econometric Theory* **1**, 97–112.

Has'minskii, R. Z. (1980). "Stochastic Stability of Differential Equations." Sijthoff & Noordhoff, The Netherlands.

Hausmann, U., and Pardoux, E. (1985). Time reversal of diffusion processes. *In* "Stochastic Differential Systems: Filtering and Control" (M. Metivier and E. Pardoux, eds.), pp. 176–183. Springer-Verlag, Berlin.

He, H. (1990). "Moment Approximation and Estimation of Diffusion Models of Asset Prices." Unpublished, Univ. of California, Berkeley.

Heath, D., Jarrow, R., and Morton, A. (1992). Bond pricing and the term structure of interest rates: A new methodology for contingent claims valuation. *Econometrica* **60**, 77–106.

Johansen, S. (1973). A central limit theorem for finite semigroups and its application to the imbedding problem for finite state Markov chains. *Zeitschrift fur Wahrscheinlichkeitstheorie und Verwandte Gebiete* **26**, 171–190.

Karatzas, I., and Shreve, S. (1988). "Brownian Motion and Stochastic Calculus." Springer-Verlag, New York.

Karlin, S., and Taylor, H. M. (1981). "A Second Course in Stochastic Processes." Academic Press, New York.

Kent, J. (1978). Time-reversible diffusions. *Advance Applications Probability* **10**, 819–835.

Krugman, P. R. (1991). Target zones and exchange rate dynamics. *Quarterly Journal of Economics* **106**, 669–682.

Levy, J. (1993). Two empirical tests of exchange rate target zones. Ph.D. Thesis, University of Chicago, Chicago.

Lewbel, A. (1995). Consistent nonparametric hypothesis testing with an application to Slutsky symmetry. *Journal of Econometrics*. **67**, 379–401.

Lo, A. (1988). Maximum likelihood estimation of generalized Ito processes with discretely sampled data. *Journal of Econometric Theory* **4**, 231–247.

Longstaff, F. A., and Schwartz, E. S. (1992). Interest rate volatility and the term structure: A two-factor general equilibrium model. *Journal of Finance* **47**, 1259–1281.

Mandl, P. (1968). "Analytical Treatment of One-dimensional Markov Processes." Academia, Publishing House of the Czechoslovak Academy of Sciences, Prague.

Millet, A., Nualart, D., and Sanz, M. (1989). Integration by parts and time reversal for diffusion processes. *Annals of Probability* **17**, 208–238.

Muckenhoupt, B. (1972). Hardy's inequality with weights. *Studia Mathematica* **44**, 31–38.

Nelson, E. (1958). The adjoint Markoff process. *Duke Mathematical Journal* **25**, 671–690.

Newey, W. K. (1990). Efficient estimation of semiparametric models via moment restrictions. *Bellcore Economics Discussion Paper* No. 65.

Newey, W. K. (1993). "The Asymptotic Variance of Semiparametric Estimators." Massachusetts Institute of Technology, Cambridge, MA.

Pazy, A. (1983). "Semigroups of Linear Operators and Applications to Partial Differential Equations." Springer-Verlag, New York.

Phillips, A. W. (1959). The estimation of parameters in systems of stochastic differential equations." *Biometrika* **59**, 67–76.

Phillips, P. C. B. (1973). The problem of identification in finite parameter continuous time models. *Journal of Econometrics* **1**, 351–362.

Powell, J. L., Stock, J. H., and Stoker, T. M. (1989). Semiparametric estimation of index coefficients. *Econometrica* **57**, 1403–1430.

Robinson, P. M. (1976). The estimation of linear differential equations with constant coefficients. *Econometrica* **44**, 751–764.

Robinson, P. M. (1989). Hypothesis testing in semiparametric and nonparametric models for econometric time series. *Review of Economic Studies* **56**, 511–534.

Rosenblatt, M. (1969). Density estimates and Markov sequences. *In* "Nonparametric Techniques in Statistical Inference" (M. L. Puri, ed.), pp. 199–210. Cambridge University Press, Cambridge, UK.

Rosenblatt, M. (1971). "Markov Processes: Structure and Asymptotic Behavior." Springer-Verlag, Berlin.

Roussas, G. G. (1969). Nonparametric estimation in Markov processes." *Annals of the Institute of Statistical Mathematics* **21**, 73–87.

Rudin, W. (1973). "Functional Analysis." McGraw-Hill, New York.

Scheinkman, J. A. (1990). Nonlinearities in economic dynamics. *Economic Journal* **100**, 33–48.

Singer, B., and Spillerman, S. (1976). The representation of social processes by Markov models. *American Journal of Sociology* **82**, 1–54.

Talenti, G. (1969). Osservazioni sopra una classe di disuguaglianze. *Rendiconti del Seminario di Matematica e Fisica di Milano* **39**, 171–185.

Wong, E. (1964). The construction of a class of stationary Markoff processes. *In* "Sixteenth Symposium in Applied Mathematics—Stochastic Processes in Mathematical Physics and Engineering" (R. Bellman, ed.), pp. 264–276. American Mathematical Society, Providence, RI.

14

NONPARAMETRIC PRICING OF INTEREST RATE DERIVATIVE SECURITIES*

YACINE AÏT-SAHALIA

Graduate School of Business
University of Chicago
Chicago, Illinois 60637

1. INTRODUCTION

Derivative securities are contingent claims whose payoffs depend upon another asset's payoff, or nontraded factor, such as an interest rate. Stock options, futures, swaps, caps, floors, bonds, callable bonds, convertibles, and bond options all fall into this category. Derivative securities are widely traded both over the counter and on exchanges, and more often than not the volume traded is much larger for the derivative securities than for the underlying assets that sustain them.

The theory of derivative security pricing relies essentially on continuous-time arbitrage arguments since the pioneering Black and Scholes (1973) paper. As further demonstrated by the work of Merton (1973, 1990) pricing derivatives in the theoretical finance literature is generally much more tractable—as well as elegant—in a continuous-time framework than through binomial or other discrete approximations. The empirical option pricing literature however has not followed suit. It is typical there to abandon the continuous-time model altogether when

* Reprinted from Aït-Sahalia, Y. A. (1996). "Nonparametric Pricing of Interest Rate Derivative Securities." *Econometrica* **64**(3), 527–560.

estimating derivative pricing models. This chapter develops tools to estimate the actual model used in the theoretical construction (which is in continuous time) by using only the data available (which are in discrete time). We will estimate the short-term interest rate process and subsequently price bonds and bond options.

The underlying process of interest $\{r_t, t \geq 0\}$ is a diffusion represented by the Itô stochastic differential equation:

$$dr_t = \mu(r_t)\,dt + \sigma(r_t)\,dW_t \tag{1.1}$$

where $\{W_t, t \geq 0\}$ is a standard Brownian motion. The functions $\mu(\cdot)$ and $\sigma^2(\cdot)$ are respectively the drift (or instantaneous mean) and the diffusion (or instantaneous variance) functions of the process. It has long been recognized in the finance literature that one of the most important features of (1.1) for derivative security pricing is the specification of the function $\sigma^2(\cdot)$.

As a consequence every model has tried to specify $\sigma^2(\cdot)$ correctly. To price interest rate derivatives, Vasicek (1977) specifies that the instantaneous volatility of the spot rate process is constant. The Cox–Ingersoll–Ross (1985b) (CIR thereafter) model of the term structure assumes that the instantaneous variance is a linear function of the level of the spot rate r, so the standard deviation is a square root. There are many other models in the literature for the instantaneous variance of the short-term interest rate, all being in general mutually exclusive (see Table 1). In the absence of any theoretical rationale for adopting one particular specification of the diffusion over another, the question must ultimately be decided by taking the models to the data.

The statistical literature contains methods appropriate for a continuous-time record of observations (e.g., Basawa and Prakasa-Rao, 1980; Florens-Zmirou, 1993). Nelson (1990) studies the limiting behavior of discrete approximations as the sampling interval goes to zero. Banon (1978) examined the estimation of the drift when the diffusion is either a constant or a known function. With discrete data, an approach based on maximum-likelihood estimation of the parameters is due to Lo (1988). This method however requires that a partial differential equation be solved numerically for each maximum-likelihood iteration, except in the few cases where it is known explicitly; see Pearson and Sun (1994) for an application to the CIR model. Duffie and Singleton (1993) propose a method of moments where sample paths are simulated for given parameter values and the moments computed (see also Gouriéroux *et al.*, 1993). Parameter estimates make the simulated moments close to the sample moments. This requires that new sample paths be simulated every time the parameters are adjusted. For a particular parametrization of $\mu(\cdot)$ and $\sigma^2(\cdot)$, this method could be applied with sample paths simulated from (1.1). Hansen

TABLE 1 Alternative Specifications of the Spot Interest Rate Process

$$dr_t = \mu(r_t)\,dt + \sigma(r_t)\,dW_t$$

Drift function $\mu(r)$	Diffusion function $\sigma(r)$	Stationary	Reference
$\beta(\alpha - r)$	σ	Yes	Vasicek (1977)
$\beta(\alpha - r)$	$\sigma r^{1/2}$	Yes	Cox *et al.* (1985b)
			Brown and Dybvig (1986)
			Gibbons and Ramaswamy (1993)
$\beta(\alpha - r)$	σr	Yes	Courtadon (1982)
$\beta(\alpha - r)$	σr^{λ}	Yes	Chan *et al.* (1992)
$\beta(\alpha - r)$	$\sqrt{\sigma + \gamma r}$	Yes	Duffie and Kan (1993)
$\beta r(\alpha - \ln(r))$	σr	Yes	Brennan and Schwartz (1979) [one-factor]
$\alpha r^{-(1-\delta)} + \beta r$	$\sigma r^{\delta/2}$	Yes	Marsh and Rosenfeld (1983)
$\alpha + \beta r + \gamma r^2$	$\sigma + \gamma r$	Yes	Constantinides (1992)
β	σ	No	Merton (1973)
0	σr	No	Dothan (1978)
0	$\sigma r^{3/2}$	No	Cox (1975); Cox *et al.* (1980)

and Scheinkman (1995) derive theoretical moment conditions characterizing the infinitesimal generators of Markov processes.

A commonly used method to estimate (1.1) consists in first parametrizing $\mu(\cdot)$ and $\sigma^2(\cdot)$, then discretizing the model in order to estimate the parameters using, for example, Hansen's (1982) General Methods of Moments (e.g., Chan *et al.*, 1992). Discretization-based methods implicitly assume that more data means more frequent data on a fixed period of observation. This hardly matches the way new data are added to the sample: We typically add today's interest rate at the end of the sample, not the 3 PM interest rate on January 15, 1992, in between two already existing observations in the sample. Even if such data were available, it is likely that market microstructure problems, such as the bid–ask spread, the discreteness of the prices observed, and the irregularity of the intraday sampling interval would complicate considerably the analysis of high frequency data compared to daily or weekly data. Furthermore, identification of the drift is generally impossible on a fixed sampling period, no matter how small the sampling interval. In contrast to this approach, we will not require that the sampling interval shrink to obtain asymptotic properties for our estimators.

Our setup has three characteristic features. First, the precise form of the diffusion function of the interest rate process is crucial to price options, as demonstrated by the extensive literature trying to capture a correct specification. Second, it is hard to form an a priori idea of the functional form of the diffusion function, as the instantaneous volatility of financial series is not observed. Third, long time-series of daily data on spot interest rates are available. These three elements together—importance of the specification, lack of a priori information regarding that specification, and availability of the data—constitute the perfect setup to try to estimate the instantaneous volatility function nonparametrically. This is the objective of this chapter. It is achieved by density matching: The instantaneous drift and diffusion functions of the short-rate process are derived to be consistent with the observed distribution of the discrete data.

The chapter is organized as follows. The first part examines the identification and estimation of the drift and diffusion functions of a continuous-time process. Section 2 identifies nonparametrically the diffusion given a restriction on the drift function. Section 3 then constructs the actual estimator and describes its properties. The second part of the chapter uses the results of the first to derive nonparametric option prices for interest rate derivative securities (section 4). Section 5 computes these prices successively for discount bonds and options on discount bonds. Section 6 concludes. Technical assumptions and details are in Appendix 1–4 while proofs are in Appendix 5.

2. NONPARAMETRIC IDENTIFICATION OF THE DIFFUSION FUNCTION

2.1. Identification via Density Matching

When estimating the CAPM, it is known empirically (see Merton, 1980) that the estimates of expected returns tend to have low precision when the observation period is finite, even though the diffusion can be estimated very precisely when the sampling interval is small. For example, suppose that a stock price were to follow $dX_t/X_t = \mu\, dt + \sigma\, dW_t$ with μ and σ constant. The maximum-likelihood estimate of μ from observations between dates 1 and n at interval Δ is the average of the log returns, $\hat{\mu} = (1/n)\sum_{t=1}^{n} \ln(X_t)$. But $\hat{\mu} = (\ln(X_n) - \ln(X_1))/n$ is independent of the sampling interval Δ and cannot be consistent as $\Delta \mapsto 0$ for fixed n. The only hope to identify the drift μ of the continuous-time process from the data consists in letting the sample period n increase. We will therefore rely on increasing n to derive identification and estimation results.

Identifying without restrictions both the drift and diffusion functions from discretely sampled data is impossible in general. In particular, without further constraints, a pair of functions (μ, σ^2) cannot be distinguished from $(a\mu, a\sigma^2)$ for any constant a, with data discretely sampled at a fixed interval. To fully identify the process, we therefore impose a restriction on the form of the drift function so that we can leave the diffusion function unrestricted. Since nothing so far is parameterized, the drift function cannot be restricted simply by fixing some parameter values. Instead, the restriction operates at one extra level of generality. It takes an otherwise completely unrestricted drift function and makes it belong to a smaller class of functions, namely a parametric class.

We will then use an essential property of stochastic differential equations. Consider a discrete normal random variable. Its distribution is obviously entirely characterized by its first two moments, mean and variance. The continuous-time process r in (1.1) in general is not normally distributed. However, because the Brownian increments are Gaussian, it turns out that under regularity conditions an analogous property will hold for stochastic differential equations: The distributions of the process (marginal and transitional densities) are entirely characterized by the first two moments of the process, here the drift and diffusion functions. By the Markov property, it suffices to consider only one set of transitions of the process; longer transitions can always be derived by iterating the shorter transitions.

In particular, the joint parameterizations of (μ, σ^2) adopted in the literature imply specific forms for the marginal and transitional densities of the process. For example, an Ornstein–Unlenbeck process $dr_t = \beta(\alpha - r_t)\,dt + \gamma\,dW_t$ generates Gaussian transitional and marginal densities. The stochastic differential equation $dr_t = \beta(\alpha - r_t)\,dt + \gamma r_t^{1/2}\,dW_t$ yields a noncentral chi-squared transitional distribution while the marginal density is a Gamma distribution (Feller, 1951). The former parameterization for the spot interest rate has been used by Vasicek (1977), while the latter has been derived in a general equilibrium framework by Cox *et al.* (1985a,b). The functional forms of the transitional densities corresponding to specifications essentially different from these two are not known explicitly [Wong (1964) summarizes the only specifications where they are linear μ and quadratic σ^2].

Our estimation approach relies on the equivalence between (μ, σ^2) and densities in the other direction: from the densities back to (μ, σ^2). To price derivatives, the drift and diffusion of the short-term interest rate process need to be estimated. But they are not observed nor can they be estimated directly. However, the densities of the process can be straightforwardly estimated from data on the short-term rate. Therefore instead of specifying parametrically both the drift and diffusion functions and then accepting whatever marginal and transitional densities are implied by these choices—as the various parametric specifications do—we start with nonparametric estimates of the densities. Then we reconstruct the drift and diffusion of the continuous-time process by matching these densities, which we can in turn use for derivative pricing.

Our construction of the pair (μ, σ^2) that will match the densities is the following. Assume that the spot interest rate process $\{r_t, t \geq 0\}$ follows (1.1) under assumptions A1–A2 in Appendix 1. These assumptions guarantee the existence and uniqueness of a strong stationary solution to (1.1), on $D = (0, \infty)$ since nominal interest rates are positive.[1] In particular they rule out the possibility that starting from any interest rate level in $(0, \infty)$ the barriers $\underline{r} = 0$ and $\bar{r} = \infty$ could be attained in finite expected time. The drift $\mu(\cdot, \theta)$ depends on an unknown parameter vector θ while the diffusion $\sigma^2(\cdot)$ is an unknown function. The available data consist of realizations of the process sampled at equally spaced discrete dates $1, \ldots, n$. The continuous record of observations between each sampling date is unobservable. The asymptotic properties of the estimators as $n \to \infty$ are derived for an expanding sampling period, i.e., the interest rate is observed over a longer period of time, not by sampling more frequently. The sampling interval is fixed at $\Delta = 1$. Let $\pi(\cdot)$ be the marginal density of the spot rate, and $p(\Delta, r_{t+\Delta} \mid r_t)$ the transition density function between two successive observations.

Consider the Kolmogorov forward equation (e.g., Karlin and Taylor, 1981, p. 219):

$$\frac{\partial p(\Delta, r_{t+\Delta} \mid r_t)}{\partial \Delta} = -\frac{\partial}{\partial r_{t+\Delta}}\left(\mu(r_{t+\Delta}, \theta) p(\Delta, r_{t+\Delta} \mid r_t)\right) + \frac{1}{2} \frac{\partial^2}{\partial r_{t+\Delta}^2}\left(\sigma^2(r_{t+\Delta}) p(\Delta, r_{t+\Delta} \mid r_t)\right) \quad (2.1)$$

In order to construct an estimator of σ^2 from the densities of the process, we use (2.1) to characterize the diffusion function. Note that by stationarity $\int_0^{+\infty} p(\Delta, r_{t+\Delta} \mid r_t)\pi(r_t)\, dr_t = \pi(r_{t+\Delta})$ has a partial derivative with respect to time equal to zero. Multiply therefore the forward equation (2.1) by the marginal density $\pi(r_t)$ and integrate through the equation with respect to the conditioning variable r_t to obtain the ordinary differential equation:

$$\frac{d^2}{dr^2}\left(\sigma^2(r)\pi(r)\right) = 2\frac{d}{dr}\left(\mu(r, \theta)\pi(r)\right) \quad (2.2)$$

where $r = r_{t+\Delta}$. This equation must be satisfied at any point r in $(0, \infty)$ and at the true parameter value θ. Integrating (2.2) twice with the boundary

[1] The Vasicek model has the undesirable feature of being distributed on $(-\infty, +\infty)$ as opposed to $(0, \infty)$.

condition $\pi(0) = 0$ now yields:

$$\sigma^2(r) = \frac{2}{\pi(r)} \int_0^r \mu(u, \theta)\pi(u)\, du \tag{2.3}$$

This equation shows that once the drift parameter vector θ has been identified, the diffusion function can be identified from the marginal distribution $\pi(\cdot)$. The identification of θ will be based on the transitional distribution. An identifying restriction on the drift is the linear mean-reverting specification $\mu(r_t, \theta) \equiv \beta(\alpha - r_t)$, $\theta \equiv (\alpha, \beta)'$. Heuristically, the interest rate is elastically attracted to its equilibrium value α at a speed $\beta\, dt$. This is consistent with the parameterization of the drift used in most spot rate models in the literature and will make comparisons with these models possible. The first step consists of deriving

$$E[r_{t+\Delta} \mid r_t] = \alpha + e^{-\beta\Delta}(r_t - \alpha) \tag{2.4}$$

To see why this must hold, let $\vartheta(r, t)$ be the solution of the backward Kolmogorov equation $\partial\vartheta(r, t)/\partial t = A\vartheta(r, t)$ with initial condition $\vartheta(r, 0) = r$, where A is the backward Kolmogorov operator: $A\vartheta(r, t) \equiv \mu(r, \theta)\partial\vartheta(r, t)/\partial r + (1/2)\sigma^2(r)\partial^2\vartheta(r, t)/\partial r^2$. Under assumptions A1–A2, this partial differential equation has the unique solution $\vartheta(r, t) = E[r_t \mid r_0 = r]$ by Dynkin's formula (e.g., Karlin and Taylor, 1981, p. 310). Now it is easy to verify directly that the function $\zeta(r, t) \equiv \alpha + e^{-\beta\Delta}(r - \alpha)$ also satisfies the equation with the same initial equation. Thus $\zeta = \vartheta$. Evaluating the equality at $t = \Delta$ and invoking stationarity gives (2.4).

This procedure can be extended to cover the case of nonlinear restrictions on the drift. In that case the only part of the identification and estimation procedure that has to be modified is (2.4); see Appendix 2. The drift parameters, once identified, can be plugged into (2.3), which remains valid for a generic drift function. We now use (2.4) to identify the parameter vector θ. Ordinary least squares (OLS) clearly identifies the parameters γ and δ in $E[r_{t+\Delta} - r_t \mid r_t] = \gamma + \delta r_t$, and therefore indirectly α and β. A discussion of efficiency is provided in the next section. Finally, the diffusion function can be completely identified from the joint and marginal density functions $p(\cdot, \cdot \mid \cdot)$ and $\pi(\cdot)$ through (2.3)–(2.4). The drift is identified from the conditional mean and given the drift the marginal density yields the diffusion function.

THEOREM 1. *Under Assumptions A1–A2, the diffusion function of the underlying spot interest rate process can be identified from its joint and marginal densities by* (2.3), *where the unknown parameter vector* θ *is identified by* (2.4).

2.2. Diffusion for Large and Small Values of the Spot Rate

The behavior for $r \to +\infty$ of the diffusion function given by Theorem 1 is not immediate since the numerator and the denominator of the ratio (2.3) tend both to zero. The resulting effect can nevertheless be characterized by L'Hôpital's Rule as:

$$\sigma^2(r) \cong_{r \to +\infty} 2\beta r \pi(r)/|\pi^{(1)}(r)| \tag{2.5}$$

where $\phi(r) \cong_{r \to +\infty} \varphi(r)$ means $\lim_{r \to +\infty} \phi(r)/\varphi(r) = 1$ and $\pi^{(1)}$ is the first derivative of π.

The asymptotic trend of the diffusion function is determined by the speed of decrease of the marginal density $\pi(\cdot)$ to zero: The faster the density goes to zero, the smaller the corresponding diffusion function becomes. Suppose that today's spot rate is high, say 20%. The stationary data on the distribution of the process essentially say that it is very unlikely for the spot rate to be so high. Consider now the spot rate level tomorrow. The increment between toady and tomorrow is given by (1.1). If a large level is reached today ($r_t = 20\%$), the mean-reverting drift term will be substantially negative ($\beta > 0$) and drive the process back toward lower levels (say $\alpha = 9\%$), which goes in the right direction. If however $\sigma(20\%)$ happens to be very large then for given realizations of the Brownian increment the stochastic term might compensate the deterministic mean-reverting drift with substantial probability. This would generate a high level of the spot rate, which would be incompatible with $\pi(\cdot)$ very close to zero above 20%.

The link between density and shape of the diffusion function can be made evident by applying successively (2.5) to the parametric family of densities which exhibit an increasingly rapid rate of decrease to zero ($\nu > 1$ and $\omega > 0$):

$$\pi(r) \equiv \xi r^{\nu-1} e^{-\omega r} \text{ generates } \sigma^2(r) \cong_{r \to +\infty} (2\beta/\omega) r \quad \text{(the CIR case)} \tag{2.6a}$$

$$\pi(r) \equiv \xi r^{\nu-1} e^{-\omega r^2} \text{ generates } \sigma^2(r) \cong_{r \to +\infty} (\beta/\omega) \quad \text{(the Gaussian case)} \tag{2.6b}$$

$$\pi(r) \equiv \xi r^{\nu-1} e^{-\omega r^3} \text{ generates } \sigma^2(r) \cong_{r \to +\infty} (2\beta/3\omega)(1/r) \quad \text{(faster decay).} \tag{2.6c}$$

The behavior near zero of the diffusion function is also given by L'Hopital's Rule as $\sigma^2(r) \cong_{r \to 0^+} 2\beta\alpha\pi(r)/|\pi^{(1)}(r)|$. So for all densities (2.6) we have $\sigma^2(r) \cong_{r \to 0^+} 2\beta\alpha r/(\nu - 1)$. Take densities polynomial in zero ($\pi(r) \cong_{r \to 0^+} \xi r^{\nu-1}$, $\nu > 1$). This encompasses most potential densities since π must be zero at zero and smooth. The result is then that every density polynomial in zero will yield a diffusion function with the same linear curvature near zero as the square root process. Thus the CIR parameterization must be a reasonable approximation for small levels of the spot rate. Because the pricing equation for interest rate derivative securities has a strong local character, these securities would tend to be priced correctly by the CIR model when interest rates are low.

3. NONPARAMETRIC ESTIMATION OF THE DIFFUSION

We propose to replace θ and $\pi(\cdot)$ in (2.3) by consistent estimators to obtain an estimate of $\sigma^2(\cdot)$. Starting with $\pi(\cdot)$, we use the interest rate data $\{r_i, i = 1, \ldots, n\}$ to form the smooth density estimator $\hat{\pi}(r) \equiv (1/nh_n)\Sigma_{i=1}^{n} K((r - r_i)/h_n)$, based on a kernel function $K(\cdot)$ and bandwidth h_n [see, e.g., Silverman (1986) for an introduction to kernel density estimation and Scott (1992) for more details]. Regularity conditions on the time-series dependence in the data (Assumption 3), the kernel (Assumption 4), and bandwidth (Assumption A5) are given in Appendix 1. The smooth estimator $\hat{\pi}(\cdot)$ can be used in the denominator of (2.3) as well as inside the integral, or only in the denominator with the density inside the integral replaced by an empirical density. The two resulting estimators of the diffusion function share the same asymptotic properties.

To estimate θ consistently, one can simply use OLS in (2.4). The efficiency of the estimator of θ will not matter for $\sigma^2(\cdot)$, because $\hat{\theta}$ will converge at speed $n^{1/2}$ while $\hat{\pi}(\cdot)$ in the denominator will force the convergence of $\hat{\sigma}^2(\cdot)$ to be slower. To obtain nevertheless a better drift estimator than OLS, useful later when pricing derivatives, consider the following procedure. We form the first step OLS estimator of $\theta = (\alpha, \beta)'$ by a one-to-one transformation from the OLS estimates of (γ, δ): $\alpha = -\gamma/\delta$ and $\beta = -\ln(1 + \delta)/\Delta$. The OLS estimator of θ can now be plugged along with the kernel estimator $\hat{\pi}(\cdot)$ into (2.3) to form a nonparametric estimator of $\sigma^2(\cdot)$.

Then the diffusion estimator can be used to correct for heteroskedasticity in the residuals from the regression (2.4). We construct the weighting matrix for the second step feasible generalized least squares (FGLS) estimation of (γ, δ) and thus (α, β). Note that no serial correlation adjustment is needed in (2.4). Any diffusion process is a Markov process so in particular $E[r_{t+\Delta} \mid r_t, r_{t-\Delta}] = E[r_{t+\Delta} \mid r_t]$. The residuals $\varepsilon_{t+\Delta} \equiv r_{t+\Delta} - (\gamma + (1 + \delta)r_t)$ from the regression $E[r_{t+\Delta} - r_t \mid r_t] = \gamma + \delta r_t$ are therefore conditionally uncorrelated; that is, $E[\varepsilon_{t+\Delta}\varepsilon_t \mid r_t, r_{t-\Delta}] = 0$. If the interest rate data were independent and identically distributed, this procedure would achieve the semi-parametric efficiency bound of Chamberlain (1987; see also Robinson, 1987, for a two-stage setup). To our knowledge, there exists no known efficiency bound for dependent data. The following result details the asymptotic properties of the diffusion estimator.

THEOREM 2. *Under Assumptions A1–A5;*

(i) *The estimator $\hat{\sigma}^2(\cdot)$ is pointwise consistent and asymptotically normal: $h_n^{1/2} n^{1/2}\{\hat{\sigma}^2(r) - \sigma^2(r)\} \xrightarrow{d} N(0, V_{\sigma^2}(r))$, with asymptotic variance*

$$V_{\sigma^2}(r) = \left\{ \int_{-\infty}^{+\infty} K(u)^2 \, du \right\} \sigma^4(r) / \pi(r). \tag{3.1}$$

(ii) *The asymptotic variance $V_{\sigma^2}(r)$ can be consistently estimated by*

$$\hat{V}_{\sigma^2}(r) = \left\{ \int_{-\infty}^{+\infty} K(u)^2 \, du \right\} \hat{\sigma}^4(r) / \hat{\pi}(r). \tag{3.2}$$

(iii) *At different points r and r' in $(0, \infty)$, $\hat{\sigma}^2(r)$ and $\hat{\sigma}^2(r')$ are asymptotically independent.*

The intuition behind the result is the following. There are three elements plugged into the right-hand side of (2.3): $\hat{\theta}$, $\hat{\pi}(\cdot)$ inside the integral and $\hat{\pi}(\cdot)$ in the denominator. The first two are responsible for $n^{1/2}$ terms, while the third generates a term converging at the slower speed $n^{1/2}$. The fast terms can be considered as fixed when computing the asymptotic distribution. Only the slowest term will matter. Part (iii) of the theorem is typical of pointwise kernel estimators (see Robinson, 1983) and is useful to know for inference purposes. The consistent estimator of the pointwise asymptotic variance makes it possible to construct pointwise confidence intervals, or later obtain estimates of the variance of derivative security prices.

4. PRICING INTEREST RATE DERIVATIVE SECURITIES

4.1. Nonparametric Plug-in

Given the underlying spot interest rate process $\{r_t, t \geq 0\}$ described by a stochastic differential equation of the type (1.1), standard arbitrage arguments (as for example in Vasicek, 1977, Section 3) determine the price $U(r, t)$ of any interest rate derivative security with time to maturity t and maturity date T when r is the spot interest rate. Let $\lambda(r)$ be the risk premium factor or market price of interest rate risk—which must be the same for all securities—$c(r, t)$ the cash flow rate paid by the security per unit of time and $g(r)$, the payoff of the derivative security at maturity. $U(r,t)$ satisfies the following partial differential equation:

$$LU(r, t) = -c(r, t) \tag{4.1}$$

where L is the parabolic differential operator:

$$LU \equiv -\partial U/\partial t + \{\sigma^2(r)/2\}(\partial^2 U/\partial r^2) + \{\mu(r,\theta) - \lambda(r)\sigma(r)\}(\partial U/\partial r) - rU. \tag{4.2}$$

This is a Cauchy problem with initial condition $U(r,0) = g(r)$. The continuous payment rate $c(\cdot,\cdot)$, the payoff $g(\cdot)$ at maturity as well as boundary condition(s) depend on the particular security considered. Assumptions A6–A8 in Appendix 3 impose regularity conditions. Consider first a discount bond with face value equal to \$100, maturing at date T. Let $U(r,t) \equiv B(r,t,T)$ be its price at date $(T-t)$ when the spot interest rate is r. This case corresponds to

$$c(r,t) = 0 \text{ (zero coupon)}, \quad g(r) = 100 \text{ for all } r \geq 0 \text{ (initial condition)},$$
$$\lim_{r\to\infty} U(r,t) = 0 \text{ for all } t \geq 0 \text{ (boundary condition)}. \tag{4.3}$$

Consider next a call option[2] on a discount bond. The call option expires at date T, has an exercise (or strike) price X, and the underlying discount bond matures at date S where $T \leq S$. Let $U(r,t) \equiv C(r,t,T;s,X)$ be the call price at time to maturity t when the spot interest rate is r. Then,

$$c(r,t) = 0, \quad g(r) = \max(0, B(r,S-T,S) - X) \text{ for all } r \geq 0,$$
$$\lim_{r\to\infty} U(r,t) = 0 \text{ for all } t \geq 0. \tag{4.4}$$

Other examples can be treated similarly. A coupon bond would be identical to (4.3), except that c is then the coupon rate received at t. An interest rate swap of r against an exogenous rate r' can be viewed as a contract paying at rate $c = r - r'$. A cap (resp. floor) is a loan at variable rate that is guaranteed to be less (resp. more) than some level $\bar{r}$ (resp. $\underline{r}$); it can be seen as a derivative with $c = \min(r,\bar{r})$ (resp. $c = \max(r,\underline{r})$). In all these cases g is the constant face value of the contract. A yield curve call option struck at X has $c = 0$ and $g(r) = \max(0, Y(n, t+n, r) - X)$ where $Y(n, T+n, r)$ is the yield to maturity at date T of an n-year bond. A binary yield option has $c = 0$ and $g(r) = 1(Y(n, T+n, r) \geq X)$. A power p yield option has $c = 0$ and $g(r) = Y(n, T+n, r)^p$. A yield curve slope call option has $c = 0$ and $g(r) = \max(0, (Y(n, T+n, r) - Y(m, T+m, r))/(n-m) - X)$, for some $n > m$.

When $\sigma(r) = \sigma$ and $\lambda(r,t) = \lambda$ are constant, (4.1)–(4.3) and (4.1)–(4.4) have known solutions [Vasicek (1977) and Jamshidian (1989), respectively]. When $\sigma(r) = \sigma r^{1/2}$ and $\lambda(r,t) = \lambda r^{1/2}/\sigma$, (4.1)–(4.3) and (4.1)–(4.4) also have known solutions (Cox *et al.*, 1985b). As was discussed

[2] It is never optimal to exercise early an American call because the underlying bond pays no coupon, so American and European calls on discount bonds have the same value.

earlier, these closed-form solutions all assume a particular parametrization of the diffusion term of the underlying spot interest rate. Since the assumptions are different, the pricing formulas differ. Hull and White (1990) compared them for a set of arbitrary values of the parameters (not estimated from actual data and time dependent) and found that the formulas can differ by as much as 15% (see their Table 1).

We estimate derivative prices by solving the pricing partial differential equation with the drift parameters replaced by the estimates $\hat{\alpha}$ and $\hat{\beta}$ and the unknown function $\sigma^2(\cdot)$ by its nonparametric estimator $\hat{\sigma}^2(\cdot)$ described in section 3. We specify that the market price of interest rate risk is constant[3] and estimate it by minimizing the squared deviations between a given yield curve and that implied by the model. The only justification for the assumptions leading to $\lambda(r,t) = \lambda_{\text{VAS}}$ or $\lambda(r,t) = \lambda_{\text{CIR}} r^{1/2}/\sigma_{\text{CIR}}$ in the Vasicek and CIR models respectively is that they yield explicit solutions given their respective choices of drift and diffusion. Here any other function for the market price of risk could be used.

4.2. The Asymptotic Distribution of Nonparametric Prices

We show next that derivative prices $\hat{U}$ still converge at the usual parametric speed $n^{1/2}$. This holds even though the σ^2 function is estimated nonparametrically and converges slower than $n^{1/2}$ (see Theorem 2). The intuition for the result is as follows. Consider the ordinary differential equation $dU(r)/dr = \sigma^2(r)U(r)$ with initial condition $U(0) = U_0$. Its solution is simply $U(r) = U_0 \exp(\int_0^r \sigma^2(x)\,dx)$. Suppose now that the coefficient function $\sigma^2(\cdot)$ of the differential equation were unknown and estimated nonparametrically by $\hat{\sigma}^2(\cdot)$ at speed $n^{1/2}h_n^{1/2}$. Estimate the solution $U(\cdot)$ by $\hat{U}(r) = U_0 \exp(\int_0^r \hat{\sigma}^2(x)\,dx)$. Now the integral $\int_0^r \hat{\sigma}^2(x)\,dx$ will converge at speed $n^{1/2}$, and so will the estimator $\hat{U}(r)$. A similar example would have the derivative price U depend upon a nonparametric estimate of the cumulative distribution function (converging at speed $n^{1/2}$) as opposed to the density (converging at speed $n^{1/2}h_n^{1/2}$).

The partial differential equation (4.1) satisfied by derivative prices is obviously substantially more complex than the simple ordinary differential equation above. However, it shares the two main insights of that example. First, the solution is also characterized as an integral over the coefficient functions. Second, the coefficient functions are similarly estimated at rate $n^{1/2}h_n^{1/2}$ (the diffusion function σ^2) or faster ($n^{1/2}$ for all the parameters α, β, and λ). Given this intuition, the first step of the proof is to obtain a

[3] Learning more about the specification of the market price of risk would require data on cross-sections of derivatives. The market price of risk could potentially be estimated nonparametrically as well by fitting a price curve exactly, along the lines of the "arbitrage-free" literature on interest rate derivatives.

representation of derivative prices as an integral over the diffusion estimator. We consider the fundamental solution $\Gamma(r,t;x)$ of (4.1).[4] This is the solution for x fixed in $(0,\infty)$ of the Cauchy problem $L\Gamma(r,t;x) = 0$ with initial condition $\Gamma(r,0;x) = \delta_{(x)}(r)$ (a mass point at x). The following lemma will provide the integral characterization needed to obtain the asymptotic distribution of the derivative prices.

LEMMA 1. *Under assumptions A6–A8, there exists a unique solution on* $(0,\infty)x(0,T]$ *to the pricing partial differential equation, and it has the form*

$$U(r,t) = \int_0^{+\infty} \Gamma(r,0;x)g(x)\,dx + \int_0^t \int_0^{+\infty} \Gamma(r,\tau;x)c(x,\tau)\,dx\,d\tau. \tag{4.5}$$

The fundamental solution $\Gamma(r,t;x)$ *is twice differential in r and x, once in t, and is a twice-differentiable function of* $\sigma^2(x)$. *It does not depend on* $\sigma^2(r)$.

The dependence of the fundamental solution $\Gamma(r,t;x)$ on $\sigma^2(x)$ is detailed in the proof of Lemma 1. As a result $U(r,t)$ will depend on σ^2 only through integrals of functions of $\sigma^2(x)\,dx$, which converge at speed $n^{1/2}$. Therefore the speed of convergence of the derivative prices will not be slowed down by the use of a nonparametric estimator of σ^2. The benefits from using a nonparametric estimator of σ^2 can be substantial in terms of avoiding misspecification of the derivative prices. This gain in robustness is achieved at no cost in terms of convergence speed for the prices.

THEOREM 3. *Under assumptions A1–A9 the nonparametric price of a derivative security satisfies*

$$n^{1/2}\{\hat{U}(r,t) - U(r,t)\} \xrightarrow{d} N(0,V(r,t)) \tag{4.6}$$

with asymptotic variance

$$V(r,t) = var\big(u_{(r,t)}(x_t)\big) + 2\sum_{k=1}^{\infty} cov\big(u_{(r,t)}(x_t), u_{(r,t)}(x_{t+k})\big),$$

where $x_t \equiv (r_t, r_{t+\Delta})$ *and* $u_{(r,t)}$ *is a cadlag function.*[5]

[4] For example, the transition density of the process over finite intervals of length Δ, $p(\Delta, r_{t+\Delta} \mid r_t)$, is the fundamental solution Γ corresponding to the backward Kolmogorov operator $-\partial U/\partial t + AU$ (i.e., the same as L in (4.2) but with $\lambda \equiv 0$). If μ and σ were parameterized in (1.1), then the parameters could be estimated by maximum-likelihood: Solve this partial differential equation numerically for Γ to obtain the transition density of the model. By the Markov property, this is sufficient to characterize the likelihood function.

[5] Cadlag = right, continuous, left, limit.

To implement Theorem 3 in practice one needs the addition of a consistent estimator of the asymptotic variance $V(r,t)$. Since the expression for $u_{(r,t)}$ is complicated and $V(r,t)$ depends upon the entire serial correlation structure of the interest rate data, the bootstrap technique can be used in practice. The validity of the bootstrap estimator of $V(r,t)$ in this context is proven in Aït-Sahalia (1992). The estimator consists of two simple steps:[6]

(1) Redraw from the original spot interest rate data. The resampling procedure redraws from blocks of contiguous observations to preserve the serial correlation existing in the original data, an idea introduced by Künsch (1989) and Liu and Singh (1992).

(2) Next estimate the drift and diffusion function associated with this new data set as in section 3 and then estimate the market price of risk and compute the resulting bond prices $\hat{U}^*(r,t)$. The bootstrap estimator $\hat{V}^*(r,t)$ of $V(r,t)$ is the sample variance of the difference of bond prices $\hat{U}^*(r,t) - \hat{U}(r,t)$.

Finally, instead of solving (4.1), derivative prices could have been computed by Monte Carlo simulations of the sample paths of the risk-neutral process $dr_t = \{\mu(r_t,\theta) - \lambda(r_t)\sigma(r_t)\}\,dt + \sigma(r_t)\,dW_t$. The sample paths, all starting at r at date $T-t$ and finishing at date T, would be simulated with the risk-neutral drift and diffusion replaced by their estimates. Rewriting (4.5) as the Feynman–Kac conditional expectation under the risk-neutral dynamics gives the prices

$$U(r,t) = E\Bigg[g(r_T)\exp\left\{-\int_{T-t}^{T} ru\,du\right\} + \int_{T-t}^{T}\exp\left\{-\int_{T-t}^{\tau} r_u\,du\right\}c(r_\tau,\tau)\,d\tau \mid r_{T-t} = r\Bigg]. \quad (4.7)$$

The price $U(r,t)$ can then be estimated by averaging the argument of the conditional expectation over the simulated sample paths.

5. DISCOUNT BOND AND BOND OPTION PRICING

5.1. The Data

The spot rate used is the seven-day Eurodollar deposit rate, bid–ask midpoint, from Bank of America. The data are daily from June 1, 1973, to February 25, 1995. Interest rates paid on other short-term financial assets

[6] In terms of computer programming, the computation of the bootstrap standard errors requires only the addition of a loop for the resampling scheme.

such as commercial paper or T-Bills typically move closely with the Eurodollar rate. A time-series plot of the data is provided in Figure 1. Choosing a seven-day rate, such as the seven-day Eurodollar, as the underlying factor for pricing derivatives is a necessary compromise between (i) literally taking an "instantaneous" rate and (ii) avoiding some of the spurious microstructure effects associated with overnight rates. For example, the second Wednesday settlement effect in the federal funds market creates a spike in the raw federal funds data that would have to be smoothed.

The rates quoted are originally bond equivalent yields (BEY). They were transformed to continuously compounded yield to maturity, deriving the current price B of the instrument from $r_{\mathrm{BEY}} = ((100 - B)/B)(365/(T - t))$ and then computing the continuously compounded yield to maturity (YTM) from $B = 100\exp(-r_{\mathrm{YTM}}(T - t))_{\mathrm{YTM}}$. Monday is taken as the first day after Friday. Whereas weekend effects have been documented extensively for stock prices, there seems to be no conclusive weekend effect in money market instruments. Descriptive statistics are provided in

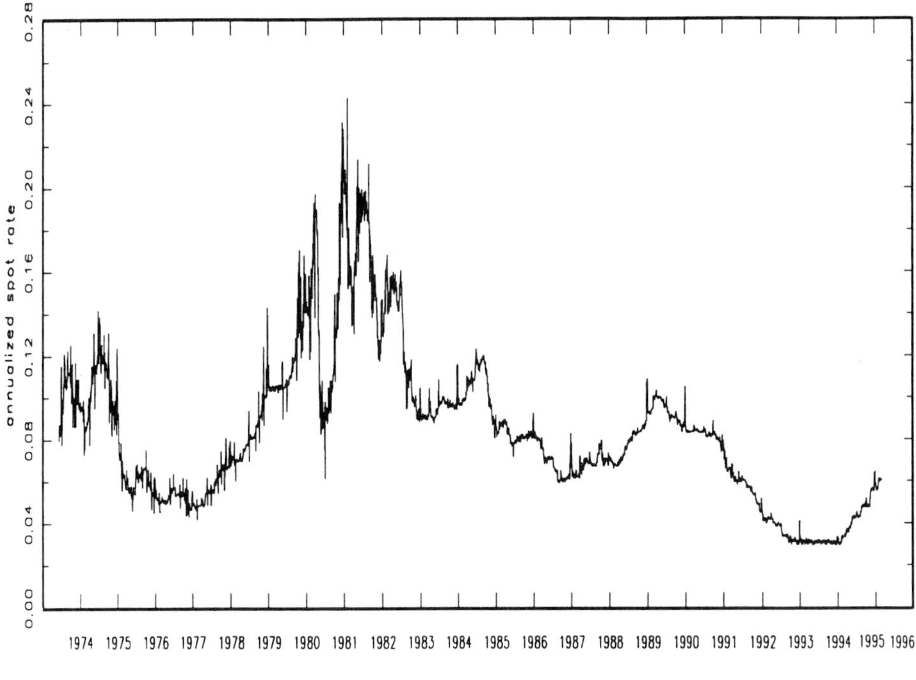

FIGURE 1 Spot interest rate.

Table 2. Since the autocorrelations of the interest rate levels decay slowly, Table 2 also reports the results of an augmented Dickey–Fuller nonstationarity test. The nonstationarity hypothesis is rejected at the 90% level. The test is known to have low power, so even a slight rejection means that stationarity of the series is likely.

Four other daily interest rate series have also been used to check the robustness of the results. First, the overnight federal funds rate (source, Telerate). Second, the one-month London Eurodollar deposit rate (average LIBOR bid–ask; source, Bank of America and Reuters). Third, the one- and three-month continuously compounded bond equivalent yields on Treasury bills (source, Bank of America). In general, the coefficient β of mean-reversion is lower when estimated on a longer-maturity proxy for the spot rate (e.g., estimated on three-month T-Bills vs. seven-day Eurodollar vs. federal funds). Also, longer maturity proxies are less volatile. Overall, all five series produce qualitatively similar shapes for σ^2.

TABLE 2 Descriptive Statistics

	Spot interest rate	First difference of spot interest rate
Mean	0.08362	−0.0000035
Standard deviation	0.03591	0.004063
Monthly ρ_1	0.9389	0.02136
ρ_2	0.8785	−0.00689
ρ_3	0.8300	−0.01658
ρ_4	0.8014	0.00242
ρ_5	0.7783	0.00858
ρ_6	0.7715	0.01573
ρ_7	0.7361	0.00056
Augmented daily Dickey–Fuller	−2.60	
H_0:	Reject at 90%	
Nonstationary	(critical value = −2.57)	

Notes:

Source: Bank of America seven-day Eurodollar (deposit rate midpoint bid-ask); Frequency: Daily; Sample period: June 1, 1973–February 25, 1995; Sample size: 5505 observations; Type: Continuously compounded yield to maturity (annualized rate).

(i) The augmented Dickey–Fuller test statistic is computed as $\hat{\tau}_\mu = \hat{\phi}/\text{ase}(\hat{\phi})$ in the model $\Delta r_l = \mu + \phi r_{t-1} + \Sigma_{j=1}^{p} \phi_j \Delta r_{t-1} + u_t$, with $p = 30$ lags (see, e.g., Harvey, 1993, section 5.4).

(ii) The justification for using the Dickey–Fuller table when the residuals are heteroskedastic and possibly serially dependent is provided by Said and Dickey (1984) and Phillips (1987).

5.2. Nonparametric Diffusion Estimation

5.2.1. Drift Estimation

The first step consists of estimating the drift coefficients α and β by OLS. Drift estimates are reported in Table 3 for daily sampling of the annualized spot rate. These estimates will be used to construct the nonparametric diffusion estimator. Given the diffusion estimator, the second-step semi-parametric FGLS estimates of the drift are computed. For both stages of drift estimates, the speed of mean reversion β is slow.

For comparison purposes, generalized method of moments (GMM) estimates of the Vasicek and CIR models are in Table 3. The GMM estimates of α, β, and σ^2 for the Vasicek and CIR models are obtained from the following four moment conditions ($\Delta = 1$ day):

$$f_t(\theta)' \equiv \left[\varepsilon_{t+\Delta}, \varepsilon_{t+\Delta} r_t, \varepsilon_{t+\Delta}^2 - E\left[\varepsilon_{t+\Delta}^2 \mid r_t \right], \left(\varepsilon_{t+\Delta}^2 - E\left[\varepsilon_{t+\Delta}^2 \mid r_t \right] \right) r_t \right] \tag{5.1}$$

where $\varepsilon_{t+\Delta} \equiv (r_{t+\Delta} - r_t) - E[(r_{t+\Delta} - r_t) \mid r_t]$ with $E[r_{t+\Delta} - r_t \mid r_t] = (1 - e^{-\beta\Delta})(\alpha - r_t)$ for both models. The exact conditional variance of interest rate changes over time intervals of length Δ is given by

TABLE 3 Parameter Estimates for the Spot Rate Process

	First-step OLS nonparametric	Second-step FGLS nonparametric	GMM	
			CIR	Vasicek
α	$8.3082 \cdot 10^{-2}$	$8.4387 \cdot 10^{-2}$	$9.0495 \cdot 10^{-2}$	$8.9102 \cdot 10^{-2}$
	(9.86)	(10.08)	(9.63)	(8.56)
β	$1.6088 \cdot 10^{-0}$	$9.7788 \cdot 10^{-1}$	$8.9218 \cdot 10^{-1}$	$8.5837 \cdot 10^{-1}$
	(2.28)	(3.03)	(3.30)	(3.27)
σ^2	Nonparametric diffusion (Figure 4)	Nonparametric diffusion (Figure 4)	$3.2742 \cdot 10^{-2}$	$2.1854 \cdot 10^{-3}$
			(19.94)	(19.51)

Notes:

(i) The estimates reported in the table are for daily sampling of the annualized one-day rate, with time measured in years.

(ii) Heteroskedasticity–robust t statistics are in parentheses. The t statistics for OLS use the White's estimator of the asymptotic variance, since in the first step the nature of the heteroskedasticity is unknown.

(iii) The two-step FGLS estimates were described in section 3. FGLS uses the first-step nonparametric diffusion function to form the weights and compute the t statistics.

$E[\varepsilon_{t+\Delta}^2 \mid r_t] = V[r_{t+\Delta} \mid r_t]$:

$$\begin{cases} \text{Vasicek:} & E\left[\varepsilon_{t+\Delta}^2 \mid r_t\right] = (\sigma^2/2\beta)(1 - e^{-2\beta\Delta}) \\ \text{CIR:} & E\left[\varepsilon_{t+\Delta}^2 \mid r_t\right] = (\sigma^2/\beta)(e^{-\beta\Delta} - e^{-2\beta\Delta})r_t \\ & \quad + (\sigma^2/2\beta)(1 - e^{-\beta\Delta})^2 \alpha \end{cases} \tag{5.2}$$

These moments correspond to transitions of length Δ and are not subject to discretization bias. Since these GMM systems are overidentified, we weighted the criterion optimally (see Hansen, 1982). Because of over-identification, the first two moments do not reduce to OLS in Table 3.

5.2.2. Marginal Density Estimation

The nonparametric kernel estimator of the density is reported in Figure 2. The kernel and optimal bandwidth are described in Appendix 1. Figure 2 also reports the Gaussian and Gamma densities corresponding to the Vasicek and CIR processes (using the respective GMM estimates of their parameters). Noticeable features of the nonparametric density estimator include its fat-tail (most high observations were recorded in the two years after the 1979 inflexion of the momentary policy). This feature is less accentuated when the subperiod 1980–1982 is not included in the sample. Whether this subperiod should ultimately be included in the sample

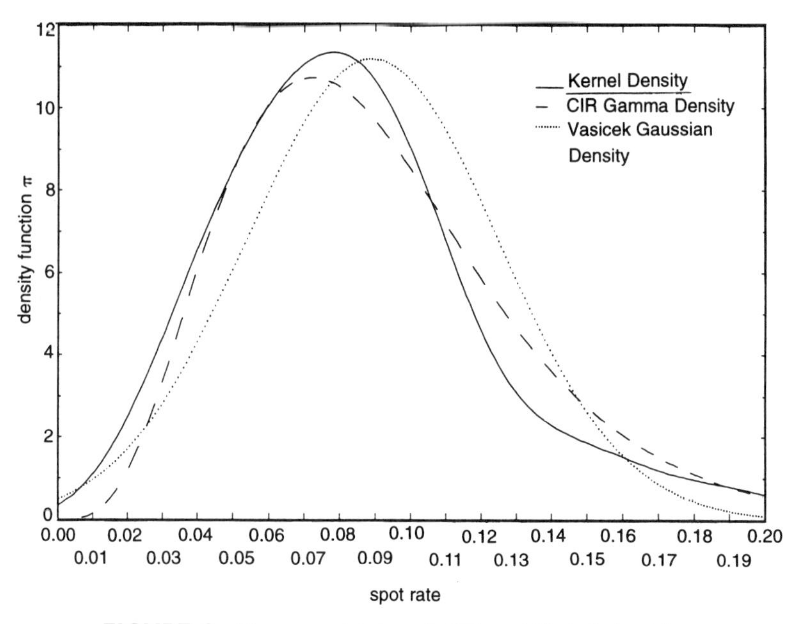

FIGURE 2 Nonparametric kernel, CIR, and Vasicek densities.

depends upon one's view regarding the likelihood that high interest rates occur again in the future. Under stationarity of the spot rate process, this subperiod provides valuable information on what can potentially happen to the spot rate. Alternatively, a model allowing for multiple interest rate regimes might be considered appropriate.

The density also exhibits a very fast decay to zero above 20%, because in the sample very few interest rate levels are recorded outside this range and the extreme observations are clustered next to the absolute smallest and largest ones (as opposed to being slowly spread on both ends of the density support). Compared to the Vasicek and CIR densities, the nonparametric density exhibits more variation and a combination of a longer tail and a faster rate of decay for large values of the interest rate. These differences will naturally be reflected in the diffusion function. A 99% pointwise confidence band for the nonparametric density is plotted in Figure 3, using the classical asymptotic distribution of the kernel density estimator. The large sample size makes it possible to estimate the density very precisely.

5.2.3. Diffusion Estimation

Given the drift parameters and nonparametric density estimator already constructed, the method of section 3 yields the nonparametric estimator of the diffusion function, reported in Figure 4. The first note-

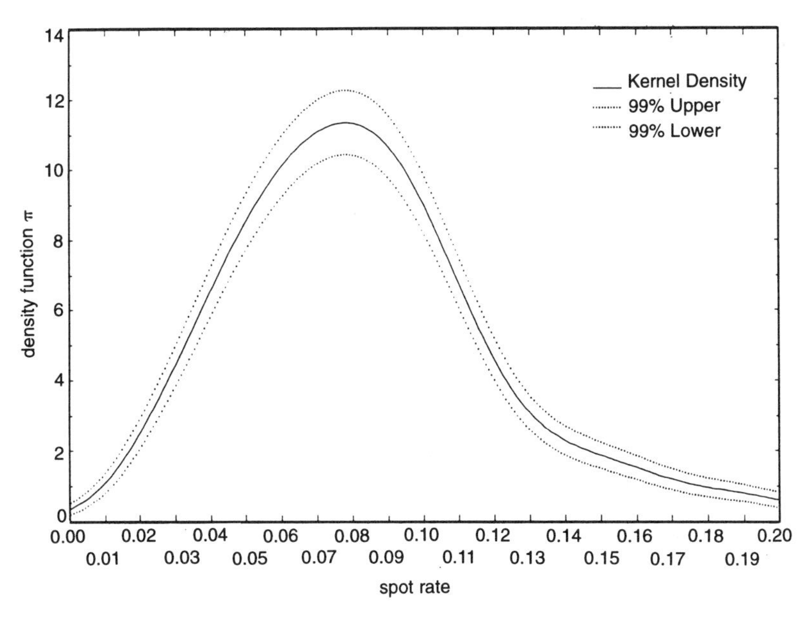

FIGURE 3 Nonparametric kernel density with 99% confidence band.

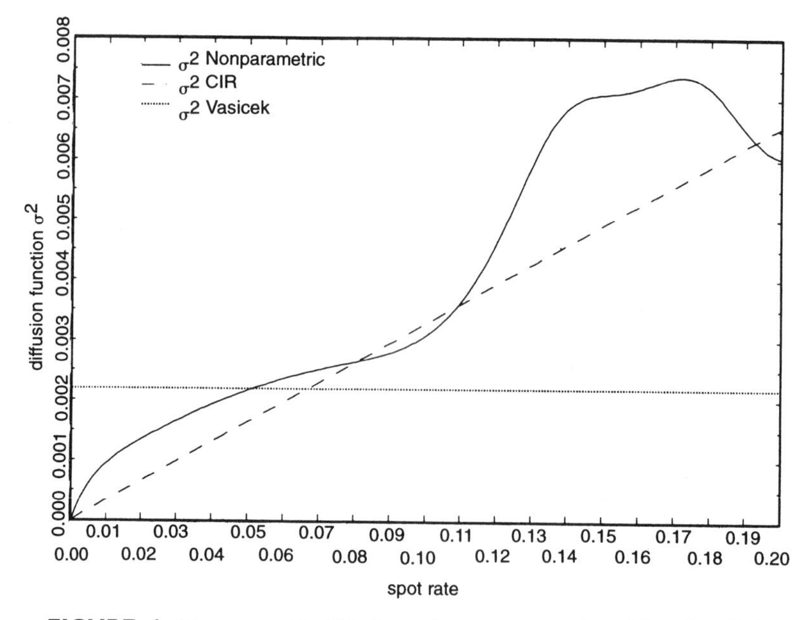

FIGURE 4 Nonparametric diffusion estimator compared to CIR and Vasicek.

worthy aspect of the graph is that σ^2 is globally an increasing function of the level of the interest rate between 0 and 14%. This gives credence to the widely expected result that high interest rates should vary more than low interest rates. Above 14%, however, the diffusion function flattens and then decreases.

At high interest rates the drift in the process (2.1) pulls the process back toward the level of $\alpha = 9\%$. The fact that the density translates into a diffusion function with this shape depends upon the specific strength of the mean-reversion effect in the drift. For example, the specification of a stronger mean-reverting drift, such as any power of the current specification, e.g., $\mu(r, \theta) \equiv \alpha_0 + \alpha_1 r + \alpha_2 r^2 + \alpha_3/r$, would make the process "more able" to pull back the spot rate toward 9% from a high level even in the presence of a larger variance (see Aït-Sahalia, 1996). A stronger drift (taking large negative values at high levels of r) would compensate for that higher potential variation and tilt the transitional density toward the left as does a medium-strength drift (e.g., linear) combined with a low diffusion. It is easy to extend the discussion in section 2.2 to see the balancing effects of a stronger (resp. weaker) mean-reverting drift with a higher (resp. lower) diffusion.

At any rate, if a model insists on specifying a linear drift $\mu(r, \theta) \equiv \beta(r - \alpha)$ as do the Vasicek, CIR, and most other models, then the diffusion function compatible with the data is given by Figure 4 and looks neither linear or flat. A formal test of misspecification of these parameterizations can be found in Aït-Sahalia (1996). A linear specification for the diffusion (CIR) is a relatively good approximation only on 0 to 9%. This is not too surprising since it was proven in section 2.2 that any density polynomial in zero would give rise to a diffusion function linear for small values of the interest rate. The nonparametrically estimated σ^2 function increases faster than r for medium values of the interest rate (9 to 14%), flattens (14 to 17%), then decreases (above 17%). The CIR diffusion instead increases steadily as a linear function of r.

Figure 5 reports pointwise confidence intervals for the nonparametric diffusion estimator. The asymptotic distribution given by Theorem 2 has the feature that the diffusion function is estimated more precisely in highly populated interest rate regions. Conversely, the confidence band tends to increase above 14%, reflecting the relative scarcity of the data. The estimates of σ^2 are stable across subperiods of the sample, with the exception that subperiods that do not include the high interest rate years 1980–1982 produce estimates of π with a truncated right tail, which in turn shift the graph of σ^2 to the left and have a lower overall level of volatility.

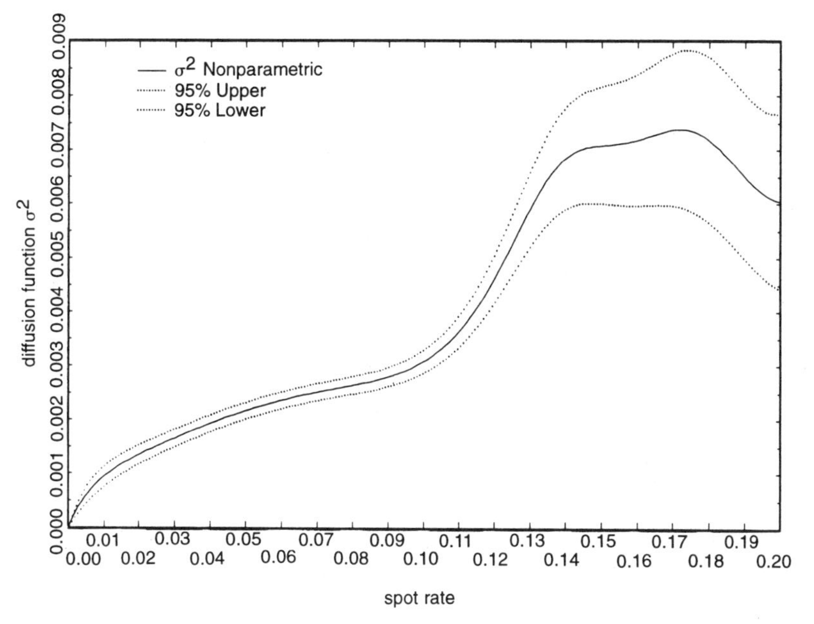

FIGURE 5 Nonparametric diffusion estimator with 95% confidence band.

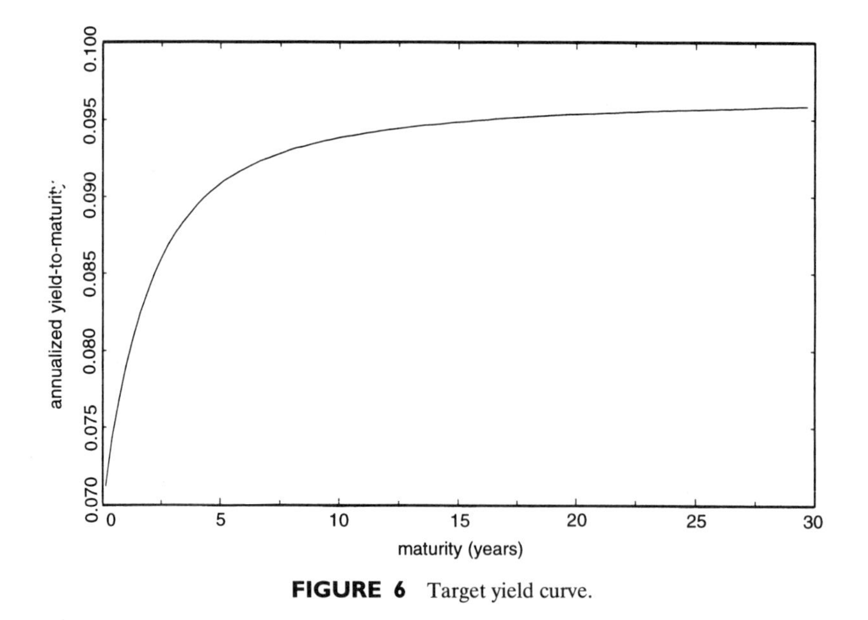

FIGURE 6 Target yield curve.

5.3. The Term Structure of Interest Rates

The market price of risk is estimated for each model by minimizing the sum of squared deviations across maturities between a given target yield curve and the yields produced by the respective model. To obtain a realistic target yield curve, we take the average yield curve over the sample period obtained from the end-of-month Reuters data. The yield curve for coupon bonds is transformed to a discount yield curve by using the standard technique of fitting third-order polynomial splines. From now on we treat this yield curve as fixed.

The target yield curve is plotted in Figure 6. Market price of risk estimates are in Table 4. Table 5 details the bond prices computed under the CIR, Vasicek, and nonparametric models. Standard errors for the nonparametric prices, derived as indicated in section 4, are also reported. As a result of fitting a common yield curve the CIR and Vasicek prices are generally within one standard deviation of the nonparametric prices. By eliminating differences in the prices of the underlying bonds, differences in option prices can be attributed to differences in the volatility of the underlying bonds, which is the determining factor.

TABLE 4 Market Price of Interest Rate Risk

	Nonparametric	CIR	Vasicek
λ	$-2.4451 \cdot 10^{-1}$	$-7.8905 \cdot 10^{-2}$	$-1.6988 \cdot 10^{-1}$
	(4.52)	(3.94)	(6.79)

Notes:

(i) The market price of risk was estimated by minimizing the squared deviations between the respective model's bond yields and those in Figure 6.

(ii) The bootstrap t statistic of the estimate is in parentheses.

TABLE 5 Nonparametric Underlying Bond Prices

Maturity	Annualized spot rate						
(years)	0.02	0.04	0.06	0.08	0.10	0.12	0.14
0.5	98.2408	97.4936	96.7065	95.9064	95.1389	94.3724	93.6150
	(0.1078)	(0.0940)	(0.0875)	(0.0709)	(0.0861)	(0.0893)	(0.0960)
	98.3089	97.5054	96.7084	95.9179	95.1339	94.3563	93.5851
	98.2859	97.4900	96.7005	95.9175	95.1408	94.3703	93.6061
1	95.5091	94.2250	92.9944	91.7708	90.5475	89.3639	88.2045
	(0.1146)	(0.0923)	(0.0839)	(0.0655)	(0.0740)	(0.0832)	(0.0992)
	95.6082	94.3132	93.0357	91.7755	90.5323	89.3060	88.0964
	95.5461	94.2721	93.0150	91.7747	90.5509	89.3435	88.1521
5	66.8409	65.4272	64.0544	62.7142	61.4002	60.1130	58.8665
	(0.1125)	(0.1088)	(0.0984)	(0.0664)	(0.0753)	(0.0888)	(0.0991)
	67.3719	65.7962	64.2574	62.7546	61.2869	59.8535	58.4537
	67.2425	65.7148	64.2218	62.7627	61.3368	59.9433	58.5814
10	41.3704	40.4903	39.5658	38.6818	37.8343	37.0836	36.3885
	(0.0920)	(0.0806)	(0.0724)	(0.0678)	(0.0622)	(0.0706)	(0.0945)
	41.5477	40.5622	39.6000	38.6606	37.7436	36.8483	35.9742
	41.4768	40.5218	39.5887	38.6771	37.7865	36.9164	36.0664
30	5.9825	5.8550	5.7211	5.5802	5.4617	5.3256	5.2232
	(0.0873)	(0.0752)	(0.0738)	(0.0705)	(0.0814)	(0.0878)	(0.0800)
	5.9788	5.8369	5.6985	5.5633	5.4313	5.3024	5.1766
	5.9759	5.8383	5.7038	5.5724	5.4441	5.3187	5.1962

Notes:

(i) All prices correspond to a face value of the bond equal to \$100. The four elements of each cell are, from top to bottom, the nonparametric price, its bootstrap standard error (in parentheses), the CIR price, and the Vasicek price.

(ii) The nonparametric prices were computed using the two-step FGLS estimates of α and β (second column of Table 3), the nonparametric diffusion estimator (Figure 4), and the market price of risk from Table 4. The CIR and Vasicek prices were computed using the GMM estimates of α, β, and σ^2 reported in Table 3 (third and fourth columns, respectively), and the market prices of interest rate risk reported in Table 4.

(iii) The bootstrap standard errors are produced by $nb = 100$ replications with blocks containing $nk = 200$ consecutive spot observations.

5.4. Bond Option Pricing

Table 6 reports call option prices on a five-year discount bond and the nonparametric prices standard errors. Option prices reflect mostly differences in the second moment of the spot rate, where differences across models are more pronounced. By Itô's Lemma, the risk-neutral bond price follows $dB/B = r\,dt + \{\sigma(r)(\partial B/\partial r)/B\}\,dZ$ when the risk-neutral spot rate has the dynamics $dr = \{\mu(r) - \lambda(r)\sigma(r)\}\,dt + \sigma(r)\,dZ$. Figure 7 reports the volatilities $\sigma(r)(\partial B/\partial r)/B$ of a five-year bond price as a function of the spot rate level r. The amount of bond volatility under each

TABLE 6 Nonparametric Call Option Prices on the Five-Year Bond

Annualized spot rate (r)	Option expiration (t, years)	Exercise price				
		0.96	0.98	1.00	1.02	1.04
0.02	0.25	3.1451	1.8624	0.7926	0.1696	0.0054
		(0.0092)	(0.0097)	(0.0035)	(0.0030)	(0.0026)
		3.1432	1.8422	0.7179	0.0966	0.0002
		3.1513	1.9061	0.9049	0.3071	0.0694
0.02	0.5	3.8179	2.5660	1.4512	0.6141	0.1543
		(0.0094)	(0.0110)	(0.0094)	(0.0089)	(0.0072)
		3.0871	2.5379	1.3910	0.5231	0.0865
		3.8196	2.5855	1.5232	0.7478	0.2950
0.02	1	5.5679	4.3117	3.1139	2.0454	1.1620
		(0.0109)	(0.0134)	(0.0155)	(0.0128)	(0.0092)
		5.5574	4.3061	3.1056	2.0077	1.0911
		5.5718	4.3133	3.1159	2.0514	1.2014
0.14	0.25	4.2449	3.1665	2.1716	1.3281	0.6962
		(0.0125)	(0.0156)	(0.0127)	(0.0148)	(0.0124)
		4.2400	3.1636	2.1670	1.3163	0.6779
		4.2211	3.0941	2.0059	1.0653	0.4243
0.14	0.5	5.9629	4.9094	3.8868	2.9515	2.0940
		(0.0175)	(0.0176)	(0.0152)	(0.0156)	(0.0139)
		5.9595	4.8964	3.8679	2.9010	2.0322
		5.9392	4.8437	3.7558	2.7009	1.7418
0.14	1	9.0293	8.0119	6.9534	5.9949	5.0125
		(0.0194)	(0.0187)	(0.0176)	(0.0175)	(0.0163)
		9.0241	8.0013	6.9857	5.9828	5.0019
		9.0063	7.9735	6.9408	5.9089	4.8809

Notes:

(i) All call option prices correspond to a face value of the discount bond equal to \$100. The exercise price is expressed as a proportion of the corresponding bond price for each model.

(ii) The four elements of each cell are, from top to bottom, the nonparametric price, its bootstrap standard error in parentheses, the closed-form CIR price, and the closed-form Vasicek price. The standard errors of the nonparametric prices were computed under the same conditions as Table 5.

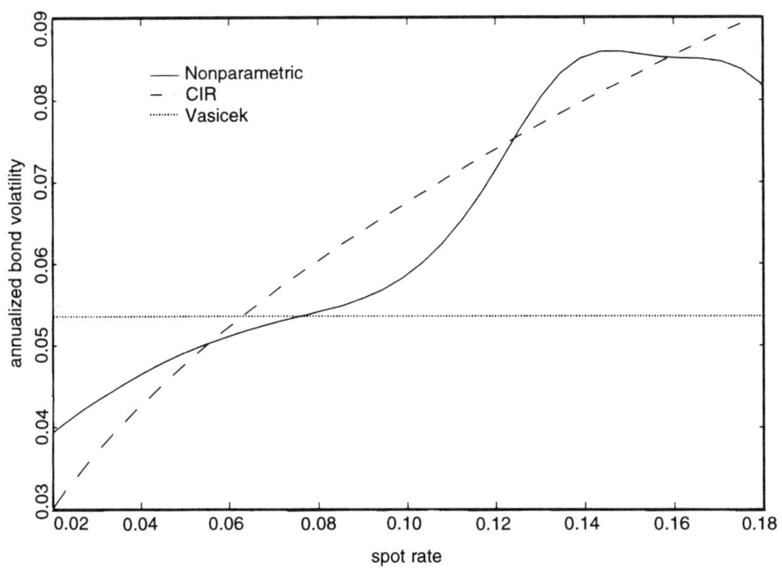

FIGURE 7 Five-year bond price volatilities.

model provides some intuition for the relative magnitude of the three prices in Table 6.

When the short rate is 2% we see that the bond price volatility from the Vasicek model is larger than that from the CIR and nonparametric models, which are close to each other. As a result, the Vasicek option prices are larger than the other two and significantly different from the nonparametric prices. Not surprisingly this effect is more visible for deep-out-of-the-money options (the right column of Table 6) and shorter maturity options (the upper rows in both parts of Table 6). Intuitively, the shorter the option, the more its price depends on the values of the parameters evaluated close to the current value of the spot rate r_t. Similarly when the short rate is 14%, the bond price volatility from the nonparametric spot process is higher than CIR, which in turn is higher than Vasicek. The ordering of the option prices is modified accordingly. Both the Vasicek and CIR prices then fall outside one or more standard deviations of the nonparametric prices.

6. CONCLUSIONS

This chapter has proposed a procedure to estimate stochastic models while preserving their basic continuous-time structure. Properties of stochastic differential equations were exploited to both identify and estimate the drift and diffusion functions of the process of interest. In particular, it was shown that the entire diffusion function can be estimated nonparametrically once a parametrization of the drift is adopted. The density-matching procedure starts from estimates of the densities of the process (which could as well be parametric) and then recovers the drift and the diffusion functions that are consistent with the density estimates. Asymptotic distributions are computed for the diffusion function of the underlying asset process and for the prices of the derivative securities. Even though the diffusion is estimated nonparametrically, and therefore converges slower than root n, the speed of convergence of the derivative prices is not penalized.

The general density-matching approach could potentially be applied to a wide variety of processes in economics and finance when it is important to identify precisely the instantaneous variance, using discrete data without discretizing the continuous-time process. The pricing of interest rate derivative securities was precisely such a problem, since option prices depend on the particular shape of the diffusion function of the underlying spot interest rate. With a linear mean-reverting drift, consistent with most of the literature, the diffusion function was found to be increasing linearly, then exponentially, before flattening and decreasing. This was shown to have implications for the pricing of derivative securities. Other relevant applications can be considered in the future. The method can easily be extended to cover multi-dimensional stochastic difference equations and therefore multi-factor models of the term structure. Hedging strategies in the option market can be assessed with the help of the nonparametric prices. A similar estimation approach for the drift and diffusion could be implemented to estimate the process followed by stock returns, or exchange rates to price foreign currency options, or identify the consequences of target zone bands.

APPENDIX 1. REGULARITY CONDITIONS FOR THE NONPARAMETRIC ESTIMATION OF THE DIFFUSION

ASSUMPTION A1.

(i) *The drift and diffusion functions have* $s \geq 2$ *continuous derivatives on* $(0, \infty)$.

(ii) $\sigma^2 > 0$ *on* $(0, \infty)$.

(iii) *The integral of* $m(v) \equiv (1/\sigma^2(v))\exp\{-\int_v^{\bar{\varepsilon}}[2\mu(u, \theta)/\sigma^2(v)]\,du\}$, *the speed measure, converges at both boundaries of* $(0, \infty)$.

(iv) *The integral of* $s(v) \equiv \exp\{\int_v^{\bar{\varepsilon}}[2\mu(u, \theta)/\sigma^2(u)]\,du\}$, *the scale measure, diverges at both boundaries of* $(0, \infty)$.

In A1(iii)–(iv), $\bar{\varepsilon}$ is fixed in $(0, \infty)$ and its particular choice is irrelevant. Under A1 (i)–(ii) the stochastic differential equation (1.1) admits a unique strong solution. Note that by the mean-value theorem A1(i) implies the following local Lipschitz and growth (LLG) conditions.

For each parameter vector θ and each compact subset of $D = (0, \infty)$ of the form $K \equiv [1/R, R]$, $R > 0$, there exist constants C_1 and C_2 such that for all x and y in K:

$$\begin{cases} |\mu(x, \theta) - \mu(y, \theta)| + |\sigma(x) - \sigma(y)| \le C_1 |x - y| \\ |\mu(x, \theta)| + |\sigma(x)| \le C_2 \{1 + |x|\} \end{cases}$$

Global Lipschitz and growth conditions are usually imposed on the drift and diffusion functions[7] to guarantee the existence and uniqueness of a strong solution to the stochastic differential equation (1.1). These global conditions fail to be satisfied for many models of interest here. It is known that the local Lipschitz condition is sufficient for pathwise uniqueness of the solution (Itô's Theorem, see, e.g., Karatzas and Shreve, 1991, Theorem 5.2.5, p. 287). Now A1(i)–(ii) guarantees the existence of a weak solution (see, e.g., Karatzas and Shreve, 1991, Theorem 5.5.15, p. 341), up possibly to an explosion time. Then the existence of a weak solution combined with pathwise uniqueness imply strong existence (a corollary to the Yamada and Watanabe Theorem, e.g., Karatzas and Shreve, 1991, Proposition 5.3.20, pp. 309–310). Therefore A1(i)–(ii) imply that (1.1) admits a unique strong solution up possibly to an explosion time.

The role of A1(iii)–(iv) is to rule out explosions in finite expected time and guarantee stationarity of the process. The solution is a Markov process with time-homogeneous transition densities (i.e., such that transition probability density $p(s, y; t, x)$ from x at time t to y at time s depends only on $t - s$, not on t and s separately). A1(iv) further guarantees that starting from any point in the interior of the state space $D = (0, \infty)$, the boundaries $x = 0$ and $\bar{x} = +\infty$ cannot be attained in finite expected time. Define the scale measure $\underline{S}(\varepsilon, \bar{\varepsilon})(b = \int_{\varepsilon}^{\bar{\varepsilon}} s(v)\, dv$. Zero cannot be attained in finite time by the spot rate if $\lim_{\varepsilon \to 0} S(\varepsilon, \bar{\varepsilon}) = +\infty$ (see Karlin and Taylor, 1981, Lemma 15.6.3 and Table 15.6.2). Infinity cannot be attained in finite time if $\lim_{R \to +\infty} S(\bar{\varepsilon}, R) = +\infty$.

An alternative condition can be formulated in terms of Lyapunov functions (see Hasminskii, 1980). Assume that there exists a twice continuously differentiable nonnegative function ϑ on D such that

[7] "Global" means here that the same constants C_1 and C_2 must be valid on the entire domain D of the diffusion. For a statement of Itô's classical result with global conditions, see, e.g., Øksendal (1992, Theorem 5.5) or Karatzas and Shreve, (1991, Theorem 5.2.5, p. 287). Even the weaker Yamada–Watanabe condition (see Karatzas and Shreve, 1991, Theorem 5.2.13, p. 291) fails to be satisfied for interesting economic models. Both would imply existence and uniqueness of a strong solution to (1.1).

$\mu(x,\theta)\partial\vartheta(x)/\partial x + (1/2)\sigma^2(x)\partial^2\vartheta(x)/\partial x^2 \leq -1$ for all x in $D = (0, +\infty)$ outside a compact interval of the form $K \equiv [1/R, R]$, $R > 0$. Also assume that $\lim_{x\to 0}\vartheta(x) = \lim_{x\to\infty}\vartheta(x) = \infty$. Since $\vartheta \geq 0$, this condition implies the weaker inequality

$$\mu(x,\theta)\partial\vartheta(x)/\partial x + (1/2)\sigma^2(x)\partial^2\vartheta(x)/\partial x^2 \leq c\vartheta(x)$$

for some (in fact any) constant $c \geq 0$. Under this weaker inequality, we can only be sure to obtain a Markov solution with time-homogeneous transition densities (Hasminskii, 1980), Theorem III.4.1 and Remark III.4.1). This is necessary but not sufficient for the process to be stationarity; for example, a Brownian motion has time-homogeneous transition densities (Gaussian) but is obviously not a stationary process. For our estimation strategy to work, we need the process solution to (1.1) to be stationary and therefore impose in A1(iii) the stronger requirement that

$$\mu(x,\theta)\partial\vartheta(x)/\partial x + (1/2)\sigma^2(x)\partial^2\vartheta(x)/\partial x^2 \leq -1$$

(use Theorem III.5.1 and footnote 9 in Hasminskii (1980), or see directly that this condition implies the existence of an integrable solution π to (2.2) for the given drift and diffusion functions). It then follows that the stationary marginal density π exists (and integrates to one by construction) and is continuously differentiable on D up to order $s \geq 2$.

ASSUMPTION A2. *The stationary density π is strictly positive on* $(0, \infty)$, *and the initial random variable r_0 is distributed as π.*

A2 makes it possible to characterize the diffusion function by (2.3) and subsequently construct the nonparametric estimator based on that expression. Under A1, the candidate stationary distribution $\pi(\cdot)$ exists. Stationarity of the process follows from initializing it with the stationary distribution itself. We now makes assumptions required to guarantee that the Central Limit Theorem to be used in the proofs of Theorems 2 and 3 holds.

ASSUMPTION A3. *The observed data sequence* $\{r_i, i = 1, \ldots, n\}$ *is a strictly stationary β-mixing sequence satisfying* $k^\delta \beta_k \xrightarrow[k\to+\infty]{} 0$ *for some fixed* $\delta > 1$.

Assumption 3 restricts the amount of dependence allowed in the sequence. Without this assumption, the Central Limit Theorem can fail. As long as $\beta_k \xrightarrow[k\to+\infty]{} 0$, the sequence is said to be absolutely regular. In order for the discretely observed process $\{r_i, i \geq 1\}$ to satisfy A3, the stochastic process $\{r_t, t \geq 0\}$ has to verify a stronger continuous-time mixing condition (see Prakasa-Rao, 1987). A condition on the drift and diffusion sufficient to ensure that the discrete observations will satisfy A3 is the following.

ASSUMPTION A3′.

$$\lim_{r \to 0 \text{ or } r \to \infty} \sigma(r)\pi(r) = 0$$

and

$$\lim_{r \to 0 \text{ or } r \to \infty} \left| \sigma(r)/\{2\mu(r,\theta) - \sigma(r)\sigma^{(1)}(r)\} \right| < \infty.$$

Geometric ergodicity follows from A3′. Indeed under A3′ Hansen and Scheinkman (1995) prove that the integral operator T (such that the infinitesimal generator A is the derivative of T) is a strong contraction. This can be used here. It follows from the strong contraction property of T that there exists λ, $0 < \lambda < 1$, such that, for each measurable f and g, $|E[f(r_i)g(r_{i+k})]| \leq \|f\| \, \|g\| \, \lambda^k$. Applying this to the indicator $f = 1_A$ and $g = 1_B$ to obtain the mixing bounds for the discrete data $2\alpha_k \leq \beta_k \leq \phi_k \leq \psi_k/2 \leq \lambda^k/2$ (where α, β, ϕ, and ψ are the various classical mixing coefficients). This is clearly a stronger mixing property of the discrete data than what is strictly required by A3. Only A3 is necessary. The next assumption is the standard regularity condition imposed on a kernel function.

ASSUMPTION A4.

(i) *The kernel* $K(\cdot)$ *is an even function, continuously differentiable up to order* r *on* R *with* $2 \leq r \leq s$, *belongs to* $L^2(R)$ *and*

$$\int_{-\infty}^{+\infty} K(x)\, dx = 1.$$

(ii) $K(\cdot)$ *is of order* r:

$$\int_{-\infty}^{+\infty} x^i K(x)\, dx = 0 \quad i = 1, \ldots, r-1$$

and

$$\int_{-\infty}^{+\infty} x^r K(x) \mid dx \neq 0, \qquad \int_{-\infty}^{+\infty} |x|^r \, |K(x)| \, dx < \infty$$

Note that in order to use effectively a kernel of order r for bias reduction, the density π must have at least that many derivatives. Hence the requirement in A1(i) that the drift and diffusion functions have $s \geq 2$ derivatives, and then the choice in A5 of $2 \leq r \leq s$. If a kernel of order $r = 2$ is used, then only $s = 2$ derivatives can be required in A1(i). More generally, when s derivatives of π exist, the optimal rate of convergence of the diffusion estimator is $n^{s/(2s+1)}$. It is achieved by a kernel of order $r = s$ (even) with bandwidth set to $h_n = O(n^{-1/(2s+1)}$. The optimal rate of convergence for π and σ^2 can only be attained with a nonvanishing bias term (meaning that the asymptotic distribution of Theorem 2 would not be centered at 0). To avoid this undesirable feature, we require A5(i)–(ii).

ASSUMPTION A5 (*Admissible Bandwidth Choices*). *As* $n \to \infty$, $h_n \to 0$ *and*

(i) *To estimate* π, $n^{1/2} h_n^{(2r+1)/2} \to 0$, *and* $n^{1/2} h_n^{1/2} \to \infty$.

(ii) *To estimate* σ^2, $n^{1/2} h_n^{(2r+1)/2} \to 0$, *and* $n^{1/2} h_n^{3/2} \to \infty$,

(iii) *To estimate derivative prices* U, $n^{1/2} h_n^r \to 0$.

A5 gives the admissible range of bandwidth choices depending upon the object to be estimated. The speed of decrease of the bandwidth h_n to 0 as n becomes large is restricted. Optimality within the respective admissible ranges is discussed later. In A5(i) the first condition ensures that the bias in the normalized density estimator is asymptotically negligible. The second condition ensures that the variance of the estimator which is of order $n^{-1} h_n^{-1}$ goes to zero. In A5(ii) the first condition serves the same purpose. Given the first, the second condition ensures that the asymptotic distribution of the diffusion estimator is indeed given by the linear term in its Taylor expansion (i.e., the remainder term in the expansion is also asymptotically negligible). As a consequence of $n^{1/2} h_n^2 \to \infty$, it follows that $n^{1/2} h_n^{1/2} \to \infty$ and therefore the variance of the estimator which is also of order $n^{-1} h_n^{-1}$ goes to zero. In A5(iii), the condition makes the bias negligible and is sufficient to make the remainder term in the Taylor expansion go to zero.

The empirical results for the density and diffusion estimators (Figures 2–5) are obtained with a Gaussian kernel $K(u) = \exp(-u^2/2)/\sqrt{2\pi}$ of order $r = 2$, so we only need to assume the existence of $s = 2$ derivatives of π. The optimal bandwidth choice (in the integrated mean squared error sense) within the range specified by either A5(i) or A5(ii) requires getting as close as possible to $n^{-1/(2r+1)}$. Choosing the bandwidth at this order would produce a bias term which does not vanish asymptotically. We therefore set the bandwidth as $h_n = c_n n^{-1/(2r+1)} = c_n n^{-1/5}$ where $c_n = c/\ln(n)$ with the constant c chosen to minimize the integrated mean square error of the diffusion estimator. The diffusion estimator can be calculated explicitly. For the actual sample size $n = 5505$, the bandwidth was $h_n = 1.6033 \cdot 10^{-2}$.

The higher order kernel $K(u) = (3/\sqrt{8\pi})(1 - u^2/3)\exp(-u^2/2)$ (order $r = 4$) with optimal bandwidth $h_n = c_n n^{-1/(1+2r)} = c_n n^{-1/9}$, as well as exponential, sinusoidal, and Epanechnikov kernels produce similar results. The results are qualitatively robust to the choice of the kernel and small changes in the bandwidth parameter around the optimal value. Selection of the bandwidth by other various forms of cross-validation also produces similar estimates. The tail behavior of the kernel does not modify substantially the tail behavior of the diffusion estimator in the interval considered (the large values of the spot rate are clustered together in the sample so that tail behavior of the density is estimated fairly accurately up to 22%).

The empirical results for the derivative prices (Tables 5–6 and Figure 7) are obtained with the same Gaussian kernel, but the optimal choice in the integrated mean square error sense is now $h_n = cn^{-1/3}$ where c is a scaling factor. The scaling factor is still set to minimize the integrated mean square error of the estimator. The precise choice of the bandwidth in this case does not modify the rate of convergence $n^{1/2}$ of derivative prices (it only affects higher order terms in the integrated mean squared error). Hence as long as A5(iii) is satisfied it is substantially less important to be close to the optimal bandwidth than for pointwise estimation of σ^2. Finally note that the optimal bandwidth choice when estimating σ^2 pointwise is larger (less smoothing) than is admissible when estimating derivative prices. This is due to respective size of the bias terms in the two cases. In particular the estimate of derivative prices would be biased if they were computed from plugging in the estimate of σ^2 computed at its pointwise bandwidth choice (A5′(i)). We summarize the actual bandwidth choice in Assumption A5′.

ASSUMPTION A5′ (*Actual Bandwidth Choice*).

(i) *To estimate* π *and* σ^2, $h_n = c_n n^{-1/(2r+1)} = c_n n^{-1/5}$ *with* $c_n = c/\ln(n)$.
(ii) *To estimate derivative prices* U, $h_n = cn^{-1/3}$.
(iii) *In both cases, the scaling factor* c *is set to minimize the asymptotic integrated mean squared error of the estimator.*

Monte Carlo evidence shows that the asymptotic distribution given by Theorem 2 is a good approximation for the actual distribution of the diffusion estimator given the same sample size as the data set. The coverage probabilities are consistently above 0.85 when simulating data with various specifications (CIR, various power functions for the diffusion $\sigma^2(r) = r^\gamma$) and computing the diffusion estimator.

APPENDIX 2: IDENTIFICATION WITH A GENERIC PARAMETERIZATION OF THE DRIFT

Let $\mu(r, \theta)$ be a parameterization of the drift. The identification restriction states that there exists a unique parameter vector θ such that the true drift is $\mu(r, \theta)$. This appendix shows how to modify (2.4) when μ is not necessarily linear. Let $\eta(r) \equiv \int_0^r \mu(u, \theta)\pi(u)\,du$. Then the equation $\partial\vartheta(r,t)/\partial t = A\vartheta(r,t)$ with initial condition $\vartheta(r,0) = r$ implies

$$\begin{aligned}\pi(r)\,\partial\vartheta/\partial t &= \mu(r,\theta)\pi(r)\,\partial\vartheta/\partial x + (1/2)\sigma^2(r)\pi(r)\,\partial^2\vartheta/\partial r^2\\ &= \{\partial\eta/\partial r\}\{\partial\vartheta/\partial r\} + \eta(r)\{\partial^2\vartheta/\partial r^2\}\end{aligned}$$

Therefore ϑ solves $\pi(r)\partial\vartheta/\partial t = \{\partial/\partial r\}[\eta(r)\partial\vartheta/\partial r]$. By Dynkin's formula again this partial differential equation has a unique solution given by $\vartheta(r,t) = E[r_t \mid r_0 = r]$. Solving this equation yields ϑ. Then θ can be estimated consistently by nonlinear least squares with this generalization of (2.4).

APPENDIX 3: REGULARITY CONDITIONS FOR THE PRICING EQUATION

The following assumptions are introduced to obtain nonparametric derivative prices and their asymptotic distribution. They are not needed to estimate the diffusion function itself.

ASSUMPTION A6. *The drift, diffusion, and density functions of the risk-neutral spot rate process,* $dr_t = \{\mu(r_t, \theta) - \lambda(r_t)\sigma(r_t)\}\, dt + \sigma(r_t)\, dW_t$, *satisfy the conditions in* $A1$–$A2$.

ASSUMPTION A7. *The derivative continuous payment rate* c *and final payoff* g *are right continuous, left limit functions and satisfy for* (r,t) *in* $(0,\infty) \times (0,T]$: $|c(r,t)| \leq N \exp(\nu r^2)$ *and* $|g(r)| \leq N \exp(\nu r^2)$ *for some* ν, $N > 0$.

ASSUMPTION A8. *The derivative price* $U(r,t)$ *satisfies the exponential boundedness condition*: $\int_0^T \int_0^{+\infty} e^{-kr^2} |U(r,t)|\, dr\, dt < \infty$ *for some* $k > 0$.

The solution to (4.1) is only unique in the class of functions satisfying A8. The bound is sharp as the solution fails to be unique in general if A8 is replaced by $\int_0^T \int_0^{+\infty} e^{-kr^{2+\varepsilon}} |U(r,t)|\, dr\, dt < \infty$ for some $\varepsilon > 0$ (see Friedman, 1975).

ASSUMPTION A9. *The market price of interest rate risk depends on a parameter* λ *and is estimated by minimizing the sum of squared deviations between a fixed set of derivative prices and the corresponding nonparametric prices.*

APPENDIX 4: SOLVING THE PRICING PARTIAL DIFFERENTIAL EQUATION

The partial differential equation (4.1) with (4.3) and then (4.4) will be solved numerically using finite differencing with a combination of an implicit scheme for the advective term $\partial U/\partial r$ and the diffusion term $\partial^2 U/\partial r^2$. Since the numerical solution is guaranteed to converge only on a compact set, we make the change of variable $s \equiv cr/(1 + cr)$ where the constant c is large enough to overrepresent interest rate levels between 0 and 20%. This increases the accuracy of the numerical procedure. Let $V(s, t) \equiv U(r, t)$, so we can replace in (4.1):

$$\partial U/\partial t = (\partial V/\partial t),\ \partial U/\partial r = (\partial V/\partial s)(ds/dr)$$

and

$$\partial^2 U/\partial r^2 = (\partial^2 V/\partial s^2)(ds/dr)^2 + (\partial V/\partial s)(d^2 s/dr^2).$$

The discretized counterpart to (4.1) is

$$\frac{v_j^m - v_j^{m-1}}{\Delta t} + p_j\left(\frac{\left(v_{j+1}^m - 2v_j^m + v_{j-1}^m\right) + \left(v_{j+1}^{m-1} - 2v_j^{m-1} + v_{j-1}^{m-1}\right)}{2(\Delta s)^2}\right)$$
$$+ q_j\left(\frac{\left(v_{j+1}^m - v_{j-1}^m\right) + \left(v_{j+1}^{m-1} - v_{j-1}^{m-1}\right)}{4\Delta s}\right) - r(s_j)\frac{v_j^m + v_j^{m-1}}{2\Delta s}$$
$$+ c\left(r(s_j), t^m\right) = 0$$

where $v_j^m \equiv V(s_j, t^m)$, and t^m and s_j are the discretized values of the time and spot interest rate transformed variable, $j = 2, \dots, J-1$ and $m = 1, \dots, M$. $r(s_j)$ is the inverted change of variable. The coefficients p and q can easily be obtained from (4.1) and the differentials above. For $j = 1$, corresponding to 0, the diffusion function is zero and this the equation becomes:

$$\frac{v_1^m - v_1^{m-1}}{\Delta t} + q_1\left(\frac{\left(v_2^m - v_1^m\right) + \left(v_2^{m-1} - v_1^{m-1}\right)}{4\Delta s}\right) + c(0, t^m) = 0.$$

For $j = J$, v_J^m is obtained explicitly from v_J^{m-1}, v_{J-1}^{m-1}, and v_{J-2}^{m-1}.

Obviously, the equation could be solved more easily with a fully explicit representation. However, explicit finite differencing of parabolic equations is known to yield estimates that do not necessarily converge to the true solution of the equation (see Ames, 1977, for a von Neumann stability analysis). We have adopted the method recommended by Ames (1977), consisting of an implicit discretization of the second-order partial derivative and an explicit discretization of the first-order derivatives. This method is known as the Crank–Nicolson scheme. Convergence to the solution as $\Delta r \to 0$ and $\Delta t \to 0$ is guaranteed. For given $\{u_j^m / j = 1, \dots, J\}$, (7.1) gives $\{u_j^{m-1} / j = 1, \dots, J\}$ as the solution of an easily solvable triangular system. The final set $\{u_j^M / j = 1, \dots, J\}$ is derived directly from the final condition (4.3) for bond pricing, and later (4.4) for bond option pricing. Then the equation is used recursively backwards to derive successively the sets $\{u_j^m / j = 1, \dots, J\}$ for $m = M-1, M-2$, etc., up to $m = 1$.

When actually pricing bonds and bond options under the CIR model and comparing the numerical solutions to the closed-form formulas, this procedure was found to be very accurate, and the approximation error to shrink as much as wanted when decreasing the step sizes.

APPENDIX 5: PROOFS

Proof of Theorem 2. The proof uses the functional delta results of Aït-Sahalia (1992). The diffusion function (2.3) is viewed as a functional of the joint cumulative distribution function F of observations sampled $\Delta = 1$ day apart, $(r_t, r_{t+\Delta})$. Compute a first-order expansion of the diffusion away from F. From (2.3),

$$\sigma^2[F+H](x) = \frac{2}{\pi[F+H](x)} \int_0^x \mu(u, \theta[F+H])\pi[F+H](u)\, du$$

The drift parameters are estimated at the parametric speed, i.e., $\theta(\cdot)$ is a regular functional. That is, there exists a linear functional $\theta^{(1)}(\cdot,\cdot) \in C^{-1}$ such that $\theta[F+H] = \theta[F] + \int_0^{+\infty} \int_0^{+\infty} \theta^{(1)}(u,v)h(u,v)\, du\, dv + O(\|H\|^2_{L(\infty,1)})$. C^{-1} is the space of cadlag functions. The norm $L(\infty, 1)$ is the sum of the supremum norms for H and its derivative h. The fact that $\theta(\cdot)$ is evaluated at the empirical distribution instead of a kernel plug-in does not change the asymptotic distribution.

Let $\delta_{(x)}$ denote a Dirac mass centered at x. The expansion for the marginal density is

$$\pi[F+H](x) = \pi[F](x) + \int_0^{+\infty} \int_0^{+\infty} \delta_{(x)}(u)h(u,v)\, du\, dv + O\left(\|H\|^2_{L(\infty,1)}\right).$$

Collecting terms

$$\sigma^2[F+H](x) = \sigma^2[F](x) + \int_{-\infty}^{+\infty} \int_{-\infty}^{+\infty} \left\{ \frac{1}{\pi(x)} \left(-\sigma^2(x)\,\delta_{(x)}(u,v) + B[F](x) \right) \right\} h(u,v)\, du\, dv + O\left(\|H\|^2_{L(\infty,1)}\right);$$

$B[F](\cdot)$ is a functional in C^{-1}, which does not influence the asymptotic distribution and therefore need be identified. It follows that $\sigma^2[\cdot](x)$ is $L(\infty,1)$ differentiable with respect to F. That is, there exists a distribution $\varphi[F](\cdot,\cdot)$ such that the functional $\sigma^2(\cdot)$ satisfies

$$\sigma^2[F+H](x) = \sigma^2[F](x) + \int_0^{+\infty} \int_0^{+\infty} \varphi[F](u,v)h(u,v)\, du\, dv + O\left(\|H\|^2_{L(\infty,1)}\right).$$

The derivative of the functional $\sigma^2(\cdot)$ is

$$\varphi[F](u,v) = \left\{ \frac{-\sigma^2[F](x)}{\pi[F](x)} \right\} \delta_{(x)}(u,v) + B(x) \in C^{-2} \setminus C^{-1}$$

where C^{-2} is the space of Dirac delta functions. Its asymptotic distribution follows directly from the functional delta method.

Proof of Lemma 1. Consider the compact set $K_\varepsilon = (\varepsilon, 1/\varepsilon)$ for a fixed $\varepsilon > 0$, and fix x in K_ε. Consider the equation $L_\varepsilon \Gamma = 0$, where $L_\varepsilon U \equiv LU + \varepsilon(\partial^2 U/\partial r^2)$. It is necessary to perturbate L slightly because $\sigma^2(r) > 0$ for all $r > 0$ but $\sigma^2(0) = 0$, thus L is parabolic on $(0, \infty) \times (0, T]$ instead of the entire domain $[0, \infty) \times (0, T]$. Under A6, L_ε is uniformly parabolic on $K_\varepsilon \times (0, T]$ and its coefficient functions satisfy the hypotheses of Friedman (1964, section I.4). We can therefore construct the fundamental solution Γ_ε corresponding to the operator L_ε by using the parametrix method. Alternatively, Γ_ε could be obtained as the limit of the Green functions associated with L_ε on $K_\varepsilon \times (0, T]$ as in Friedman (1976, Lemma 15.1.1; see also Freidlin, 1985).

Form the parametrix solution for any (r, t) in $K_\varepsilon \times (0, T]$: $\Gamma_\varepsilon(r, t; x) = Z(r, t; x) + \int_0^t \int_\varepsilon^{1/\varepsilon} Z(r, t - s; y)\Phi_\varepsilon(y, s; x)\, dy\, ds$, where we have introduced

$$Z(r, t; x) \equiv \left(1/\sqrt{2\pi t}\,\sigma(x)\right) \exp\left\{-(r - x)^2/2t\sigma^2(x)\right\}$$

and $\Phi_\varepsilon(y, s; x) \equiv \sum_{p=1}^{\infty} (L_\varepsilon Z)_p(y, s; x)$. The differential operator $(L_\varepsilon \cdot)_p$ is defined recursively by $(L_\varepsilon Z)_1 = L_\varepsilon Z$ and

$$(L_\varepsilon Z)_{p+1}(y, s; x) = \int_0^s \int_\varepsilon^{1/\varepsilon} [L_\varepsilon Z(y, s - \tau; z)] \times \left[(L_\varepsilon Z)_p(z, \tau; x)\right] dz\, d\tau.$$

By Theorem I.8 in Friedman (1964) the infinite series in Φ_ε converges and the parametrix solution is well-defined.

Next it is clear that the parametrix expression is a fundamental solution of (4.1) on $K_\varepsilon \times (0, T]$. Write

$$\begin{aligned} L_\varepsilon \Gamma_\varepsilon(r, t; x) &= L_\varepsilon Z(r, t; x) \\ &\quad + L_\varepsilon \left\{ \int_0^t \int_\varepsilon^{1/\varepsilon} Z(r, t - s; y)\Phi_\varepsilon(y, s; x)\, dy\, ds \right\} \\ &= L_\varepsilon Z(r, t; x) - \Phi_\varepsilon(r, t; x) \\ &\quad + \int_0^t \int_\varepsilon^{1/\varepsilon} L_\varepsilon Z(r, t - s; y)\Phi_\varepsilon(y, s; x)\, dy\, ds \end{aligned}$$

Thus for Γ_ε to be a fundamental solution, Φ_ε must satisfy the Volterra integral equation:

$$\begin{aligned} \Phi_\varepsilon(r, t; x) &= L_\varepsilon Z(r, t; x) \\ &\quad + \int_0^t \int_\varepsilon^{1/\varepsilon} L_\varepsilon Z(r, t - s; y)\Phi_\varepsilon(y, s; x)\, dy\, ds. \end{aligned}$$

By construction the Volterra equation is satisfied by $\sum_{p=1}^{\infty} (L_\varepsilon Z)_p(r, t; x)$.

Then as in Theorem 15.1.2 in Friedman (1976) we obtain that as ε goes to zero $\Gamma_\varepsilon(r,t;x)$ converges to a limit $\Gamma(r,t;x)$, together with the first two r derivatives, x derivatives, and the first t derivative uniformly for all r and x in $(0,\infty)$ and t in $(0,T]$. Furthermore the limit $\Gamma(r,t;x)$ satisfies (4.1) on $(0,\infty)\times(0,T]$ as in Corollary 15.1.3 in Friedman (1976). The fundamental solution has the form $\Gamma(r,t;x)=\gamma(r,t,x,\sigma^2(x),\mu(x))$ with γ continuously differentiable in each of its arguments (see the parametrix form). The only difference here with the results in Chapter 15 in Friedman (1976) is that the parabolic operator L contains the term $(-rU)$. The proof is otherwise identical. Finally the fundamental solution is unique in the class of nonexplosive solutions (assumption A8). By Theorem I.16 in Friedman (1964), for each $\varepsilon>0$, Γ_ε is unique on $K_\varepsilon\times(0,T]$. Any two solutions on $(0,\infty)\times(0,T]$ would have to coincide on $K_\varepsilon\times(0,T]$.

Proof of Theorem 3. We first consider the case where the market price of risk is known, or without lack of generality $\lambda\equiv 0$. Then the coefficients of the differential operator L are functionals of the joint cumulative distribution function F of the data $(r_t, r_{t+\Delta})$ so the differential operator is $L[F]$. Then consider the equations followed by the derivative prices associated to F and $F+H$, respectively: $L[F]U[F](r,t)=-c(r,t)$ with initial condition $U[F](r,0)=g[F](r)$, and $L[F+H]U[F+H](r,t)=-d(r,t)$ with initial condition $U[F+h](r,0)=g[F+H](r)$. The final payoff $g(\cdot)$ depends on F because it could be the price of a derivative previously computed; for example, the price of the underlying bond when valuing a bond option.

Lemma 1 characterizes the fundamental solution Γ in the form $\Gamma(r,t;x)=\gamma(r,t,x,\sigma^2(x),\mu(x))$ with γ continuously differentiable in each of its arguments. From this characterization, and the (linear in H) expansions of the drift and diffusion functionals in Theorem 2, it follows that $\Delta U[F,H]\equiv U[F+H]-U[F]$ admits the expansion ($L(\infty,0)$ is the standard supremum norm for H):

$$
\begin{aligned}
\Delta U[F,H](r,t) &= \int_0^{+\infty}\Gamma[F+H](r,0;x)g[F+H](x)\,dx \\
&\quad -\int_0^{+\infty}\Gamma[F](r,0;x)g[F](x)\,dx \\
&\quad +\int_0^t\int_0^{+\infty}\{\Gamma[F+H]-\Gamma[F]\}(r,\tau;x)\,c(x,\tau)\,dx\,d\tau \\
&= \int_0^{+\infty}\int_0^{+\infty}u[F](r,t,x,y)h(x,y)\,dx\,dy+O\big(\|H\|^2_{L(\infty,0)}\big)
\end{aligned}
$$

Suppose that for the derivative under consideration the final payoff $g(\cdot)$ is independent of F (e.g., a bond for which $g=1$, or an interest rate option $g=\max(0,r-k)$). Then the functional $F\to U[F](r,t)$ for fixed

(r,t) in $(0,\infty)\times(0,T]$ is $L(\infty,1)$-differentiable with respect to F with a continuous functional derivative. Its functional derivative $u(\cdot)$ is continuous as can be seen from collecting terms from the linearizations of $\mu(\cdot)$ and $\sigma^2(\cdot)$ (which were computed in Theorem 2) and the partial derivatives of $\gamma(r,t,x,\sigma^2(x),\mu(x))$ with respect to σ^2 and μ. If now the derivative's payoff g depends on F, typically through the previously computed price of a more primitive derivative, then the expansion of $g(\cdot)$ also matters. An example of this situation is a bond call option where $g(F)$ is the maximum of zero and the price of the underlying bond (dependent on F) net of the strike price. By what precedes, the bond price admits a first-order expansion in F with derivative in C^{-1} (since its payoff is independent of F). Under A7 the final payoff for the derivative also satisfies this requirement. The functional derivative $u(\cdot)$ then has additional terms which reflect this additional dependence on F.

We next show that the introduction of the market price of risk does not modify the rate of convergence (only the asymptotic variance). This holds provided that λ is estimated at a speed[8] that is at least as fast as that of σ^2, in particular when the estimator of λ converges at speed $n^{1/2}$. This will hold as a further special case when the market price of risk is estimated by matching the prices a cross-section of derivatives ($\{U_i\}$, $i=1,\ldots,J$). Indeed under A9 set $\hat{\lambda}=\arg\min_{\lambda\in L}\Sigma_{j=1}^J\{\hat{U}_\lambda(r,t_i)-U_i\}^2$ by minimizing the squared deviations between the derivative prices ($\{U_i\}$) and those predicted by the model ($\hat{U}_\lambda(r,t_i)$ evaluated at λ).

Note that under A9 we assume that we are matching a fixed set of derivative prices and therefore do not have to account for the variability of prices being matched.[9] Recall that the solution U_λ has μ replaced by $\mu-\lambda\sigma$. Furthermore, the characterization of the fundamental solution has $\Gamma(r,t;x)=\gamma(r,t,x,\sigma^2(x),\mu(x))$ with γ continuously differentiable in each of its arguments. Therefore $\hat{U}_\lambda(r,t_i)$ is continuously differentiable in λ and $\lambda(\cdot)$ is $L(\infty,1)$ differentiable as a functional of F with derivative in C^{-1} by the implicit functional theorem: View λ as the implicit solution of the first-order condition $\Sigma_{j=1}^J\{\partial\hat{U}_\lambda(r,t_i)/\partial\lambda\}\{\hat{U}_\lambda(r,t_i)-U_i\}=0$. As a consequence, this particular $\hat{\lambda}$ converges at speed $n^{1/2}$ and $\hat{U}$ also converges at speed $n^{1/2}$. Collect all the linear terms in the first-order expansion of U and let $u_{(r,t)}$ be the derivative (in C^{-1} by what precedes)

[8] If, however, λ were estimated at a slower speed than that of σ^2 (e.g., its derivative were in $C^{-3}\setminus C^{-2}$, the space of first derivatives of Dirac masses) then $U[\cdot,\cdot]$ would no longer converge at speed $n^{1/2}$.

[9] If we were matching a random set of derivative prices, such as the term structure on a particular day, then derivatives prices would depend not only on F but also on G, the joint cdf of the prices being matched. For example, G would be the joint cdf of discount bond prices at the fixed maturities $t_1,\ldots,t_J$. A formal proof in that case goes beyond the scope of this chapter.

of U. A complete expression for $u_{(r,t)}$ can be computed explicitly but is not needed since the asymptotic variance of the prices will be obtained by the bootstrap method.

ACKNOWLEDGMENTS

This chapter is part of my Ph.D. dissertation at the MIT Department of Economics. I am very grateful to my advisors Jerry Hausman, Andrew Lo, and Whitney Newey, as well as John Cox, Lars Hansen, Jose Scheinkman, and Daniel Stroock for suggestions. The comments of a coeditor and three referees led to considerable improvement of the chapter. I thank seminar participants at Berkeley, British Columbia, Carnegie-Mellon, Chicago, Columbia, Harvard, LSE, Michigan, MIT, Northwestern, Princeton, Stanford, Toulouse, ULB, Washington University, Wharton, and Yale for comments. Financial support from Ecole Polytechnique and MIT Fellowships is also gratefully acknowledged. All errors are mine.

REFERENCES

Aït-Sahalia, Y. (1992). "The Delta and Bootstrap Methods for Nonparametric Kernel Functionals," Mimeo. Massachusetts Institute of Technology, Cambridge, MA.

Aït-Sahalia, Y. (1996). Testing continuous-time models of the spot interest rate. *Review of Financial Studies*.

Ames, W. F. (1977). "Numerical Methods for Partial Differential Equations." Academic Press, New York.

Banon, G. (1978). Nonparametric identification for diffusion processes. *SIAM Journal of Control and Optimization* **16**, 380–395.

Basawa, I. V., and Prakasa-Rao, B. L. S. (1980). "Statistical Inference for Stochastic Processes." Academic Press, New York.

Black, F., and Scholes, M. (1973). "The pricing of options and corporate liabilities. *Journal of Political Economy* **3**, 133–155.

Brennan, M. J., and Schwartz, E. S. (1979). A continuous-time approach to the pricing of bonds. *Journal of Banking and Financial* **3**, 133–155.

Brown, S. J., and Dybvig, P. H. (1986). The empirical implications of the Cox, Ingersoll, Ross theory of the term structure of interest rates. *Journal of Finance* **41**(3), 617–630.

Chamberlain, G. (1987). Asymptotic efficiency in estimation with conditional moment restrictions. *Journal of Econometrics* **34**, 305–334.

Chan, K. C., Karolyi, G. A., Longstaff, F. A., and Sanders, A. B. (1992). An empirical comparison of alternative models of the short-term interest rate. *Journal of Finance* **47**(3), 1209–1227.

Constantinides, G. M. (1992). A theory of the nominal term structure of interest rates. *Review of Financial Studies* **5**(4), 531–552.

Courtadon, G. (1982). The pricing of options on default-free bonds." *Journal of Financial and Quantitative Analysis* **17**(1), 75–100.

Cox, J. C. (1975). "Notes on Option Pricing I: Constant Elasticity of Diffusions," Mimeo. Stanford University, Stanford, CA.

Cox, J. C., Ingersoll, J. E., and Ross, S. A. (1980). An analysis of variable rate load contracts. *Journal of Finance* **35**(2), 389–403.

Cox, J. C., Ingersoll, J. E., and Ross, S. A. (1985a). An intertemporal general equilibrium model of asset prices. *Econometrica* **53**(2), 363–384.

Cox, J. C., Ingersoll, J. E., and Ross, S. A. (1985b). A theory of the term structure of interest rates. *Econometrica* **53**(2), 385–407.

Dothan, L. U. (1978). On the term structure of interest rates. *Journal of Financial Economics* **6**, 59–69.

Duffie, D., and Kan, R. (1993). "A Yield Factor Model of Interest Rates," Mimeo. Stanford University, Stanford, CA.

Duffie, D., and Singleton, K. (1993). Simulated moments estimation of Markov models of asset prices. *Econometrica* **61**(4), 929–952.

Feller, W. (1951). Two singular diffusion problems. *Annals of Mathematics* **54**(2), 173–182.

Florens-Zmirou, D. (1993). On estimating the diffusion coefficient from discrete observations. *Journal of Applied Probability* **30**. 790–804.

Freidlin, M. (1985). "Functional Integration and Partial Differential Equations," Annals of Mathematics Studies, No. 109. Princeton University Press, Princeton, NJ.

Friedman, A. (1964), "Partial Differential Equations of Parabolic Type." Prentice-Hall, Englewood Cliff, NJ.

Friedman, A. (1975). "Stochastic Differential Equations and Applications." Vol. 1. Academic Press, New York.

Friedman, A. (1976). "Stochastic Differential Equations and Applications." Vol. 2. Academic Press, New York.

Gibbons, M. R., and Ramaswamy, K. (1993). A test of the Cox, Ingersoll and Ross model of the term structure. *Review of Financial Studies* **6**(3), 619–658.

Gouriéroux, C., Monfort, A., and Renault, E. (1993). Indirect inference. *Journal of Applied Econometrics* **8**, S85–S118.

Hansen, L. P. (1982). Large sample properties of generalized method of moments estimators. *Econometrica* **50**, 1029–1054.

Hansen, L. P., and Scheinkman, J. A. (1995). Back to the future: Generating moment implications for continuous time Markov processes. *Econometrica* **63**, 767–804.

Harvey, A. C. (1993). "Time Series Model." 2nd ed. MIT Press, Cambridge, MA.

Hasminskii, R. Z. (1980). "Stochastic Stability of Differential Equations." Sijthoff & Noordhoff, Alphen aan den Rijn, The Netherlands.

Hull, J., and White, A. (1990). Pricing interest-rate-derivative securities. *Review of Financial Studies* **3**(4), 573–592.

Jamshidian, F. (1989). An exact bond option formula. *Journal of Finance* **44**(1), 205–209.

Karatzas, I., and Shreve, S. E. (1991). "Brownian Motion and Stochastic Calculus," 2nd ed., Springer-Verlag, New York.

Karlin, S., and Taylor, H. M. (1981). "A Second Course in Stochastic Processes." Academic Press, New York.

Künsch, H. R. (1989). The jacknife and the bootstrap for general stationary observations. *Annals of Statistics* **17**(3), 1217–1241.

Liu, R. Y., and Singh, K. (1992). Moving blocks jacknife and bootstrap capture weak dependence. *In* "Exploring the Limits of Bootstrap" (R. LePage and Y. Billard, eds.). Wiley, New York.

Lo, A. W. (1988). Maximum likelihood estimation of generalized Itô processes with discretely sampled data. *Econometrica Theory* **4**, 231–247.

Marsh, T. A., and Rosenfeld, E. R. (1983). Stochastic processes for interest rates and equilibrium bond prices. *Journal of Finance*, **38**(2), 635–646.

Merton, R. C. (1973). Theory of rational option pricing. *Bell Journal of Economics and Management Science* **4**, 141–183.

Merton, R. C. (1980). On estimating the expected return on the market: An exploratory investigation. *Journal of Financial Economics* **8**, 323–361.

Merton, R. C. (1990). "Continuous-Time Finance." Blackwell, Cambridge, MA.

Nelson, D. B. (1990). ARCH models as diffusion approximations. *Journal of Econometrics* **45**(1/2), 7–38.

Øksendal, B. (1992). "Stochastic Differential Equations," 3rd ed. Springer-Verlag, New York.

Pearson, N. D., and Sun, T.-S. (1994). Exploiting the conditional density in estimating the term structure: An application to the Cox, Ingersoll, and Ross model. *Journal of Finance* **49**(4), 1279–1304.

Phillips, P. C. B. (1987). Time series regression with a unit root. *Econometrica* **55**(2), 277–310.

Prakasa-Rao, B. L. S. (1987). "On Mixing for Flows of σ-Algebras," Tech. Rep. No. 95. University of California, Davis.

Robinson, P. M. (1983). Nonparametric estimators for time series. *Journal of Time Series Analysis* **4**(3), 185–207.

Robinson, P. M. (1987). Asymptotically efficient estimation in the presence of heteroskedasticity of unknown form. *Econometrica* **55**(4), 875–891.

Said, E. S., and Dickey, D. A. (1984). Testing for unit roots in autoregressive-moving average models of unknown order. *Biometrika* **71**(3), 599–607.

Scott, D. W. (1992). "Multivariate Density Estimation: Theory, Practice and Visualization." Wiley, New York.

Silverman, B. W. (1986). "Density Estimation for Statistical and Data Analysis." Chapman & Hall, London.

Vasicek, O. (1977). An equilibrium characterization of the term structure. *Journal of Financial Economics* **5**, 177–188.

Wong, E. (1964). The construction of a class of stationary Markov processes. *In* "Sixteenth Symposium in Applied Mathematics—Stochastic Processes in mathematical Physics and Engineering" (R. Bellman, ed.), pp. 264–276. American Mathematical Society, Providence, RI.

INDEX

Page references followed by *f*, *n*, or *t* indicate figures, notes, or tables, respectively.